RECENT ADVANCES IN REMOTE SENSING AND GEOINFORMATION PROCESSING
FOR LAND DEGRADATION ASSESSMENT

International Society for Photogrammetry and Remote Sensing (ISPRS) Book Series

Book Series Editor

Paul Aplin
School of Geography
The University of Nottingham
Nottingham, UK

information from imagery

Recent Advances in Remote Sensing and Geoinformation Processing for Land Degradation Assessment

Editors

Achim Röder & Joachim Hill

Remote Sensing Department, University of Trier, Germany

CRC Press
Taylor & Francis Group
Boca Raton London New York Leiden

CRC Press is an imprint of the
Taylor & Francis Group, an **informa** business

A BALKEMA BOOK

CRC Press/Balkema is an imprint of the Taylor & Francis Group, an informa business

© 2009 Taylor & Francis Group, London, UK

Typeset by Vikatan Publishing Solutions (P) Ltd., Chennai, India
Printed and bound in Great Britain by TJ International Ltd, Padstow, Cornwall

Published by: CRC Press/Balkema
          P.O. Box 447, 2300 AK Leiden, The Netherlands
          e-mail: Pub.NL@taylorandfrancis.com
          www.crcpress.com – www.taylorandfrancis.co.uk – www.balkema.nl

British Library Cataloguing in Publication Data

Library of Congress Cataloging-in-Publication Data
Recent Advances in Remote Sensing and Geoinformation Processing for Land Degradation Assessment/editors, Achim Röder & Joachim Hill.
    p.cm. -- (International Society for Photogrammetry and Remote Sensing (ISPRS) Book Series; v. 8)
    Includes bibliographical references and index.
    ISBN: 978-0-415-39769-8 (handcover: alk. paper) -- ISBN: 978-0-203-87544-5 (e-book)
1. Land degradation -- Remote sensing. 2. Remote sensing. I. Röder, Achim. II. Hill, Joachim. III. Title.

GE140.R425 2009
551.41028--dc22

                                                                        2009002727

ISBN: 978-0-415-39769-8 (hbk)
ISBN: 978-0-203-87544-5 (e-book)
ISSN: 1572-3348

Recent Advances in Remote Sensing and Geoinformation Processing
for Land Degradation Assessment – Röder & Hill (eds)
© 2009 Taylor & Francis Group, London, ISBN 978-0-415-39769-8

Table of contents

Acknowledgements

The editors of this volume would like to acknowledge the authors who have contributed their energy and work to provide this state-of-the-art analysis and demonstration of applications of geospatial information technology in land degradation and desertification research. We are particularly grateful to all members of the review panel for giving their valuable time to review the manuscripts: Adrian Chappell (University of Salford, Manchester/UK), Gabriel del Barrio (Consejo Superior de Investigaciones Cientificas-Estacion Experimental de Zonas Aridas, Almeria/Spain), Ferdinand Bonn (CARTEL, Université de Sherbrooke/Canada, R.I.P.), Guy Engelen (Vlaamse Instelling voor Technologisch Onderzoek, Mol/Belgium), Richard Escadafal (Centre d'Etudes Spatiales de la Biosphère, Toulouse/France), Ulf Helldén (GeoBiosphere Science Centre, Lund/Sweden), Joachim Hill (Remote Sensing Department, University of Trier, Germany), Bernard Lacaze (Centre National de Recherche Scientifique-PRODIG, Paris/France), Eric Lambin (University of Louvain, Louvain-la-Neuve/Belgium), Keith McCloy (Danish Institute for Agricultural Science, Copenhagen/Denmark), Achim Röder (Remote Sensing Department, University of Trier, Germany), Stefan Sommer (Joint Research Centre of the European Commission, Ispra/Italy), Compton J. Tucker (NASA/Goddard Space Flight Center, Greenbelt/USA), Thomas Udelhoven (Centre Recherche Gabriel Lippmann, Belval/Luxembourg), Susan Ustin (University of California, Davis/USA), Riccardo Valentini (Department of Forest Science and Environment, University of Tuscia, Viterbo/Italy).

For many years, the European Commission has been supporting this field of research through numerous research and coordination projects. More specifically, it has provided the financial means to organise the '1st International Conference on Remote Sensing and Geoinformation Processing in the Assessment and Monitoring of Land Degradation and Desertification (RGLDD)', which formed the starting point for this compilation. This support is gratefully acknowledged. In particular, we would like to mention the responsible scientific project officers, past and present, Mr. Denis Peter and Mrs. Maria Yeroyanni, for their continued support.

Finally, we wish to thank Paul Aplin, ISPRS Book Series editor in charge of this volume, for invaluable counsel and support throughout the preparation process.

Achim Röder & Joachim Hill
Trier, Germany
November 2008

Recent Advances in Remote Sensing and Geoinformation Processing
for Land Degradation Assessment – Röder & Hill (eds)
© 2009 Taylor & Francis Group, London, ISBN 978-0-415-39769-8

Contributors

Clive T. Agnew
School of Environment and Development, University of Manchester, Manchester, UK

Amir Arnon
Agricultural Research Organization, Bet-Dagan, Israel. E-mail: amir@volcani.agri.gov.il

Hidetoshi Asai
Kyoto University, Kyoto, Japan

Ophélie Aussedat
European Commission—DG Joint Research Centre, Institute for Environment and Sustainability, Global Environment Monitoring Unit, Via E. Fermi, 2749, I-21027 Ispra (VA), Italy. E-mail: ophelie.aussedat@jrc.it

Gloria Bordogna
IDPA CNR, Institute for the Dynamics of Environmental Processes, Bergamo, Italy

Mirco Boschetti
IREA CNR, Institute for Electromagnetic Sensing of the Environment, Milano, Italy

Pietro Alessandro Brivio
IREA CNR, Institute for Electromagnetic Sensing of the Environment, Milano, Italy. E-mail: brivio.pa@irea.cnr.it

Paola Carrara
IREA CNR, Institute for Electromagnetic Sensing of the Environment, Milano, Italy

David Celis
International Center for Agricultural Research in the Dry Areas, (ICARDA), Aleppo, Syria

Adrian Chappell
Centre for Environmental Systems Research, University of Salford, Manchester, UK. E-mail: a.chappell@salford.ac.uk

Abdelghani Chehbouni
Centre d'Etudes Spatiales de la Biosphère BP 31055 cedex Toulouse, France. E-mail: ghani@cesbio.cnes.fr

Barnaby J.F. Clark
Department of Geography, University of Helsinki, Finland. E-mail: barnaby.clark@helsinki.fi

Sergio Contreras
Estación Experimental de Zonas Áridas (CSIC), Almería, Spain. E-mail: sergio@eeza.csic.es

Eddy De Pauw
International Center for Agricultural Research in the Dry Areas, (ICARDA), Aleppo, Syria. E-mail: e.de-pauw@cgiar.org

Gabriel del Barrio

Estación Experimental de Zonas Aridas, Consejo Superior de Investigaciones Cientificas, Almeria, Spain. E-mail: gabriel@eeza.csic.es

Francisco Domingo

Estación Experimental de Zonas Áridas (CSIC), Almería, Spain.
E-mail: poveda@eeza.csic.es

Linkham Dounagsavanh

National Agriculture and Forest Research Institute, Vientiane, Laos PDR

Jamal Ezzahar

Physics Department LMFE, Faculty of Sciences Semlalia, Marrakesh, Morocco

Karsten Friedrich

Institut für Meteorologie, Freie Universität Berlin, Carl-Heinrich-Becker-Weg 6–10, 12165 Berlin, Germany. E-mail: friedkar@zedat.fu-berlin.de

Jaime Garatuza-Payan

ITSON, Ciudad Obregon, Sonora, Mexico

Mónica García

Estación Experimental de Zonas Áridas (CSIC), Almería, Spain.
E-mail: monica@eeza.csic.es

Helmut Geist

School of Geosciences, University of Aberdeen, Aberdeen, United Kingdom.
E-mail: h.geist@abdn.ac.uk

Uri Gilad

The Remote Sensing Laboratory, Jacob Blaustein Institutes for Desert Research, Ben Gurion University of the Negev, 84990, Israel. E-mail: gileadu@bgu.ac.il

Nadine Gobron

European Commission—DG Joint Research Centre, Institute for Environment and Sustainability, Global Environment Monitoring Unit, Via E. Fermi, 2749, I-21027 Ispra (VA), Italy. E-mail: nadine.gobron@jrc.it

Joachim Hill

University of Trier, FB VI Geography/Geosciences, Remote Sensing Department, Campus-II, D-54286 Trier, Germany. E-mail: hillj@uni-trier.de

Takeshi Horie

Kyoto University, Kyoto, Japan

Yvon Carmen Hountondji

Faculty of Agronomy, University of Parakou, BP: 123 Parakou, Benin.
E-mail: yvon.hountondji@gmail.com

Yoshio Inoue

National Institute for Agro-Environmental Sciences, Tsukuba, Japan.
E-mail: yinoue@affrc.go.jp

Stéphane Jacquemoud

Institut de Physique du Globe de Paris, Géodésie & Gravimétrie, 4 Place Jussieu, 75252 Paris Cedex 5, France. E-mail: jacquemoud@ipgp.jussieu.fr

Thomas Jarmer

Technion—Faculty of Civil and Environmental Engineering, Transportation
and Geo-Information Engineering Unit, Haifa 32000, Israel. E-mail: jarmer@technion.ac.il

Arnon Karnieli

The Remote Sensing Laboratory, Jacob Blaustein Institutes for Desert Research,
Ben Gurion University of the Negev, 84990, Israel. E-mail: karnieli@bgu.ac.il

Yoshiyuki Kiyono

Forestry and Forest Products Research Institute, Tsukuba, Japan.
E-mail: kiono@ffpri.affrc.go.jp

Konstantin König

Institute for Ecology, Evolution and Diversity, Goethe-University, Frankfurt am Main,
Germany. E-mail: k.koenig@em.uni-frankfurt.de

Dirk Koslowsky

Institut für Meteorologie, Freie Universität Berlin, Carl-Heinrich-Becker-Weg 6–10,
12165 Berlin, Germany. E-mail: kosze@zedat.fu-berlin.de

Tobias Kuemmerle

Department of Geomatics, Humboldt University Berlin, Germany.
E-mail: kuemmerle@geo.hu-berlin.de

Eric F. Lambin

University of Louvain, Louvain-la-Neuve, Belgium. E-mail: eric.lambin@uclouvain.be

Hanoch Lavée

Department of Geography, Bar-Ilan University, Ramat Gan 52900, Israel.
E-mail: laveeh@mail.biu.ac.il

John F. Leys

Department of Natural Resources, Gunnedah, NSW, Australia

Lin Li

Department of Geology, Indiana University-Purdue University, Indianapolis, IN USA.
E-mail: ll3@iupui.edu

Grant H. McTainsh

Australian Rivers Institute, The Griffith School of Environment, Griffith University,
Nathan, Queensland, Australia

Wolfgang Mehl

European Commission, Joint Research Centre, Institute for Environment and
Sustainability (IES), I-21020 Ispra (Va), Italy. E-mail: wolfgang.mehl@jrc.it

Jonas V. Müller

Royal Botanic Gardens, Kew, United Kingdom. E-mail: j.mueller@kew.org

Mark Mulligan

Environmental Monitoring and Modelling Research Group, Department of Geography,
King's College London, Strand, London, WC2R 2LS, UK. E-mail: mark.mulligan@kcl.ac.uk

Jacques Nicolas

Environmental Sciences Management Department, University of Liege, Avenue de
Longwy 185, B-6700 Arlon, Belgium. E-mail: j.nicolas@ulg.ac.be

Yukinori Ochiai
Forestry and Forest Products Research Institute, Tsukuba, Japan

Albert Olioso
INRA-CSE, Avignon, France. E-mail: Albert.Olioso@avignon.inra.fr

Pierre Ozer
Environmental Sciences Management Department, University of Liege, Avenue de Longwy 185, B-6700 Arlon, Belgium. E-mail: pozer@ulg.ac.be

Alicia Palacios-Orueta
Dpto. Silvopascicultura. U.D. Edafología ETSI Montes. Universidad Politécnica de Madrid, Spain. E-mail: apalacios@montes.upm.es

Vasilios P. Papanastasis
Faculty of Forestry and Natural Environment, Laboratory of Range Ecology, Aristotle University Thessaloniki, Greece. E-mail: vpapan@for.auth.gr

Petri K.E. Pellikka
Department of Geography, University of Helsinki, Finland. E-mail: petri.pellikka@helsinki.fi

Avi Perevolotsky
Agricultural Research Organization, Bet-Dagan, Israel. E-mail: avi@volcani.agri.gov.il

Bernard Pinty
European Commission—DG Joint Research Centre, Institute for Environment and Sustainability, Global Environment Monitoring Unit, Via E. Fermi, 2749, I-21027 Ispra (VA), Italy. E-mail: bernard.pinty@jrc.it

Juan Puigdefábregas
Estación Experimental de Zonas Áridas (CSIC), Almería, Spain. E-mail: puigdefa@eeza.csic.es

Jiaguo Qi
Michigan State University, East Lansing, USA. E-mail: qi@msu.edu

James F. Reynolds
Department of Biology, Duke University, Durham, USA. E-mail: james.f.reynolds@duke.edu

Achim Röder
University of Trier, FB VI Geography/Geosciences, Remote Sensing Department, Campus-II, D-54286 Trier, Germany. E-mail: roeder@uni-trier.de

Julio-César Rodriguez
IMADES, Hermosillo, Sonora, Mexico

Kazuki Saito
Kyoto University, Kyoto, Japan

Pariente Sarah
Department of Geography, Bar-Ilan University, Ramat Gan 52900, Israel. E-mail: pariens@mail.biu.ac.il

Marco Schmidt
Research Institute Senckenberg, Frankfurt am Main, Germany. E-mail: mschmidt@senckenberg.de

Rakefet Shafran-Nathan
Ben-Gurion University of the Negev, Beer-Sheva 84105, Israel. E-mail: shafranr@bgu.ac.il

Tatsuhiko Shiraiwa
Kyoto University, Kyoto, Japan

Maxim Shoshany
Geo-Information Engineering, Faculty of Civil & Environmental Engineering,
Technion Institute of Technology, Haifa 32000, Israel. E-mail: maximsh@tx.technion.ac.il

Nestor Sokpon
Faculty of Agronomy, University of Parakou, BP: 123 Parakou, Benin

D. Mark Stafford-Smith
CSIRO Sustainable Ecosystems, Centre for Arid Zone Research, CSIRO, Alice Springs,
Australia. E-mail: mark.staffordsmith@csiro.au

Craig Strong
Australian Rivers Institute, The Griffith School of Environment, Griffith University, Nathan,
Queensland, Australia

Daniela Stroppiana
IREA CNR, Institute for Electromagnetic Sensing of the Environment, Milano, Italy

Tal Svoray
Department of Geography and Environmental Development, Ben Gurion University of the
Negev, Beer-Sheva 84105, Israel. E-mail: tsvoray@bgu.ac.il

Malcolm Taberner
European Commission—DG Joint Research Centre, Institute for Environment
and Sustainability, Global Environment Monitoring Unit, Via E. Fermi, 2749,
I-21027 Ispra (VA), Italy. E-mail: malcolm.taberner@jrc.it

Georgios M. Tsiourlis
Research Institute, Laboratory of Ecology, National Agricultural Research Foundation
(NAGREF), Vasilika-Thessaloniki, Greece. E-mail: gmtsiou@fri.gr

Thomas Udelhoven
Centre de Recherche Public Gabriel Lippmann, Belveaux, Luxembourg.
E-mail: udelhove@lippmann.lu

Eugene David Ungar
Agricultural Research Organization, Bet-Dagan, Israel. E-mail: eugene@volcani.agri.gov.il

Susan Ustin
California Space Institute Center of Excellence, University of California Davis, CA 95616,
USA. E-mail: slustin@ucdavis.edu

Michel M. Verstraete
European Commission—DG Joint Research Centre, Institute for Environment
and Sustainability, Global Environment Monitoring Unit, Via E. Fermi, 2749,
I-21027 Ispra (VA), Italy. E-mail: michel.verstraete@jrc.it

Christopher J. Watts
UNISON, Hermosillo, Sonora, Mexico

Michael L. Whiting
California Space Institute Center of Excellence, University of California Davis, CA 95616,
USA. E-mail: mwhiting@ucdavis.edu

Ted M. Zobeck
USDA, Cropping Systems Research Laboratory, Lubbock, Texas, USA

Recent Advances in Remote Sensing and Geoinformation Processing
for Land Degradation Assessment – Röder & Hill (eds)
© *2009 Taylor & Francis Group, London, ISBN 978-0-415-39769-8*

Remote sensing and geoinformation processing in land degradation assessment—an introduction

A. Röder & J. Hill
Remote Sensing Department, FB VI Geography/Geosciences,
University of Trier, Germany

The history of Earth is a history of changing environmental and human systems. Yet, the rates of transformation processes observed in past decades are causing increasing concern. The Millenium Ecosystem Assessment reports more rapid and extensive human-induced ecosystem changes over the past 50 years than in any comparable period of time. These changes have brought about increasing wealth to a small part of the world's population at the expense of the ability of ecosystems to continuously provide a diverse range of goods and services. Among the multiple processes of change, desertification and land degradation are of particular importance, as they threaten the livelihoods of some 2 billion people living in the world's drylands. While land degradation refers to the reduction or loss of the biological or economic productivity of drylands, desertification is defined by the U.N. Convention to Combat Desertification (UNCCD) as "land degradation in arid, semiarid and dry subhumid areas resulting from various factors, including climatic variations and human activities". Beside the immediate impact on people's livelihoods, desertification is connected with other pressing global problems, as it is accompanied by the loss of biological diversity, and contributes to global climate change through the loss of carbon sequestration capacity and an increase in land-surface albedo.

As a result of these alarming facts, the struggle against land degradation and desertification has found its way to the agenda of decision-makers and politicians on national and global scales, as well as land managers responsible at the local level. This is reflected in numerous initiatives on global and regional levels, such as the UNCCD and FAO's Land Degradation in Drylands Assessment initiative (LADA), or networks such as AridNet (A Research Network for Studies of Global Desertification) or the European DesertNet (European Network on Global Desertification Research). At the same time, desertification and land degradation are considered key processes in adjacent initiatives, such as the LUCC (Land Use/Land Cover Change) Project and its successor, the Global Land Project (GLP). Recently, the Global Earth Observation System of Systems (GEOSS) has been launched to streamline efforts for establishing a consistent and standardised framework for observing the Earth with regard to different societal benefit areas. Within this framework, desertification is listed as one of the fields where action is most urgently required. This initiative takes up one of the key requests issued by stakeholders, which is the call for accurate, spatially explicit information on the extent and magnitude of desertification, along with an assessment of the effects of management interventions.

In this context, remote sensing data are of outstanding value. With the history of operational earth observation sensors spanning back over three decades they allow to retrospectively analyse the state and development of ecosystems on different scales and with different spatial coverage. Remote sensing data adhere to the principles of repetitiveness, objectivity and consistency, which are prerequisites in the frame of monitoring and surveillance. Thus, they provide essential base information for integrated, region- or problem-specific approaches combining satellite data products with specific interpretation and modelling approaches.

The European Union has long been committed to support actions against desertification at different levels. This has been reflected in past and present Research Framework Programmes, and included both experimental and applied research in the field, as well as numerous horizontal, coordinating and supporting actions. In this context, DG Research commissioned the '1st International Conference on Remote Sensing and Geoinformation Processing in the Assessment and Monitoring of Land Degradation and Desertification (RGLDD)', which attracted global attendance of scientists and stakeholders to the city of Trier, Germany, in 2005. Contributions reflected both recent innovations in sensor design and methodological developments, as well as the wide range of applied studies targeted at specific local conditions. Given the wide interest and the relevance of the topic, it was decided to prepare a book that demonstrates the benefits of geospatial data interpretation in the context of desertification and land degradation, and reflects both traditional and innovative data processing, interpretation and modelling approaches.

The book is organized in four major parts. The first provides a conceptual background on information extraction, spatial modelling and data integration in land degradation assessment. The second part focuses on options for mapping large areas and the implications resulting from the selection of data sources. The third part examines specific biophysical indicators that can be related to land degradation and desertification processes. Part four casts a spotlight on the potential of integrating remotely sensed information with auxiliary data sets to address specific local problems.

PART 1   SETTING THE SCENE: PRINCIPLES IN REMOTE SENSING
AND SPATIAL MODELLING FOR LAND DEGRADATION ASSESSMENT

In the first chapter, Lambin and colleagues provide the conceptual background for integrating biophysical and socio-economic data, one of the most important issues in the context of dryland degradation. They illustrate the challenges in linking data at different levels of aggregation and propose an eight-step methodology for place-based research on desertification. This aspect is complemented by Mulligan with a discussion of the concept of modelling land degradation and desertification by integrating the socio-economic and biophysical domains. In addition, the author provides an overview of European funded research in this field. On a more technical side, the chapters by Ustin et al. and Chehbouni et al. focus on the more general aspect of deriving biophysical parameters that are linked to desertification processes. Ustin et al. provide an exhaustive overview of remote sensing-based options to produce vegetation- and soil-related information from remote sensing data using spectral features and model representations of radiation interaction with surfaces of interest. Chehbouni et al. introduce approaches to provide spatially explicit estimates of evapotranspiration to contribute to an efficient water management in dryland agriculture and discuss different measurement setups with respect to their quantitative accuracy.

PART 2   THE GLOBAL PERSPECTIVE: STRATEGIES FOR LARGE AREA MAPPING

Many global initiatives request standardised information for regional to continental scales. Accordingly, the contributions of part 2 of the book introduce data archives suitable for such tasks, investigate methodologies to extract meaningful information from these, and discuss some of the pitfalls arising from the influence of external processes, such as fluctuations in precipitation. Gobron et al. present an algorithm for sensor-independent derivation of the fraction of absorbed photosynthetically active radiance (FAPAR) developed at the European Commission's Joint Research Centre and showcases some examples for Europe and the Sahel. Complementing the archive being developed at the JRC, Friedrich and colleagues present the MEDOKADS (Mediterranean Extended Daily One Km Archive Data Set) archive hosted at the Free University Berlin, which provides recalibrated one km NOAA AVHRR Pathfinder data for Mediterranean Europe and Northern Africa. Udelhoven & Hill demonstrate state-of-the-art time series analysis using the MEDOKADS archive, illustrated with examples from Syria. The next papers address the issue of large area mapping. For the Central and West Asia and North Africa (CWANA) region, Celis & De Pauw present an approach to land cover change mapping using monthly NDVI-composites derived from NOAA-AVHRR time series and a subsequent decision-tree classification. For the non-Saharian part of the African continent, Brivio et al. introduce an environmental anomaly indicator which highlights departure of environmental conditions from a reference state. Using a fuzzy logic approach, it integrates multiple input factors derived from different sources, including remotely sensed information on growing cycles. In the last two papers of part 2, special consideration is given to the role of climate and precipitation driving environmental processes. Chappell and Agnew review the influence of the way meteorological data are recorded and interpolated and quantify uncertainty associated with spatial interpolation, which obviously affects conclusions on the postulated desiccation patterns in the west African Sahel. For the same area, Hountondji et al. investigate relations between rainfall and vegetation dynamics and report a differentiated pattern where seemingly stable areas are contrasted by regions where desertification processes appear to be ongoing.

PART 3   TAKING A CLOSER LOOK: BIOPHYSICAL INDICATORS OF VEGETATION AND SOILS

Part 3 of the book contains six chapters focussing on the derivation of vegetation- and soil-related biophysical indicators for specific areas. It is commenced by Shoshany with a review on the potential of linking advanced remote sensing techniques, such as multi-date Spectral Mixture Analysis, with field-based surveys of plant species distribution and biomass, which is exemplified for transects along climatic gradients. It is followed by a study of König and colleagues on ecological niche modelling using phytosociological samples and Landsat-based satellite imagery to provide spatially explicit information on phytodiversity, thus addressing the link between trends in biodiversity and land degradation processes. Leaf area index as an important state variable in dryland ecosystems is the target of the chapter by Hill et al. They review potentials and limitations of combining satellite imagery from different sensors to parameterize a canopy reflectance model for woody LAI retrieval in sparsely vegetated, heterogeneous rangeland areas. Focusing on the soil domain, Jarmer et al. introduce a methodology combining spectroscopic measurements and chemical analysis of soil samples with C.I.E. colour coordinates to predict soil inorganic carbon. Subsequently, they use the derived relation to map soil inorganic carbon for a test area in the Judean desert using Landsat imagery. Part 3 of the book is concluded by a chapter of García and colleagues who introduce a method to map land degradation status in water-limited ecosystems based on changes in the surface energy balance, where they combine vegetation-related information and the non-evaporative fraction calculated from remote sensing and meteorological field data.

PART 4   STORIES BEHIND PIXELS: PROCESS-BASED ASSESSMENT OF GEOSPATIAL DATA

The last part of the book provides a platform for concepts and applications that emphasize the integration of remote sensing and geoinformation processing for specific questions related to local land use. Röder et al. place particular emphasis on the combination of time series analysis of vegetation cover derived from Landsat images with terrain models derived from digital elevation data and cost-surface based modelling of livestock grazing behaviour to map temporal and spatial trends in grazing-affected rangelands of Northern Greece. Closely connected, Svoray et al. investigate grazing behaviour at a high level of detail using GPS collar data which they combine with a predictive habitat distribution model incorporating forage production. This is an important step towards spatially explicit prediction of grazing pressure, which is an important driver of land degradation. In the last grazing-related chapter, Karnieli et al. demonstrate the use of geostatistical methods, such as kriging, to map local patterns of resource degradation in a rangeland region in Kazachstan characterized by a distinct piospheric pattern. Part 4 of the book is concluded by two case studies illustrating different image interpretation concepts. Clark and Pellikka assess and map landscape change processes in the Taita Hills region in Kenya, which is characterized by an interface of natural, settlement and agricultural areas with considerable change dynamics. They use SPOT XS data from different dates to compare object-oriented classification following multi-resolution segmentation vs. maximum-likelihood classifications and map land use change for a period of 16 years. Eventually, Inoue and colleagues present a study in a highly vulnerable mountain ecosystem in northern Laos where they investigate the connection between carbon stock capacity, food security and sustainable use of forestry resources in an ecosystem traditionally characterized by slash-and-burn culture. They combine a multi-temporal Landsat data set with spectral surface reflectance measurements and field measurements of biomass, soil CO_2 fluxes, and soil carbon content to characterize the potential impact of different management priorities and land use decisions on the goods and services provided by this ecosystem.

This book addresses specialists working at the interface of remote sensing and geoinformation processing with a focus on desertification and land degradation. Yet, it is also intended as a point of reference for stakeholders without deeper experience in spatial sciences who seek to better understand potentials and limitations of spatial technologies in their field. Its scope and the wide range of methods and topics reflect the multi-disciplinary and multi-faceted character of the underlying problem, which can not be analysed as an isolated problem. Rather, land degradation and desertification need to be perceived in their inter-connectedness with other global issues, such as global climate change, the preservation of biodiversity and the maintenance of ecosystem's capacity to deliver vital goods and services. Against this background, the contributions in this volume make a strong case for state-of-the-art analysis of geospatial data, which are continuously gaining importance in facing present and future challenges of managing the Earth's resources in a sustainable way.

Part 1
Setting the scene: principles
in remote sensing and spatial scene modelling
for land degradation assessment

Recent Advances in Remote Sensing and Geoinformation Processing
for Land Degradation Assessment – Röder & Hill (eds)
© 2009 Taylor & Francis Group, London, ISBN 978-0-415-39769-8

Coupled human-environment system approaches to desertification: Linking people to pixels

E.F. Lambin
University of Louvain, Louvain-la-Neuve, Belgium

H. Geist
School of Geosciences—Geography and Environment, University of Aberdeen

J.F. Reynolds
*Division of Environmental Science and Policy, Nicholas School of the Environment and Earth Science
and Department of Biology, Duke University, Durham, USA*

D. Mark Stafford-Smith
CSIRO Sustainable Ecosystems, Centre for Arid Zone Research, CSIRO, Alice Springs, Australia

ABSTRACT: The first part of this paper provides the rationale for approaching desertification by integrating biophysical and socio-economic data. We start with a brief overview of the causes and consequences of land degradation in global drylands, with an emphasis on the interaction between human and natural dimensions of the problem. We then introduce a new paradigm on desertification that accounts for the need to adopt an integrated approach to the problem. The second part discusses methodologies to achieve that integration by linking remote sensing data with household survey data. The combination of remote sensing and socio-economic data is increasingly used to better assess land-use changes but little work has been conducted in the context of drylands. Different approaches have been followed: socio-economic surveys to explain or supplement observed patterns of land-cover change; overlay analyses of spatially-explicit socio-economic and land cover data; proximity analyses; interpretations of spatial patterns of land use in terms of land use practices; and joint statistical analyses of spatially-explicit household survey and land cover data. After a review of the literature, we propose an eight-step methodology to conduct integrated, place-based research on desertification.

1 INTRODUCTION

The definition of desertification used by the Convention to Combat Desertification (CCD) makes it clear that whilst biophysical components of ecosystems and their properties are involved (e.g. soil erosion, loss of vegetation), the interpretation of change as 'loss' is dependent upon the integration of these components within the context of the socio-economic activities of human beings. The CCD states that land degradation is the reduction or loss of the biological and economic productivity and complexity of terrestrial ecosystems, including soils, vegetation, other biota, and the ecological, biogeochemical, and hydrological processes that operate therein. The CCD's definition of desertification explicitly focuses on the linkages between humans and their environments that affect human welfare in arid and semi-arid regions: land degradation in arid, semi-arid and dry sub-humid areas resulting from various factors, including climatic variations and human activities.

Unfortunately, the CCD definition of desertification is not amenable to easy quantification, especially as a single number or synthetic index. Most estimates of desertification are derived solely from either biophysical factors (e.g. soil erosion, loss of plant cover, change in albedo) or socio-economic factors (decreased production, economic loss, population movements, etc), but rarely both types simultaneously. Much of the confusion surrounding rates of desertification and regions affected could be eliminated by focusing on a small number of critical variables that contribute to an understanding of the *cause*, rather than *effect*, of desertification (Stafford Smith & Reynolds 2002).

This paper is divided in two parts: a first part provides the rationale for approaching desertification by integrating biophysical and socio-economic data, and a second part discusses methodologies to achieve that integration, by linking remote sensing data with household survey data. Firstly, we provide a brief overview of the causes and consequences of land degradation in global drylands, with an emphasis on the interaction between human and natural dimensions of the problem. We then introduce a new paradigm on desertification that accounts for the need to adopt an integrated approach to the problem. Secondly, we discuss methods to link remote sensing and socio-economic data to better assess land-use changes, first in general and then in the context of drylands. After a review of the literature, we propose an eight-step methodology to conduct integrated, place-based research on desertification.

2   THE CAUSES OF DESERTIFICATION

Land degradation in drylands is still poorly documented and its causes are hardly understood (Lambin et al. 2003). On the one hand, proponents of single-factor causation suggest various primary causes, such as irrational or unwise land mismanagement by nomadic pastoralists and growing populations in fragile semi-arid ecosystems. Central to this understanding is the notion of 'man-made deserts', i.e. the human-driven, irreversible extension of desert landforms (Le Houérou 2002). On the other hand, desertification has been attributed to multiple causative factors that are specific to each locality, revealing no distinct pattern (Dregne 2002, Warren 2002). There has also been a great deal of debate on whether the causes of desertification lie in the socio-economic or biophysical spheres (human-induced land degradation *versus* climate-driven desiccation).

Geist & Lambin (2004) carried out a worldwide review of the causes of desertification based on 132 carefully selected case studies. Their aim was to generate a general understanding of the proximate causes and underlying driving forces of desertification, including cross-scalar interactions of causes and feedbacks, while preserving the descriptive richness of these case studies. Results showed that desertification is driven by a limited suit of recurrent core variables, with identifiable regional patterns of causal factor synergies.

At the proximate level, desertification is best explained by the combination of multiple social and biophysical factors, rather than by single variables. Dominating the broad clusters of proximate factors is the combination of agricultural activities, increased aridity, extension of infrastructure, and wood extraction (or related extractional activities), with clear regional variations. In particular, agricultural activities and increased aridity are associated together.

Agricultural activities include extensive grazing, nomadic pastoralism and annual cropping. Livestock production activities slightly outweigh crop production as a cause of desertification, but both activities remain intricately interlinked in most of the cases. In comparison with pastoralism, agricultural expansion into marginal rangeland areas during wet periods often leaves farmers more seriously exposed to hazard when drought returns (Glantz 1994). Agricultural expansion on areas previously used for pastoral activities can also lead to overstocking on the remaining, reduced rangeland. In addition, it can trigger soil degradation at sites that are not suitable, in particular, for permanent agriculture. Increased aridity is a robust proximate cause of desertification, through greater rainfall variability and prolonged droughts.

The extension of infrastructure associated with desertification is frequent in cases from Asia, Africa, and Australia mainly. Desertification is mostly linked to the development of water-related infrastructure for cropland irrigation and pasture development (reservoirs, dams, canals, boreholes). This in turn leads to a decrease in livestock mobility (Niamir-Fuller 1999). Water droughts are replaced with feed droughts. In the Asian and African cases, the build-up of irrigation infrastructure is associated with expanding human settlements, following an increase in food production and food security.

At the underlying level, desertification is also best explained by regionally distinct combinations of multiple, coupled social and biophysical factors, and drivers acting synergistically rather than by single-factor causation. A recurrent and robust broad factor combination implies the interplay of climatic factors leading to reduced rainfall, agricultural growth policies, newly introduced land use technologies, and malfunct land tenure arrangements which are no longer suited to contemporary dryland ecosystem management.

Climatic factors, mainly associated with a decrease in rainfall are prominent underlying driving forces of desertification (Puigdefábregas 1998). They operate either through the (indirect) impact of rainfall variability via changes in land use or by directly impacting upon land cover in the form of prolonged droughts. Technological innovations are associated with desertification as frequently as are deficiencies of technological applications. Innovations do mainly comprise improvements in land and water management through motor pumps and boreholes, or through the construction of hydrotechnical installations such as dams, reservoirs, canals, collectors, and artificial drainage networks. When applied, these developments often cause high water losses due to poor maintenance of the infrastructure. The disaster of the Aral Sea is an extreme case of such a perturbation (Saiko & Zonn 2000).

Among institutional and policy factors underlying desertification, modern policies and institutions are equally involved as are traditional institutions. Growth-oriented agricultural policies including measures such as land (re)distribution, agrarian reforms, modern sector development projects, diffusion of agricultural intensification methods, and market liberalization policies are as important underlying causes of desertification as are institutional aspects of traditional land tenure such as equal sharing of land and splintering of herds due to traditional succession law, thus reducing flexibility in management and increasing the pressure upon constant land units. The collapse of necessary conditions for effective maintenance of common property grazing regimes—e.g. reduced capacity to prevent encroachment by other land uses—is another cause. The introduction of new land tenure systems, be it private (individual) or state (collective) management, is another important factor associated with desertification. Uncertain land tenure may arise from the overlapping of conflicting property-rights regimes, often leading to violent conflicts about land, thus reducing the adaptive capacity of herding and farming populations.

Often, the growth or increased economic influence of urban population triggers out-migration of poor cultivators and/or herders from high potential, peri-urban zones onto marginal dryland sites. Consequently, the sometimes rapid increases in the size of local human populations are often linked to in-migration of cultivators onto rangelands or in regions with large-scale irrigation schemes, or of herders onto hitherto unused, marginal sites, with the consequence of rising population densities there.

It is mostly the interactions between multiple causal factors that lead to desertification. A frequent pattern of causal interactions stems from the necessity for water-related infrastructures that are associated with the expansion of irrigated croplands and pastures. Typically, newly introduced irrigation infrastructures induce accelerated in-migration of farm workers into drylands, and often stir more commercial-industrial developments as well as the growth of human settlements and related service economies. Irrigation infrastructure is often nested in a system of larger infrastructure extension related to regional economic growth. Commonly, road extension paves the way for the subsequent extension of irrigation, and (semi)urban land uses. In the developing world, underlying these proximate factors are national policies aimed at consolidating territorial control over remote, marginal areas, and attaining self-sufficiency in food and clothing, with rice and cotton being the key irrigated crops.

3   CONSEQUENCES OF DESERTIFICATION

Desertification is only a problem if it threatens the short- or long-term welfare of users of affected ecosystems. Actually, the definition of what is a significant change in environmental conditions depends on the uses to which the environment is put (Lynam & Stafford Smith 2003). Since natural resources are only one of many elements that make up livelihoods, which exist in very complex environmental, social and political milieux, the impacts of desertification can only be evaluated in their social context (Warren 2002). Empirical measures of land degradation must therefore be linked to socio-economic variables that represent how people collect information to evaluate threats to their natural resources, human behaviours and motivation with respect to resource management, and the capacity of land managers to respond to potential threats through their access to technology, social and economic resources and their risk-avoidance strategies (Lambin et al. 2003, Warren et al. 2001).

From the socio-economic point of view, most consequences of desertification (especially in pastoral systems) are a direct result of the decline in 'productivity' or the capacity of the land to support plant growth and animal production. During early stages of desertification such losses are compensated by the social resilience of the local human populations, especially in developing countries, or by economic inputs from government (Vogel & Smith 2002). However, when certain thresholds are crossed, social resilience or government subsidies may not be enough to compensate for the loss of productivity, and this fuels a battery of socio-economic changes that range from modifications in trade promoted by lower agricultural production to large population migrations (Fernandez 2002). Therefore, in any assessment of desertification, the level of degradation of the biophysical environment cannot be separated from the capacity of the affected community to mobilize internal or external resources to respond to and cope with the degradation, and recover the system. The concept of resilience establishes the link between risks of degradation of the human-environment system and impacts on human well-being. Biophysical and socio-economic indicators need always to be associated for any meaningful evaluation of the severity, for a given stakeholder, of land degradation.

4   INTEGRATED APPROACH TO DESERTIFICATION

The simultaneous assessment of biophysical and socio-economic drivers and consequences of desertification has been recognized as one of the most challenging topics for further research. Stafford Smith & Reynolds (2002) proposed the Dahlem Desertification Paradigm (DDP) that is unique in two ways: it attempts to capture the multitude of interrelationships within human-environment systems that cause desertification within a single, synthetic framework; and it is testable, which ensures that it can be revised and improved upon. The DDP embraces a hierarchical view of land degradation and highlight key linkages between socio-economic and biophysical systems at different scales.

The monitoring of desertification is an increasingly important development in the management of dryland areas. The establishment of long-term and rigorous monitoring programs is an effective way to assess the status of natural resources and the evolution of desertification processes. Increased research is dedicated to the development of easily accessible monitoring methods based on simple soil and vegetation indicators, and combining ground-based methods with remote sensing data (Pyke et al. 2002). The challenge, however, is to monitor not just biophysical indicators but the coupled human-environment system as a whole. This requires methods to link biophysical data to socio-economic data on land managers, to capture the varying capacity of local agents to cope and respond to a decline in land productivity.

Informal monitoring by land managers tends to focus on those factors that they perceive readily and that vary from day to day as a result of the variable external environment—i.e. "fast" variables such as rainfall, grass growth, animal condition, and market prices (Lynam & Stafford Smith 2003). However, there is a growing acceptance that a small number of "slow" variables act as the critical determinants in human-environment systems (Gunderson & Holling 2002). Monitoring systems of desertification should therefore be designed around these slow variables which warn for impending fundamental change when some thresholds are crossed—e.g. shrub encroachment, cropland expansion, change in land tenure, opening of new export markets for agricultural products. In arid and semi-arid ecosystems, the noise in the fast variables hides any signal in the critical slow variables (Lynam & Stafford Smith 2003). Hence the need for long time series of observation and appropriate methods to isolate the slow trends from the fast fluctuations.

5 LINKING PEOPLE TO PIXELS

Recent studies demonstrated the complementarity of remote sensing and socio-economic survey data for understanding causes, processes and impacts of land-use/land-cover changes. Most studies linking socio-economic with remote sensing data are aimed at better understanding human-environment relationships and predicting the impact on land resources of changes in land use. Beyond a simple understanding of drivers of land-use change, such studies also highlight dynamic interactions between land-use changes and processes of landscape transformation (e.g. soil erosion, water availability, modifications in climate patterns induced by land-cover changes, spread of vector-borne diseases, etc.). This can lead to assessments or scenarios of sustainability of land use and vulnerability (or resilience) of communities, based on models of human-environment interactions.

Creating a direct link between spatially-explicit land cover information, as derived by remote sensing, and information on land-use change processes requires the development of new methods and models which are merging landscape data with data on human behaviour (Liverman et al. 1998). Several studies combined socio-economic household survey data and remote sensing data to better understand processes of land-use change (Guyer & Lambin 1993, Sussmann et al. 1994, Behrens et al. 1994, Entwisle et al. 1998, Moran & Brondizio 1998). One of the major challenges of merging these data from heterogeneous sources concerns the definition of the appropriate spatial observation units, i.e. the appropriate level of aggregation of information derived from the domains of social phenomena and natural environment (Liverman et al. 1998). While conceptually straightforward, developing the appropriate linkages between household-level and remote sensing datasets can be difficult to implement operationally (Entwisle et al. 1998). In remote sensing, the spatial unit of observation is not directly associated to any unit of observation in the social sciences, e.g. individuals, households or communities.

Most of the studies linking remote sensing observations and socio-economic data have been performed at the scale of administrative units (e.g. Green & Sussman 1990, Geoghegan et al. 2001). For instance, Wood & Skole (1998) and Pfaff (1996), in their studies of deforestation in the Brazilian Amazon, have aggregated the land-cover data to conform to the administrative units (*municipio* or county level, in these cases). In this approach, the dependent variable (deforestation derived from remote sensing data) refers precisely to the independent variables (e.g. socio-demographic data obtained through the national census and transport infrastructure density), and vice versa. Entwisle et al. (1998) attempted to link population dynamics, derived from field surveys, to land-use/land-cover change data in Thailand. They noted the difficulty of relating remotely-sensed patterns of land-cover change with field observations of land-use change since people live in nucleated villages away from their fields, and since households cultivate multiple non-contiguous plots. For this reason, the integration of the two datasets was performed at the village level.

Aggregated to the village-level, household data offer an additional perspective to the remotely sensed land-cover dynamics. The village profiles provide a crosscheck on the dynamics observed by remote sensing, and can be related to remotely sensed landscape variables (Entwisle et al. 1998, Mertens et al. 2000). However, the aggregation of land-cover change data to a coarse resolution leads to a loss of information as it obscures the variability within the units (Wood & Skole 1998) and the role of the heterogeneity between actors (i.e. social networks, leadership status, role of education, differing access to resources, knowledge and land, etc.) in driving land-use changes. Moreover, this aggregation can introduce "ecological fallacies" in the interpretation of statistical correlations. When the measures of statistical association are calculated across administrative units, the data do not correspond to the level of the decision units (Wood & Skole 1998).

For these reasons, more recent research efforts have attempted to integrate remote-sensing observations and field surveys at finer levels of aggregation, i.e. at the scale of households. This property level linkage between remote sensing and socio-economic data requires extremely precise fieldwork to georeference every plot used (i.e. travelling with the household to every plot to collect GPS coordinates, or identifying the user of every plot based on detailed maps) and to ensure a high geometric accuracy when matching pixels with plots. This is achieved at a certain cost. McCracken et al. (1999) successfully analysed land-cover changes at the property level, because each household lived on a well-defined plot and had little impact on adjacent land. Individual household data allow for a better understanding of the land-use practices within each village, as most land-use decisions are made by individuals and households.

While a village-level linkage of socio-economic and remote sensing data often includes the collection of some data at the household level, only the boundaries of the land used at the village-level—rather than at the level of individual plots—is georeferenced. This is done either by assuming a maximum travel distance to the plots from the houses or by identifying, for each village, the boundary of land use on a map with a key informant. The optimal level of analysis of a joint remote sensing-socio-economic study depends on the research question. It involves a trade-off between the information that can be extracted with a reasonable level of accuracy and the cost of field data collection. Working at the finest possible level is not always the optimal strategy. However, aggregated level analyses tend to oversimplify explanation and modelling of the phenomenon by focussing on average behaviour rather than on the heterogeneity between actors.

Rindfuss et al. (2003) provide an in-depth discussion of the methodological and practical problems in designing a study linking household and remotely sensed data, including issues of ephemeral households, boundaries of farms and villages, prospective or retrospective study designs, differences in sampling between land-to-people or people-to-land approaches, connection with households through land use or land ownership, plot and pixel size mismatch, sample size, etc. One of the most difficult problems is the difficulty to protect the identity of surveyed households, for obvious confidentiality reasons, once their plots have been georeferenced. Releasing survey data with geographic coordinates associated with the plots used or owned by every household is tantamount to disclosing the identity of the respondent households (Rindfuss et al. 2003).

6 THE CHALLENGES OF LINKING PEOPLE TO PIXELS IN DRYLANDS

Land use in drylands is characterised by a number of features that make the fine scale linking of socio-economic and remote sensing data particularly difficult: extensive agriculture (including pastoralism) practiced over large plots with fuzzy boundaries and communal ownership; often complex and flexible tenure arrangements regulating land access across space and time; mixture of natural vegetation and managed rangelands or croplands (e.g. tree cultivation); multifunctional landscapes (e.g. grazing in crop fields after the harvest); mobility of households and their livestock in transhumant or nomadic farming systems: households move across the landscape in search of forage and water, in the face of environmental uncertainty and seasonal fluctuations; large inter-annual variability in climatic conditions; and risk-avoidance land use strategies that can only be understood by adopting a long-term perspective, cutting across several climatic cycles.

For these reasons, most previous studies linking people to pixels at the household level have been conducted in moist tropical forest environments. However, a few pioneer studies have taken place in rangelands. They have combined socio-economic with remote sensing data following different approaches. These approaches are presented in order of increasing level of integration of the socio-economic and remote sensing data sources. The selection of studies below suffers from an African bias, given the background of the lead author.

6.1 *Socio-economic survey to explain or supplement observed patterns of land-cover change*

The following studies included analyses of remote sensing and socio-economic data, but these two data sources were not linked and jointly processed. Rather, they were part of two separated but complementary components of a research project. Ringrose et al. (1996a) compared remote sensing data to map vegetation recovery to drought over a ten-year period in Botswana. Semi-structured interviews and group discussions were then conducted to assess the impacts of the observed changes on local people and to evaluate their perception of changes. People were keenly aware of declining productivity but mostly attributed this to drought rather than to intensive grazing and groundwater use. In south-western Burkina Faso, Gray (1999) has measured agricultural expansion and land degradation at the landscape scale, based on a time series of aerial photographs. A household survey explained how farmers intensify their production system to prevent a reduction of soil quality in their fields. Abbot & Homewood (1999) measured a decline in closed canopy miombo woodland in a protected area in Malawi. Surveys of fuelwood use by different actors revealed that domestic use has a much lower impact on woodlands than wood collection for commercial fish smoking, highlighting the importance of disaggregating different wood use practices by different agents. Mbow et al. (2000) compared the spatial and temporal distribution patterns of fires in central Senegal (as detected using remote sensing techniques) to local population's perceptions of the causes and consequences of fires (as revealed by interviews). They showed that fire use practices are closely linked to livestock grazing and crop production, and that the local population has a high degree of awareness about the application of fire. In Inner Mongolia, Jiang (2003) showed how remote sensing analysis offers additional stories about landscape changes that supplement ethnographic research. Analysis of landscape-scale spatial patterns and diverse trends of change—in this case, increase of both sandy land and highly productive land covers—provided a cross-check with ethnographic data on cultural change, and was combined with these data to enrich understanding of culture-landscape interactions.

6.2 Overlay of spatially-explicit socio-economic and land cover data, and proximity analyses

These studies computed the spatial overlap in the distributions of variables related to land use and land cover, or the proximity of landscape attributes to point (settlements, water points) or line (roads, rivers) features, using functionalities of geographic information systems (GIS). As such, spatial overlaps do not allow making causal inference—a jump that is too often made in an unwarranted way (Turner 2003). Several authors have proposed methodologies to assess and map desertification based on the overlay in a GIS of biophysical and human indicators combined with simple ecosystem models (Grunblatt 1992). Typical variables include rainfall, soil type, elevation and slope, vegetation cover, livestock density, and human population density. Key to the integration of these variables in a desertification risk map is the weighting assigned to different layers. The models used to combine variables tend to make simplistic assumptions on land management strategies. In a more advanced version of this GIS-based approach, Mallawaarachchi et al. (1996) combined economic and ecological models within a spatially consistent framework to track the socio-economic implications associated with soil erosion and evaluate the opportunity costs of land degradation. Amissah-Arthur et al. (2000) combined data on soil quality, land use intensity, and population density to measure the carrying capacity and vulnerability to land degradation of landscapes in Niger. Crop fields and fallows were mapped by classification of remote sensing data. The area of influence of villages was defined by assuming a maximum travel distance from settlements for farming-related activities. This study documented a process of land abandonment on some soil types and cropland expansion on marginal land and at increasing distances from villages.

In an assessment of desertification around deep wells in the Ferlo, Senegal, Hanan et al. (1991) have overlaid a map of deep wells and their area of influence on satellite measurements of a vegetation index, as a proxy for primary production. They found no consistent relationship between primary production and proximity to a well along a series of transects from these deep wells. They concluded that grazing around wells has not created "desert patches". By contrast, Ringrose et al. (1996b) found in southern Botswana that degraded vegetation was concentrated around villages and boreholes. Land cover classes containing mainly woody weeds, sparse vegetation cover and large areas of bare soil—which could all be identified on remotely sensed data—were associated spatially with cattle grazing and bush product harvesting. Rasmussen et al. (1999) mapped areas suitable for winter grazing in Mongolia by combining a digital elevation model with a land-cover map derived from a satellite image. They overlaid the location of family winter settlements and the areas being grazed, and calculated stocking rates by household. They demonstrated the uneven distribution of grazing pressure.

6.3 *Interpretation of spatial patterns of land use in terms of land use practices*

The following studies provided a broader context or generalized to a larger region relationships and changes observed by more localized fieldwork. They did so by interpreting spatial patterns of land use in terms of land use practices or processes of land-use change. For example, mechanized farming is associated with larger, more geometric and more homogenous fields compared to plots under traditional agro-forestry practices. Certain categories of land-use changes tend to fragment the landscape (e.g. expansion of smallholder farming, land privatisation) while others increase landscape homogeneity (e.g. mechanized cultivation or ranching over large areas). These studies therefore exploited the correlation between remotely-sensed spatial patterns and farming system attributes, in an implicit model inversion procedure. Lambin (1988) found statistically that, in Burkina Faso, spatial patterns interpreted from Landsat MSS data correlate more closely with ethnic groups and farming systems than with any other physical or cultural landscape variables. Each ethnic group had a clear landscape-scale footprint visible on remote sensing data. At a finer spatial scale, Guyer & Lambin (1993) succeeded in discriminating and quantifying the total area and proportion of tractor-cleared and hand-cleared fields in a region of Nigeria by using several shape criteria derived from multispectral SPOT data. These authors also computed the crop-fallow cycle and the importance of the land reserve using a remote sensing-based land-cover map. Millington et al. (1999) used multi-date remotely-sensed imagery to construct a typology of field (large *versus* small, rain-fed *versus* irrigated, permanent *versus* temporary) and to analyse their dynamics in northern Jordan. Semi-structured interviews with farmers uncovered the reasoning behind changing farming systems and cultivation practices. Access to wells and synergy between livestock and cultivation systems were important factors as part of complex land colonization processes.

6.4 *Joint statistical analysis of spatially-explicit household survey and land cover data*

This last group of studies collected socio-economic and landscape data at the level of households, georeferenced all these data, and integrated these in a joint statistical analysis. They achieved a higher level of integration of socio-economic and remote sensing data compared to the previous set of studies. Reenberg et al. (1998) studied field expansion in the desert margin in Burkina Faso by explaining patterns of land use change (observed from a long time series of aerial photos and SPOT satellite images). Their explanation is based on a household survey aimed at understanding parameters of agricultural decisions. Land use information obtained during interviews was georeferenced and then linked with the successive land-use maps and with field observations associated with GPS measurements. Their descriptive analysis, supported by these quantitative data, highlighted an expansion of cultivation that is influenced spatially by climate variations and is associated with a reallocation of fields to different soil types. Thompson et al. (2002) analysed production systems amongst a Maasai pastoral population adjacent to the Masai Mara National Reserve in Kenya, where land-use systems have recently diversified (i.e. large scale mechanised cultivation and tourism in addition to traditional pastoral livestock keeping). A set of landscape, accessibility and socio-economic variables measured at the household level were combined in a generalised logit model to assess the determinants of land-use strategies adopted by different households. The model resulted in accessibility to the park and the nearest market centre as the variable with most explanatory power, followed by landscape attributes. Leadership, education and wealth were also significantly linked to mechanised cultivation as a choice of land use. Results showed that decision-making on land-use is influenced by both location and socio-economic factors. Households were often involved in several land use activities that have an impact on the environment in a wide range around the homestead, so linking households to changes in land cover as measured by remote sensing was difficult, except for mechanised farming.

Galvin et al. (2001) explored the effects of the 1997 drought and El Niño rains on Maasai herders in northern Tanzania. They used satellite NDVI data from the AVHRR sensor to scale up from the household to the region, using analyses of household responses to climate variability. These responses, as revealed by in-depth household interviews using ethnographic research techniques, were related to regional vegetation patterns. They thus created a method to generalize to regions the effects of changes in rainfall patterns on pastoralist households. At a finer spatial scale, a few authors have collected spatial data on grazing itineraries of herders and linked these to household-level survey data on social determinants of grazing management modes and grazing intensity in West Africa (Turner 2003) and East Africa (Coppolillo 2000, 2001). The later study demonstrated considerable household-level variation in herding practice and in resultant levels of cattle productivity, which is not apparent at the population level. In Kajiado, Kenya, Burnsilver et al. (2003) analysed household-level socio-economic data, daily spatial data on pastoral resource use, and landscape heterogeneity and seasonal variability data derived from 1km resolution time series of a vegetation index (NDVI) and from other biophysical data. Socio-economic data related to household strategies, herd size and composition, marketing strategies, and spatial movements. Pathways of daily grazing movements—collected by following the main cattle herds with a hand-held GPS—were overlaid on landscape heterogeneity indices. The results confirmed that ecological heterogeneity is crucial for the maintenance of pastoralism in highly variable environments.

7 AN EIGHT STEP METHODOLOGY

While the need for an integration of biophysical and socio-economic variables to assess desertification status is well illustrated in the studies reviewed above, is it possible to suggest general guidelines on how to achieve this in practice? Based on past experiences, one could propose a comprehensive methodology to assess desertification in an integrated way for a given place. This methodology is inductive as it starts with a description of patterns of land change and proceeds towards an understanding of processes and causal links in land-use change. It is based on previous empirical studies that have generally only focused on some of the steps outlined below. This methodology has therefore never been applied in such a comprehensive way so far. It is an attempt to "put the pieces" together to gain the methodological capacity to implement a truly integrated human-environment approach to case studies in drylands as advocated above. It thus attempts to allow researchers to move from the theory of integrated desertification assessment to the practice of place-based empirical and integrated studies of desertification. Note that this sequence of analytical steps probably applies to any land-use change issue and is not specific to drylands. The successive steps are:

1. Land cover mapping and assessment of current state of ecosystem, using a combination of continuous spatial coverage based on remote sensing and cartographic data, and measurements of ecosystem attributes in a sample of locations;
2. Reconstruction of past changes in land cover and ecosystem attributes based on retrospective data, also combining remote sensing, cartographic and field data, at several spatial and temporal scales (plot- to landscape-levels; recent past to historical reconstruction); Change in the slow (biophysical and socio-economic) variables is of main interest here;
3. Evaluation of the significance of observed change for the welfare and livelihood of the multiple users of the ecosystem, through an assessment of the potential impact of the changes on the key goods and services delivered by the ecosystem for specific stakeholders;
4. Spatial association analysis between observed patterns of change, and natural and cultural landscape variables and/or census socio-economic data at the level of administrative units, using GIS tools (overlay and proximity analysis) and multivariate analysis; This step aims at understanding the proximate (or direct) causes of change;

5. Causal analysis based on in-depth and spatially-explicit household survey and macro-level analysis, to understand how agents make decisions about land use and resource management and how these decisions impact ecosystems; This step aims at understanding the underlying (or indirect) causes of change; It is based on statistical methods to link people (or agents) to pixels through household surveys and georeferencing of the land parcels used and/or owned by these households, as described in the previous sections;
6. Investigation of coping strategies of agents to observed and predicted changes, including the portfolio of technological, institutional and social responses, to assess the resilience of the socio-ecological system;
7. Simulation of the dynamic of the socio-ecological system—e.g. using multi-agent simulations—to predict possible future trajectories of the system; This exercise produces more realistic outcome if it is conducted in a participatory way, involving representatives of key stakeholders. There is an emerging body of literature on multi-agent simulations and participatory approaches in drylands, e.g. in the African Sahel (Becu et al. 2003, Bah et al. 2006). These studies do not yet integrate remote sensing data but this would be an easy step to make to better represent landscape heterogeneity and its dynamic;
8. Based on the previous steps, assessment of desertification risk and possible policy interventions to increase the resilience of the socio-ecological system.

8   CONCLUSION

The above paper has first provided compelling reasons for an integration of biophysical and socio-economic variables to assess desertification status. This is related to the nature, causes and consequences of desertification. It then discussed new methodologies linking remote sensing data to household survey data to achieve this goal. These methodologies have still to be widely developed and applied to land use in drylands, which poses particularly difficult challenges. Based on the studies reviewed in this chapter, we outlined a sequence of steps to achieve in practice an integration of biophysical and socio-economic variables to assess desertification status for specific localities. This approach integrates geospatial and socio-economic data at the level of agents.

Understanding complex human-environment dynamics in drylands requires moving beyond simple map overlay approaches. Remote sensing can add to the broader human ecological analysis and spatial analysis must be informed by the more complete understanding of causal connections uncovered by detailed field work (Turner 2003). Remote sensing is just one step in a broader, more comprehensive methodology.

ACKNOWLEDGEMENT

This paper has greatly benefited from ideas developed within several workshops of the Land-Use and Cover Change (LUCC) project.

REFERENCES

Abbot, J.I.O. & Homewood, K. 1999. A history of change: causes of miombo woodland decline in a protected area in Malawi. *Journal of Applied Ecology* 36:422–433.

Amissah-Arthur, A., Mougenot, B. & Loireau, M. 2000. Assessing farmland dynamics and land degradation on Sahelian landscapes using remotely sensed and socio-economic data. *International Journal of Geographical Information Science* 14:583–599.

Bah, A., Toure, I., Le Page, C., Ickowicz, A. & Diop, A.T. 2006. An agent-based model to understand the multiple uses of land and resources around drillings in Sahel. *Mathematical and Computer Modelling* 44(5–6):513–534.

Becu, N., Bousquet, F., Barreteau, O., Perez, P. & Walker, A. 2003. A methodology for eliciting and modelling stakeholders' representations with agent based modelling. *Lecture Notes in Artificial Inteligence* 2927:131–148.

Behrens, C.A., Baksh, M.G. & Mothes, M. 1994. A regional analysis of Bari land use intensification and its impact on landscape heterogeneity. *Human Ecology* 22(3):279–316.

BurnSilver, S.B., Boone, R.B. & Galvin, K.A. 2003. Linking pastoralists to a heterogeneous landscape. In J. Fox, R.R. Rindfuss, S.J. Walsh & V. Mishra (eds), *People and the Environment: Approaches for linking household and community surveys to remote sensing and GIS*. Boston: Kluwer Academic Publishers.

Coppolillo, P.B. 2000. The landscape ecology of pastoral herding: Spatial analyses of land use and livestock production in East Africa. *Human Ecology* 28:527–560.

Coppolillo, P.B. 2001. Central-place analysis and modeling of landscape-scale resource use in an East African agropastoral system. *Landscape Ecology* 16:205–219.

Dregne, H. 2002. Land Degradation in the Drylands. *Arid Land Research and Management* 16:99–132.

Entwisle, B., Walsh, S.J., Rindfuss, R.R. & Chamratrithirong A. 1998. Land-use/land-cover and population dynamics, Nang Rong, Thailand. In D. Liverman, E.F. Moran, R.R. Rindfuss & P.C. Stern (eds), *People and pixels: linking remote sensing and social science*: 121–144. Washington D.C.: National Acadamy Press.

Fernandez, R.J., Archer, E.R.M., Ash, A.J., Dowlatabadi, H., Hiernaux, P.H.Y., Reynolds, J.F., Vogel, C.H., Walker, B.H. & Wiegand., T. 2002. Degradation and recovery in socio-ecological systems: A view from the household/farm level. In J.F. Reynolds & D.M. Stafford Smith (eds), *Global Desertification: Do Humans Cause Deserts?* (Dahlem Workshop Report 88), 297–323. Berlin: Dahlem University Press.

Galvin, K.A., Boone, R.B., Smith, N.M. & Lynn, S.J. 2001. Impacts of climate variability on East African pastoralists: linking social science and remote sensing. *Climate Research* 19:161–172.

Geist, H.J. & Lambin., E.F. 2004. Dynamic causal patterns of desertification. *Bioscience* 54:817–829.

Geoghegan, J., Villar, S.C., Klepeis, P., Mendoza, P.M., Ogneva-Himmelberger, Y., Chowdhury, R.R., Turner, B.L. & Vance, C. 2001. Modelling tropical deforestation in the southern Yucatan peninsular region: comparing survey and satellite data. *Agriculture, Ecosystems and Environment* 85:24–46.

Glantz, M.H. 1994. Drought follows the plow: Cultivating marginal areas. Cambridge, New York: Cambridge University Press.

Gray, L.C. 1999. Is land being degraded? A multi-scale investigation of landscape change in southwestern Burkina Faso. *Land Degradation Development* 10:329–343.

Green, G.M. & Sussman, R.W. 1990. Deforestation history of the eastern rain forests of Madagascar from satellite images. *Science* 2:212–215.

Grunblatt, J., Ottichilo, W.K. & Sinange., R.K. 1992. A GIS approach to desertification assessment and mapping. *Journal of Arid Environments* 23:81–102.

Guyer, J. & Lambin, E.F. 1993. Land use in an urban hinterland: Ethnography and remote sensing in the study of African intensification. *American Anthropologist* 95(4):839–859.

Hanan, N.P., Prevost, Y., Diouf, A. & Diallo, O. 1991. Assessment of desertification around deep wells in the Sahel using satellite imagery. *Journal of Applied Ecology* 28:173–186.

Gunderson, L.H. & Holling, C.S. 2002. *Panarchy: understanding transformations in human and natural systems*. Washington, D.C.: Island Press.

Jiang, H. 2003. Stories remote sensing images can tell: integrating remote sensing analysis with ethnographic research in the study of cultural landscapes. *Human Ecology* 31(2):215–232.

Lambin, E.F. 1988. L'apport de la télédétection dans l'étude des systèmes agraires d'Afrique. *Africa* 58:337–352.

Lambin, E.F., Geist, H.J. & Lepers, E. 2003. Dynamics of land-use and land-cover change in tropical regions. *Annual Review of Environment and Resources* 28:205–241.

Lambin, E.F. 2005. Conditions for sustainability of human-environment systems: information, motivation, and capacity. *Global Environmental Change*. In press.

Le Houérou, H.N. 2002. Man-Made Deserts: Desertization Processes and Threats. *Arid Land Research and Management* 16:1–36.

Liverman, D., Moran, E.F., Rindfuss, R.R. & Stern, P.C. (eds) 1998. *People and pixels: linking remote sensing and social science*. Washington, D.C.: National Academy Press.

Lynam, T. & Stafford Smith, M. 2003. Monitoring in a complex world: seeking slow variables, a scaled focus and speedier learning. In N. Allsopp, A.R. Palmer, S.J. Milton, K.P. Kirkman, G.I.H. Kerley, C.R. Hurt & C.J. Brown (eds), *Proceedings of the VIIth International Rangelands Congress*: 617–629. Durban.

Mallawaarachchi, T., Walker, P.A., Young, M.D., Smyth, R.E. & Lynch, H.S. 1996. GIS-based integrated modeling systems for natural resource management. *Agricultural Systems* 50:169–189.

Mbow, C., Nielsen, T.T. & Rasmussen, K. 2000. Savanna fires in East-Central Senegal: Distribution patterns, resource management and perceptions. *Human Ecology* 28:561–583.

McCracken, S.D., Brondizio, E.S., Nelson, D., Moran, E.F., Siqueira, A.D. & Rodriguez-Pedraza, C. 1999. Remote sensing and GIS at farm property level: demography and deforestation in the Brazilian Amazon. *Photogrammetric Engineering & Remote Sensing* 65:1311–1320.

Mertens, B., Sunderlin, W.D., Ndoye, O. & Lambin, E.F. 2000. Impact of macroeconomic change on deforestation in South Cameroon: integration of household survey and remotely-sensed data. *World Development* 28:983–999.

Millington, A., al-Hussein, S. & Dutton, R. 1999. Population dynamics, socioeconomic change and land colonization in northern Jordan, with special reference to the Badia Research and Development Project area. *Applied Geography* 19:363–384.

Moran, E.F. & Brondizio, E. 1998. Land-use change after deforestation in Amazonia. In D. Liverman, E.F. Moran, R.R. Rindfuss & P.C. Stern (eds), *People and pixels: linking remote sensing and social science*: 94–120. Washington, D.C.: National Acadamy Press.

Niamir-Fuller, M. 1999. *Managing mobility in African rangelands: The legitimization of transhumance.* London: Intermediate Technology Publications.

Pfaff, A.S.P. 1996. What drives deforestation in the Brazilian Amazon: evidence from satellite and socioeconomic data. *World Bank Policy Research Working paper* no.1772. Washington, DC.: World Bank.

Puigdefábregas, J. 1998. Ecological impacts of global change on drylands and their implications for desertification. *Land Degradation & Development* 9:393–406.

Pyke, D.A., Herrick, J.E., Shaver, P.L. & Pellant, M. 2002. Rangeland health attributes and indicators for qualitative assessment. *Journal of Range Management* 55:584–597.

Rasmussen, M.S., James, R., Adiyasuren, T., Khishigsuren, P., Naranchimeg, B., Gankhuyag, R. & Baasanjargal, B. 1999. Supporting Mongolian pastoralists by using GIS to identify grazing limitations and opportunities from livestock census and remote sensing data. *Geojournal* 47:563–571.

Reenberg, A., Nielsen, T.L. & Rasmussen, K. 1998. Field expansion and reallocation in the Sahel—land use pattern dynamics in a fluctuating biophysical and socio-economic environment. *Global Environmental Change* 8(4):309–327.

Rindfuss, R.R., Walsh, S.J., Mishra, V., Fox, J. & Dolcemascolo, G.P. 2003. Linking household and remotely sensed data: Methodological and practical problems. In J. Fox, R.R. Rindfuss, S.J. Walsh & V. Mishra (eds), *People and the environment: Approaches for linking household and community surveys to remote sensing and GIS*: 1–29. Boston, Dordrecht, London: Kluwer Academic Publishers.

Ringrose, S., Chanda, R., Nkambwe, M. & Sefe, F. 1996a. Environmental change in the mid-Boteti area of North-Central Botswana: Biophysical processes and human perceptions. *Environmental Management* 20(3):397–410.

Ringrose, S., Vanderpost, C. & Matheson, W. 1996b. The use of integrated remotely sensed and GIS data to determine causes of vegetation cover change in southern Botswana. *Applied Geography* 16(3):225–242.

Saiko, T.A. & Zonn, I.S. 2000. Irrigation expansion and dynamics of desertification in the Circum-Aral region of Central Asia. *Applied Geography* 20:349–367.

Stafford Smith, D.M. & Reynolds, J.F. 2002. The Dahlem Desertification Paradigm: A new approach to an old problem. In J.F. Reynolds & M. Stafford Smith (eds), *Global Desertification: Do Humans Cause Deserts?* Berlin: Dahlem University Press.

Sussman, R.W., Green, M.G. & Sussman, L.K. 1994. Satellite imagery, human ecology, anthropology and deforestation in Madagascar. *Human Ecology* 22(3):333–354.

Turner, D.T. 2003. Methodological reflections on the use of remote sensing and geographic information science in human ecological research. *Human Ecology* 31(2):255–279.

Thompson, M., Serneels, S. & Lambin, E.F. 2002. Land use strategies in the Mara Ecosystem: A spatial analysis linking socio-economic data with landscape variables. In S.J. Walsh & K.A. Crews-Meyer (eds), *Linking People, Place, and Policy. A GIScience approach*: 39–68. Boston: Kluwer Academic Publishers.

Vogel, C.H. & Smith, J. 2002. Building social resilience in arid ecosystems. In J.F. Reynolds & M. Stafford Smith (eds), *Global Desertification: Do Humans Cause Deserts?*: 149–166. Berlin: Dahlem University Press.

Warren, A., Batterbury, S. & Osbahr, H. 2001. Soil erosion in the West African Sahel: a review and an application of a "local political ecology" approach in South West Niger. *Global Environmental Change* 11:79–95.

Warren, A. 2002. Land degradation is contextual. *Land Degradation & Development* 13:449–459.

Wood, C.H. & Skole, D. 1998. Linking satellite, census, and survey data to study deforestation in the Brazilian Amazon. In D. Liverman, E.F. Moran, R.R. Rindfuss & P.C. (eds), *Stern People and pixels: linking remote sensing and social science*:70–93. Washington, D.C.: National Acadamy Press.

Recent Advances in Remote Sensing and Geoinformation Processing
for Land Degradation Assessment – Röder & Hill (eds)
© 2009 Taylor & Francis Group, London, ISBN 978-0-415-39769-8

Remote sensing based assessment of biophysical indicators for land degradation and desertification

S.L. Ustin, A. Palacios-Orueta & M.L. Whiting
California Space Institute Center of Excellence, University of California Davis, CA, USA

S. Jacquemoud
Institut de Physique du Globe de Paris, Géodésie & Gravimétrie, Paris Cedex 5, France

L. Li
Department of Geology, Indiana University-Purdue University, Indianapolis, IN USA

ABSTRACT: Desert ecosystems spanning moisture conditions from dry grasslands to barren hyper-arid landscapes are the largest terrestrial biome with more than 40% of the terrestrial land-mass. Remote sensing data provide an efficient cost-effective means to assess biophysical indicators of land degradation and desertification, providing that essential ecosystem properties can be monitored. We review the spectral characteristics of plants and soils that are detectable using optical sensors and methods to identify and quantify properties that have potential for monitoring arid ecosystem processes. Vegetation indexes have little sensitivity at low leaf area, particularly when the soil background is highly variable, as is characteristic of many arid systems. Additionally, accumulated dry plant material on the soil surface challenges measurement. Although the absorption characteristics of the major biochemical constituents of plants and soils are generally understood, the methods to retrieve this information from reflectance data and to understand the significance of how the structural organization alters the absorption features remains an area of active research. The overlapping absorption features of plants and soils preclude direct assessment of many biogeochemicals of interest. New biophysical methods that take the full spectral shape into account, including the effect of one compound on the spectral absorption of another, are needed to reduce uncertainty in their estimates. As a result, despite significant progress in developing fundamental understanding of ecosystem processes and optical properties, more research is needed before fully predictable quantitative methods are available.

Keywords: spectral indexes, biophysical indicators, radiative transfer, reflectance models

1 INTRODUCTION: DESERTS AND DESERTIFICATION

Desert and dryland ecosystems, including hyperarid, arid, semiarid, and dry subhumid areas are the largest terrestrial biome, about 41% of the terrestrial landmass (MEA 2005). Deserts are categorized by precipitation with extremely arid lands having at least 12 consecutive months without precipitation, arid lands having less than 250 mm annual rainfall, and semiarid lands (typically grasslands or savanna) having mean annual precipitation between 250 and 500 millimeters (Meigs 1953). Deserts have evolved into their modern distributions (Fig. 1) largely in response to Quaternary climate conditions, with local environmental factors superimposed. Present expansion of deserts is attributed to land degradation from land use, land use change, and climate change (MEA 2005).

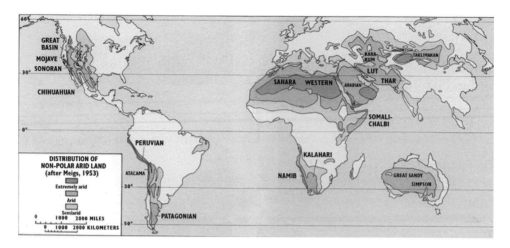

Figure 1. Global distribution of arid ecosystems (MEA 2005) (see colour plate page 369).

There is growing evidence that human activities are creating unprecedented rates of ecosystem change such that novel combinations of climate and disturbance factors may perhaps exceed the capacity of current ecosystems to adapt. Desert ecosystems are typically fragile and even small perturbations can have long-lasting impacts on the distribution of vegetation and on functionality of the landscape (Schlesinger et al. 1990, Okin et al. 2004). Vegetation change may have positive feedbacks to climate that accelerate change (Scheffer et al. 2005). Conversely, within the historic range of climate, vegetation distributions tend to be stable, despite large inter-annual variability (Beatley 1980).

Desertification is defined as "land degradation in arid, semiarid, and dry subhumid areas resulting from various factors, including climatic variations and human activities" (MEA 2005), where land degradation is defined as reduction or loss of biological or economic productivity. Dregne (1986) identified three processes that promote desertification: 1) deterioration of vegetation cover from overgrazing, wood cutting and burning; 2) wind and water erosion from improper land management; and 3) salinization from improper irrigation management. Desertification affects all continents (except Antarctica), and 10–20% of global arid systems are believed to be degraded. Because nearly one-third of the human population lives in arid regions, the threat of desertification ranks among the most important environmental problems and has significant impacts on meeting human well-being.

Many interacting factors affect ecosystem sustainability and resistance to degradation (Fig. 2). Maintenance of biotic composition, diversity, and cover is essential to sustainable productivity. In addition to plant factors, maintenance of soil quality processes, including soil texture, nutrient, and biogeochemical cycles, soil microorganisms and biological crust, and resistance to wind and water borne erosion is also essential to sustaining the health of the ecosystem. Changes in vegetation characteristics and shifts in phenological cycles may accelerate impacts from climate variability and land use. To protect arid lands from accelerated desertification and reduce the potential for feedbacks that may accelerate global climate changes, it is important to understand these integrated ecosystem responses.

Deserts and semi-arid ecosystems are by definition limited by water. Low but highly variable precipitation, often with unpredictable periodicity, dominates the responses of desert ecosystems (Reynolds et al. 2004, Scanlon et al. 2005). Fractional cover, biomass and leaf area are generally proportional to precipitation and available soil moisture (Beatley 1974). Life forms shift depending on the predictability of precipitation, with grasses more abundant under predictable precipitation regimes and shrublands dominating when rainfall becomes more irregular in timing and amount (Schlesinger et al. 1990, Smith et al. 2000, Scanlon et al. 2005). A primary driver of desertification is loss of vegetative cover which creates a positive feedback to lower

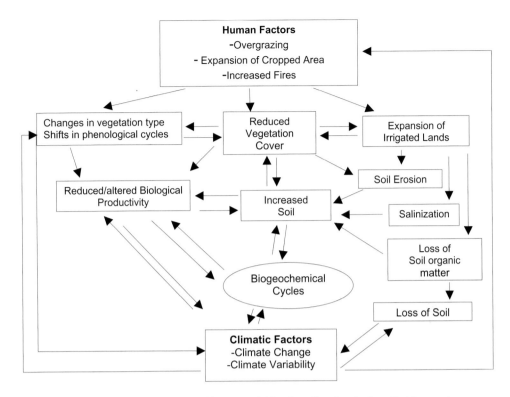

Figure 2.   Interactions among climate and human activities that affect functioning of arid ecosystems.

precipitation (Charney et al. 1977). Desert ecosystems are often secondarily limited by nutrients, particularly nitrogen (Smith et al. 2000). In contrast, nitrogen deposition along with CO_2 enrichment is likely to enhance growth and affect other ecosystem attributes like species composition, growth form, and phenological patterns (Smith et al. 1997). Enhanced growth may also alter delicately balanced desert ecosystem functions—changing nutrient, fire and water cycles. Conditions that favor succession may promote invasion by exotic species. A landmark study by Schlesinger et al. (1990) showed a progressive pattern of desertification occurred as soil nutrients became less uniformly distributed as an ecosystem transformed from grassland to desert shrubland, that lead to unproductive barren soil between islands of higher fertility. How deserts will respond to multiple simultaneous environmental changes becomes highly uncertain. An increase in inter-annual climate variability (e.g., El Niño precipitation) could under multiple stressor conditions, exceed a threshold and produce a step change in ecosystem structure and function (Scheffer et al. 2005).

The first few centimeters of soil are the most fertile and most susceptible to rain and wind erosion. Improved ability to quantify and monitor soil surface properties, especially soil chemistry, can provide key indicators of desertification. Knowledge of the distribution of secondary clay and organic matter will contribute to understanding plant and soil responses in arid regions and also contribute to monitoring desertification. Mapping of clay and organic matter will improve estimates of soil carbon, a key parameter for developing an understanding of the impacts of altered rainfall distributions, deposition of pollutants, and increases in greenhouse gas concentrations. Of particular concern for desertification from climate change is the formation and dissolution of pedogenic inorganic carbon. Despite the fact that the global inorganic carbon pool is comparable in size to the pool of atmospheric CO_2 and to soil organic carbon, there has been little research on the impact of changing hydrologic processes on this pool of inorganic carbon.

2 RETRIEVING BIOPHYSICAL SPECTRAL INFORMATION

2.1 *Biophysical information from living plants*

Desert species have developed convergences in the types of leaf structures that have evolved. Similarly, growth forms converge being typically limited to grasses and drought-tolerant shrubs. Plant cover is the most obvious indicator of precipitation and soil moisture status. Numerous studies have shown a strong linear relationship between plant cover, biomass, and leaf area in response to water availability (Beatley 1980). Typically, leaf traits include small thick leaves, with thick cuticles and pubescence, sunken stomates, and small mesophyll cells with thick cell walls and small air spaces. The reflectance and transmittance of leaves is a function of both the concentration of light absorbing compounds (chlorophyll, water, dry plant matter, etc.) and the surface/internal scattering of light that is not efficiently absorbed (Fig. 3). Leaf absorbance is reduced by pubescence and waxes, and leaves appear more reflective after the hairs have dried out (Ehleringer & Björman 1978, Ehleringer & Mooney 1978). As the properties of xeromorphic plant leaves change, e.g., they become more schlerophyllous (drought hardened) or desiccated, and reflectance and transmittance changes in predictable directions (Rotondi et al. 2003). These wavelength specific absorption and scattering produces potentially diagnostic optical patterns of vegetation conditions in deserts. A number of analytical methods have been developed for directly estimating biochemical composition and structural characteristics. An extensive review of the literature and available optical models are available online at (http://www.multimania.com/opticleaf/).

There is strong interest today in developing techniques to detect and quantify individual pigments (Ustin et al. 2004). The absorption of light by photosynthetic pigments dominates green leaf properties (Carter 1991) in the visible (VIS) spectrum (400–700 nm), with a minimum at 550 nm (Fig. 4a–c). Leaf inversion models, e.g., PROSPECT (Jacquemout & Baret 1990, Jacquemoud et al. 1996, 2000) predict total pigment concentration ($\mu g\,cm^{-2}$), by assuming that they are entirely composed of chlorophyll a and b (Fig. 4a). Characterization of in vitro absorption coefficients for other photosynthetic pigments, e.g., xanthophylls, anthocyanins, caroteins and even chlorophyll b could further improve leaf radiative transfer models (Fig. 4b). In vitro absorption spectra for chlorophylls and several accessory photosynthetic pigments (e.g., carotenoids, xantophylls) are available (Lichtenthaler 1987). However, the pigment spectrum shifts with the solvent used to extract them, so the in vivo configuration of plant pigments remains uncertain. The overlapping wavelengths of pigment absorptions make individual identification challenging.

Carotenoids are a family of red, orange, or yellow pigments, which effectively expand the wavelength range for energy absorption by photosynthesis. In photosynthesis, the energy absorbed by carotenoids is transferred to a chlorophyll a photoreceptor. Xanthophylls, a class of carotenoids, are

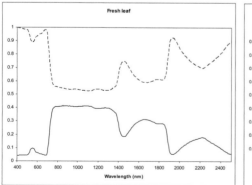
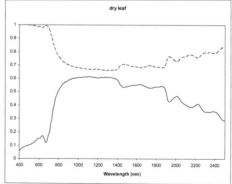

Figure 3.   Reflectance (solid line) and transmittance (dashed line) of (left) fresh leaf and (right) dry leaf of a semiarid species, *Quercus pubescens*. Note that the leaf in Figure 3b was dried rapidly since strong chlorophyll absorption and red-edge features remain in the 400–800 nm region.

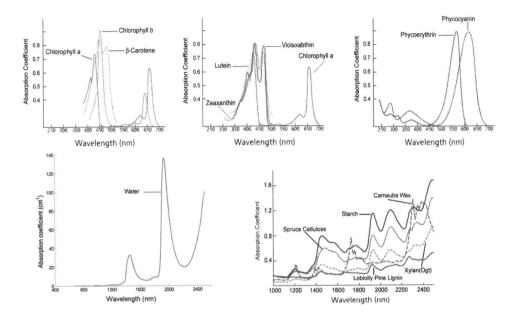

Figure 4. Specific absorption coefficients for a) chlorophyll a and b and β-carotene, b) chlorophyll a and xanthopyll pigments: leutin, violaxanthin and zeathanthin, c) phycoerythrin and phycocyanin, d) water (after Kou et al. 1993) and e) five common biochemical compounds found in leaves (see colour plate page 369).

involved in photoregulation of light by dissipating excess absorbed energy and avoiding oxidative damage to the photocenter (Dey & Harborne 1997). Under excess light conditions, violaxanthin is converted to zeaxanthin in the xanthophyll cycle, which due to its absorption characteristics, increases leaf reflectance. Gamon, Peñuelas, and colleagues (Garnon et al. 1992, 1993, 1997, 1999, Peñuelas et al. 1995a, b, 1997a, b) have extensively investigated the use of these spectral differences to develop a normalized ratio termed the Photosynthetic Reflectance Index (PRI), which is sensitive to changing xanthophylls cycle pigments. They successfully used this method to estimate photosynthetic efficiency. Stylinski et al. (2000) showed that the PRI was well correlated to photosynthetic electron transport capacity in *Quercus pubescens* exposed to long term elevated CO_2 enrichment. Subsequently Stylinski et al. (2002) showed that the PRI could track seasonal changes in carotenoid pigments and photosynthetic activity of mature semiarid evergreen shrubs.

Anthocyanins are a class of water-soluble red pigments which are not involved in photosynthesis but occur widely in flowers, fruits and leaves. They have a single absorption peak around 529 nm. Anthocyanins may also protect the photosynthetic system from excess light and/or to deter herbivory. Anthocyanins are common in leaves and stems, although their red to purple color may be masked by chlorophylls, carotenoids, or pubescence (Neill et al. 1999, Lee et al. 2002). Under stress (e.g., high temperatures) or during early leaf development (Hoch et al. 2001) they become more visible, consequently, measurement of leaf anthocyanins may be another indicator of physiological state (Curran et al. 1991, Peñuelas et al. 1999).

Phycobilins, specifically phycoerythrin (red) and phycocyanin (blue) are water-soluble photosynthetic pigments that are only found in the cytoplasm of Cyanobacteria (formerly termed blue-green algae) and chloroplasts of Rhodophyta (red algae). Because cyanobacteria are the nitrogen fixing component in lichens, these are important constituents of biological soil crust. The absorption coefficients for these pigments are shown in Figure 4c which suggests that it is possible to spectrally identify and distinguish green and cynaophytic lichen classes based on pigment composition differences, if not additional lichen and algae types.

The near-infrared (NIR, 700–1100 nm) region has limited biochemical absorptions which are contributed by compounds characteristic of dry leaves, primarily cellulose, lignin and other structural carbohydrates. Reflectance in this region is affected by multiple scattering from internal leaf

19

structure, including the fraction of air spaces, and refraction at air-water interfaces. As intercellular air spaces and cell sizes decrease, overall leaf reflectance or albedo declines. Eller and Willi (1977) showed that pubescence acted mainly in the visible range, because the increase in reflectance in the infrared was offset by a decrease in transmittance.

Reflectance and transmittance in the shortwave-infrared (SWIR, 1100–2500 nm) are dominated by water absorption in green leaves (Fig. 4d). The specific absorption coefficients for pure liquid water were observed by Curcio and Petty (1951) and have been extensively studied (see Prahl 2001). The primary absorptions by water in leaf reflectance occur at 1450, 1940, and 2500 nm with important secondary absorptions at 980 nm, and 1240 nm (Carter 1991). The water absorptions result from the following vibrational modes: $v1$ (H–O–H symmetric stretch mode transition), $v2$ (H–O–H bending mode transition), and $v3$ (H–O–H asymmetric stretch mode transition). The absorption feature at 970 nm is attributed to a $2v1 + v3$ combination, at 1200 nm to a $v1 + v2 + v3$ combination, at 1450 nm to a $v1 + v3$ combination, and at 1900 nm to a $v2 + v3$ combination. The fundamental water absorption is at 2800 nm.

The spectral signatures of biochemical absorptions in the SWIR region are complex. Even when substances are composed of well-characterized repeating units such as starch, sugar, and cellulose, their molecular weights vary, while others families of biochemical substances are not precisely defined, vary between species, or have not been isolated with their precise molecular structure intact (e.g., proteins and lignins). In dry leaves, reflectance and transmittance (Fig. 4d, e) are controlled by dry carbon-based compounds like cellulose, lignin, and sugars, and nitrogen-based molecules such as proteins and enzymes. For example, nitrogen (N–H bonds) has a first harmonic overtone at 1510 nm and a series of combination bands at 1980, 2060, and 2180 nm (Wessman 1990). Because many plant and soil macromolecules contain common chemical bonds (–H, C–N, C–O, and N–O bonds, and C=O, O=H, and N=H) they create overlapping absorption features (Barton et al. 1992, Grossman et al. 1996) that have resisted specific biochemical identification.

2.2 Biophysical information from dry plants

Remote sensing studies often group dry plant material as a class, termed non-photosynthetic vegetation (NPV), which is all of the non-green plant parts, including dry leaves, bark, and wood. NPV are highly spectrally variable, especially when considering a range of species (Elvidge 1990, Asner 1998, Roberts et al. 2004). In dry leaves, reflectance forms a more continuous monotonic spectral shape from 400 to 1500 nm. If some green foliage remains then a minor red-edge can be observed. The red-edge is a narrow spectral region between 700–725 nm at the long-wavelength edge of the chlorophyll absorption feature. It is absent in rocks, soils, and most dry plant materials. High spectral resolution studies have concentrated on identifying dry plant materials which are indistinguishable in broadband data. Baret et al. (1992) showed that the position of the red-edge inflexion point was little influenced by the soil background or by the atmosphere. Elvidge et al. (1993) noted that the problem in detecting trace quantities (<10% cover) of green vegetation in arid and semiarid region could be solved by measuring the red-edge. Anecdotally, because the red-edge is considered a biosignature of green vegetation, it has become a new avenue of research for astronomers searching for life on extrasolar planets (Seager et al. 2005)

Lignin and cellulose are usually present in plant residues in an intimate mixture, and have fundamental absorptions in the SWIR infrared, which produce several combination and overtone bands between 2000 and 2500 nm (Fig. 4e). Lignin has a strong absorption in the ultraviolet at 280 nm with an absorption wing across VIS and NIR, observable after pigments have degraded in dry leaves. The combination of cellulose and lignin have two diagnostic bands around 2090 and 2300 nm (Fig. 4d) which are not observable in live green vegetation (Elvidge 1990).

Nagler et al. (2000) developed a Cellulose Absorption Index (CAI) using band depth at 2100 nm. Most soils do not absorb at this wavelength while plant residues do. Additionally, they found that CAI changed with soil moisture and that CAI of wet litter was significantly brighter (higher reflectance) than CAI of wet soil, and therefore, distinguishable. Since lignin is more resistant to decay than cellulose, as plant residues decompose they become enriched in lignin and the ultraviolet absorption becomes more distinct. Thus, reflectance curves change significantly during decomposition.

As non-photosynthetic vegetation (NPV) decomposes, significant mixing with soil occurs until eventually, cellulose decomposes and only lignin residues remain. Over time, absorption features from decomposition resistant components come to dominant the spectrum. Thus in soil, lignin features become more developed over time. Surface residues continue to be highly variable in reflectance until decomposition progresses to the point that it becomes indistinguishable from soil (Baumgardner et al. 1985).

One of the most complex remote sensing tasks in arid or semiarid environments is to identify NPV at various levels of cover and decomposition. When working with image data the soil background is critical to detecting variable amounts of NPV. At low NPV, the soil is the main contributor to temporal and spatial variability. While soil and NPV have a similar magnitude and spectral shape over the visible-NIR region (Fig. 5), residues have narrow-band absorptions at some wavelengths, making them sufficiently distinct that they can be discriminated from soils. Wanjura and Bilbro (1986) found that residue reflectance increased from 400 to 1300 nm, and decreased sharply at longer wavelengths while soil reflectance continued to increase to 1750 nm and then decreased slowly at longer wavelengths. Aase and Tanaka (1991) found that residue reflectance was greater over wet soil than dry soil. Various methods have been successful in direct estimates of crop residues (Wanjura et al. 1986, Whiting et al. 1987, Abers et al. 1990, Biard & Baret 1997, Nagler et al. 2000, Streck et al. 2002, Vina et al. 2003). Since dry residues absorb more strongly in the SWIR than soils, multispectral residue indexes are based on the relationships at two wavelengths. McNairn and Protz (1993) developed the Normalized Difference Index (NDI) but it is less sensitive in light colored and sandy soils, which reduces its value in arid and semiarid regions. Biard and Baret (1997) transformed the NDI into a Soil Adjusted Corn Residue Index (SACRI) which reduced the index sensitivity to soil texture. They found a near linear relationship up to 20% crop residue cover, suggesting that this type of index could be applied in arid areas where vegetation cover is sparse and plant litter and soil are significant components of reflectance. A typical threshold for

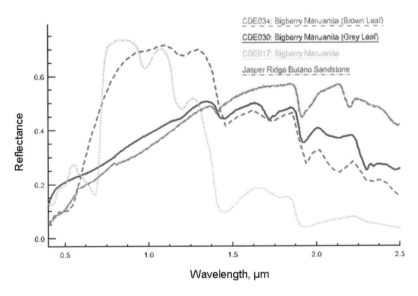

Figure 5. Typical spectra of fresh green leaf from bigberry Manzanita (┊), two stages of dry leaf weathering (┊, ┊), and bare soil (- - - - bulano sandstone) from Jasper Ridge Biological Preserve, California, showing spectral differences that can be used to differentiate them. Note that the *Quercus pubescens* leaf in Figure 3b was dried rapidly since a strong chlorophyll absorption remains in the 400–700 nm region. The dry grass residue here lacks a red-edge feature (see colour plate page 370).

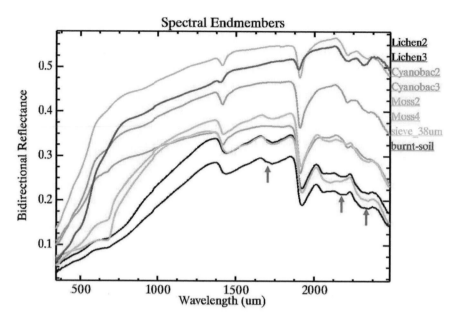

Spectral Endmembers

Lichen2
Lichen3
Cyanobac2
Cyanobac3
Moss2
Moss4
sieve_38um
burnt-soil

Bidirectional Reflectance

Wavelength (um)

Figure 6.   Components of biological soil crust from a Mojave desert habitat at the Nevada Desert Test Site, USA. Sieved and burnt (SOM exhausted) soil has higher reflectance than lichen, cyanobacteria, or mosses (P. Valko, unpublished results). Red arrows denote wavelengths where organic matter of biological curst are reported to show absorption features (see colour plate page 370).

minimum residue to prevent soil erosion is often set at 15%, suggesting that this index is valuable for detecting potential erosion.

2.3   Biophysical information from biological soil crusts

Biological soil crusts (BSC) occur on every continent (2001a) and are a major biological component of arid ecosystems. Although they occur only the first few millimeters of the soil surface they have a key role in arid ecosystem dynamics. BSC typically account for up to 70% of the live cover in arid deserts, therefore account for much of the surface reflectance. BSC are physical and chemical associations between soil particles and microphytic mixtures of lichen, moss, fungi, cyanobacteria, and free-living algae (Belnap 2001a, b). BSC markedly contribute to the high spectral variability in arid regions depending on their active or dormant states and their composition (Fig. 6). BSC provide critical ecological functions, including stabilizing soil, nitrogen fixation, water and nutrient retention, carbon fixation, and a seed bed for germination (Belnap et al. 2001a, Belnap & Gardner 1993, Harper & Belnap 2001). BSC are typically darker than surrounding bare soil (Fig. 6) thus loss increases surface albedo (Belnap & Elridge 2001). In contrast to BSC, abiotic physical soil crusts and chemical weathering processes are light colored, have low permeability, and are generally much thinner, often less than 1 mm. Bechtel et al. (2002) and Buffoni-Hall et al. (2002) showed that lichen have low transmittance, making it difficult to estimate spectral properties of the underlying rocks or soils.

BSC react rapidly even to low moisture inputs making their spectral responses highly variable in time. Building on Ager and Milton (1971), Zhang et al. (2005) used normalized spectral mixture analysis to accurately ($r = 0.90$) separate rocks from lichen in a sub-arctic environment, a technique that might be extended to arid environments. Karnieli et al. (1999) showed that the main spectral features were due to chlorophyll at 670 nm and organic matter at 1720, 2180, and 2309 nm (see red arrows on Figure 6). Bechtel et al. (2002) used ratios of 2132/2198 nm and 2232/2198 nm to differentiate lichens from rocks. When wet, they observed NDVI as high as 0.3 comparable to vascular plants in arid regions. Rees et al. (2004) used principal component analysis, and found

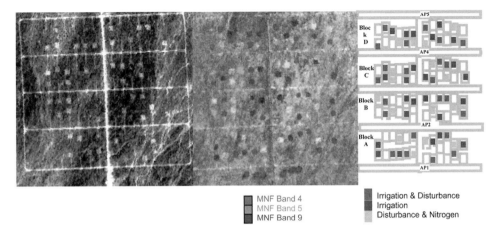

MNF Band 4	Irrigation & Disturbance
MNF Band 5	Irrigation
MNF Band 9	Disturbance & Nitrogen

Figure 7. False color Airborne Advanced Visible Infrared Imaging Spectrometer (AVIRIS) image (left), and minimum noise fraction image (MNF) acquired 9 July 2003 at ~3 m spatial resolution over the Mojave Global Change Facility experimental plots, Nevada, USA (P. Valko, unpublished results). Each of the 96 plots is 14 m × 14 m. Treatments include added monthly summer irrigation equivalent to three extra monsoon storm events, two levels of nitrogen enrichment simulating dry deposition (added in the fall before winter precipitation period, and crust disturbance simulating effect of over-grazing, performed annually. Treatments began summer 2001 (see colour plate page 371).

maximum variability for arctic lichens was between 700 and 1450 nm although our data (Fig. 6) show greatest variability between 1.3–2.5 μm.

Some authors suggest that BSC may be early indicators of ecosystem change because of their rapid response to environmental change (Jeffries et al. 1993a, b, Lange et al. 2001). BSC are sensitive to physical disturbance and to climate variability, including sensitivity to increased atmospheric CO_2 concentrations (Belnap & Elridge 2001). BSC have significantly greater lichen and moss diversity in winter rain dominated deserts than summer rain deserts with equal precipitation. The rapid wet and dry cycles that are typical of summer precipitation regimes generally increase net carbon losses and produce negligible nitrogen fixation (Jeffries et al. 1993a, b). Lange et al. (1998) reports that CO_2 fixation in the cyanolichen, Collema tenax is negatively impacted by extremes of both hydration and desiccation. It is likely that increased temperatures will reduce lichen cover with secondary effects on decreased soil stability, decreased soil nitrogen, and decreased soil carbon.

The unique assemblages of biota associated with BSC create detectable spectral characteristics (Fig. 6) (Rees et al. 2004, Karnieli & Tsoar 1995, Karnieli & Safaris 1996, Karnieli et al. 1996, 1999, Karnieli 1997, Hill et al. 1998). Figure 7 shows spectral changes in a Mojave desert scrub (Larrea tridentata, *Ambrosia dumosa*, and *Lycium spp.*) ecosystem in southern Nevada, after four years of treatments in a long-term manipulation experiment. The Mojave Global Change Facility (MGCF) experiment is designed to test the cumulative and interactive effects of increased summer precipitation, increased nitrogen deposition, and increased mechanical disturbance from grazing began in 2001. (http://www.unlv.edu/climate_change_research)

2.4 Soil structural crust

In contrast to BSC, soil structural or physical crust is a thin layer (1 mm) formed by kinetic energy of drop impact over bare soils during rainstorms. The rainfall drops on the bare soil rearranges a thin surface layer by physical disintegration of the soil particles and dispersion of soil clays which migrate with water in the first few mm of soil. As a result of crust formation, infiltration decreases markedly and runoff and erosion is accentuated (McIntyre 1958, Goldshleger et al. 2002). Studies show that effects are directly related to the rainfall energy so this is an important degradation process in arid regions where severe storm events bring the most precipitation and soils have a low vegetation cover. In contrast to BSC and desert varnish formation, crust formation is fast and highly dynamic, happening nearly instantaneously after a storm.

Soil crusts alter reflectance because of increased albedo, due to smaller particle size distribution (de Jong 1992, Goldshleger et al. 2001, Eshel et al. 2004), orientation of the particles (Courault et al. 1993), and enhancement of the 2200 nm absorption band by sorting montmorillonite and illite clays onto the surface. Generally there is no significant change in the carbonate 2330 nm band, because it becomes concentrated in the silt or sand fractions. Since soil crust contains higher clay content than the bulk soil, it can bias soil classifications (Goldshleger et al. 2002). Albedo and absorptions at 1833 and 2143 nm (no specific absorption feature) have the best correlation with raindrop energy and infiltration rate, indicating that scattering processes govern these effects (Goldshleger et al. 2002, Ben-Dor et al. 2003) with different soils (clay vs. sandy) producing different spectral behaviors (Goldshleger et al. 2004). Goldshleger et al. (2004) tracked crust formation using reflectance at 1700 nm as an indicator of albedo and the first derivative of absorption at 2200 nm, which represented absorption of clay minerals. They found that a clay-particle relationship was a better indicator than particle size alone.

2.5 Desert pavement and varnish

Although little studied for their spectral characteristics, desert pavements are important for controlling erosion and ecosystem stability and represent an early stage of biologic weathering in arid landscapes. Desert pavements (Fig. 8) are ancient mosaics of rock fragments embedded in sediments (Cooke & Warren 1973) that occur on a wide range of undisturbed arid and semiarid landforms. The presence of desert pavement indicates that an area has been undisturbed for an extended geologic time (10,000s of years) (Dorn & Oberlander 1981). Generally, as pavement ages the rock surfaces become smoother with few protruding fragments. Exposed surfaces are coated with thicker varnishes, which are black or brown coatings formed by bacteria interacting with clay minerals (~70%) and iron or manganese oxides (Potter & Rossman 1979, Dorn & Oberlander 1981, 1982). Spectrally, desert pavement and varnish lower surface albedo relative to younger soil and rock surfaces. Loss of the pavement increases albedo by exposing native fine materials, gravel and cobbles. Disturbance may turn over pavement and expose less weathered (oxidized) rocks of different colors, or which have a different mineralogy (less clay and manganese for example). Varnish can be mechanically removed by wind and scraping or water erosion, exposing the less weathered gravel surfaces.

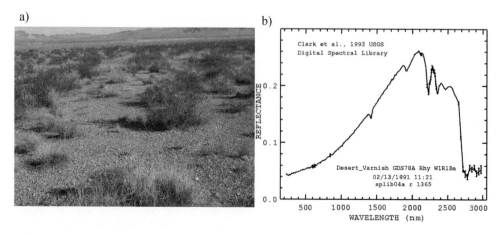

Figure 8. a) Desert pavement in Mojave Global Change Facility site, near Mercury, Nevada, USA. b) Spectrum of coating on desert pavement composed of rhyolite and quartz cobbles (desert varnish.gds78A) http://speclab.cr.usgs.gov (see colour plate page 371).

2.6 Biophysical information from soils

The weathering process that soils undergo in arid and semiarid regions are related primarily to high temperatures and low and/or irregular availability of water. Important soil surface processes are due to the combination of irregular precipitation events, shallow soil development, and scarce vegetative cover which create low infiltration capacity and high runoff. Thus, salt dissolution combined with high evaporation can precipitate alkaline salts to form hardpans and caliche layers at the wetting front. Sparse vegetation produces little organic matter with high spatial heterogeneity. Biological and structural soil crusts have very different properties in terms of soil health and spectral properties. The combination of all these processes produces highly heterogeneous soils, which requires synoptic coverage over large regions with sufficient spatial and spectral resolution to quantify soil patterns.

Soil has less spectral structure than decomposing dry plant material (Fig. 9). Since it is composed of well mixed minerals and organic matter, absorption features result from overlapping bands with no strong narrow-band absorption features such as those found in pure minerals. Consequently, soil spectra vary slowly with time (Baumgardner et al. 1985). The VIS and NIR reflectance is driven by a wide absorption wing that extends from the ultraviolet, due to electronic transitions of metals, mainly iron. In contrast, the general shape and specific absorption bands of the SWIR are mainly driven by water and minerals that contain water and hydroxyl ions.

Many soils only differ in brightness in red and NIR, which form a linear relationship that is represented in spectral space as a soil brightness vector (Richardson & Wiegand 1977). The position of a given soil on the soil line depends on factors that affect the albedo such as shadows, moisture content, and roughness (Elvidge & Chen 1995, Baret et al. 1993). The slope of the line depends on the relative position between the R and NIR bands and therefore depends on the intrinsic soil properties such as iron or organic matter content. Also it can be affected by atmosphere and sensor properties such that it becomes soil and scene-dependent.

The width of the soil line expresses variable soil properties in these wavelengths (e.g., moisture or organic matter). To identify the source of the variability and quantify this information requires measuring additional bands. The soil line has been used to estimate vegetation cover by measuring the reflectance of a surface that is perpendicular to a soil line defined by bare soil pixels. The problem in using this index in arid environments is that the high soil heterogeneity makes it difficult to define a unique soil line. To overcome this limitation, Fox et al. (2004) proposed an automated soil line identification routine. The soil line only represents a two dimensional spectral space so it only shows the general shape of the spectrum in the VIS-NIR space. Most soil biological and physicochemical properties have information contained in a large number of other wavelengths that make it necessary to use more complex methodologies to study soil properties.

The most widely used semi-analytical technique to determine mineral absorption strength is normalization based on the background continuum (Kokaly & Clark 1999, van der Meer 2004). The continuum technique calculates the normalized band depths by interpolating reflectance between two local reflectance maxima (the shoulders of the absorption feature) then calculates the band depth, area, and asymmetry within the wavelength interval, relative to the maxima. The mineral type is defined by the wavelength positions at the minima and the feature asymmetry (van der Meer 2004). However other studies (Palacios-Orueta & Ustin 1998) have demonstrated that spectral shape results from all interactions that affect absorption and scattering. For example, Whiting et al. (2006) found water broadened and shifted the wavelength positions of specific absorption features to longer wavelengths. These shifts make correlating a particular wavelength with a particular mineral chemistry and absorption depth inconsistent.

The most complex analytical task is to distinguish bare soils and dry residues at various levels of cover and decomposition (Figs. 5–9). The soil background is the main contributor to variable surface reflectance in areas with variable and/or low amounts of residue. Wanjura and Bilbro (1986) found that residue reflectance increased with wavelength to 1300 nm, then it decreased sharply, while soil reflectance continued to increase to 1750 nm, and then decreased slowly at longer wavelengths (Figs. 5, 9). Aase and Tanaka (1991) observed that variable soil moisture looked similar to different amounts of straw on dark soil as the effect of moisture was to lower albedo across the SWIR.

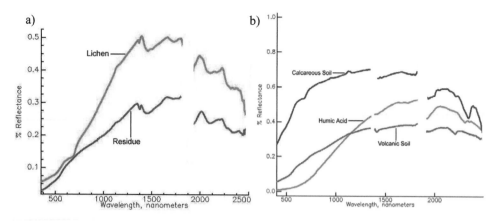

Figure 9. a) Contrasting spectral shapes between biologically active lichen and dry plant residues. b) Respiration resistant humic acid soil component and two contrasting soils, derived from calcareous and volcanic parent material (see colour plate page 372).

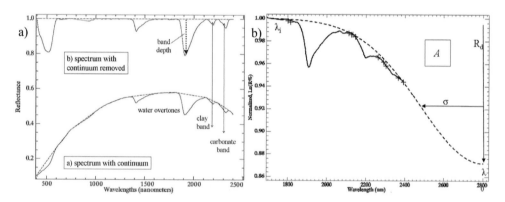

Figure 10. a) Upper, soil spectrum relative to the continuum, which normalizes absorption features. Lower, soil reflectance spectrum with continuum shown as dashed line. b) Soil Moisture Gaussian Model, fit to a soil spectrum from water fundamental at 2800 nm. The shoulders of the absorption features used for calculating the continuum are shown as crosses (Whiting et al. 2004). The Gaussian curve is fit from the wavelength (λ) absorption maximum to λi, Rd is depth at λ, σ is the full width half max, and A is the area.

Moisture is the main source of temporal variability in soil reflectance. Soil reflectance is greatly affected by slight differences in water content (Clark 1999), reducing the spectral contrast needed for identifying minerals and organic matter absorption depths (Bowers & Hanks 1965). The change in band depth is non-linear with water content (Lobell & Asner 2002), which requires knowing the light absorption characteristics of the soil for all pixels in an image. Several researchers have successfully modeled soil moisture using 1450 and 1900 nm bands (Dalal & Henry 1986, Lobell & Asner 2002, Liu et al. 2002). Unfortunately, these absorption bands also absorb water vapor in the atmosphere, confounding soil quantification in image data. Whiting et al. (2004) extrapolated the general shape of the SWIR spectrum to the fundamental water absorption at 2800 nm by fitting the convex hull to a Gaussian function and reported that the area under the Gaussian curve is correlated with soil moisture content ($r^2 = 0.90$). The Soil Moisture Gaussian Model (SMGM) was highly correlated to the soil water content, up to the water holding capacity of the soil without dependence on specific water absorption bands. Organic matter is closely related to soil quality and degradation processes in natural vegetation and agricultural lands. It is not only an indicator of soil degradation, but also regulates

26

other biogeochemical processes. Aranda & Oyonarte (2005) found lowest quality soil organic matter under degraded shrubs suggesting that degraded vegetation may lead to soil degradation. Both low content of OM and poor quality are characteristic of arid and semiarid environments. Organic matter has been studied by Henderson et al. (1992) and Coleman and Montgomery (1987) with the main effect observed in the VIS-NIR, causing high albedo and a convex hull shape for soils with low organic matter content (Palacios-Orueta & Ustin 1996). Figure 11a shows three soils with high, low and intermediate amounts of organic matter. Demattê et al. (2003) observed a decrease in reflectance between 600 and 2500 nm when organic matter content increased after applying an organic fertilizer. Chen et al. (2000) mapped organic matter at the field scale using aerial photography, obtaining high accuracy in bare soil. However this task becomes more complicated in natural areas with variable vegetation, high soil variability, and low organic matter in arid areas.

Soil salinity is one of the main consequences of soil degradation. Salinization not only affects the growth of vegetation, but destruction of soil structure due to clay dispersion and subsequent decrease in infiltration and aeration. Reflectance of saline soils results from the spectral properties of salt and surface roughness, presence of crust, soil color and moisture, which have a combined effect on albedo (Metternicht & Zinck 2003). A good review of spectral characteristics of soil salinity is found in Metternicht & Zinck (2003). Most soil salts (i.e., halite, calcium carbonate, sodium sulfate and gypsum) are highly reflective in the VIS-NIR and show distinct absorption features in the SWIR. Hydrated salts like gypsum have a significant decrease in reflectance toward longer wavelengths and sharp absorptions at the water bands, while salts like halite are highly reflective in this region of the spectrum. As a result most saline areas have high albedo except highly alkaline soils which are dark black due to distillation of organic matter. Roughness and moisture are two additional factors affecting albedo. For example the presence of saline crust decreases roughness and increases albedo, while moisture dissolves salt crusts so albedo decreases. Since these properties are highly dynamic their variability makes assessment difficult. While high salt concentrations are easily identifiable, low levels present more difficulty for detection because spectral features are weak and the presence of salt-tolerant vegetation dominates the soil signal. Hypersaline soils have little or no vegetation on them but may have BSC if moisture is available.

Roughness is the primary factor affecting soil reflectance and it is also highly variable in arid soils. Surface roughness is influenced by the size and distribution of soil aggregates and by the density and distribution of any vegetation cover. The effect of roughness on albedo is on the distribution of shadows (Matthias et al. 2000), which affect the spectral contrast and reduce discrimination of spectral features. However, several studies have demonstrated that bi-directional reflectance distribution is distinct on desert surfaces (Deering et al. 1990). Thus, Shoshany (1993) demonstrated that most soils in the Australian desert were anisotropic with a dominant backscattering component. Cierniewski & Karnielli (2002) simulated reflectance on four virtual surfaces (desert pavement, BSC, sands and playa) under different illumination conditions and showed that forward scattering and backscattering properties significantly affected spectral contrasts between surfaces. For example, contrast between sand and playa was greatest in the backscattering directions for a specific view and zenith angles, while for crust and desert pavement, the contrast was best in the forward scattering direction.

Erosion is important in arid regions. Common indicators of erosion are a decrease of organic matter, appearance of lower soil horizons at the surface, and presence of rocks. Eroded soil surfaces can have high levels of iron oxides or carbonates from the B or even C horizon. Whiting et al. (2006) mapped carbonates in the Tomelloso area of Spain where the underlying carbonate layer (caliche) was exposed on the surface. Figure 11b shows the spectral soil signature from the surface soil samples of an eroded terrace and from a floodplain. The eroded soil has high albedo and a strong carbonate absorption feature, while the floodplain soil has stronger clay absorption. De Jong (1992) used imaging spectroscopy to map erosion based on iron and carbonates and Palacios-Orueta et al. (1999) used Foreground and Background analysis on AVIRIS data to map organic matter and iron. In a spatial context, eroded soils can occupy broad extensions or form linear features like gullies or rills which are difficult to identify without high spatial resolution. Metternich & Fermont (1998) mapped regional patterns of soil surface erosion using spectral mixture analysis on Landsat data.

27

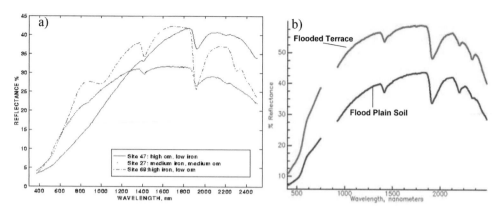

Figure 11.   a) Three soils showing effect of organic matter on soil reflectance. b) Soil spectra from an eroded calcareous terrace and an alluvial floodplain (see colour plate page 372).

As with spectral features of plants, any overlapping biogeochemical absorption will contribute to the total absorption causing interpretation errors (Mustard 1993). Whiting et al. (2004) found that classical continuum analysis under-estimated the band depth of carbonate and clay absorptions if the effect of adjacent mineral absorptions and water on soil reflectance were not quantified. Given the hydrophilic nature of clays, their model provided a robust methodology that could be used to obtain quantitative estimates of soil minerals. Overlapping adjacent absorption features affect reflectance along the shoulders between the absorptions (Clark & Roush 1984, Sunshine et al. 1990, Mustard 1992), lowering reflectance and reducing the apparent depth of the absorption. This error in band depth significantly reduces the correlation between measured and predicted mineral content as moisture and other minerals increase. The short wavelength side of the clay absorption band at 2200 nm is affected by broadening of the water overtone at 1900 nm and an absorption at 2100 nm by dry plant material (Whiting et al. 2006). Quantifying soil moisture and adjacent materials allows a correct interpretation of clay and carbonate band depths. Including this model in materials that have overlapping absorption features should improve their quantification.

3   MAPPING PLANT AND SOIL PROPERTIES USING
   BIOPHYSIOLOGICAL INDEXES

A wide range of spectral indexes have been developed that take advantage of specific absorption features of plants and soils relative to spectral bands where the materials do not absorb (Huete et al. 1997). These methods are quasi-quantitative in that specific chemicals absorb energy at specific wavelengths, thus we know a priori where these features occur. However, the absorption depth is not linear with concentration and the presence of other materials with overlapping absorption features will confound quantification. Some plant and soil chemicals of interest are actually a family of related molecules (e.g., organic matter) rather than a specific molecule (e.g., H_2O) which results in inconsistent identification of the wavelength location of absorption features, thus limiting index methods.

Vegetation indexes are sensitive to vegetation cover and provide an estimate of the "greenness" of an area. Newer indexes identify specific pigments, water content, cellulose, lignin, and nitrogen. Tables 2 and 3 list 36 widely used indexes of four basic types: pigment, foliar water, foliar chemistry, and soil indexes (Table 1). Not all of these indexes are independent. Obviously the multiple indexes estimating chlorophyll content will be correlated. Additionally, correlations between physiological characteristics lead to further correlations between indexes. For example, the water indexes and nitrogen index are highly correlated to NDVI. This is because nitrogen increases with water

Table 1. Soil Indexes Developed as Biophysical Indicators.

Index[a]	Formula	Details	Source[b]
NDI	$(R_{840} - R_{1650})/(R_{840} + R_{1650})$	Discriminates soil and dry matter	1
SACRI	$\dfrac{\alpha(R_{840} - \alpha R_{1650} - \beta)}{R_{1650} - \alpha R_{840} - \alpha\beta}$	Improved NDI discrimination	2
CRIM			3
	TM Indices	**HRV Indices**	
BI	$BI = \sqrt{(TM1^2 + TM2^2 + TM3^2)/3}$	$BI = \sqrt{(XS1^2 + XS2^2)/2}$	4
SI	$SI = (TM3 - TM1)/(TM3 + TM1)$		5
HI	$HI = (2 \times TM3 - TM2 - TM1)/(TM2 - TM1)$		5
CI	$CI = (TM3 - TM2)/(TM3 + TM2)$	$CI = (XS2 - XS1)/(XS2 + XS1)$	6, 7
RI	$RI = TM3^2/(TM1 \times TM2)$	$RI = XS2^2/XS1^3$	7
RI		$RI = XS2^2/XS1^4$	4

[a] **NDI**: Normalized Index; **SACRI**: Soil Adjusted Crop Residue Index; **CRIM**: Crop Residue Index Multiband; **BI**: Brightness Index—average soil reflectance; **SI**: Saturation Index—spectral slope; **HI**: Hue Index—dominant wavelength; **CI**: Coloration Index—hematite/hematite+geothite ratio; **RI**: Redness Index—Hematite Content.

[b] 1, McNairn & Protz 1993; 2, 3, Biard & Baret 1997; 4, Mathieu et al. 1998; 5, Escadafal et al. 1994; 6, Escadafal & Huete 1991; 7, Madeira et al. 1997.

as does photosynthetic machinery, resulting in higher NDVI. Conditions where indexes become uncorrelated or partially uncorrelated may provide evidence for stress conditions.

3.1 *Vegetation indexes*

Vegetation indexes (VI) take advantage of the strong contrast between red and NIR reflectance produced by green foliage (related to vigor and amount of vegetation). In desert environments, VIs must be able to discriminate vegetation at low cover, from zero up to about 50%, and to extract information about vegetation properties, while minimizing soil, atmosphere, and sun and view angle effects. In arid environments soil is the most significant contribution to scene variance therefore, a correction for soil background effects is necessary.

3.2 *Pigment indexes*

Pigment indexes focus on photosynthetic pigments, primarily chlorophyll, the dominant photosynthetic pigment of green vegetation (Jacquemoud & Baret 1990, Jacquemoud et al. 1996). However, some like the PRI identify xanthophylls cycle pigments, which are important in regulating light absorption in high light intensity environments (Gamon et al. 1992, 1993, 1994). Leaves possess a range of accessory photosynthetic pigments, including carotenoids and xanthophylls that serve functions to regulate and increase photosynthetic performance.

3.3 *Foliar water indexes*

Water absorbs strongly throughout the SWIR wavelengths (Carter 1991). Most narrow-band indexes use the 970 or 1240 nm water absorption features, although broad band indexes have been used (e.g., Landsat bands 4, 5 and 7).

3.4 *Foliar chemistry indexes*

Several leaf chemistry indexes have been proposed. The normalized difference nitrogen index (NDNI) was successfully used by Serrano et al. (Serrano et al. 2002) on AVIRIS data to estimate nitrogen concentration in a semiarid shrubland. Nonetheless, other studies have found it difficult

Table 2.   Vegetation Indexes Developed as Biophysical Indicators.

Index[a]	Formula	Details	Source[b]
Pigment			
SR	R_{NIR}/R_R	Index of green vegetation cover. Wavelengths, depending on sensor, e.g., NIR = 845 nm, R = 665 nm.	1
NDVI	$(R_{NIR} - R_R)/(R_{NIR} + R_R)$	Index of green vegetation cover. Wavelengths, depending on sensor., e.g., NIR = 845 nm, R = 665 nm.	1
mNDVI	$(R_{750} - R_{705})/(R_{750} + R_{705})$	Leaf chlorophyll content	2
SGR	$\sum\limits_{n=500}^{599} R_n$	Index of green vegetation cover.	2
PRI	$(R_{531} - R_{570})/(R_{531} + R_{570})$	Xanthophyll light response $\sim$ photosynthetic efficiency. Sensitive to carotenoid/chlorophyll ratio	3
RGR	$(R_{600-699})/(R_{500-599})$	Anthocyanins/chlorophyll	2
NPCI	$(R_{680} - R_{430})/(R_{680} + R_{430})$	Total pigments/chlorophyll	4
SRPI	R_{430}/R_{680}	Carotenoid/chlorophyll a content	5
NPQI	$(R_{415} - R_{435})/(R_{415} + R_{435})$	Chlorophyll degradation, detects early stress	5
SIPI	$(R_{800} - R_{445})/(R_{800} - R_{680})$	Carotenoid/chlorophyll a concentrations	5
PI1	R_{695}/R_{420}	Plant stress status	5
PI2	R_{695}/R_{760}	Plant stress status	5
PI3	R_{440}/R_{690}	Vegetation health index, chlorophyll fluorescence ratios	6
PI4	R_{440}/R_{740}	Vegetation health, chlorophyll fluorescence ratios	6
Water			
NDWI	$(R_{860} - R_{1240})/(R_{860} + R_{1240})$	Leaf water content	7
WBI	R_{900}/R_{970}	Leaf water content	8
Foliar chemistry			
NDNI	$[\log(R_{1680}/R_{1510})]/[\log(^1/R_{1680}R_{1510})]$	Foliar nitrogen concentration	9
NDLI	$[\log(R_{1680}/R_{1754})]/[\log(^1/R_{1680}R_{1754})]$	Foliar lignin concentration	9
CAI	$0.5(R_{2020} + R_{2220}) - R_{2100}$	Cellulose & lignin absorption features, discriminates plant litter from soils	10

[a]**SR**: Simple Ratio; **NDVI**: Normalized Difference Vegetation Index; **mNDVI**: Modified NDVI; **SGR**: Summed green reflectance; **PRI**: Photochemical Reflectance Index; **RGR**: Red/Green ratio; **NPCI**: Normalized Pigments; **CRI**: Chlorophyll Ratio Index; **SRPI**: Simple Ratio Pigment Index; **NPQI**: Normalized Phaeophytinization Index; **SIPI**: Structure Intensive Pigment Index; **PI**: Pigment Index; **NDWI**: Normalized Difference Water Index; **WBI**: Water Band Index; **NDNI**: Normalized Difference Nitrogen Index; **NDLI**: Normalized Difference Lignin Index; **CAI**: Cellulose Absorption Index.

[b]1,Tucker 1979; 2, Fuentes et al. 2001; 3, Gamon et al. 1992; 4, Peñuelas et al. 1995a; 5, Zarco-Tejada 1998; 6, Lichtenthaler et al. 1996; 7, Gao 1996; 8, Peñuelas et al. 1997b; 9, Serrano et al. 2002; 10, Nagler et al. 2000

Table 3. Leaf chemistry measured in the LOPEX dataset (Jacquemoud et al. 1996) and chemistry predictions from the PROSPECT model for an average of five *Quercus pubescens* leaf measurements.

Leaf properties	LOPEX chemistry	Fresh leaf predictions	Dry leaf prediction
N		1.38	2.97
Cab (μg/cm^2)	42.7	42.2	28.1
Brown pigments (μg/cm^2)	–	30.2	105.7
Cw (g/cm^2)	0.00825	0.00750	0.00063
Cm (g/cm^2)	0.00604	0.00604	0.00604

to extract this data from two band indexes. Piñzon et al. (1998) used singular value decomposition to accurately estimate leaf nitrogen from leaf samples in laboratory data. Later, Smith et al. (2002) used partial least squares to relate field measured leaf nitrogen to canopy nitrogen in AVIRIS hyperspectral data. The cellulose absorbance index (CAI; Nagler et al. 2000), estimates the cellulose content of senescent plant matter, emphasizing the distinctions between soil and plant litter. This index may be useful in arid and semi-arid regions where green vegetation is seasonally sparse and much of the biomass is in undecomposed plant litter and stems.

3.5 *Soil indexes*

To observe desert conditions and desertification it is critical to map soils with none or low amounts of vegetation. The soil line, related to soil brightness in red and NIR bands, has already been introduced. Although soil spectra lack strong absorption bands, several soil indexes (Table 1) related to soil color have been developed (Madeira et al. 1997, Mathieu et al. 1998). The latter found high correlation between the Helmholtz definition of color and Landsat Thematic Mapper (TM) indexes in an arid environment. Redness indexes are important for characterizing soil weathering and oxidation. Color is important for assessing soil quality and is closely related to surface processes and types of chemical weathering. Soil color is defined by parent material, weathering, and topography which are related to soil biophysical properties, therefore, color might be used as an indicator of erosion or deposition or as an indicator of surface organic matter. If soil variability is high, as in many desert environments, there will not be a unique soil line and the width of the soil line becomes critical for defining soil properties.

4 LEAF REFLECTANCE AND TRANSMITTANCE MODELS

Over the past decade, sophisticated radiative transfer models have been developed to account for leaf properties and changes in reflectance/transmittance due to changing leaf biochemistry and structure. The simplest models consider the leaf as a single scattering and absorbing plane-parallel layer while the most complicated models consider the full three-dimensional structure and biochemistry of the cells and tissues that form the leaf. At a minimum, physically realistic models require information about the refractive index and the specific absorption coefficients of leaf constituents (Fig. 4). Computer-based leaf models can be categorized into different classes, arranged in order of increasing complexity (Fig. 12).

4.1 *Plate models*

Figure 12a: Allen et al. (1969) were the first to represent a compact leaf as an absorbing plate with rough surfaces producing diffusion. This approach was extended to non-compact leaves by regarding them as layers of plates separated by N-1 air spaces (Allen et al. 1970). The parameter N provides an internal structure for scattering. The widely used PROSPECT model (Leaf Optical Properties Spectra) (Jacquemoud & Baret 1990) was designed to accurately simulate the hemispherical reflectance and transmittance of various types of plant leaves (fresh monocot and

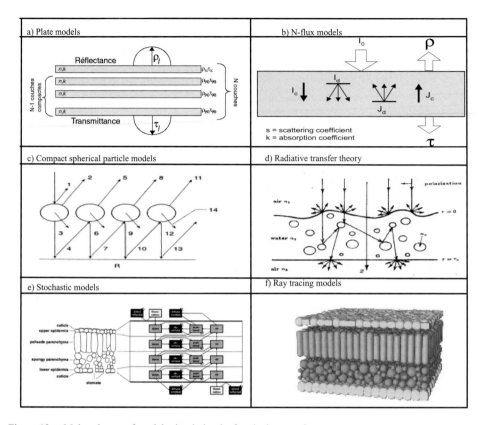

Figure 12.   Major classes of models simulating leaf optical properties.

dicot leaves, senescent and dry leaves) over the solar spectrum from 400 to 2500 nm. The original model was improved significantly by optimization of input parameters (Jacquemoud et al. 1996, Fourty et al. 1996, Baret & Fourty 1997, Fourty & Baret 1998, Jacquemoud et al. 2000). Its four input parameters today are the leaf structure parameter, chlorophyll a+b concentration, equivalent water thickness, and dry matter content. Recent studies (Ceccato et al. 2001, Bacour et al. 2002) have quantified the contribution of these parameters to the PROSPECT model outputs, as well as their interactions (Fig. 13), demonstrating that most foliar absorptions are accounted for in the current model, with the gray line the sum of these contributions. Clearly there is potential to improve the model in both the VIS and SWIR regions.

PROSPECT provides good estimates of leaf water content (equivalent water thickness) in fresh leaves, illustrated in Table 2 using data from the LOPEX experiment (Jacquemoud et al. 1996). Likewise estimate of dry matter from dry leaves is well predicted although the prediction of dry matter content in fresh leaves is poorly predicted. For the *Quercus pubescens* example shown in Figure 3, Table 2 provides a comparison between measured and predicted chemistry. Chuvieco et al. (2005) have empirically solved this problem for wildfire risk in Mediterranean ecosystems by using multi-temporal data and combining the dry biomass estimate from the driest annual period with water content from real-time estimates. Because total ecosystem biomass changes slowly over time, this method provides better estimates than direct model inversions.

4.2   N-flux models

Figure 12b: Derived from the Kubelka-Munk theory these models consider the leaf as a slab of diffusing and absorbing materials. Different parameters are allowed in each layer and the model

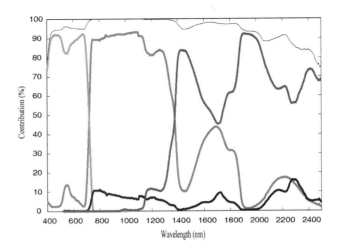

Figure 13. Contribution to leaf transmittance simulated by PROSPECT for chlorophyll concentration Cab (green), water content Cw (blue), dry matter content Cm (brown) and the structure parameter N (red) (Pavan, unpublished) (see colour plate page 373).

estimates leaf reflectance and transmittance. Leaf biochemistry was introduced by Conel et al. (1993) who used a two-flux model to evaluate the influence of water, protein, cellulose, lignin, and starch on leaf reflectance, however their model was not validated.

4.3   Compact spherical particle models

Figure 12c: Dawson et al. (1998) adapted Melamed's theory of light interaction with suspended powders and designed the LIBERTY model (Leaf Incorporating Biochemistry Exhibiting Reflectance and Transmittance Yields) specifically to calculate the optical properties of both dried and fresh stacked conifer (particularly pine) needles. By treating the leaf as an aggregation of cells, with multiple scattering, output reflectance and transmittance are a function of three structural parameters (cell diameter, intercellular air space, and leaf thickness) and the combined absorption coefficients for chlorophyll, water, lignin and cellulose, and nitrogen.

4.4   Radiative transfer equation

Figure 12d: Few leaf models directly use the radiative transfer equation because essential information about internal leaf structure and biochemical distributions is lacking, leading to major simplifications. Ma et al. (1990) described the leaf as a slab of water with an irregular surface containing randomly distributed spherical particles. Ganapol et al. (1998) developed LEAFMOD (Leaf Experimental Absorptivity Feasibility MODel), which models the leaf as a homogeneous mixture of biochemicals which scatter and absorb light. The homogeneity simplification ignores the obvious cellular organization of leaf chemistry. Nonetheless, Johnson (2001) used this model to predict leaf nitrogen from several laboratory datasets.

4.5   Stochastic models

Figure 12e: Tucker & Garatt (1977) proposed a stochastic model in which radiation transfer is simulated by a Markov chain. The leaf is partitioned into palisade parenchyma and spongy mesophyll tissues with four radiation states (solar, reflected, absorbed, and transmitted) with defined the transition probabilities. The SLOP (Stochastic model for Leaf Optical Properties) model (Maier et al. 1999) is a recent improvement in which the leaf is partitioned into four tissues.

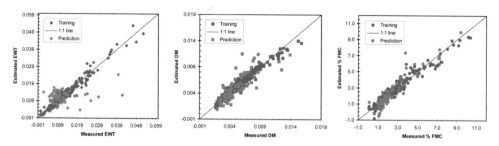

Figure 14. a) Prediction of leaf water content from fresh leaves b) leaf dry matter from dry leaves, and c) leaf dry matter from fresh leaves using a generic algorithm-partial least square regression, GA-PLS (red dots) on LOPEX leaf data for calibration and validation (blue dots). Leaf reflectance and transmission data and biochemistry from the LOPEX dataset (Hosgood et al. 1994). The mean reflectance of the samples at each spectral band is subtracted before running GA-PLS (see colour plate page 373).

4.6 Ray tracing models

Figure 12f: Only ray tracing techniques account for the full three dimensional complexity of internal leaf structure as it appears in a photomicrograph. This technique requires a detailed description of individual cells and their unique arrangement inside tissues and the optical properties of leaf materials (cell walls, cytoplasm, pigments, air cavities, etc.) must be defined. Using the laws of reflection, refraction, and absorption, the propagation of individual photons incident on the leaf surface can be simulated. Once a sufficient number of rays have been simulated, statistically valid estimates of the radiation transfer in a leaf can be predicted.

Govaerts et al. (1996) used a three-dimensional ray tracing model, RAYTRAN (1998), and developed a virtual three-dimensional leaf, to characterize the light environment, including absorption, scattering and transmission, within and between cells. Cells of variable size, cell wall thicknesses, chemistry, and air spaces were modeled to simulate realistic leaf tissues and the implications for absorption profiles, light harvesting, and photosynthesis was successfully investigated (Ustin et al. 2001). The model simulated the leaf anatomy of a typical mesic dorsiventral dicotyledonous leaf. In general, we lack sufficient information about leaf anatomy and biochemistry to apply this model to a wider range of morphological types, e.g., xeromorphic leaves. However, if anatomical and morphological data were available, it would be possible to simulate a range of environmental conditions and improve our understanding of biophysical measurements in arid ecosystems.

4.7 Next generation of leaf models

Despite decades of research, much more work is required before we will accurately model leaf optical properties. Progress on the next generation of optical models requires improvements in understanding detailed cell and leaf anatomy for leaves having adaptations to different environmental conditions. Perhaps new approaches to modeling may lead to fundamental improvements, e.g., the ABM (Algorithmic BDF Model) or FSM (Foliar Scattering Model) which was used to study the interaction of light with plant leaves for image synthesis applications (Baranoski et al. 2004). Further, more understanding of the relationships between structure and function at the leaf level are clearly needed to drive model improvements. Lastly, better optical characterization of more biochemical compounds is needed to expand the range of biochemistry that can be detected. It is not demonstrated that specific leaf photosynthetic pigments (chlorophyll a and b, carotenoids and xanthophylls) can be identified and quantified and yet this information would significantly improve understanding of the biological controls on photosynthesis. One reason for needing more information is that a direct interpretation of chlorophyll only estimates potential carbon fixation and does not measure actual state of photosynthetic activation.

A small but measurable fraction of the radiation reflected and transmitted by leaves is actually fluorescence emission (Zarco-Tejada et al. 2003a, Dobrowski et al. 2005) by polyphenols in the

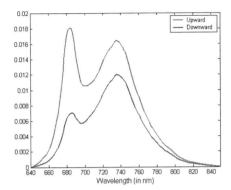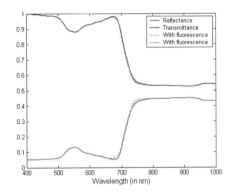

Figure 15. Simulation of the adaxial and abaxial chlorophyll fluorescence (on the left) and of the reflectance and transmittance with and without fluorescence (on the right) using fluormodleaf. Structure parameter: N = 1.5, chlorophyll a + b content: Cab = 33 μg cm^{-2}, equivalent water thickness: Cw = 0.025 cm, dry matter content: Cm = 0.01 g cm^{-2}, fluorescence quantum yield: Φ = 0.04, temperature: T = 20°C, species: green bean, PSII/SPI ratio: Sto = 2.0 (see colour plate page 374).

blue and chlorophyll a in the red and NIR as shown in Figure 15. Passive sensing of chlorophyll fluorescence in the oxygen bands emerges as a promising approach for monitoring vegetation health. However, the interpretation of this signal depends both on the leaf environment (incoming light quality and intensity, temperature, etc.) and on intrinsic parameters (physiological state, species, biochemical composition, etc.). In the FluorMOD project (Development of a Vegetation Fluorescence Canopy Model), a leaf fluorescence model has been developed that predicts reflectance and transmittance of a fresh leaf including chlorophyll fluorescence (Miller et al. 2005), Pedrós et al. 2006). Beside the classical input variables for PROSPECT, fluorescence quantum yield, PSII/PSI ratio, temperature, species, and incident PAR are required. Figure 15 shows the first results of the fluormodleaf for a typical dicot leaf. Although the model is still at an early stage of validation, these results are encouraging.

Many studies have demonstrated a direct relationship between photosynthetic rates, light absorbance, chlorophyll content, leaf nitrogen, and dry matter production are all related to leaf area (e.g., Gamon et al. 1993, Esteve et al. 1998, Alt et al. 2000). There appears to be a linear dependence of maximum photosynthetic capacity on leaf nitrogen (Field & Mooney 1986) which is partially species specific. Leaf nitrogen exhibits a linear relationship with specific leaf mass, the reciprocal of specific leaf area (1/specific leaf area) (Garnier et al. 1999), and a property inversely related to dry biomass estimated by PROSPECT.

As described earlier, hairy and/or waxy leaves are typical of plants in arid regions. Specific surface optical properties of these leaves are poorly known. For instance, a thick cuticle, which may protect leaves from desiccation or insects acts as a mirror, preferentially reflecting light in the specular direction. In contrast, hairs tend to scatter light in all directions. This duality between specular and Lambertian surfaces is characteristic of the leaf bidirectionnal reflectance distribution function (BRDF). Leaf BRDF was recently acquired for several species using a goniophotometer at 400 wavelengths and 98 viewing angles (Bousquet et al. 2005). These measurements (Fig. 16) have been successfully estimated with a specular reflection model (Cook-Torance) coupled with prospect to simulate the diffuse fraction.

4.8   Extending biophysical measurements to canopy models

Heterogeneous and open plant canopies, characteristic of desert environments, preclude the use of simple 1-D radiative transfer models that rely on an assumption of homogenous turbid media (Luquet et al. 1998). On such surfaces, the interactions between light and the various objects (for instance, shrubs, grasses, BSC, bare soil), which are distributed in clumped patches rather

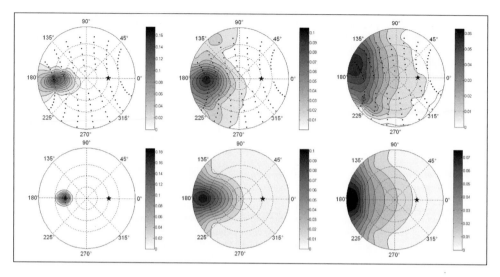

Figure 16. Measured (top) and modeled (bottom) BRDF for θs = 41° at wavelengths of minimum reflection (after Bousquet et al. 2005).

than uniformly distributed, generate a higher level of complexity that requires 3-D models (Pinty et al. 1996). Esteve et al. (1998) used the DART (Discrete Anisotropic Radiative Transfer) model (Gastellu-Etchegorry et al. 1996), to simulate radiative transfer in heterogeneous arid 3-D landscape comprised of trees, shrubs, grass, soil, etc. They inverted DART on a temporal series of Landsat MultiSpectral Scanner (MSS) acquired in Burkina Faso, and determined the woody cover allowing them to follow its degradation over time. Zarco-Tejada et al. (2003) followed changing leaf water content over the summer drought in semiarid southern California using a MODIS time series analyzed with a linked PROSPECT and SAILH canopy model. Some information about surface heterogeneity, which may change fast when land degradation occurs, can also be inferred using multi-angular reflectance data. Widlowski et al. (2001) generated various 3-D vegetation canopy representations, from homogeneous to heterogeneous, at the nominal ground resolution of the MISR instrument and simulated their reflectance using RAYTRAN. They showed that the reflectance anisotropy could be obtained by inversion of a simple parametric model.

5   SOIL REFLECTANCE MODELS

Most soil mapping has been related to the fraction of exposed soil or more precisely, the fraction of green vegetation cover, assuming that non-vegetated surfaces are soil. To date the identification of soil biogeochemical components from reflectance measurements have been largely limited to regression models that are valid only for specific locations. Greater use of multivariate statistical models, e.g., neural nets or support vector machines, will provide techniques that increase accuracy of soil quantities and portability across larger regions. Wider use of classification and regression tree (CART) models, partial least-squares, and general additive model (GAM) methods may also improve local regression models by including more spatial data in the analysis, and provide the templates for applying methods in other soil regions. These regression models may also benefit from the inclusion of hierarchical partitioning the surface reflectance components as a means of stratifying the image region to improve the local spectral model, just as much as landscape partitioning improves the accuracy.

   The greater task, although highly promising for producing more robust models, is in developing radiative transfer models that can identify the presence and abundance of soil components in the context of intimate mixtures and surface geometry. Knowledge gained in recent years in extending radiative transfer modeling to a wider range of vegetation and soil backgrounds will restart soil modeling efforts begun in 1990's. The identification of absorption coefficients for soil minerals, organic matter, and water are needed to develop radiative transfer models that will measure mineral abundance through spectral curve fitting.

6   CONCLUSION

The geographic extent and remote locations of arid ecosystems and the harsh climate character-istic of arid ecosystems create challenges to effectively monitoring desertification. Furthermore monitoring must be done over long periods to fully observe the stochastic behavior of deserts in a globally changing context. The potential to use remote sensing for cost-effective monitoring has long been recognized, although the extensive heterogeneity of geologic parent material and soils combined with low vegetative cover, rapid biologic responses to wet and dry rainfall pulses, has limited our ability to fully monitor the condition of these ecosystems. Furthermore, infrequent precipitation leads to extended dormancy with short unpredictable periods of biological activity. We reviewed the spectral characteristics of plants and soils that are detectable using optical sensors and current methods to identify and quantify properties that have potential for monitoring arid ecosystem processes. The wealth of existing information on the reflectance and transmittance of desert plants, soils, and other materials, combined with new modeling and index approaches will lead to improved capability for monitoring global desert environments.

A wide range of plant biochemicals are detectable or potentially detectable in hyperspectral data, e.g., chlorophylls, photosynthetic accessory pigments, water, and soil constituents e.g., clays, carbonate, iron, and organic matter. Biophysical methods may improve detection and monitoring of inter-plant spaces composed of bare soil crust, biological soil crust, or dead plant material. The expanded use of absorption features in the VIS and SWIR has produced a large number of spectral index-based methods to identify specific compounds or properties. As we move toward more quantitative predictive models, it is clear that the overlapping structure of many absorption features of plants and soils precludes direct assessment of their concentration. New biophysical methods that take the full spectral shape into account, including the effect of one compound on the spectral absorption of another, are needed to reduce uncertainty in their estimates. Over the past decade several radiative transfer models have been developed based on physical absorption and scattering processes using principles of spectroscopy and information over the entire spectrum. Other models take advantage of new types of data like passive fluorescence or thermal-infrared emissions or bidirectional reflectance.

ACKNOWLEDGEMENTS

A significant part of this review was done while the first author was on sabbatic leave at the Department of Geography and Environmental Sciences, University of Auckland, New Zealand. The authors wish to thank Margaret Andrew for her help in assembling the index information and references for Table I and Dr. David Riaño for help in preparing some figures. We wish to thank Van Lay for help in formatting the document and references and preparation of other figures.

REFERENCES

Aase, J.K. & Tanaka, D.L. 1991. Reflectances from four wheat residue cover densities as influenced by three soil backgrounds. *Agronomy Journal* 83: 753–757.

Abers, J.D., Wessman, C.A., Peterson, D.L., Melillo, J.M. & Fownes, J.H. 1990. Remote sensing of litter and soil organic matter decompsition in forest ecosystems. In Hobbs, R.J. & Mooney, H.A. (eds.), *Remote Sensing of Biosphere Functioning*: 87–103. New York: Springer-Verlag.

Ager, C.M. & Milton, N.M. 1987. Spectral reflectance of lichens and their effects on the reflectance of rock substrates. *Geophysics* 52: 898–906.

Allen, W.A., Gausman, H.W. & Richardson, A.J. 1970. Mean effective constants of cotton leaves. *Journal of Optical Society of America* 60: 542–547.

Allen, W.A., Gausman, H.W., Richardson, A.J. & Thomas, J.R. 1969. Interaction of isotropic light with a compact leaf. *Journal of Optical Society of America* 59: 1376–1379.

Alt, C., Stutzel, H. & Kage, H. 2000. Optimal nitrogen content and photosynthesis in cauliflower (Brassica oleracea L. botrytis). Scaling up from leaf to whole plant. *Annals of Botany* 85: 779–787.

Aranda, V. & Oyonarte, C. 2005. Effect of vegetation with different evolution degree on soil organic matter in a semi-arid environment (Cabo de Gata-Nijar Natural Park, SE Spain). *Journal of Arid Environments* 62: 631–647.

Asner, G.P. 1998. Biophysical and biochemical sources of variability in canopy reflectance. *Remote Sensing of Environment* 64: 234–253.

Bacour, C., Jacquemoud, S., Tourbier, Y., Dechambre, M. & Frangi, J.P. 2002. Design and analysis of numerical experiments to compare four canopy reflectance models. *Remote Sensing of Environment* 79: 72–83.

Baranoski, G.V.G. & Rokne, J.G. 2004. Light Interaction with Plants. A Computer Graphics Perspective, Horwood Publishing.

Baret, F. & Fourty, T.H. 1997. Estimation of leaf water content and specific leaf weight from reflectance and transmittance measurements. *Agronomie* 17: 455–464.

Baret, F., Jacquemoud, S. & Hanocq, J.F. 1993. The soil line concept in remote sensing. *Remote Sensing Reviews* 7: 65–82.

Baret, F., Jacquemoud, S., Guyot, G. & Leprieur, C. 1992. Modeled analysis of the biophysical nature of spectral shifts and comparison with information content of broad bands. *Remote Sensing of Environment* 41: 133–142.

Barton, F.E., II, Himmelsbach, D.S., Duckworth, J.H. & Smith, M.J. 1992. Two-dimensional vibration spectroscopy: correlation of mid- and near-infrared regions. *Applied Spectroscopy* 46(3): 420–429.

Baumgardner, M.F., Silva, L.F., Biehl, L.L. & Stoner, E.R. 1985. Reflectance properties of soils. *Advances in Agronomy* 38: 1–44.

Beatley, J.C. 1974. Effects of rainfall and temperature on the distribution and behavior of Larrea tridentata (creosote-bush) in the Mojave Desert of Nevada. *Ecology* 55: 245–261.

Beatley, J.C. 1980. Fluctuations and Stability in Climax Shrub and Woodland Vegetation of the Mojave, Great Basin and Transition Deserts of Southern Nevada. Israel. *Journal of Botany* 28(3–4): 149–168.

Bechtel, R., Rivard, B. & Sanchez-Azofeifa, A. 2002. Spectral properties of foliose and crustose lichens based on laboratory experiments. *Remote Sensing of Environment* 82(2–3): 389–396.

Belnap, J. & Eldridge, D.J. 2001. Disturbance and recovery of biological soil crusts. Belnap, J. & Lange, O.L. (eds.), *Biological Soil Crusts: Structure, Function, and Management*: 363–384. Berlin: Springer-Verlag.

Belnap, J. & Gardner, J.S. 1993. Soil microstructure in soils of the Colorado Plateau: the role of the cyanobacterium Microcoleus vaginatus. *Great Basin Naturalist* 53: 40–47.

Belnap, J. 2001a. Comparative structure of physical and biological soil crusts. In Belnap, J. & Lange, O.L. (eds.), *Biological Soil Crusts: Structure, Function, and Management*: 177–192. Berlin: Springer-Verlag.

Belnap, J. 2001b. Biological soil crust and wind erosion. In Belnap, J. & Lange, O.L. (eds.), *Biological Soil Crusts: Structure, Function, and Management*: 339–348. Berlin: Springer-Verlag.

Belnap, J., Budel, B. & Lange, O.L. 2001a. Biological soil crusts: Characteristics and distribution, In Belnap, J. & Lange, O.L. (eds.), *Biological Soil Crusts: Structure, Function, and Management*: 3–30. Berlin: Springer-Verlag.

Ben-Dor, E., Goldlshleger, N., Benyamini, Y., Agassi, M. & Blumberg, D.G. 2003. The spectral reflectance properties of soil structural crusts in the 1.2- to 2.5-mu m spectral region. *Soil Science Society of America Journal* 67: 289–299.

Biard, F. & Baret, F. 1997. Crop residue estimation using multiband reflectance. *Remote Sensing of Environment* 59: 530–536.

Bousquet, L., Lacherade, S., Jacquemoud, S. & Moya, I. 2005. Leaf BRDF measurement and model for specular and diffuse component differentiation. *Remote Sensing of Environment* 98: 201–211.

Bowers, S.A. & Hanks, R.J. 1965. Reflection of radiant energy from soils. *Soil Science* 100: 130–138.

Buffoni Hall, R.S., Bornman, J.F. & Björn, L.O. 2002. UV-induced changes in pigment content and light penetration in the fruticose lichen Cladonia arbuscula ssp. mitis. *Journal of Photochemistry and Photobiology* 66: 13–20.

Carter, G.A. 1991. Primary and secondary effects of water content on the spectral reflectance of leaves. *American Journal of Botany* 78(7): 916–924.

Ceccato, P., Flasse, S., Tarantola, S., Jacquemoud, S. & Grégoire, J.M. 2001. Detecting vegetation water content using reflectance in the optical domain. *Remote Sensing of Environment* 77: 22–33.

Charney, J., Quirk, W.J., Chow, S.H. & Kornfield, J. 1977. Comparative-study of effects of albedo change on drought in semi-arid regions. *Journal of Atmospheric Science* 34: 1366–1385.

Chen, F., Kissel, D.E., West, L.T. & Adkins, W. 2000. Field-scale mapping of surface soil organic carbon using remotely sensed imagery. *Soil Science Society of America Journal* 64: 746–753.

Chuvieco, E., Ventura, G., Martin, M.P. & Gomez, I. 2005. Assessment of multitemporal compositing techniques of MODIS and AVHRR images for burned land mapping. *Remote Sensing of Environment* 94 (4): 450–462.

Cierniewski, J. & Karnielli, A. 2002. Virtual surfaces simulating the bi-directional reflectance of semi-arid soils. *International Journal of Remote Sensing* 23: 4019–4037.

Clark, R.N. & Roush, T.L. 1984. Reflectance spectroscopy: Quantitative analysis techniques for remote sensing applications. *Journal of Geophysical Research* 89: 6329–6340.

Clark, R.N. 1999. Spectroscopy of rocks and minerals, and principles of spectroscopy. In Rencz, A.N. (ed.), *Remote Sensing for the Earth Sciences: Manual of Remote Sensing*: 3–58. New York: John Wiley & Sons, Inc.

Coleman, T.L. & Montgomery, O.L. 1987. Soil moisture, organic matter, and iron content effect on the spectral characteristics of selected vertisols and alfisols in Alabama. *Photogrammetric Engineering and Remote Sensing* 53: 1659–1663.

Conel, J.E. & Van Den Bosch, J., Grove, C.I. 1993. Application of a two-stream radiative transfer model for leaf lignin and cellulose concentrations from spectral reflectance measurements. Parts 1 & 2. In Green, R.O. (ed.), *Proceedings 4th Annual JPL Airborne Geoscience Workshop. Vol. 1. AVIRIS Workshop 25–29 October 1993, Washington (DC)*, NASA-JPL Publication 93–26: 39–51.

Cooke, R.U. & Warren, A. 1973. Geomorphology in deserts. London: B.T. Batsford Ltd.

Courault, D., P. Bertuzzi, P. & Girard, M.C. 1993. Monitoring surface changes of bare soils due to slaking using spectral measurements. *Soil Science Society of America Journal* 57: 1595–1601.

Curcio, J.A. & Petty, C.C. 1951. The near infrared absorption spectrum of liquid water. *Journal Optical Society of America* 41(5): 302–304.

Curran, P.J., Dungan, J.L., Macler, B.A. & Plummer, S.E. 1991. The effect of a red leaf pigment on the relationship between red edge and chlorophyll concentration. *Remote Sensing of Environment* 35(1): 69–76.

Dalal, R.C. & Henry, R.J. 1986. Simultaneous determination of moisture, organic carbon and total nitrogen by near infrared reflectance spectroscopy. *Soil Science Society of America Journal* 50: 120–123.

Dawson, T.P., Curran, P.J. & Plummer, S.E. 1998. LIBERTY—Modelling the effects of leaf biochemical concentration on reflectance spectra. *Remote Sensing of Environment* 65: 50–60.

de Jong, S.M. 1992. The analysis of spectroscopical data to map soil types and soil crusts of Mediterranean eroded soils. *Soil Technology* 5: 199–211.

Deering, D.W., Eck, T.F. & Otterman, J. 1990. Bidirectional reflectances of selected desert surfaces and their three-parameter soil characterization. *Agricultural and Forest Meteorology* 52: 71–93.

Demattê, J.A.M., Pereira, H.S., Nanni, M.R., Cooper, M. & Fiorio, P.R. 2003. Soil chemical alterations promoted by fertilizer application assessed by spectral reflectance. *Soil Science* 168: 730–747.

Dey, P.M. & Harborne, J.B. 1997. *Plant Biochemistry*. Academic Press Inc.

Dobrowski, S.Z., Pushnik, J.C., Zarco-Tejeda, P.J. & Ustin, S.L. 2005. Simple reflectance indicesindexes track heat and water stressed induced changes in steady state chlorophyll fluorescence at the canopy scale. *Remote Sensing of Environment* 97: 403–414.

Dorn, R.I. & Oberlander, T.M. 1981. Microbial origin of desert varnish. *Science* 213: 1245–1247.

Dorn, R.I. & Oberlander, T.M. 1982. Rock Varnish. *Progress in Physical Geography* 6: 317–367.

Dregne, H.E. 1986: Desertification of arid lands. In El-Baz, F. & Hassan, M.H.A. (eds.), *Physics of Desertification*: 473. Dordrecht: Martinus Nijhoff.

Ehleringer, J.R. & BjÖrkman, O. 1978. Pubescence and leaf spectral characteristics in a desert shrub, Encelia farinosa. *Oecologia* 36: 151–162.

Ehleringer, J.R. & Mooney, H.A. 1978. Leaf hairs: effects on physiological activity and adaptive value to a desert shrub. *Oecologia* 37: 183–200.

Eller, B.M. & Willi, P. 1977. The significance of leaf pubescence for the absorption of global radiation by Tussilago farfara L. *Oecologia* 29: 179–187.

Elvidge, C.D. & Chen, Z.K. 1995. Comparison of Broad-Band and Narrow-Band Red and Near-Infrared Vegetation Indices. *Remote Sensing of Environment* 54: 38–48.

Elvidge, C.D., Chen, Z.K. & Groeneveld, D.P. 1993. Detection of trace quantities of green vegetation in 1990 AVIRIS data. *Remote Sensing of Environment* 44: 271–279.

Elvidge, C.L. 1990. Visible and NIR reflectances characteristics of dry plant materials. *International Journal of Remote Sensing* 2: 1775–1795.

Escadafal, R. & Huete, A. 1991. Etude des propriétés spectrales des sols arides appliquée à l'amélioration des indices de végétation obtenus par télédétection. *C. R. Academy of Science Paris* 312: 1385–1391.

Escadafal, R., Belghith, A. & Ben Moussa, H. 1994. Indices spectraux pour la degradation des milieux naturels en Tunisie aride. *6éme Symposium International Measures Physiques et Signatures en Télédétection*. Val d'Isère, France : ISPRS-CNES.

Eshel, G., Levy, G.J. & Singer, M.J. 2004. Spectral reflectance properties of crusted soils under solar illumination. *Soil Science Society of America Journal* 68(6): 1982–1991.

Esteve, P., Fontes, J. & Gastellu-Etchegorry, J.P. 1998. Tropical dry ecosystems modelling and monitoring from space. *Ecological Modelling* 108: 175–188.

Field, C. & Mooney, H.A. 1986. The photosynthesis-nitrogen relationship in wild plants. In Givinsh, T.J. (ed.), *On the economy of form and function:* 25–55. Cambridge: Cambridge University Press.

Fourty, T.H., Baret, F., Jacquemoud, S., Schmuck, G. & Verdebout, J. 1996. Optical properties of dry plant leaves with explicit description of their biochemical composition: direct and inverse problems. *Remote Sensing of Environment* 56(2): 104–117.

Fourty, T.H. & Baret, F. 1998. On spectral estimates of fresh leaf biochemistry. *International Journal of Remote Sensing* 19: 1283–1297.

Fox, G.A., Sabbagh, G.J., Searcy, S.W. & Yang, C. 2004. An Automated Soil Line Identification Routine for Remotely Sensed Images. *Soil Science Society of American Journal* 68: 1326–1331.

Fuentes, D.A., Gamon, J.A., Qiu, H.L., Sims, D.A. & Roberts, D.A. 2001. Mapping Canadian boreal forest vegetation using pigment and water absorption features derived from the AVIRIS sensor. *Journal of Geophysical Research* 106: 33565–33577.

Gamon, J.A., Field, C.B., Roberts, D.A., Ustin, S.L. & Valentini, R. 1993. Functional patterns in an annual grassland during an AVIRIS overflight. *Remote Sensing of Environment* 44: 239–253.

Gamon, J.A. & Qiu, H-L. 1999. Ecological applications of remote sensing at multiple scales. In Pugnaire, F.I. & Valladares, F. (eds.), *Handbook of Functional Plant Ecology*: 805–266. New York: Marcel Dekker, Inc.

Gamon, J.A., Peñuelas, J. & Field, C.B. 1992. A narrow-waveband spectral index that tracks diurnal changes in photosynthetic efficiency. *Remote Sensing of Environment* 41: 35–44.

Gamon, J.A., Serrano, L. & Surfus, J.S. 1997. The photochemical reflectance index: an optical indicator of photosynthetic radiation use efficiency across species, functional types, and nutrient levels. *Oecologia* 112: 492–501.

Ganapol, B., Johnson, L., Hammer, P., Hlavka, C. & Peterson, D. 1998. LEAFMOD: a new within-leaf radiative transfer model. *Remote Sensing of Environment* 6: 182–193.

Gao, B.-C. 1996. NDWI—A normalized difference water index for remote sensing of vegetation liquid water from space. *Remote Sensing of Environment* 58: 257–266.

Garnier, E., Salager, J.L., Laurent, G. & Sonie, L. 1999. Relationships between photosynthesis, nitrogen and leaf structure in 14 grass species and their dependence on the basis of expression. *New Phytologist* 143: 119–129.

Gastellu-Etchegorry, J.-P., Demarez, V., Pinel, V. & Zagolski, F. 1996. Modeling radiative transfer in heterogeneous 3-D vegetation canopies. *Remote Sensing of Environment* 58: 131–156.

Goldshleger, N., Ben-Dor, E., Benyamini, Y., Agassi, M. & Blumberg, D.G. 2001. Characterization of soil's structural crust by spectral reflectance in the SWIR region. *Terra Nova* 13: 12–17.

Goldshleger, N., Ben-Dor, E., Benyamini, Y., Agassi, M. 2004. Soil reflectance as a tool for assessing physical crust arrangement of four typical soils in Israel. *Soil Science* 69: 677–687.

Goldshleger, N., Ben-Dor, E., Benyamini, Y., Blumberg, D. & Agassi, M. 2002. Spectral properties and hydraulic conductance of soil crusts formed by raindrop impact. *International Journal of Remote Sensing* 23: 3909–3920.

Govaerts, Y.M. & Verstraete, M.M. 1998. Raytran: A Monte Carlo ray-tracing model to compute light scattering in three-dimensional heterogeneous media. *IEEE Transactions in Geoscience and Remote Sensing* 36(2): 493–505.

Govaerts, Y.M., Jacquemoud, S., Verstraete, M.M. & Ustin, S.L. 1996. Three-dimensional radiation transfer modeling in a dicotyledon leaf. *Applied Optics* 35(33): 6585–6598.

Grossman, Y.L., Ustin, S.L., Sanderson, E., Jacquemoud, S., Schmuck, S. & Verdebout, J. 1996. Critique of stepwise multiple linear regression for the extraction of leaf biochemistry information from leaf reflectance data. *Remote Sensing of Environment* 56: 182–193.

Harper, K.T. & Belnap, J. 2001. The influence of biological soil crusts on mineral uptake by associated vascular plants. *Journal of Arid Environments* 47: 347–357.

Henderson, T.L., Baumgardner, M.F., Franzmeier, D.P., Stott, D.E. & Coster, D.C. 1992. High dimensional reflectance analysis of soil organic matter. *Soil Science Society of America Journal* 56: 865–872.

Hill, J., Hostert, P., Tsiourlis, G., Kasapidis, P., Udelhoven, T. & Diemer, C. 1998. Monitoring 20 years of increased grazing impact on the Greek island of Crete with earth observation satellites. *Journal of Arid Environments* 39(2): 165–178.

Hoch, W.A., Zeldin, E.L. & McCown, B.H. 2001. Physiological significance of anthocyanins during autumnal leaf senescence. *Tree Physiology* 21: 1–8.

Hosgood, B., Jacquemoud, S., Andreoli, G., Verdebout, J., Pedrini, A. & Schmuck, G. 1994. *The JRC Leaf Optical Properties Experiment (LOPEX'93)*, European Commission, Directorate-General XIII, Telecommunications, Information Market and Exploitation of Research, L-2920, Luxembourg. CL-NA-16095-EN-C.

Huete, A.R., Liu, H.Q., Batchily, K. & van Leeuwen, W. 1997. A comparison of vegetation indices global set of TM images for EOS-MODIS. *Remote Sensing of Environment* 59: 440–451.

Jacquemoud, S. & Baret, F. 1990. PROSPECT: a model of leaf optical properties spectra. *Remote Sensing of Environment* 34: 75–91.

Jacquemoud, S., Bacour, C., Poilve, H. & Frangi, J.-P. 2000. Comparison of four radiative transfer models to simulate plant canopies reflectance–Direct and inverse mode. *Remote Sensing of Environment* 74: 471–481.

Jacquemoud, S., Ustin, S.L., Verdebout, J., Schmuck, G., Andreoli, G. & Hosgood, B. 1996. Estimating leaf biochemistry using the PROSPECT leaf optical properties model. *Remote Sensing of Environment* 56: 194–202.

Jeffries, D.L., Link, S.O. & Klopatek, J.M. 1993a. CO_2 fluxes of cryptogamic crusts. 1. Response to resaturation. *New Phytologist* 125: 163–173.

Jeffries, D.L., Link, S.O. & Klopatek, J.M. 1993b. O_2 fluxes of cryptogamic crusts. 2. Response to dehydration. *New Phytologist* 125: 391–396.

Johnson, L.F. 2001. Nitrogen influence on fresh-leaf NIR spectra. *Remote Sensing of Environment* 78: 314–320.

Karnieli, A., Shachak, M., Tsoar, H., Zaady, E., Kaufman, Y., Danin, A. & Porter, W. 1996. The effect of microphytes on the spectral reflectance of vegetation in semiarid regions. *Remote Sensing of Environment* 57(2): 88–96.

Karnieli, A. & Sarafis, V. 1996. Reflectance spectrophotometry of cyanobacteria within soil crusts—A diagnostic tool. *International Journal of Remote Sensing* 17(8): 1609–1614.

Karnieli, A. & Tsoar, H. 1995. Spectral reflectance of biogenic crust developed on desert dune sand along the Israel-Egypt border. *International Journal of Remote Sensing* 16(9): 1741–1741.

Karnieli, A. 1997. Development and implementation of spectral crust index over dune sands. *International Journal of Remote Sensing* 18(6): 1207–1220.

Karnieli, A., Kidron, G.J., Glaesser, C. & Ben-Dor, E. 1999. Spectral characteristics of cyanobacteria soil crust in semiarid environments. *Remote Sensing of Environment* 69(1): 67–75.

Kokaly, R.F. & Clark, R.N. 1999. Spectroscopic determination of leaf biochemistry using band-depth analysis of absorption features and stepwise multiple linear regression. *Remote Sensing of Environment* 67(3): 267–287.

Kou, L., Labrie, D. & Chylek, P. 1993. Refractive indices of water and ice in the 0.65 2.5 µm spectral range, *Applied Optics* 32: 3531–3540.

Lange, O.L., Belnap, J. & Reichenberger, H. 1998. Photosynthesis of the cyanobacterial soil-crust lichen Collema tenax from arid lands in southern Utah, USA: Role of water content on light and temperature responses of CO_2 exchange. *Functional Ecology* 12: 195–202.

Lange, O.L., Green, T.G.A. & Heber, U. 2001. Hydration-dependent photosynthetic production of lichens: what do laboratory studies tell us about field performance? *Journal of Experimental Botany* 52: 2033–2042.

Lee, D.W. & Gould, K.S. 2002. Why leaves turn red: pigments called anthocyanins probably protect leaves from light damage by direct shielding and by scavenging free radicals. *American Scientist* 90: 524–528.

Lichtenthaler, H.K. 1987. Chlorophylls and carotenoids: pigments of photosynthetic biomembranes. *Methods Enzymology* 148: 350–382.

Lichtenthaler, H.K., Lang, M., Sowinska, M., Heisel, F. & Miehe, J.A. 1996. Detection of vegetation stress via a new high resolution fluorescence imaging system. *Journal of Plant Physiology* 148: 599–612.

Liu, W., Baret, F., Gu, X., Tong, Q., Zheng, L. & Zhang, B. 2002. Relating soil surface moisture to reflectance. *Remote Sensing of Environment* 81: 238–246.

Lobell, D.B. & Asner, G.P. 2002. Moisture effects on soil reflectance. *Soil Science Society of America Journal* 66: 722–727.

Luquet, D., Begue, A., Dauzat, J., Nouvellon, Y. & Rey, H. 1998. Effect of the vegetation clumping on the BRDF of a semi-arid grassland: comparison of the SAIL model and ray tracing method applied to a 3D computerized vegetation canopy. In *Proceedings of Geoscience and Remote Sensing Symposium (IGARSS '98), 6–10 July 1998*: 791–793.

Ma, Q., Ishimaru, A., Phu, P. & Kuga, Y. 1990. Transmission, reflection, and depolarization of an optical wave for a single leaf. *IEEE Transactions in Geoscience and Remote Sensing* 28(5): 865–872.

Madeira, J., Bedidi, A., Cervelle, B., Pouget, M. & Flay, N. 1997. Visible spectrometric indices of hematite (Hm) and goethite (Gt) content in lateritic soils: the application of a Thematic Mapper (TM) image for soil-mapping in Brasilia, Brazil. *International Journal of Remote Sensing* 18(13): 2835–2852.

Maier, S.W., Lüdeker, W. & Günther, K.P. 1999. SLOP: A revised version of the stochastic model for leaf optical properties. *Remote Sensing of Environment* 68(3): 273–280.

Mathieu, R., Pouget, M., Cervelle, B. & Escadafal, R. 1998. Relationships between satellite-based radiometric indices simulated using laboratory reflectance data and typic soil color of an arid environment. *Remote Sensing of Environment* 66: 17–28.

Matthias, A.D., Fimbres, A., Sano, E.E., Post, D.F., Accioly, L., Batchily, A.K. & Ferreira, L.G. 2000. Surface Roughness Effects on Soil Albedo. *Soil Science Society of America Journal* 64: 1035–1041.

McIntyre, D.S. 1958. Permeability measurement of soil crust formed by raindrop impact. *Soil Science* 85: 158–189.

McNairn, H. & Protz, R. 1993: Mapping corn residue cover on agricultural fields in Oxford County, Ontario, using Thematic Mapper. *Canadian Journal of Remote Sensing* 19: 152–159.

Meigs, P. 1953. World distribution of arid and semi-arid homoclimates. *Reviews of research on arid zone hydrology* 203–209. Paris: United Nations Educational, Scientific, and Cultural Organization, Arid Zone Programme-1.

Metternicht, G.I. & Fermont, A. 1998. Estimating erosion surface features by linear mixture modeling. Remote Sensing of Environment 64: 254–265.

Metternicht, G.I. & Zinck, J.A. 2003. Remote sensing of soil salinity: potentials and constraints. *Remote Sensing of Environment* 85: 1–20.

Millennium Ecosystem Assessment 2005. *Ecosystems and Human Well-Being: Desertification Synthesis.* Washington D.C.: World Resource Institute.

Miller, J.R., Berger, M., Goulas, Y., Jacquemoud, S., Louis, J., Moise, N., Mohammed, G., Moreno, J., Moya, I., Pedrós, R., Verhoef, W. & Zarco-Tejada, P.J. 2005. *Development of a Vegetation Fluorescence Canopy Model*, ESTEC Contract No. 16365/02/NL/FF, Final Report, May 2005 (http://www.ias.csic.es/fluormod/)

Mustard, J.F. 1992. Chemical analysis of actinolite from reflectance spectra. *American Mineralogist* 77: 345–358.

Nagler, P.L., Daughtry, C.S.T. & Goward, S.N. 2000. Plant litter and soil reflectance. *Remote Sensing of Environment* 71: 207–215.

Neill, S. & Gould, K.S. 1999. Optical properties of leaves in relation to anthocyanin concentration and distribution. *Canadian Journal of Botany* 77(12): 1777–1782.

Noy-Meir, I. 1973. Desert ecosystems: Environment and producers. *Annual Review of Ecology and Systematics* 4: 25–41.

Okin, G.S., Mahowald, N., Chadwick, O.A. & Artaxo, P. 2004. Impact of desert dust on the biogeochemistry of phosphorus in terrestrial ecosystems. *Global Biogeochemical Cycles* 18(2): Art. No. GB2005.

Palacios-Orueta, A. & Ustin, S.L. 1996. Multivariate classification of soil spectra. *Remote Sensing of Environment* 57: 108–118.

Palacios-Orueta, A. & Ustin, S.L. 1998. Remote sensing of soil properties in the Santa Monica mountains: I. Spectral analysis. *Remote Sensing of Environment* 65: 170–183.

Palacios-Orueta, A., Piñzon, J.E., Ustin, S.L. & Roberts, D.A. 1999. Remote sensing of soil properties in the Santa Monica Mountains. II. Hierarchical foreground and background analysis. *Remote Sensing of Environment* 68: 138–151.

Pedrós, R., Goulas, Y., Jacquemoud, S., Louis, J. & Moya, I. 2006. A new leaf fluorescence model. Part 1: Fluorescence excitation & Part 2: Fluorescence emission, *Remote Sensing of Environment* (forthcoming).

Peñuelas, J., Baret, F. & Filella, I. 1995a. Semi-empirical indexes to assess carotenoids/chlorophyll a ratio from leaf spectral reflectance. *Photosynthetica* 31: 221–230.

Peñuelas, J., Filella, I. & Gamon, J.A. 1995b. Assessment of photosynthetic radiation-use efficiency with spectral reflectance. *New Phytologist* 131: 291–296.

Peñuelas, J., Filella, I., Gamon, J.A. & Field, C.B. 1997a. Assessing photosynthetic radiation-use efficiency of emergent aquatic vegetation from spectral reflectance. *Aquatic Botany* 58: 307–315.

Peñuelas, J., Pinol, J., Oqaya, R. & Filella, I. 1997b. Estimation of plant water concentration by the reflectance Water Index WI (R900/R970). *International Journal of Remote Sensing* 18: 2869–2875.

Pinty, B., Verstraete, M.M., Iaquinta, J. & Gobron, N. 1996. Advanced modelling and inversion techniques for the quantitative characterization of desertification. In Hill, J. & Peter, D. (eds.), *Proc. The use of remote sensing for land degradation and desertification monitoring in the Mediterranean basin, EN, Valencia (Spain), 13–15 June 1994*. European Commission, Vol. EUR 16732, pp. 79–93.

Pinzóon, J.E., Ustin, S.L., Castañneda, C.M. & Smith, M.O. 1998. Investigation of leaf biochemistry by hierarchical foreground/background analysis. *IEEE Transactions in Geoscience and Remote Sensing* 36: 1–15.

Potter, R.M. & Rossman, G.R. 1979. Mineralogy of manganese dendrites and coatings. *American Mineralogist* 64: 1219–1226.

Prahl, S. 2001. Optical absorption of water. Oregon Medical Laser Center. *http://omlc.ogi.edu/spectra/water/*

Rees, W.G., Tutubalina, O.V. & Golubeva, E.I. 2004. Reflectance spectra of subarctic lichens between 400 and 2400 nm. *Remote Sensing of Environment* 90(3): 281–292.

Reynolds, J.F., Kemp, P.R., Ogle, K. & Fernandez, R.J. 2004. Modifying the pulse-reserve paradigm for deserts of North America: precipitation pulses, soil water, and plant responses. *Oecologia* 141: 194–210.

Richardson, A.J. & Wiegand, C.L. 1977. Distinguishing Vegetation from Soil Background Information. *Photogrammetric Engineering and Remote Sensing* 43: 1541–1552.

Roberts, D.A., Ustin, S.L., Ogunjemiyo, S., Greenberg, J., Dobrowski, S.Z., Chen, J. & Hinckley, T.M. 2004. Spectral and structural measures of northwest forest vegetation at leaf to landscape scale. *Ecosystems* 7: 545–562.

Rotondi, A., Rossi, F., Asunis, C. & Cesaraccio, C. 2003. Leaf xeromorphic adaptations of some plants of a coastal Mediterranean macchia ecosystem. *Journal of Mediterranean Ecology* 4: 25–35.

Scanlon, B.R., Levitt, D.G., Reedy, R.C., Keese, K.E. & Sully, M.J. 2005. Ecological controls on water-cycle response to climate variability in deserts. *Proceedings National Academy of Science* 102: 6033–6038.

Scheffer, M., Holmgren, M., Brovkin, V. & Claussen, M. 2005. Synergy between small- and large-scale feedbacks of vegetation on the water cycle. *Global Change Biology* 11: 1003–1012.

Schlesinger, W.H., Reynolds, J.F., Cunningham, G.L., Huenneke, L.F., Jarrell, W.M., Virginia, R.A. & Whitford, W.G. 1990. Biological Feedbacks in Global Desertification. *Science* 247(4946): 1043–1048.

Seager, S., Turner, E.L., Schafer, J. & Ford, E.B. 2005. Vegetation's red edge: a possible spectroscopic biosignature of extraterrestrial plants. *Astrobiology* 5: 372–390.

Serrano, L., Peñuelas, J. & Ustin, S.L. 2002. Remote sensing of nitrogen and lignin in Mediterranean vegetation from AVIRIS data: decomposing biochemical from structural signals. *Remote Sensing of Environment* 81: 355–364.

Shoshany, M. 1993. Roughness-reflectance relationship of bare desert terrain, an empirical study. *Remote Sensing of Environment* 45: 15–27.

Sims, D.A. & Gamon, J.A. 1999. Estimating anthocyanin, chlorophyll, and carotenoid concentrations using hyperspectral reflectance. *Poster, Ecological Society of America*, Annual Meeting.

Smith, M.-L., Ollinger, S.V., Martin, M.E., Aber, J.D., Halett, R.A. & Goodale, C.L. 2002. Direct estimation of aboveground forest productivity through hyperspectral remote sensing of canopy nitrogen. *Ecological Applications* 12: 1286–1302.

Smith, S.D., Huxman, T.E., Zitzer, S.F., Charlet, T.N., Housman, D.C., Coleman, J.S., Fenstermaker, L.K., Seemann, J.R. & Nowak, R.S. 2000. Elevated CO_2 increases productivity and invasive species success in an arid ecosystem. *Nature* 408: 79–81.

Smith, S.D., Monsoon, R.K. & Anderson, J.E. 1997. Physiological Ecology of North American Desert Plants. Berlin: Springer.

Streck, N.A., Rundquist, D. & Connot, J. 2002. Estimating residual wheat dry matter from remote sensing measurements. *Photogrammetric Engineering and Remote Sensing* 68: 1193–1201.

Stylinski, C.D., Gamon, J.A. & Oechel, W.C. 2002. Seasonal patterns of reflectance indices, carotenoid pigments and photosynthesis of evergreen chaparral species. *Oecologia* 131(3): 366–374.

Stylinski, C.D., Oechel, W.C., Gamon, J.A., Tissue, D.T., Miglietta, F. & Raschi, A. 2000. Effects of lifelong [CO_2] enrichment on carboxylation and light utilization of *Quercus pubescens* Willd. examined with gas exchange, biochemistry and optical techniques. *Plant, Cell and Environment* 23(12): 1353–1362.

Sunshine, J.M., Pieters, C.M. & Pratt, S.F. 1990. Deconvolution of mineral absorption bands—an improved approach. *Journal of Geophysical Research-Solid Earth and Planets* 95: 6955–6966.

Tucker, C.J. & Garratt, M.W. 1977. Leaf optical properties as a stochastic process. *Applied Optics* 16(3): 635–642.

Tucker, C.J. 1979. Red and photographic infrared linear combinations for monitoring vegetation. *Remote Sensing of Environment* 8: 127–150.

Ustin, S.L., Jacquemoud, S. & Govaerts, Y.M. 2001. Simulation of photon transport in a three-dimensional leaf: Implication for photosynthesis. *Plant Cell and Environment* 24: 1095–1103.

Ustin, S.L., Roberts, D.A., Gamon, J.A., Asner, G.P. & Green, R.O. 2004. Using Imaging Spectroscopy to Study Ecosystem Processes and Properties. *Bioscience* 54: 523–534.

Van der Meer, F. 2004. Analysis of spectral absorption features in hyperspectral imagery. *International Journal of Applied Earth Observation Geoinformatics* 5: 55–68.

Vina, A., Peters, A.J. & Ji, L. 2003. Use of multispectral ikonos imagery for discriminating between conventional and conservation agricultural tillage practices. *Photogrammetric Engineering and Remote Sensing* 69: 537–544.

Wanjura, D.F. & Bilbro, J.D. 1986. Ground cover and weathering effects on reflectances of three crop residues. *Agronomy Journal* 78: 694–698.

Wessman, C.A. 1990. Evaluation of canopy biochemistry. In Hobbs, R.J. and Mooney, H.A. (eds.), *Remote Sensing of Biosphere Functioning*: 135–156. New York: Springer-Verlag.

Whiting, M.L. 2004. Effects of overlapping absorptions on soil mineral content estimates. In *Soil moisture model to improve mineral abundance estimates from hyperspectral data*. Ph.D. Dissertation, University of California.

Whiting, M.L., DeGloria, S.D., Benson, A.S. & Wall, S.L. 1987. Estimating conservation tillage residue using aerial photography. *Journal of Soil Water Conservation* 42: 130–132.

Whiting, M.L., Li, L. & Ustin, S.L. 2004. Predicting Water Content Using Gaussian Model on Soil Spectra. *Remote Sensing of Environment* 89: 535–552.

Whiting, M.L., Palacios-Orueta, A., Li, L. & Ustin, S.L. 2006. Soil moisture and stratified landscape parameters improve soil mineral mapping with hyperspectral imagery. (in preperation).

Widlowski, J.L., Pinty, B., Gobron, N., Verstraete, M.M. & Davis, A.B. 2001. Characterization of surface heterogeneity detected at the MISR/TERRA subpixel scale. *Geophysical Research Letters* 28(24): 4639–4642.

Zarco-Tejada, P.J., Pushnik, J.C., Dobrowski, S. & Ustin, S.L. 2003a. Steady-state Chlorophyll a fluorescence detection from canopy derivative reflectance and double-peak effects. *Remote Sensing of Environment* 84: 283–294.

Zarco-Tejada, P.J. 1998. Optical indexes as bioindicators of forest sustainability. *Graduate Programme in Earth and Space Science*. Toronto: York University.

Zarco-Tejada, P.J., Rueda, C.A. & Ustin, S.L. 2003. Water content estimation in vegetation with MODIS reflectance data and model inversion methods. *Remote Sensing of Environment* 85: 109–124.

Zhang, J.K., Rivard, B. & Sanchez-Azofeifa, A. 2005. Spectral unmixing of normalized reflectance data for the deconvolution of lichen and rock mixtures. *Remote Sensing of Environment* 95: 57–66.

*Recent Advances in Remote Sensing and Geoinformation Processing
for Land Degradation Assessment – Röder & Hill (eds)*
© *2009 Taylor & Francis Group, London, ISBN 978-0-415-39769-8*

Integrated environmental modelling to characterise processes of land degradation and desertification for policy support

M. Mulligan

*Environmental Monitoring and Modelling Research Group, Department of Geography,
King's College London, London, UK*

ABSTRACT: This paper reviews the role of integrated environmental modelling in the better understanding of land degradation and desertification specifically for improved policy formulation to prevent, mitigate or adapt to its consequences. Integrated environmental modelling attempts to couple the socio-economic drivers of desertification with the biophysical desertification processes and their outcomes. This is best achieved as a bi-directional feedback in which (a) the socio-economic drivers force the biophysical desertification processes (alongside external biophysical drivers such as aridification), (b) the desertification processes lead to some development of land or environmental condition which can be either stasis, aggradation or degradation and (c) the socio-economic system responds to those changes which then further changes the socio-economic drivers, and so on in a feedback loop.

Integrated environmental modelling is a relatively new approach with most models to date being confined to either the biophysical or the socio-economic systems, rarely both. Indeed the majority of models used in desertification cover only part of the biophysical system (for example. soil erosion models) or part of the socio-economic system (for example land use change models) whereas desertification is actually a suite of multiple, interacting processes and factors (aridification, soil erosion, biological decline, tourism, agricultural change...). A defining characteristic of integrated environmental models is that they are—necessarily—spatial in nature and thus take account of spatially varying biophysical and socio-economic states, patterns and processes alongside the important lateral flows which connect natural and economic landscapes. They are also focused on policy relevant scales and those tend to be larger than the experimental scales of most process models. This spatiality and scale necessitates the availability of significant spatial data for the parameterisation of climate, soils, vegetation, water resources, landscape and land use, population and infrastructure. This makes integrated environmental modelling highly dependent (for model parameterisation and validation) upon inputs from remote sensing and geoinformation processing. This is especially true for desertification because desertification processes operate at fine grains yet the policies which drive (or prevent) them are usually implemented regionally.

In this paper my emphasis is on desertification in the European Mediterranean and in particular on the science developed under the successive European Commission (EC) funded studies on desertification under frameworks II–VI (1990–2010). I briefly review the main processes of land degradation and the science behind the assessment of land degradation indicating the role of measurement, modelling and remote sensing in the assessment process. This is followed by a more detailed review of recent desertification modelling efforts on the path towards integrated environmental modelling. I then discuss the data needs of integrated assessment models (IAMs) for desertification and the potential role of remote sensing in providing such data before presenting a case study of one of the most advanced Integrated Environmental Models for the assessment of Desertification: the MedAction model. The model is described briefly and the role of remote sensing in providing data for model application and validation discussed. Finally a case study on the application of the MedAction integrated model for better understanding the impact of climate and policy changes on desertification in the Guadalentin Basin, SE Spain are discussed with particular emphasis on the feedbacks between socio-economic and biophysical processes and on the assemblage of observed environmental complexity from simple processes distributed over spatially complex landscapes.

1 INTRODUCTION AND AIMS

Land degradation takes many forms and occurs in many different environments. Where land degradation occurs in dry sub-humid, semi-arid and arid zones and is characterised by a diminution of the biological potential of the land caused by climatic aridification and/or the impact of human land or water use, then it is often called 'desertification'. Desertification is not a process but rather is the outcome of a series of processes that act together to transform environmental conditions from productive to less productive (and thus more desert-like). Desertification is not often a simple transformation to desert as might occur at an existing desert margin, for example. Rather, desertification most often occurs in the context of a spatially complex mosaic of patches of aggradation, stasis and degradation but where the patches of degradation have the greater spatial extent or impact on society. The balance between the state of these patches and the location of them relative to human need from the land will determine the perception (and impact) of the state of desertification. Patches may be degrading because of their physical properties (for example south facing with high evaporation losses; low hydraulic conductivity or high upslope area leading to high runoff and erosion), because of human use of the patch (unsuitable land use and land management practices) or because of impacts imported from other (connected) patches (for example fire, flood, reduced runon, wind-eroded sediment deposition). Indeed this lateral connection is another important feature of desertification, which adds another level of complexity in terms of understanding on-site and off-site controls (usually upstream[1]) and effects of desertification (usually downstream). Desertification is the outcome of processes that may be slow and gradual (such as aridification, soil erosion and exhaustion, productivity decline) or much more rapid but episodic extreme events (for example flooding, erosion and mass movements, wildfire) and thus rates of desertification may be highly spatially and temporally variable (Mulligan 1998).

Desertification engages with human society because it affects the reliability of natural resources on which large, complex and demanding societies have come to depend. Since it is unpleasant (inconvenient, costly and at its worst impossible) for those societies to adapt farming and living conditions to cope with resource change, desertification can be a threat to human health and wellbeing. Desertification also engages with society because human impacts in the landscape are often a driver of desertification. So, humans can be part of the cause and are certainly prone to the effects of desertification.

The purpose of this paper is to examine desertification with a particular emphasis on the European Mediterranean and a particular focus on the spatial complexity of the processes leading to it and the extent of coupling and feedback between the biophysical and human components. My aims here are to:

a. describe more of the (process, spatial and temporal) complexity of the processes which produce desertification.
b. review available models for understanding European desertification in an integrated way (including human and biophysical components, their linkages and feedbacks) over space and time with some examples of the state-of-the-art systems.
c. indicate the role of remote sensing in the provision of data and understanding for the state of the art integrated assessment models for desertification assessment and policy support.
d. indicate areas in which further conceptual and technical development are necessary to improve the science, the data and the models.

---

[1] 'Stream' here is considered analogous to 'network' so the connections may be upstream in the case of fluvial processes, upwind in the case of Aeolian processes or up or downstream along some other relevant network, which may even be socio-economic.

I will first discuss the science and rationale for desertification assessment before presenting a short review of the variety of processes of desertification that should be taken into account in any particular assessment. The bulk of this paper reviews the available models for desertification and the data requirements for the successful application of these models, indicating the role of remote sensing in the provision of these data spatially. I conclude with a case study of a state-of-the-art integrated desertification model (the MedAction policy support system) and its application. The model is described briefly and the role of remote sensing in providing data for model application and validation is discussed. A case study on the application of the MedAction integrated model for better understanding the impact of climate and policy changes on desertification in the Guadalentin Basin, SE Spain are discussed with particular emphasis on the feedbacks between socio-economic and biophysical processes and on the assemblage of observed environmental complexity from simple processes distributed over spatially complex landscapes. This leads to a concluding discussion on necessary developments for the improvement of integrated assessment models in desertification policy support and the role of remote sensing in this.

2   LAND DEGRADATION AND DESERTIFICATION

In this text I will use the terms land degradation and desertification interchangeably since desertification is a subset of land degradation applied according to the definition given in section 1. My use of the term land degradation (which I prefer to desertification because of its more accurate and less alarmist description of the issue) should be understood to refer only to the many types of dryland land degradation that may be considered desertification for those who prefer the latter term). In other words when I say land degradation here I mean reduction in the biological potential of the land and consequent effects in dry environments as a result of combined climate and human impact.

2.1   *Defining land condition and degradation*

2.1.1   *Measures*
In order for one to assess land degradation one must first have some non-destructive measures of land condition that can be repeated over time and compared in order to first define and then measure degradation, stasis or aggradation. The measures of land condition that one chooses will depend upon the impacts of land degradation that one is interested in. An ecologist would probably use measures such as species richness, natural vegetation cover, change in extinction rates. A farmer would be more interested in crop yields, artificial (water energy and nutrient) input requirements and the outcome of these two—agricultural profit. A water resource manager would focus more on water resource reliability, quality and quantity in the main subsurface, surface and riparian 'reservoirs'. Thus all of these 'policy makers' will have different views on the most important measures of land condition and also different views of the direction in which change in land condition is degradation and aggradation.

2.1.2   *My degradation, your aggradation*
For example, reduction in natural forest cover over time may be an important measure of land condition to the ecologist and the water resource manager, but the reduction is a degradation to the ecologist but an aggradation to the water resource manager (because of reduced evaporation from shrublands compared with closed forests[2]). So degradation means different things to different 'policy makers'. Moreover, as is clear from the case of the consequences of forest loss for the water

---

[2] Though this ignores the potential water resource benefits of forest over shrubland in terms of potentially increased baseflows, better flow regulation, reduced erosion and thus increased water quality. Thus the water resource manager must determine whether the balance of these effects makes reduced forest cover a hydrological aggradation or degradation.

resource manager, a simple change in condition may have many opposing effects (see footnote) some of which are beneficial to the water resource manager and others of which are not.

2.1.3 *Scientific uncertainty*
Our scientific understanding of the processes is such that even determining whether a particular change in land condition represents a degradation or not for a particular policy maker, like the water resource manager, may not always be possible. Of course some measures of land condition are much more clearly either degradation or aggradation: reduced agricultural productivity is a degradation for the agriculturalist because it leads to reduced incomes and profits—clearly. However, reduced agricultural productivity may be an aggradation to the ecologist, especially where reduced productivity leads to agricultural abandonment and natural regeneration. So, even in this very first stage of desertification assessment there can be difficulties in defining (a) appropriate measures of land condition and (b) which changes in those measures constitute an overall degradation. This is partly because of the scientific uncertainty in understanding the controls on desertification and partly because of the variety of stakeholders operating in any one landscape and their very different views of what constitutes an adverse change.

2.1.4 *Part of the problem*
In providing science-based policy or decision support one sometimes has a single stakeholder in mind (a water resource manager, tourism enterprise, agricultural stakeholder or environmental group). The danger of building systems for a limited set of stakeholders are that the policy actions that may seem to be appropriate in preventing, adapting to or mitigating the impacts of desertification on the single or few stakeholders may have negative consequences for other stakeholders. Thus as much as possible science needs to move away from partial descriptions of the environmental and human interactions in desertification, towards a more holistic treatment of multiple processes, multiple stakeholders. Since the same human-environment interactions can have very different outcomes in different spatial locations (because of their different properties) and at different times (because of the impact of cyclical changes in drivers such as climate and global markets) spatial and temporal variation must also be accounted for in these systems.

Such an approach requires a multidisciplinary effort that brings together data and process descriptions from a variety of disciplines describing the relevant human-environment interactions and integrating them with appropriate feedback loops in a scientifically meaningful and technically plausible way. As we will see, such an approach is very difficult to achieve outside a modelling framework.

2.2 The assessment of land degradation

2.2.1 *The purpose of assessment*
Avoiding the deleterious impacts of desertification involves either prevention or adaptation or mitigation. The primary purpose of desertification assessment is to sectorally and spatially target efforts at prevention, adaptation and mitigation. Assessment assists by providing spatially targeted information on desertification hazard that can be used to assist land use and land management planning. Thus the ideal desertification assessment must provide the appropriate information for the relevant strategy. No system exists to date which will do all of this especially because different strategies for overcoming desertification (prevention, adaptation or mitigation) require different types of information.

2.2.2 *Assessment for prevention*
The prevention of desertification involves monitoring changes in land condition, identifying the cause of any degradation and adapting land and water use practices and intensities in order to put less pressure on the more degrading or more sensitive areas of the landscape. In other words preventing desertification by removing its drivers. This may mean changing land use to require less irrigation, having lower cropping densities, using more native or drought-tolerant varieties and accepting lower yields. It may also mean changing land management practices to encourage lower water use and protection against soil erosion by, for example, contour ploughing or low tillage, retention of

rock fragments (Van Wesemael et al. 2000), low inputs. Thus an assessment system that is focused on prevention needs to be able to forecast the cumulative impact of current practices in particular landscape positions, examine the cost of the resulting desertification and be capable of simulating the impact of land use and land management scenarios designed to reduce desertification. The cost of implementing these scenarios should be less than the benefit of reducing desertification that results from their implementation. It is then down to the regional and national financial bureaucracy to find mechanisms for redistributing costs and benefits between the different stakeholders such that the costs are incurred in an equitable manner relative to the benefit winners.

2.2.3   Assessment for adaptation

Adaptation is quite different to prevention. It does not focus on halting desertification but rather on adapting the use of the land to the changing land condition. Adaptation is particularly important where desertification is forced by externally driven (for example by global climate change) and is thus unpreventable by the local or regional stakeholders or where prevention is otherwise not possible or the degradation not so serious as to warrant preventative measures. The requirements of an assessment system for adaptation are less focused on scenario testing and assessment of long term cumulative impacts and more focused on comparative analysis of different land use and land management practices in order to know which is possible (sustainable) under a degrading environment. Thus, the focus is not on preventing desertification through land use and land management change but on finding the use-management combination that provides the greatest benefits (without necessarily accelerating desertification) and continuing to adapt those practices as the land condition changes. It is thus an optimization exercise rather than a comparative cost:benefit exercise. The key difference between prevention and adaptation is that prevention attempts to find a stable use-management combination which is sustainable in the long term and avoids land degradation whereas adaptation recognises that degradation cannot be prevented and provides an adaptive scenario which gets the most from the land as degradation proceeds, always in the hope that stasis or aggrading conditions will return.

2.2.4   Assessment for mitigation

The focus of mitigation is in the development of land use and management practices that lessen the actual or potential negative effects of desertification so that, although degradation is allowed to take place, land use practices focus, not on adapting practices to the degrading potential of the land but rather on developing land management strategies that help recover land condition and reduce the signs of desertification. Mitigation is the application of 'technical fixes' for desertification. Such strategies might include the development of irrigation schemes, the construction of reservoirs and erosion dams, the transfer of water from wetter regions to drier ones and the development of high input agricultural techniques that import production resources from outside the degrading region (e.g. greenhouse crops). Assessment for mitigation focuses on understanding the dominant processes of desertification in a region and simulating the success of different mitigation techniques in reducing the deleterious impacts. Assessment systems may examine user interventions such as terracing and the construction of check dams and compare the cost of construction and maintenance of these features with the benefits gained from them.

2.2.5   The science of assessment

When we talk of desertification assessment we must distinguish between the assessment of desertification intensity and extent and the assessment of desertification causality. Most of the scientific effort to date has focused on the former with various assessments of desertification extent both globally (Mabutt 1984, Dregne 1998) and regionally, especially in the Mediterranean (Mairota et al. 1997, Geeson et al. 2001). The global assessments have often taken the form of maps from the World Desertification Map (1977) prepared by UNEP, FAO, UNESCO and WMO for the United Nations Conference on Desertification (UNCOD), through the UNEP (1998) Plan of action to combat desertification (PACD) map to the assessment of UNEP/ISRIC[3] (1990) and the World

---

[3] International Soil Reference and Information Centre.

Desertification Atlas (UNEP, 1992), now in it's second edition (Middleton 1997). Most of these efforts combined national expert opinions and field-collected data with mapping and later GIS techniques and some simple multi-criteria analysis. These efforts continue in the various desertification early warning systems (EWS) such as the Degradation Early Warning System (DEWS)[4] and The European Land Degradation Monitoring System[5]. Some of these latter efforts make extensive use of satellite remote sensing as well as ground-collected data. They also continue in the various indicator approaches to desertification assessment summarized by Geeson (2001).

Any indicator of change should be quantitative and objective, sensitive to the change of interest, few in number and readily measured at the necessary scale (Dregne 1998). Most indicators are focused on desertification assessment rather than causality and most are indirect since not all symptoms of desertification (e.g. reduction in the biological potential of the land) are always the consequence of desertification. Moreover, indicators, like most environmental variables are rarely constant in time so that a sufficiently long record of observation is necessary to separate an important trend in such an indicator from natural background variability. Different indicators will be representative of different processes of desertification and will usually be most representative at well-defined scale. Moving from the complex 'landscape' of desertification to the few indicators necessary for operational use is a rather subjective and challenging process.

In the arena of Mediterranean Desertification, the European Commission has supported research on natural disasters since the 1980s via the Framework Programmes for research and technological development. This research has included a focus on desertification processes in the semi-arid European Mediterranean. In the late 1980s and early 1990s this research focused on scientific projects which collected the necessary baseline field data to better understand the state of desertification in threatened regions. By the early-mid 1990s this research was much more focused on developing an understanding of the processes of desertification and, later in the 1990s on developing sophisticated research models for understanding hillslope to catchment scale desertification process and causality. The science indicated the complexity of Mediterranean desertification in terms of cause and of process. In the early 2000s two main approaches to interfacing this complexity with the policy domain were devised. The first focused on the generation of simple desertification indicators and the second focused on making these complex process models much more relevant to the policy domain and to policy support.

In order to select and apply appropriate actions to counter desertification we must also assess desertification causality since maps of desertification extent cannot be used by policy makers in the implementation of control measures. This is not to say that maps are not the correct tool: the spatially variable nature of desertification processes and thus of the controls necessary to counter them means maps are appropriate but rather than focusing on where desertification occurs, they need to focus on why it occurs there. This transition from where to why has occurred as desertification has progressed from a poorly known issue in need of attention into one of the best understood of the global environmental changes in which the focus is on remedy rather than diagnosis (policy rather than science). Understanding the controls on desertification in a particular area requires assessment to focus on the causal processes of desertification. The direct cause of desertification is usually (mis) management of the land in a way which is inappropriate for its climate but this is, in turn, a consequence of various indirect causes which may include: population pressure, land tenure or lack thereof, international markets for crops, inappropriate land management or agricultural policies, drought, poverty, inadequate agricultural extension services, lack of research (Dregne 1998). Thus desertification assessment tools that aim to inform policy by tackling the question *'why does desertification occur here?'* must continue the use of ground based and remote sensing data, continue to provide spatial assessments but also incorporate an understanding of physical processes as well as direct and indirect causes. This is best achieved by integrating environmental and socio-economic modelling with spatial information systems driven by remote sensing data and

---

[4] http://www.geog.umd.edu/LGRSS/Projects/degradation.html
[5] ftp://ftp.fao.org/agl/agll/lada/montana.pdf

focused on policy relevant scales. This provides a formal integration of process, data and context and may be used to develop policy-focused indicators of desertification in a more robust manner than can be achieved by direct human interpretation of the data and science.

2.2.6 *Processes of land degradation*

We saw in the previous section that desertification usually has a simple direct cause (human mismanagement of land) but that this can be driven by all manner of complex indirect causes, many of which lie in the policy domain. Human mismanagement of land can occur because inappropriate management is applied or because land (and climate) conditions degrade (or vary cyclicly) and thus render previously appropriate management, inappropriate. We can identify a range of different types of physical degradation that result from either external change or mismanagement or both. We can classify these types of degradation according to the physical process domain in which they lie, with the main examples listed below:

1. Climatic degradation
    i. Aridification (drying of the climate)
    ii. Climatic intensification (changes in the frequency and magnitude of extreme events)
2. Hydrological degradation
    i. Changes to river flow regimes
        a. Increased flashiness
        b. Reduced dry season flows
        c. Reduced water quality
    ii. Land aridification
3. Biological degradation
    i. Reduction in productivity
    ii. Reduction in standing biomass and patchification
    iii. Loss of biodiversity
    iv. Shift in dominant species
4. Pedological degradation
    i. Soil erosion
    ii. Nutrient loss
    iii. Organic matter loss

Of course these processes are tightly interlinked and the interaction of climate and human activity is likely to bring about changes in many of these, some of which are conditional upon changes in others of them. In order to better understand these multi-process linkages in desertification and to understand them within the context of a specific environment, because desertification processes and thus control are often site-specific we must focus on computer simulation.

2.3   Geospatial modelling of land degradation

Before examining in detail the role of modeling in desertification, let us look more generally at the purpose of modeling as a research activity in the environmental sciences.

2.3.1   *Definition of modelling*

To produce a model is to produce a simplification of reality. A model can be used to formalise understanding gained through data collection or theoretical advance and to explore the properties of that understanding. Models can take the form of conceptual models which formalise theories in a pictorial, illustrative, rule-based or tabular, non-computational form, physical models—which are hardware representations of processes or mathematical models which use the formal language of mathematics to represent processes and the relationships between them. Within the context of scientific research, models are tools for simplifying, formalising and testing theories as well as for implementing projections of particular scenarios for future change. Models are often used to integrate different research activities, in particular theoretical cogitation, delphic knowledge, field and laboratory-based experimentation and monitoring as well as earth observation. Within the context of desertification research there appear to be two basic types of modelling strategy. The first strategy sees models as single equations to represent a specific process whilst the second sees models as software tools for integrating the results of interdisciplinary, multi-institution research efforts. As well as integrating the research activity itself, models are also used to explore the consequences of theoretical developments that are made as a result of the research and are therefore used as exploratory tools for developing and presenting research results.

The development of models can be a primary research objective as well a post-research integrative tool since modelling can be used as a means of:

i.   *Understanding the system*. Particularly where research includes a number of simple processes connected in complex ways (as desertification) or where the research deals with understanding complex processes, models are a useful way to simplify the systems and explore their sensitivity to different parameters or manipulations,
ii. *Testing of hypotheses*. Models can be used to formally test hypotheses under particular environmental conditions that are representative of the same or different locations in the past, present or future.
iii. *Prediction and scenario development*. Models can be used to develop predictions based on current knowledge of the nature of the impact of present environmental changes or scenarios of future environmental changes.

2.3.2   Ubiquitous modelling

Modelling is not so different from theory or data. Data are—in fact—models (simplified representations of process, time and space compared with reality) and so all sensors form a model of reality. All scientists employ some form of mental or conceptual model of the system they work with and a mathematical model is just a formalisation of this. Thus there is little difference between mathematical modelling and many other scientific endeavours. Those who distrust models either distrust the science behind them, the fact that models are more portable than other forms of science (and thus could be employed by non-experts, even out of context), or what people (sometimes) choose to do with them (e.g. purporting to 'predict' the future) and the perceived authority given to those predictions by the fact that they originate from one of these 'mysterious' models.

2.3.3   The process of modelling

A very brief review of the process of mathematical modelling is necessary to contextualise the material ahead. Mathematical models are developed from some conceptual understanding of the parameters affecting a particular process and the relationships between processes in a multi-process system. Models are conceptualised on the basis of field and laboratory experimentation or through theoretical induction. Once conceptualised the formal process of modelling involves some or all of the following stages:

1. *Model development*. In the case of simple empirical equations this may be a rather short process of mathematical formalisation of the relationship between two or more variables. In the case of more complex multi-process models, model development may involve the construction of a detailed conceptual model, which is then formalized, process by process, relationship by relationship, in computer code. The modeller pays particular attention to the links between processes in the model.
2. *Model parameterisation*. The relationships between state variables represent processes and are defined by parameters. Model parameterisation is the process of measuring and supplying the values for these parameters so that input variables can be converted to output variables. For simple equations parameterisation may be simply the elucidation of relationships between variables through field or laboratory experimentation but for more complex models, parameterisation may require a high cost and exhaustive field and remote sensing campaign.
3. *Model calibration*. Most empirical and physically based models are not sufficiently generalised as to provide accurate results under and wide range of conditions. As a result, these models must be calibrated, that is, measured parameter values must be optimised to produce the best

agreement between modelled and measured output for a particular variable (the critical output variable for the present objective). Calibration is also usually required where a model is applied to an area or time outside the conditions of its development.

4. *Model verification and validation*. One of the most important, but often neglected, aspects of model development is testing the results of the modelling process. Verification is the process of going through model output testing the realism of the output variables against known behaviour. Verification tests that the model behaviour is appropriate. Validation is the application of the model to a study case for which independently measured data for the input and output variables are available. Validation is the numerical comparison of the model output fields with their measured equivalent over a particular simulation period. Validation tests the predictive accuracy of a model for a particular set of variables.

5. *Model sensitivity analysis*. An analysis of the sensitivity of model outputs to perturbation in the model parameters can (a) assist in the understanding of the sensitivity of the real system and (b) indicate which parameters are important and which are not. This, in turn, can guide simplification of the model and can also indicate the significance of field parameters and thus the precision with which they must be measured. Sensitivity analysis is an often-neglected tool in understanding the model itself and the system that the model represents.

6. *Simulations and scenarios*. The main purpose of the model is, of course, to allow simulation of the real-world process. A well understood, calibrated and validated model can be applied as a tool for (a) understanding the controls on some past change through comparison of modelled versus measured data, (b) simulation of some future scenario for change or (c) application to 'what if?' type scenarios in order to better understand the system under consideration.

2.3.4   *Why modelling?*

Modelling is increasingly important in the assessment of desertification and the support of desertification policy. This is largely because desertification is a landscape scale phenomenon and thus not easily tackled by the other main forms of scientific enquiry: laboratory experimentation and field monitoring. Humans interact with desertification at all scales but policy is usually implemented at so-called 'policy relevant' scales that are usually regional, national or international. Modelling is the only practicable approach to desertification policy support at those scales in the same way that remote sensing is the only practicable approach to desertification monitoring. The (vegetation-hydrology) interactions in desertification are often too complex for conceptual modelling approaches. The time and space scales are too great for field experimental approaches (using, for example, spatial environmental analogues). Since we would probably like to mitigate against or adapt systems to cope with any adverse effects of desertification, the 'wait and see what happens' approach is not usually an option either.

However, in complex systems like the Mediterranean environment our understanding of the systems is fragmented and over-simplified, our datasets for model parameterisation and validation are poor over the long term and at the landscape-scale, and our understanding of the links and interactions between socio-economic, climatic, land surface and sub-surface processes is elementary. So, can our models be any good? In many cases they are not, but usually they are the best tool that we have to tackle issues such as desertification, and they are of value in formalising our understanding of the process, applying these formalised concepts to measured data, understanding the 'processes' underlying the patterns that we observe, combining knowledge from many (researchers, disciplines, institutions), exploring the results of process interactions over time, space and variation and finally sharing and communicating that understanding. Thus models should be seen, not as crystal balls predicting the future, but as tools to help us to understand the outcomes of interacting processes and for investigating the reasons that these outcomes are as they are: i.e. the causality.

2.3.5   *Simple versus complex models*

Environmental models come in all levels of complexity. The best of two equally predictive models is always the simplest if we follow the principal of parsimony or Ockham's razor. This is probably better expressed as the best model is that which achieves its objective most simply. This is not the same as saying the simplest model is always the best. . .sometimes a particular objective requires a complex model. Different objectives require vastly different levels of model complexity. The objective of modeling is sometimes prediction (of empirical events or states), sometimes explanation (of processes, events or states) but increasingly exploration of data, concepts and processes and their interaction. For prediction the simplest model can often be as good a predictor of a variables trajectory as a complex model and so simplicity (parsimony) is paramount. For explanation it is important that the process descriptions are as accurate and complete as possible and this may require an added level of complexity. For exploration the model needs to be as complete a representation of processes and their interactions as necessary and this requires the most complex parsimonious model of the three cases since the model is trying to replicate the significant complexity of the real system. If the measure of a models worth is its predictive capacity then Ockam's razor ("Given two equally predictive theories, choose the simpler") clearly applies but when the measure is the accuracy of representation of the internal workings of the system then since the real system will always be more complex than the model, the best model will be the more complex one if it is also accurate. But should only be as complex as necessary, and no more. Though "simplicity is the ultimate sophistication" (Leonardo da Vinci) nature is inherently complex and thus even simplicity, with respect to nature, is rather complex. Complexity in nature arises in many ways, some of which are inherently simple. Complexity can arise from the (a) existence of complex processes, (b) the existence of simple processes with complex outcomes, (c) from the accumulation of repeated simple processes over spatial or temporal variation, or (d) from the interaction of multiple simple processes. (a) and (b) are best truly complex whereas (c) and (d) are better termed 'complicated' since they are not simple. So simplicity and complexity are not necessarily opposite ends of a spectrum but simple and complicated are. A more thorough representation of environmental complexity might include:

a. Process complexity—the sophistication and detail of the description of processes,
b. Spatial complexity—the spatial extent and grain of variation (and any lateral flows) represented,
c. Temporal complexity—the temporal horizon and resolution of system dynamics,
d. Inclusivity (complication)—the number of processes included,
e. Integration—the extent to which the important feedback loops between processes are closed.

Environmental researchers have tended to concentrate on (a) whereas (b)–(e) are probably more important in natural systems, especially at spatial and temporal scales outside of those used in experimentation.

2.4   Model integration

For a long time the focus in science has been on reductionism, splitting nature into pieces in order to understand how the parts work. That has led to significant progress in many areas but it does not allow us to understand whole systems well—we need to be able to do that in managing desertification and the environment generally. Now, modelling and computers are enabling us, for the first time, to put these pieces together and understand their interactions and emergent properties as a (process, spatial and temporal) whole and this capability is critical to understanding response to environmental change. Integrated modelling (or Integrated Assessment Modelling, IAM) attempts integration across: issues, scales, stakeholders, disciplines (biophysical and socio-economic), and models. IAM helps in exploring the behaviour of bi-directionally coupled human-environment systems and helps to highlight the previously unforeseen consequences of scenarios of change and policy or other interventions. Integration is important because the links and feedbacks between processes can be as important for system behaviour as the processes themselves, because it facilitates the production of decisions and policies that are well tested with respect to their long term sustainability and that do not have deleterious unforeseen and unintended consequences in other policy domains, because nature is complex (especially with humans around).

We have seen that desertification is a bewildering series of processes connected by a complex web of feedbacks loops. The human brain is not well adapted to understanding or tracking the outcomes of these processes and feedbacks (even at only one point in space and time). Miller (1956) showed that a person can store 7 ± 2 items (numbers, faces, words, processes) in their short term memory under optimal conditions. In the presence of distractions even 3 items can become difficult and may disappear in 2–18 seconds (Peterson & Peterson 1959, Marsch et al. 1997). Thus the outcome of a complex set of processes can be more robustly examined computationally than mentally. The real mental effort is in the understanding of system components and linkages that are required in the development of the model rather than the routine execution of that model. A model based on the best available knowledge and assembled in a careful and well-tested manner can be a useful aid in the decision making process. These IAMs are thus 'thinking tools' that can be used to shed light on problems that otherwise cannot be managed by the human brain alone.

In this paper integrated modelling is used to refer to (spatial) modelling systems which couple models from a variety of disciplines in the science and social science domains with a view to understanding the outcomes of their interaction and the impact of scenarios for change (e.g. climate change) and policy interventions (e.g. land use planning) upon them. This integrated modelling promises a great deal in terms of better understanding complex environmental problems but also provides a series of important challenges which must be solved before its full potential can be realized. Model integration itself is a challenge. Engelen (2000) points out that there are very few operational definitions or procedures for model integration available from the scientific literature. Appropriate model integration requires attention to scientific integration early on (which models should be included?, which parameters and variables should be passed between them?, how should differences in model philosophy, time and space scale be dealt with?). In addition, one has to tackle the technical integration aspects (how should the integrated model be assembled and run?). Other challenges posed by integrated modelling include:

a. *Parameterisation*—integrated models by their very nature tend to have a large number of parameters. If the integrated model is to be used for decision support (as is often the case) then model parameterisation at policy relevant scales becomes a significant challenge and the need for integration with satellite remote sensing and geoinformation processing becomes all the more important.

b. *Calibration*—even if individual models operate well in isolation, they may need re-calibration (or redesign) when coupled with other models. Calibration of an integrated model is complex and time-consuming because model integration focuses on tight coupling and thus parameter interdependence is common.

c. *Uncertainty and Error propagation*—though there is some evidence that errors in model components become more evident when the models are integrated rather than run in isolation [2], this may not always be the case and understanding error propagation in a complex, highly connected model is a significant intellectual and computational challenge, especially for DSS where input data are lean and end users are in need of some estimates of the uncertainty of model outcomes.

d. *Validation*—validation is always difficult, never more so than in multi-component and projective models. The validation problem stems in part from lack of available data but also from the need to ensure that validation is a true measure of model performance against some 'reality' rather than a measure of differences in scale, method and time of parameter validation data compared with the model representation of the same parameter, as is often the case. The focus of validation in integrated models has to be primarily on verification of model behaviour against expectations and secondly in the validation of critical subcomponents rather than attempts at validating the whole integrated model per se. Validation at the latter level takes the form of critical appraisal rather than validation per se.

e. *Understanding the outcomes*—the more complex the model the more difficult it is to ascertain the series of process interactions that led to a particular outcome. As models become more inclusive they gain exploratory power but lose interpretability. There are to date few tools to assist in the 'mining' of model results and the interpretation of complex model outputs. In the end the few simple indicators of desertification causality required in the policy domain will need to be generated from model output in the same way that indicators of desertification extent will need to be generated from remote sensing output. Well developed protocols and methodologies for this do not exist.

f. *Documentation*—the larger and more includive the model, the greater the task of fully documenting its routines, assumptions and codes and the more difficult the task of making this documentation accessible and interpretable to the models user community in both the scientific and the policy domain.

There are many advances that need to be made in terms of computational power, data availability and integration, multiparameter calibration, error analysis, verification and validation, interpretive tools and indicators and documentation technologies before IAMs will be on an equal footing with simpler or single process approaches to modeling. We will later highlight how these specific problems have been tackled in the MedAction PSS but remains a long way to go. Consequently, IAMs and any encompassing decision support capacity should be used for explorative rather than predictive purposes. Simple (empirical) models can are usually better for prediction, complex or integrated and physically based models are better for understanding and exploring the system. The basic process for development of an IAM is as follows:

a. Manageable slices of science are produced tested and wrapped into modules.
b. The modules are linked appropriately with parameters, state variables and other controlling or controlled process modules.
c. A scientific integration ensures the modules are scientifically compatible (conceptually and philosophically).
d. A technical integration ensures the modules connect appropriately in terms of data passing at the right times and spaces (and in the right units!).
e. A spatial database is added.
f. Appropriate temporal scenarios (outside control of policymaker) and policy interventions are added.
g. The computer deals with the (short term) memory and logic leaving the scientists to deal with the science.

2.5 'Research' versus 'policy' models

Models are usually built by scientists and social scientists. In many cases their purpose is to advance understanding through research though they are often useful for, and applied to, policy or decision support, particularly where research funding agencies require this as a condition of funding and this has certainly been the focus of EU desertification research since 2000. The characteristics of a model designed for research may be somewhat different to those required for a model developed specifically for policy application. Thus models which begin as research models often evolve into what might be called policy models—models use in decision or policy support. Of course there are no clear boundaries that separate research versus policy models and most models lie somewhere on a continuum from research to policy use. Table 1 identifies some of the key differences in emphasis between what might be called research-centred models and policy focused models. The major difference between a research and policy model are in their objectives. Research problems are usually extremely well defined as a testable hypothesis that the model addresses whereas policy problems are rather ill-defined and may be of the type: *what is the optimum land use pattern in the area for sustainable production?; how can we improve nature conservation in the area?; what should we do to adapt to desertification in the area?; how might we improve the livelihoods of the communities in the area?*. These kinds of questions require a quite different approach to modelling than a hypothesis does. The approach required by ill-defined questions is an integrative approach. The model is thus designed to be more of a general description of the workings of the system ("an instrument for exploration representing part of a complex reality with some degree of certainty", (Engelen 2000)) rather than a specific description of the processes required to confirm or refute a particular hypothesis.

Table 1.   The characteristics of research and policy models.

Research models	Policy models
Research problem well defined as hypothesis which model addresses	Policy problem ill-defined, model more generalized
Accurate representation of processes	Adequate representation of processes
Complexity and (time and space) resolution reflect processes	Complexity and (time and space) resolution reflect data
Accurate representation of spatial variability	Adequate representation (existing data)
Sectoral and detailed	Less detailed but multi-sectoral (integrated or holistic)
Scientifically innovative	Scientifically proven
Raises more questions than answers	Provides simple(?), definitive(?) answers
Interesting and worthwhile in its own right	Interesting and worthwhile only through its output
Process centred	Input/output centred
Numbers validatable	Outcomes validatable
As complex as necessary	As simple as possible
As fast as possible	Faster (no more than a few minutes running)
Data hungry if necessary	Data lean

Research models are used primarily to understand how a system functions and thus the description of processes should be as accurate as possible for the particular objective. Policy models on the other hand require only the accuracy of process description which is necessary to achieve the policy objective, and no more. Similarly model complexity and resolution for research models often reflect the complexity and resolution of the processes whereas for policy models they reflect the complexity and resolution of available data. Research models are usually scientifically innovative (much of the value of the model is its description of new science) whereas policy models need to be scientifically proven such that they can be trusted in application. Research models usually raise more questions than answers whereas policy-focused models are required to give simple, clear (and true!) answers and the value of a policy model is solely within its output whereas research models have value in their own right (as descriptions of the system of interest). Thus policy models tend to be input/output centred and research models, process centred. All models should be the simplest possible description of the process required for the objective but this is a particularly stringent requirement for policy models that are both data and runtime limited.

2.6   Models as decision and policy support systems

Policy questions have reached a level of complexity that can no longer be dealt with by politicians alone. High-level technicians play an ever-increasing role, and, the significant advances in hardware and software technologies have equipped them with "very powerful multi-media calculators" [5]. In the last decade there has been considerable growth in the number of governmental organizations who develop these rather sophisticated model-based information systems to support the policy making process. This growth is fuelled by the growing awareness that policy-making should be based on an integrated approach (Engelen 2000) in order to develop policies that are most sustainable and that are not prone to negative unintended consequences in other sectors than the sector of interest. The systems that policy-makers deal with rarely exist in isolation, rather they are parts of a larger entity and interact in feedback loops with other local and global systems and with other spaces and times. Even weak connectivity between system components can have important consequences for the behaviour of the system as a whole (see May, 1974). Policy makers have to manage these fragile systems that exhibit a very rich behaviour, not least because of the presence of intelligent (human) actors which steer the system in the direction of their interest (Engelen 2000, Wainwright & Mulligan 2004) and require technical support systems which summarise knowledge (data and understanding) of their system and help them to better anticipate the effects of their interventions within the system as fully as possible (Engelen 2000). The development of such support systems has been fuelled by the requirement for integrated natural resources management (Campbell & Sayer 2003) and integrated assessment (Gough et al. 1998), by the availability of research funding to provide the support tools that it requires, by the massive advances in comput-ing and analytical power of the last two decades and the increasing availability of high resolution remote sensing and Geographical Information Systems (GIS) data that make spatial modelling for policy support a reality.

Policy-focused models are often also called decision support systems (DSS) or policy support system (PSS). DSS refers to a wide range of support tools from maps through spreadsheets to the kind of spatial IAMs which are the focus of this paper. PSS are DSS applied specifically to supporting the policy domain. A DSS has aims to aid a decision maker in addressing unstructured or semi-structured decisions through realizing one or more of the following DSS objectives:

a. To assist with problem recognition (clarifying or better specifying the problem). This is an important function of DSS and often their most important contribution.
b. To supplement one or more of a decision maker's abilities e.g. knowledge collection, knowledge derivation, analysis (what if?), and formulation of potential plans for analysis or action.
c. To facilitate one or more of the decision-making phases e.g. intelligence (providing relevant information), design (identifying or analysing alternatives), choice (which alternative to choose).

In addition, within a desertification policy context a DSS should:

a. SIMPLIFY—distill complex but good data and science into usable models or simple rules.
b. INTEGRATE—integrate research results from very different disciplines in a common and formal language (mathematics).
c. COMMUNICATE—hide complex science from the end user and link scientists with policy advisors.
e. BE FLEXIBLE—be flexible in the analysis of scenarios for change and policy options.
f. BE INTERACTIVE—be interactive, fast, easy to understand and cater for a short attention span.
g. PROVIDE—to provide the end user with the information they want at the scale they like when they like.

The value of a DSS depends upon the nature of the DSS itself, the nature of the decision maker and the decision context (for example the data availability, range of stakeholders). Where these conditions are suitable the deployment of a DSS can augment a decision maker's innate knowledge handling abilities, can solve problems that the decision maker alone would not even attempt or that would consume significant decision-maker time and resources due to their complexity. Even for relatively simple problems, a DSS may be able to reach solutions faster and/or more reliably than a decision maker. Even if a problem is too complex to be solved using a DSS, the DSS could be used to stimulate the decision maker's thoughts about the problems through exploratory retrieval, analysis, by solving a "similar" problem which triggers insight about the present problem or by providing more compelling evidence to justify or support the decision maker's position. Thus a DSS can also be a learning (and teaching) tool. Finally, one of the most valuable aspects of DSS is the activity of constructing them (with end user or decision maker involvement) which may reveal new ways of thinking about the decision domain and may partially formalize aspects of decision making. DSS are no panacea. They have a number of limitations especially in tackling problems such as desertification. Two of these are outside of the scientific and technical area of DSS:

a. they only provide information to support decisions, those decisions are then made within a social, cultural and political context which may or may not allow the decisions to mature into policies and
b. policies need to be acted upon in a sustained manner if they are to succeed. The remaining limitations are scientific and technical:

   a. DSS are unable to replicate some innate human knowledge management skills/talents such as scientific reasoning,

b. DSS may be too specific, that is, many models and DSSs may be needed for a single decision so a means of coordinating them is required,

c. A DSS may not match the decision makers mode of expression or perception, making communication difficult,

d. DSS cannot overcome a faulty decision maker or one who is not making science-based decision/policy,

e. a DSS is constrained by the knowledge it possesses (the data, the models and the analysis and visualization tools),

f. No amount of slick presentation can overcome scientific or technical flaws in a DSS, one has to remain always critical of the output of DSS (as with any information system) and not become over-dependent upon them for making decisions,

g. Models are never finished, they should continuously evolve as the knowledge and experience of the modeler and the computational capacity evolves. The same also applies to DSS.

2.7 Building a DSS

Engelen (2000) defines the basic structure of a DSS as containing four major components, (a) the database (model parameters, GIS and remote sensing based spatial data, timeseries data), (b) the model base (process descriptions and connections), (c) the tool base (analytical techniques and functions), (d) the user interface which hides the complexity of the first three components and presents the results to the decision maker in a way that they can understand. There are two very different ways in which DSS can be constructed. In some cases existing research models are tweaked and have user interfaces added or are integrated loosely with other models within a DSS shell (this so-called soft integration was the approach taken in the MODULUS (Engelen 2000, Oxley et al. 2000). In other cases scientific models are redesigned from scratch as policy models, taking note of the requirements outlined in Table 1 (this so-called hard integration was the approach adopted in the MedAction PSS (van Delden et al. 2004). The former is often necessary where there already exists a bank of research models representing a substantial investment which, with a little further funding, can interface better with the policy domain, though the latter is usually much more successful in that regard.

Soft-integration is a matter of choosing appropriate models which share a common modeling philosophy and which together simulate the processes and spatio-temporal domains of interest. These models are analysed in order to know where there points of connection are, how one might link them temporally and spatially with the model database and with other models and which variables they should obtain from the database and which from other models, where, when and in what units. Integration is then a technical matter of either recoding or software 'wrapping' the existing models and having them run together with the appropriate database, toolbase and user interface and verifying that the system works. Such a process often produces a rather imprecisely aimed, buggy, slow and cumbersome system since the models are developed with different objectives in mind and may not follow the requirements identified in Table 1 very well, if at all.

Hard integration is altogether more challenging and rewarding. It involves the redevelopment of models as policy models through changing their: spatial and temporal domains, data requirements, output focus, level of connectivity and integration and general philosophy and then tightly integrating them within a time and spatial domain, with a database, toolbase and appropriate user interface. The models would simulate the relevant processes, allow user exploration of the system and its dynamics and display outcomes as readily interpretable indicators. This approach produces a much more policy-relevant model but is a significant investment of time and effort.

2.8 The recent history of Modelling Mediterranean Desertification

Mulligan (2004) presents a review of EC-DGXII (Environment and Climate) funded research into modelling desertification under frameworks III and IV (1990–1997) of the EC research and technological development funding. It is fair to say that before 1990 (EC frameworks I and II) there was relatively little research that focused on modelling desertification in the Mediterranean since most activities concentrated on better understanding the processes of desertification through field monitoring, remote sensing and experimentation. The modelling which did exist was largely theoretical modelling which was focused on understanding the system dynamics of desertification rather than having a particular 'policy-focus' per se (Thornes 1990, Lavee et al. 1991).

Through frameworks III and IV the majority of the modelling effort was single process modeling or modelling a few processes in a narrowly defined field (e.g. regionalising climate variables, climate change, land-atmosphere exchange, the hydrological budget, overland flow and soil erosion, fluvial and catchment response, soil and water pollution, fire, crops and agriculture, natural vegetation, socioeconomic processes, land use change and crop choice), see appendix 1. At this time very few projects attempted integrated desertification modelling. Those projects are listed in Table 2. Though the models produced (Kirkby et al. 1996, Kirkby et al. 1998, Kirkby 1998) were integrated in the sense that they incorporated a range of desertification processes, they were usually not highly policy focused and are what might be called scientific models.

It was only under frameworks VI and now VII that policy-focused integrated desertification assessment models were produced, notably within the context of the MODULUS, MedAction and now DESURVEY projects, all of which were multi-process, achieved a level of integration between the human and biophysical aspects of desertification (i.e. were cross-domain) and incorporated as many of the relevant desertification processes as was possible (they were integrated). A simplified timeline of the changing approaches to desertification modeling is given in Table 3.

This evolution of desertification models reflects the following driving factors: (a) improved knowledge of desertification processes, (b) an appreciation of the importance of process interaction and feedback and of spatial and temporal variability, (c) an increasing EC focus on policy relevance, policy support and human dimensions of desertification, (d) an increase in the size and multidisciplinarity of projects, and (e) the fact that modelling, like remote sensing, is fundamentally limited by the tools available for the activity. This means that major improvements in the ability to model desertification processes rely on technological innovation as well as theoretical innovation. In this way the major advances under Frameworks III and IV have come about, at least in part, by advances in computing (hardware and software) technology and by advances in instrumental and remote sensing techniques for model parameterisation and validation and in database and GIS techniques for data management.

Table 2.   EU funded efforts at producing integrated models for Mediterranean Desertification.

Integrated Physical and Biological Models

Dynamic modelling of DRUs, land use change, water resources allocation (MEDALUS II, project 3, EV5V-CT92-0165)

MEDALUS I EPOC-CT90-0014-(SMA) The MEDALUS slope catena model, surface and subsurface hydrology, implementation of the SHE model

ERMES EV5V-0023 Modelling the ecosystem impact of climate change

Integrated Physical, Biological, Cultural and Socio-economic Models

The ARCHAEOMEDES PROJECT—(EV5V-91-0021) the role of modelling, human ecodynamics, the multiscalar modelling framework.

Mediterranean Desertification and Land Use—MEDALUS I (EPOC-CT90-0014-(SMA)

ICALPE MEDIMONT Analysis of environmental and socio-economic parameters in

Mediterranean mountain regions (EV 5V CT91 0045) Social Economic regional approach

ERMES (EV5V CT91 0023) Modelling and exploring the impact of climate change on ecosystem degradation, hydrology and land use along a transect across the Mediterranean

Table 3. A timeline of approaches to modeling Mediterranean desertification.

1990–1995
Simple indicators and GIS overlay (MEDALUS I)
Conceptual modelling (MEDALUS, ARCHEAOMEDES)
One domain, one process models (e.g. soil erosion, MEDALUS, evapotranspiration, EFEDA)

1995–2000
One domain, multiprocess models (e.g. MEDALUS hillslope model)
Cross-domain, multiprocess model (e.g. MODULUS)

2000-present
Cross-domain integrated, policy focused multiprocess models (e.g. MedAction PSS)
Cross-domain, integrated, policy focused, multiprocess and operational (surveillance)
models (e.g. DESURVEY)

I will now cover in more detail the most recent (and current) developments in integrated desertification modeling that took place in three EU projects in which the author was/is a partner, starting with the MODULUS[6] project, then MedAction[7] and finally the ongoing DESURVEY[8] project. In MODULUS the research team essentially took a wide variety of models written by different research groups in different previous EU projects (ERMES, ARCHEOMEDES, EFEDA) and coded in different languages (Pascal, C, LISP...), chose the most suitable from them, making the necessary modifications for better integration and wrapped them in ACTIVE-X wrappers specifying the connections between models and passing data between them at runtime within the Research Institute for Knowledge Systems (RIKS) GEONAMICA© DSS system. This approach allowed existing models to be used but (a) does not make those models policy models, they tend to remain rather heavy research models, (b) produces a rather slow, clunky product that is not well streamlined for policy application, (c) requires a lot of work to ensure that the model philosophies are similar/compatible and that there is a good scientific rationale for their wrapping and integration in this way. The MODULUS system proved that this kind of scientific and technical integration was indeed possible but produced no more than a prototype DSS which is useful as a scientific testbed IAM but is probably not very attractive to as a DSS. The system was applied to desertification processes in the Mariba Baixa of SE Spain and in the Argolid of Greece (Oxley et al. 2002, Oxley et al. 2003, Oxley et al. 2004).

In MedAction we took a quite different approach and redeveloped all of the models from scratch in hard code. The models were based on knowledge developed from previous desertification research activities and models but were re-developed to be less like 'scientific models' as more like 'policy models' though there is, of course, a continuum between them. They were tightly coupled, fine tuned, made to work with routinely available data (particularly from remote sensing), tightly integrated with feedback loops closed and coupled with the appropriate database, toolbase and GUI. The PSS development was driven by regional policy makers in the region and used model derived indicators as a tool for better communication of model outcomes to policy makers. The advantages of this approach over that of MODULUS are that the end product is more likely to be useful to a policy analyst, it is likely to be more streamlined, simpler, better targeted at the specific policy issue and better rounded in terms of the robustness of the underlying model philosophy and the quality of the process integration (especially across the biophysical/socio-economic divide). The model was applied to desertification issues in the Guadalentin Basin of SE Spain (van Delden et al. 2004).

---

[6] http://www.riks.nl/projects/MODULUS
[7] http://www.riks.nl/projects/MedAction
[8] http://www.desurvey.net/

2.9   The Medaction PSS

Here we describe the basics of the MedAction PSS before presenting an example application of the tool. The purpose of the MedAction PSS was to make use of the best models of Mediterranean desertification to assist technical policy analysts in:

a. understanding the processes shaping the region and their linkages,
b. identifying and anticipating current and future problems in the region,
c. designing policy measures to mitigate the problems and assessing their effectiveness,
d. evaluating different alternatives and selecting a candidate for implementation.

This was achieved with tightly coupled process models with a regional extent, 100 m grain, timescales from sub-minute to annual and a 30 year time horizon, starting in 2000. The model includes 6 core modules (see Figure 1 climate and weather, hydrology and soil, vegetation, water management, land use and farmer's decisions) within which a series of submodules exist, for example hydrology and soils includes submodules for hydrology, sedimentation and salinisation. Within a submodule are a series of submodels for example the hydrology submodule contains models for the soil moisture balance, runoff generation, streamflow and groundwater. Each sub-model is made up of a series of processes. For the water balance sub-model these are evapotranspiration, infiltration, soil drainage, throughflow and continuity. The process is defined by a number of parameters and state variables that are either generated by another process in the same or different submodel, submodule or even module or derive from the spatial or temporal databases. The process is run for each relevant raster cell of the spatial database, conditioned by interactions with other processes and with inputs from the temporal database at the appropriate temporal resolution (for example rainfall inputs). The running model provides a variety of spatial (map-based), temporal (graph based) and statistical (spreadsheet based) output of scientific interest and a series of moving indicators of environmental sensitivity of policy relevance. The policy analyst interacts with the system via these indicators but also via a series of policy options and scenarios which they may apply to the system.

The MedAction system couples external and internal biophysical processes with external and internal human processes. External biophysical processes include climate and weather which must be supplied as either historic data files or are downscaled from GCM-derived grid box projections of climate change for three GCMs (HADCM2, ECHAM and GFDL, IS92 scenarios). The downscaling makes use of local station data in order to increase the spatial and temporal resolution of supplied historic daily data or projected monthly data. Climate and weather thus contain a series of weather generators designed to improve the spatial and temporal realism of routinely available meteorological data in line with the needs of process based hydrological models in environments where desertification can be an event-based process (Mulligan 1998) and thus it is important to properly represent storm dynamics. Internal biophysical processes are all of those within the spatial realm of the model and which do not have to be supplied as a boundary condition. These include all of the hydrological processes (surface and subsurface), soil wash erosion and sedimentation, vegetation growth, development and succession for crops and natural vegetation and soil salinisation. The external socio-economic components include external markets for crops, agricultural incentives, water 'imports' and the various policy options (water pricing, terracing, crop planning and irrigation). These must all be supplied to the model as data. The internal socioeconomic components include water demands and usage, water resources allocation, land use (change), profit and crop choice and dynamic land suitability.

The detail for the individual biophysical models is described fully in Mulligan (2004b) and for the socio-economic models in van Delden et al. (2004). A short summary is given here. The biophysical models were developed and tested first in the PCRASTER GIS[9] before being ported to the Geonamica system.

---

[9] www.pcraster.nl

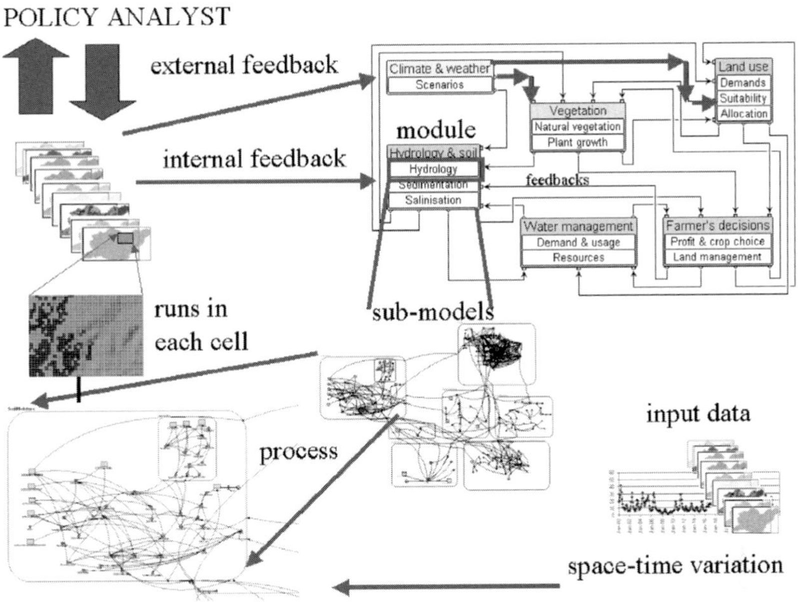

Figure 1. A Basic outline of the structure of the MedAction PSS (see colour plate page 374).

2.9.1 *Climate and weather*

The climate and weather module simulates spatially distributed sub-hourly rainfall using a timestep in inverse proportion to the rainfall intensity (down to a few seconds) which recognizes the importance of high intensity events for runoff generation. All other inputs are integrated twice daily, at sunset and sunrise, to give a mean value during the daytime and during the nighttime, which are considered appropriate timesteps and different process domains. These inputs include solar, net and phosynthetically active radiation, air temperature. The data are based on GCM scenario inputs or historic data from a single base station and are spatialised and downscaled in temporal resolution using data from a nearby automatic weather station (AWS) for a single year. Data are spatialised using inputs from additional stations around the catchment along with appropriate climate covariates such as altitude and distance to coast. Further details can be found in Mulligan (2004b).

2.9.2 *Hydrology*

The hydrology module integrates at sunrise and sunset during periods of no rain but at the resolution determined by the rainfall intensity (the 'bucket-tip' timestep) during rainfall periods. It has submodels for interception, evapotranspiration, soil sealing, infiltration and runoff, throughflow, and drainage in the soil compartment, groundwater recharge, groundwater flow and return flow in the groundwater compartment and runoff accumulation, transmission loss and river evaporation in the fluvial compartment. Water is routed downriver using a kinematic wave or a simpler cascading scheme. Further details can be found in Mulligan (2004b).

2.9.3 *Soil erosion and deposition*

Since one of the major desertification issues in the Guadalentin basin is soil erosion, the erosion and sedimentation module includes a wash erosion model which simulates erosion based on the soil properties, slope gradient, land cover and stream power [37] along with a sediment transport and deposition model after Kirkby [9]. This model both reduces soil thicknesses (and breaks surface seals) in areas where erosion occurs but also deposits soil on slopes, in streams, rivers, checkdams and reservoirs which has implications for water storage and associated policy options for dredging and check dam construction and clearing. Further details can be found in Mulligan (2004b).

2.9.4 *Salinisation*

The salinisation module uses salt concentrations from the various water sources in the region (rainfall, river, groundwater, desalinised sea water and water from the segura-tajo transfer as well as salt concentrations in the soil to calculate the accumulation of salinity in soils by evaporation and the washout of this salinity with infiltration water (eventually to the aquifer and back) and with runoff water (to the rivers and also, eventually to the aquifer). Further details can be found in Mulligan (2004b).

2.9.5 *Water resource allocation, demands and use*

The water resources module calculates the water balance of the aquifer, any reservoirs and stocks of desalinated seawater and allocates this water spatially based on its availability and any current policy restrictions. The water demands module calculates the requirements for water by each agricultura; sector on the basis of the unit demands and the spatial extent of the sector. Actual water use is calculated taking into account the policy restrictions and water prices from the different sources. Further details can be found in van Delden et al. (2004).

2.9.6 *Land use and dynamic suitability*

The land use model is based on a constrained cellular automata model and allocated land according to the demand by different sectors, neighbourhood attraction and repulsion by particular land uses and the land use specific suitability of the land, any policy zoning and accessibility requirements (White and Engelen, 1997). Land use suitability is dynamic, based on the changing conditions of soil salinity, soil moisture, available soil thickness, all of which respond to desertification processes as well as air temperature which responds to climate change and slope gradient which is rather static in the absence of a terracing policy. In this way land use determines the trajectory of environmental processes which in turn set the state of the land suitability for further land use change in a tightly coupled dynamic feedback. Further details can be found in van Delden et al. (2004).

2.9.7 *Plant growth and natural vegetation*

The plant growth model uses a production efficiency approach (Monteith 1977, Prince 1991, Prince et al. 1994) to calculate net biomass growth on the basis of environmental conditions. This growth is then partitioned to the relevant plant parts including yield using a resource dependent partitioning algorithm. On this basis plant structural properties (biomass, leaf area index, vegetation cover fraction) can be determined and in part determine the productivity of the plant in the next timesteps. At the biophysical level natural vegetation and crops are represented as a series of functional types: nonirrigated tree (e.g. Pine, Olive), irrigated tree (e.g. Citrus), dwarf shrub (e.g. Quercus), nonirrigated grass (e.g. Stipa) and irrigated grass (e.g. wheat). These classes are determined from the EEA CORINE land cover data which is used to initialize land use in the model. Further details can be found in Mulligan (2004b). The natural vegetation and land management modules use a much more detailed specification of natural vegetation and crop types which are mapped onto these for biophysical purposes. The natural vegetation model represents the processes of succession at the community level on the basis of the vegetation maturity (height) and the various grazing and firing conditions. The module also permits vegetation dispersal through seed diffusion (Mazzoleni et al. 1998, McIntosh et al. 2001).

2.9.8 *Land management*

The land management module is one of the main interfaces with policy analysts and incorporates procedures on crop planning, terracing and irrigation. The planning model allows a user to define the timing of sowing, ploughing and harvesting for each crop type. The irrigation model allows the user to define a minimum soil moisture which, if reached under a crop, will start crop irrigation and determines from which source the water will come and how much will be available for use, based on the approach adopted in MODULUS (Oxley et al. 2000). The terracing model allows users to construct terraces which has profound implications for aspects of the hydrology. Further details can be found in van Delden et al. (2004).

3   EXPLORING DESERTIFICATION WITH THE MEDACTION PSS

The MedAction PSS has been extensively tested and verified (Mulligan 2004b, van Delden et al. 2004). Though some of the individual models have been verified and validated in isolation (Mulligan 1996a, Mulligan 1996b) outputs from the whole IAM have not been validated against historic data as yet—though, as discussed earlier, this is less relevant for explorative models. Here we present some case studies which explore the dynamics of desertification using the MedAction PSS. We will conduct two experiments, the first examines the implications of projected climate change on the integrated environment system for the Guadalentin and the second examines the impacts of switching off the inputs of water from the Rio Tajo-Segura transfer. Each experiment is an equilibrium experiment in which the model is run twice, once as a baseline and then again with the perturbation. The model results (averaged over the catchment and the 30 year simulation period) are compared between the two sets of equilibrium output.

3.1   *Exploring climate change in the Guadalentin*

In this 30 yr simulation we run a baseline scenario based on current climate and another based on the downscaled IS92 Climate change scenario of the ECHAM GCM ($+0.48°C$ over the period, rainfall change is $+1.87$ mm/month or 22.5 mm/year (around 7.5% in the lowlands) so an overall slight wetting). After running the two simulations, all of the 392 output variables were analysed and those changing by more than ±5% between the simulations on average across the catchment and period were set aside. 254 out of 392 (65%) of variables changed by more than 5% in response to this scenario. This process allows us to examine the model outcomes in a more holistic way than would be possible by analyzing specific processes or variables in isolated. By looking at the suite of variables that have changed the greatest we can generate a narrative of change from the model outcomes. The variables increasing by more than 5% included: dryland vineyard productivity and cover (150%), matorral productivity and cover (300%), dryland olives and fruit productivity and cover (100%), dryland almonds productivity and cover (60%), profit for dryland fruit (86%), greenhouse vegetables (39%), vineyards (34%) and olives (21%), evaporative interception losses (50%), aquifer demands for rural residential (14%), reservoir demands for industry and commerce (7.5%), urban residential (7.5%) and tourism (9.5%), soil moisture (6%), runoff (13%) and river discharge (13%). Variables decreasing by more than 5% included: desalinated sea water demands in tourism, rural residential and expats (all 1-00%), water shortage in the tourism sector (-70 to -90%), sea water irrigation for pretty much all crops, aquifer salinity (-43%) and soil salinity (-9%), biomass and productivity for most irrigated crops (Olives, Almonds and Fruit, -34 to -42%), aquifer recharge (-30%), reservoir (-19%) and checkdam (-15%) sedimentation, drip irrigation (-13.6%) and aquifer demands for agriculture (-14%), soil erosion (12.2%) and productivity and cover for irrigated Almonds, Fruit, Olives, Cirus and Cereal (-14 to -10%). The net area under natural vegetation fell by 28 ha over the period to be replaced by agriculture.

The mild climate change thus precipitated a reduction in the productivity of irrigated crops (which tend to be based on the warmer, drier lowland plains) and an increase in the productivity of dryland crops (which tend to occur on the cooler, wetter hillslopes) with resulting profit reductions for irrigated crops and thus a shift to non-irrigated or reduced irrigation crops. There is also and a general 'blooming' of agricultural and natural biomass with the extra water availability. Although soil moisture and runoff slightly increase, soil erosion and sedimentation decrease, probably as a result of the increases in vegetation cover. Moreover salinisation also decreases because of the extra water inputs. The reductions in erosion are consistent with the train of feedbacks that the climate change precipitates: increased water availability ▶ reduced dependence on irrigation ▶ growth of the vegetation cover ▶ reduced erosion. None of the other profits fell or rose sufficiently to be counted in this analysis so that although large changes in the landscape occurred, the economic consequences were not serious in a positive or negative way regionally and in the long term.

The three GCMs used (HADCM2, ECHAM and GFDL) all show marginal changes in rainfall, often slight increases for the SE Spanish grid box so the previous analysis is the most likely outcome. We can, however, produce an aridification scenario. This was done with a temperature increase of +1°C over the 30 year period and a rainfall decrease of 4 mm/month (47 mm/year or 12.4% of the rainfall). As before we compared this aridification scenario with the baseline scenario for all 392 model output variables and 249 variables (64%) showed a change of more than ±5%. Variables that increased by more than 5% as a result of this scenario included: desalinated sea water demands for expatriates (+390%), desalinated sea water irrigation for irrigated greenhouse vegetables (+149%), olives (+181%) and almonds (+183%), vegetables (+61%), fruit (+72%), citrus (+76%) and vineyards (+77%), water shortage for the tourism sector (+134%) and thus desalinated sea water demands for tourism (+110%), rural residential (+128%) and agriculture (+47%). Reservoir demands for expats (+91%), rural residential (+14%) and agriculture (+15%) also increase as do extraction from desalination sea water (+30%) and reservoirs (+15%). The shortage of water for urban residential increases by 11% and by 5% for industry and commerce. Aquifer salinity increases by 10%. Variables that decreased by more than 5% as a consequence of the scenario included: biomass and profit for dryland, greenhouse and irrigated fruit and vegetables. Profit for greenhouse vegetables fell by −116%, by −70% for irrigated fruit, by −50% for irrigated vegetables. Biomass and profit also fell for dryland olives (−44%) and vineyards (−40%) as well as matorral and forest communities. Interception losses fell by −36% as a result of reduced rainfall and reduced vegetation cover. Profit for olives fell by −30%, for irrigated citrus by −20%, irrigated almonds (−16%), dryland vineyards (−16%), irrigated vineyards (−13%) and this leads to reduced aquifer irrigation for these crops (around −15%) and thus lower demands on the aquifer. Even with less rainfall, the fall in biomass production and profitability and consequent land use changes led to a reduction in the shortage of water for agriculture of −27%! The lower rainfall also causes lower runoff and river discharge (−16%) and thus −10% less check-dam and slope sedimentation, −19% less river sedimentation and transmission loss to the aquifer (−12%) and −7% less water in the reservoirs. There is also lower transpiration (−15.8%) in part also because of the reduced plant covers. This leads to −22.5% less recharge, −19% less aquifer replenishment and 8% less infiltration and soil moisture. The net area under agriculture fell by 1518 ha over the period to be replaced by natural vegetation.

So, the aridification scenario showed an overall decrease in biomass and productivity for both natural vegetation and crops (especially irrigated ones but also for dryland crops), leading to much-reduced profitability, greater dependence on desalinized sea water and reservoirs (also for the tourism sector). Reduced recharge led to higher levels of aquifer salinity. The resulting biomass decline, reduced transpiration losses reduced agricultural hectareage and increased natural vegetation hectareage leads to a reduction in the agricultural water deficit in the region! Less water leads to less runoff and erosion and reduced soil moisture, groundwater recharge but higher salinity. This is more like the typical desertification scenario and indicates the significant dependence of the agricultural and tourism economies in this region to the available rainfall. Since rainfall in the Mediterranean is extremely temporally variable at all timescales (Mulligan 1996b) it is a rather shaky foundation upon which to build enormous rainfall-dependent agricultural and touristic economies.

3.2 Turning off the Tajo-Segura supply to the Guadalentin

The Tajo-Segura transfer (TST) or Trasvase (a man made aqueduct) carries water from the Tagus basin to the Segura Basin (mainly to support tourism and irrigation). The baseline simulation of the MedAction PSS has inputs to the Guadalentin from the Trasvase of some 2 Mm^3/month. In this experiment, two 30 yr simulations were performed: a baseline simulation with the 2 Mm^3/month of Trasvase water and a simulation in which the inputs from the Trasvase were set to zero. Once again we compared the results for all 392 output variables averaged across the catchment and the simulation period and identified those that had changed by more than ±5%. 79 out of 392 (20%) variables changed by more than 5% in response to this scenario. Variables increasing by more than 5% included: biomass of matorral (+25%), demands for desalinated sea water for expatriates (74%), rural residential (55%), tourism (28%), desalinated sea water irrigation for almonds (55%), greenhouse vegetables (53%) and olives (52%), fruit (11%), citrus (12%) and vines (14%), water shortage in the tourism sector (58%), checkdam and reservoir sedimentation (7%). Variables that decreased by more than 5% included: average profit for greenhouse vegetables (−39%), and dryland fruit (−12%), evaporative interception losses (−8%), agricultural water shortage (−7%), desalinated sea water irrigation for vegetables (−6%), reservoir demands for rural residential (−5%). The remaining 370+ variables changes by less than 5%. The net area under agriculture increased by 1225 ha over the period replacing natural vegetation.

These are altogether surprising results but are all rather logical within the context of the system being simulated. It is not surprising that water shortages for the tourism sector should increase with the switch off because the tourism sector require high quality water (with low salt concentrations) and this can only be obtained from the TST or from desalination (hence the observed increase in use) in much of the region. The switch off also led to some land abandonment (for crops entirely irrigated by TST water, which are far away from or too saline sensitive for groundwater sources) as evidence by the increase in matorral biomass. In a few areas where profitability is very high and other suitable water sources are not available irrigation continues with desalinized sea water, hence the increase in irrigation using desalinised water, especially for almonds, greenhouse vegetables and olives, though the magnitude of these increases is small. The fact that interception evaporation declines significantly suggests a shift from spray irrigation (in which interception losses are high) to drip irrigation (in which they are much lower), but may also result from lower leaf areas where dryland (secano) has replaced irrigated agriculture. The increased soil moisture combined with the land use changes may be responsible for the observed increase in reservoir sedimentation. The overall agricultural water shortage decreased with the switch off because the overall demand for irrigation water declined as a result of land abandonment and increased efficiency of use. A decrease in profits for greenhouse vegetables and dryland fruits is observed and a resulting decrease in desalinized irrigation for greenhouse vegetables.

Once again an apparently disastrous scenario with respect to desertification, the switching off of external water supplies to a region, turns out to have few negative impacts in the long term and at the regional scale. The reduction in the availability of TST water forces greater use of desalinized water but without large scale changes in profitability, except in the greenhouse sector. In all but the most profitable TST irrigated crops land use change or abandonment occurred without dramatic economic consequences for the region, resulting in an overall decrease in water deficits (because of reduced agricultural demand). Changes in land management are also evidenced by the shift from spray to drip irrigation practices and reduced evaporative losses as a result. Where there is a negative impact (reservoir sedimentation) it not in the process domain expected for this policy option. So, the train of feedbacks became reduced TST inputs ► increase in water shortage for tourism ► greater reliance on desalinisation ► irrigation efficiency improvements or land abandonment ► reduced agricultural demand ► reduced water deficits ► slightly increased reservoir sedimentation.

3.3   Leave it to the feedbacks

It is clear from the analyses above that models with feedback loops closed tend to behave less dramatically than partial and unclosed models. If we would run a soil erosion model in isolation on any of the scenarios above we would get a quite different picture of the outcomes of that scenario than we get from this analysis. Where feedback loops are closed, systems react to change and the reaction usually dampens the change. After all, most of the feedbacks that exist in natural systems are negative feedbacks. Where positive feedbacks exist they usually become negative at a point or else they tend to destroy the system in which they exist (hurricanes, earthquakes) and are thus short lived. Positive feedbacks that are sustained in the longer term tend to be artificially sustained by human activity (for example artificially sustained agriculture on eroding soils or in drying climates). Thus in many cases the expected consequence of the input scenario is not realized. . .but other consequences are, for example reduced water deficits as a result of switching off the Tajo supply.

In addition to many negative feedbacks the MedAction model (as the real world) incorporates a high degree of **connectivity** and, as a result, the consequences of the scenarios can be seen in very different domains from those in which the scenario directly operates. So, switching off the

Tajo supply leads to increased reservoir sedimentation; a wetter climate leads to less erosion and a drier climate leads to less erosion.

The more of the processes we include the more complicated it becomes to interpret the outputs of models and the causal chains that have led to those outcomes but also the greater the opportunity for identifying unexpected or counterproductive outcomes of particular policies in the domain in which they are aimed or in other domains. The complication of the MedAction PSS is not necessary for identifying desertification but is for understanding precisely why it is occurring and building successful policies that will manage it without causing (too many) other (possibly worse) problems. I hope that I have been able to convince the reader that complex models also have their place in dealing with issues as multi-factoral, spatially variable and interactive as desertification. It is not the modeller's aim to make the models complex. They are complex because the reality of desertification is complex and, looking at one or a few processes in isolation will give only a partial view of the potential trajectories of desertification is a region. Worse still, this partial view will tend to be more dramatic than the reality because the important negative feedbacks are ignored.

3.4   ...and the policy makers?

As support systems these tools do not provide all the answers. Any model (even these) are partial and there may also be processes, feedbacks or actors that are not included but that may very well reverse the success of a particular policy to adapt to, prevent or mitigate desertification. These models can assist in highlighting 'dangerous' or negative interaction, can help to identify unforeseen consequences of apparently benign interactions, but are most useful as learning tools to help those working in science and policy to view the connectedness and behavioural dynamics and interactions of landscapes and their occupants. As we all know change can be more to do with politics than policies. Whilst the kind of tool outlined here can help to compare the long-term and multi-domain impact of policies, policies will only be implemented if leaving things as they are is not a feasible alternative, if the stakeholders and thus their politicians will accept it and if the (short-term) benefits of tackling desertification are greater than the long term losses of not doing so—and are accounted for as such.

3.5   Developmental needs for improved IAM

If these tools are to be used operationally at least the following developments are required some of which will require significant development in the analysis but also the dissemination of the remote sensing products discussed elsewhere in this volume. Remote sensing has a lot to offer these regional scale integrated models but to-date there has been relatively little interaction between the remote sensing community and the modelling community at this scale (much more has occurred at the global scale driven by various of the global change initiatives). The following basic data need to be improved, particular in terms of the representation of spatial variability:

a. Spatio-temporal climate fields (solar radiation, temperature, precipitation)
b. Better (and higher spatial resolution) LUCC (and land management) products
c. More spatial data on soil properties (thickness, bulk density, texture, stoniness)
d. Validation products (soil erosion, vegetation change timeseries, moisture dynamics timeseries)

In addition more and deeper exploration of the dynamics of complex spatial models is necessary so as to better understand their behaviour and that of the processes that they represent. In order for this to happen the systems themselves require significant theoretical and software developments in order to facilitate more efficient mining through the model results. These tools include:

a. Error and uncertainty tracing tools capable of representing model results alongside a measure of cumulative error,
b. Analytical tools focused on interpreting detailed model output and converting it to narratives which represent the behaviour of the system much more simply than by multitude graphs, maps and data tables.

ACKNOWLEDGEMENTS

The MODULUS project was carried out by a consortium including: Research Institute for Knowledge Systems b.v., International Ecotechnology Research Centre, Universitá di Napoli 'Federico II', The Spatial Modelling Centre on Human Dimensions of Environmental Change, Agricultural University of Athens and The University of Edinburgh Institute of Ecology and Resource Management who supplied models and data, which is gratefully acknowledged. MedAction was collaborative between King's College London and Research Institute for Knowledge Systems b.v. Data were also supplied by local partners in the Guadalentin and by various partners in the MEDALUS project.

REFERENCES

Campbell, B.M. & Sayer, J.A. 2003. *Integrated Natural Resources Management: Linking Productivity, the Environment and Development*. Cabi Publishing.

de Kok, J.L., Engelen, G., White, R. & Wind, H.G. 2001. Modelling land-use change in a decision-support system for coastal-zone management. *Environmental Modelling and Assessment* 6: 123–132.

Dregne, H.E. & Chou, N.-T. 1992. Global desertification dimensions and costs. In: *Degradation and restoration of arid lands*. Lubbock: Texas Tech. University.

Dregne, H.E. 1998. Desertification assessment and control. *UNU Desertification Series No. 1. New Technologies To Combat Desertification* Proceedings of the International Symposium held in Tehran, Iran 12–15 October 1998. http://www.unu.edu/env/workshops/iran-1/.

Engelen, G. (ed.) 2000. *MODULUS: A spatial modelling tool for integrated environmental decision making*. Vol. 2, Ch. 6, Final report, EU-DGXII (contract ENV4-CT97-0685): Brussels.

Geeson, N. 2001. Review of other indicator systems. Deliverable 1.1a (ii) *Desertlinks: combating desertification in Mediterranean Europe linking science with stakeholders*. http://www.kcl.ac.uk/projects/desertlinks.

Geeson, N.A., Brandt, C.J. & Thornes, J.B. 2002. *Mediterranean Desertification: A Mosaic of Processes and Responses*. Chichester: Wiley.

Gough, C., Castells, N. & Funtowics, S. 1998. Integrated assessment; an emerging methodology for complex issues. *Environmental Modelling and Assessment* 3: 19–29.

Kirkby, M.J. (1976) Hillslope hydrology. John Wiley and Sons.

Kirkby, M.J., Baird, A.J., Diamond, S.M., Lockwood, J.G., McMahon, M.L., Mitchell, P.L., Shao, J., Sheehy, J.E., Thornes, J.B. & Woodward, F.I. 1996. The MEDALUS slope catena model: a physically based process model for hydrology, ecology and land degradation interactions. In Thornes, J.B. & Brandt, J. (eds.), *Mediterranean Desertification and Land Use*: 303–354, Chichester: John Wiley.

Kirkby, M.J., Abrahart, R., McMahon, M.D., Shao, J. & Thornes, J.B. 1998. MEDALUS soil erosion models for global change. *Geomorphology* 24: 35–49.

Kirkby, M.J. 1998. Modelling across scales: the MEDALUS family of models. In Boardman, J. & Favis-Mortlock, D. (Eds.), Modelling Soil Erosion by Water, Springer, Berlin, pp. 161–73.

Lavee, H., Imeson, A., Pariente, S. & Benyamini, Y. 1991. The response of soils to simulated rainfall along a climatological gradient in an arid and semi-arid region. *Catena Suppl.* 19: 19–37.

Mabbutt, JA. 1984. New Global Assessment of the Status and Trends of Desertification. *Environmental Conservation* 11 (2): 103–113.

Mairota, P., Thornes, J.B. & Geeson, N.A. 1997. *Atlas of Mediterranean Desertification*, Wiley.

Marsh, R.L., Sebrechts, M.M., Hicks, J.L. & Landau, J.D. 1997. Processing strategies and secondary memory in very rapid forgetting. *Memory & Cognition*, 25: 173–181.

May, R.M. 1974. *Stability and complexity in model ecosystems*. Princeton University Press.

Mazzoleni, S., Legg, C., Strumia, S. & Migliozzi, A. (Eds.), 1998. *ModMED, Modelling Vegetation Dynamics and Degradation in Mediterranean Ecosystems*, EU-DGXII (ENV4-CT95-0139): Brussels.

McIntosh, B.S., Legg, C.J., Csontos, P., Arianoutsou, M. & Mazzoleni, S. 2001. Rule-based modelling. In Mazzoleni, S. & Legg, C.J. (Eds.), *ModMED: Modelling Mediterranean Ecosystem Dynamics, ModMED III Project, EU-DGXII* (ENV4-CT97-0680): Brussels.

Middleton, N., Thomas, D.S.G. & UNEP, 1997. *World Atlas of Desertification. Second Edition.* Oxford University Press.

Miller, G.A. 1956. The magical number seven, plus or minus two: Some limits on our capacity for processing information. *Psychological Review* 63: 81–97.

Monteith, J.L. 1977. Climate and the Efficiency of Crop Production in Britain. *Philosophical Transactions of the Royal Society of London Series B*. 281: 277–294.

Mulligan, M. 1996a. *Modelling the hydrology of vegetation competition in a degraded semi-arid environment.* Unpublished PhD thesis, King's College London, University of London.

Mulligan, M. 1996b: Modelling the complexity of landscape response to climatic variability in semi arid environments, in Anderson, M.G.A. & Brooks, S.M. (eds.), *Advances in Hillslope Processes:*1099–1149, Chichester: Wiley.

Mulligan, M. 1998. Modelling the Geomorphological Impact of Climatic Variability and Extreme Events in a Semi-Arid Environment. *Geomorphology* 24: 59–78.

Mulligan, M. 2004. A review of European Union funded research into modelling Mediterranean desertification. *Advances in Environmental Monitoring and Modelling* 1 (4): 1–78.

Mulligan, M. 2004b. MedAction, Development, testing and application of the climate, hydrology and vegetation components of a Desertification Policy Support System. *Final report for work undertaken as part of 'MedAction: Policies to combat desertification in the Northern Mediterranean region'*, EU-DGXII (contract EVK2-2000-22032): Brussels.

Noilhan, J. 1996. *Co-ordinator summary report of the EFEDA modelling group VI* (EV5V-CT93-0269). CNRM France.

Oxley, T., Jeffrey, P. & Lemon, M. 2002. Policy relevant modelling: relationships between water, land use and farmer decision processes. *Integrated Assessment* 3 (1): 30–49.

Oxley, T. & Lemon, M. 2003. From social-enquiry to decision support tools: towards an integrative method in the Mediterranean rural environment. *Journal of Arid Environments* 54: 595–617.

Oxley, T., McIntosh, B., Winder, N., Mulligan, M., Engelen, G. 2004: Integrated modelling and decision-support tools: a Mediterranean example. *Environmental Modelling & Software* 19 (11): 999–1010.

Oxley, T., McIntosh, B.S., Mulligan, M. & De Roode, R. 2000. Adaptation, integration and application of an integrative modelling framework. In Engelen, G. (ed), *MODULUS: A spatial modelling tool for integrated environmental decision making. Vol. 2, Ch. 6, Final report*, EU-DGXII (contract ENV4-CT97-0685): Brussels.

Peterson, L.R. & Peterson, M.J. 1959. Short-term retention of individual verbal items. *Journal of Experimental Psychology* 58: 193–198.

Prince, S.D. 1991. A Model of Regional Primary Production for Use With Coarse Resolution Satellite Data. *International Journal of Remote Sensing*. 12: 1313–1330.

Prince, S.D., Justice, C.O. & Moore III, B. 1994. Monitoring and modelling of terrestrial net primary production. *IGBP-DIS Working Paper: 8*. IGBP.

Rosema, A., Roebeling, R., Peters, S., Garrido, S., de Bruin, H.A.R., Saraber, M.J.M., Roozekrans, J.N., Garcia, F., Kok, K., Stolle, F. & Bronsveld, M.C. 1994. *Assessment and monitoring of desertification in the Mediterranean area (ASMODE), final report, summary*. Project EV5V-CT91-0029. (summary available).

Thornes, J.B. 1990. The interaction of erosional and vegetational dynamics in land degradation: spatial outcomes. In Thornes, J.B. (ed), *Vegetation and Erosion*. Chichester: John Wiley & Sons.

van Delden, H., Luja, P. & Engelen, G. 2006. Integration of multi-scale dynamic spatial models of socio-economic and physical processes for river basin management, *submitted Environmental Modelling and Software.*

van Delden, H., Luja, P. & Engelen, G. 2004. *MedAction PSS. Final report for work undertaken as part of 'MedAction: Policies to combat desertification in the Northern Mediterranean region'*, EU-DGXII (contract EVK2-2000-22032), Brussels.

van Wesemael, B., Mulligan, M. & Poesen, J. 2000. Spatial patterns of soil water balance on intensively cultivated hillslopes in a semi-arid environment: the impact of rock fragments and soil thickness. *Hydrological Processes* 14 (10): 1811–1828.

Wainwright, J. & Mulligan, M. 2004. Environmental modelling: finding simplicity in complexity. Chichester: Wiley & Sons.

White, R. & Engelen, G. 1997. Cellular automata as the basis of integrated dynamic regional modelling. *Environment and Planning B: Planning and Design* 24: 235–246.

APPENDIX 1

A list of EC funded projects modelling aspects of desertification (1990–1997)

Regionalising Climatic Variables
ASMODE (EV5V-CT91-0029) EFEDA II (EV5V-CT93-0284)

Forecasting Climatic Variability (Drought)
MEDALUS EPOC-CT90-0014 (SMA), Temporal rainfall analysis EV5V-CT93-0257

Modelling Climatic Change
MEDALUS I EPOC-CT90-0014-(SMA)—general circulation model based scenarios
MEDALUS II Project 2 (EV5V-0164) Downscaling of GCM scenarios
MEDALUS II Project 2 (EV5V-CT92-0164) Impact of the greenhouse effect for extreme climatic events and agricultural and water impacts

Modelling Land Surface—Atmosphere Exchange
EFEDA (EPOC-CT 90-0030) Calibration of the BERLIN SVAT scheme for natural vegetation
EFEDA II Hydrology Group EV5V-CT93-0282 PATTERN model
EFEDA II Vegetation and soil physics group EV5V-CT93-0272 Sispat model
MEDALUS EPOC-CT90-0014-(SMA) Evapo-transpiration modelling.
DEMON (EV5V-CT90-0035) GIS based evapo-transpiration modelling.
An integrated monitoring and modelling study of desertification and climatic change impacts in the Messara Valley of Crete (EV5V-CT94-0466)

Modelling the Hydrological Budget
MEDALUS EPOC-CT90-0014-(SMA) Evapo-transpiration modelling.
EFEDA II (EPOC-CT90-0284)
(EV5V-CT94-0466) An integrated monitoring and modelling study of desertification and climatic changes in the Messara Valley of Crete
EV5V CT94 0487 (AGUAS)
EPOC CT90 0028 (TSTS) Drought effects on vegetation and soil degradation in Mediterranean countries.

Modelling Water Resources
MEDALUS II, project 3, EV5V-CT92-0165
EFEDA (EPOC-CT 90-0030) Hydrological modelling of groundwater in Castilla La Mancha (IH)
EFEDA II Hydrology Group EV5V-CT93-0282 Groundwater Modelling, regionalised land surface modelling, modelling recharge (REM), modelling soil profile drainage
An integrated monitoring and modelling study of desertification and climate change impacts in the Messara Valley of Crete (EV5V-CT94-0466)

Modelling Overland Flow and Soil Erosion
PROJECT DM2E Models of the impact of waste litter treatment on soil erosion and overland flow (EV5V-CT-91-0039), models of runoff and soil erosion for representative catchments
Climate change, soil erosion and slope instability in selected agricultural areas of Italy and southern Britain (EV5V-CT93-0257)
DEMON (EV5V-CT90-0035) Erosion Hazard Modelling
Mediterranean Desertification and Land Use—MEDALUS I (EPOC-CT90-0014-SMA)
Climate change, soil erosion and slope instability in selected agricultural areas of Italy and southern Britain (EV5V-CT93-0257)

Modelling Fluvial and Catchment Response
EFEDA II Hydrology Group EV5V-CT93-0282—monthly precipitation runoff relationships
EFEDA (EPOC-CT 90-0030) Modelling of flow for ephemeral streams (IH)
Mediterranean Desertification and Land Use—MEDALUS I (EPOC-CT90-0014-SMA)
DM2E (EV5V-CT91-0039)

An integrated monitoring and modelling study of desertification and climatic change impacts in the Messara Valley of Crete (EV5V-CT94-0466)
EV5V CT91 0045 (MEDIMONT)

Modelling Soil and Water Pollution
DM2E (EV5V-CT91-0039)

Modelling Fire
Modelling vegetation Dynamics and Degradation in Mediterranean Ecosystems (EV5V-CT94-0489)—fire modelling in Mediterranean ecosystems
PHOENIX—A GIS decision support system for the prevention of desertification resulting from forest fires: the rule base (EV5V-0025)

Crops and Agriculture
WEPP modelling of crop growth an erosion for the South Downs EV5V-CT93-0257

Semi-natural Vegetation
EV5V-CT94-0489 (MODMED Modelling vegetation Dynamics and Degradation in Mediterranean Ecosystems
(EV5V-CT94-0489)—computer modelling of vegetation dynamics
Drought effects on forest vegetation in central Greece EPOC CT-90-0028 (TSTS) [drought/growth index models]

Modelling Cultural and Socio-economic Processes
EV5V-0486—Environmental perception and policy making—cultural and natural heritage and the preservation of degradation-sensitive environments in southern Europe
EV5V-0487—new interpretative modelling of the long term dynamics of human settlement and land use

Modelling Land Use Change and Crop choice
EV5V-0486—Environmental perception and policy making—cultural and natural heritage and the preservation of degradation-sensitive environments in southern Europe

Recent Advances in Remote Sensing and Geoinformation Processing
for Land Degradation Assessment – Röder & Hill (eds)
© 2009 Taylor & Francis Group, London, ISBN 978-0-415-39769-8

Estimating area-averaged surface fluxes over contrasted agricultural patchwork in a semi-arid region

A. Chehbouni
Centre d'Etudes Spatiales de la Biosphère BP cedex Toulouse, France

J. Ezzahar
Physics Department LMFE, Faculty of Sciences Semlalia, Marrakesh, Morocco

C. J. Watts
UNISON, Hermosillo, Sonora, Mexico

J.-C. Rodriguez
IMADES, Hermosillo, Sonora, Mexico

J. Garatuza-Payan
ITSON, Ciudad Obregon, Sonora, Mexico

ABSTRACT: Population growth has resulted in intense demands on the quantity and quality of water resources worldwide. The sustainability of water resources in the 21st Century will depend on our ability to correctly manage water resources systems under a more variable future climate. Semi-arid regions are in particular jeopardy, experiencing rates of population development that exceed those of other climatic regions and are highly sensitive to increasing anthropogenic pressures, variations in climate, and the disruptions associated with long-term climate change. In arid and semi-arid regions, irrigated agriculture consumes about 80 to 90% of total available water. Therefore, a sound and efficient irrigation practice is an important step for achieving sustainable management of water resources in these regions. In this regard, a better understanding of the water balance is essential for exploring water-saving techniques. One of the most important components of water balance in semi-arid areas is the evapotranspiration (ET). Measuring ET over large and heterogeneous surfaces is possible through the deployment of a network point sampling devices such as Eddy Correlation systems which are expensive and require a well trained staff to operate and maintain them. Alternatively, one can use scintillometer to derive area-averaged sensible heat flux and then obtain latent heat or ET as the residual terms of the energy balance equation since available energy can be easily estimated through a combination of ground and remotely sensed data. In this study, an experimental setup has been designed to investigate the effectiveness of the large aperture scintilllometer (LAS) to obtain area-averaged sensible and latent heat flux along a transect of about 1.8 km made up of three contrasting agricultural fields: cotton, wheat and chickpeas. The comparison against reference area-averaged fluxes measured by eddy correlation systems (EC) shows that the LAS based fluxes are overestimated by about 15%. In an attempt to explain such behavior, several reasons are proposed: a) The LAS measurement was made below the blending height which may violate the requirement for the applicability of Monin-Obukhov Similarity Theory (MOST); b) The fact the sensitivity of the LAS along the optical path is not uniform but follows a bell-shaped curve; c) The division and the contrast between the fields may provide a source of additional turbulence, the LAS does not distinguish between upward and downward eddies, both contribute to the recorded signal; d) The eddy correlation underestimates sensible and latent heat fluxes since their sum is smaller than the observed available energy (the no closure of the energy balance) . This later explanation is supported by the fact that the correspondence between measured and simulated heat flux improved when the fluxes were adjusted to force closure of the energy balance. Therefore, the scintillometer technique can be considered as an effective tool to estimate surface fluxes over large and heterogeneous surfaces.

1   INTRODUCTION

Population growth has resulted in intense demands on the quantity and quality of water resources worldwide. The sustainability of water resources in the 21st Century will depend critically on our ability to correctly manage water resources systems under a more variable future climate. Semi-arid regions are in particular jeopardy, experiencing rates of development that exceed those of other climatic regions and are highly sensitive to increasing anthropogenic pressures and variations in climate. In arid and semi-arid regions, irrigated agriculture consumes about 80 to 90% of total available water. Therefore, a sound and efficient irrigation practice is an important step for achieving sustainable management of water resources in these regions. In this regard, a better understanding of the water balance is essential for exploring water-saving techniques. One of the most important components of water balance in semi-arid area is the evapotranspiration (ET) which needs to be estimated for management purposes over large and heterogeneous surfaces (at the irrigation district scale).

On the other hand Soil-Vegetation-Atmosphere-Transfer schemes (SVATs) are now widely used in meteorological models to describe energy and mass partitioning at the earth surface. This is a major step towards interpreting changes in land use and land cover in terms of their impact on the balances of mass and energy at different space-time scales. However, the credibility and the realism of the model simulations might be questioned if the impact of sub-grid scale hetero-geneity encountered in the real world is not properly taken into account. Substantial progress has been made in the development of aggregation schemes to estimate area-average surface fluxes over heterogeneous surfaces (Koster & Suarez 1992, Sellers et al. 1997; Noilhan & Lacarrere 1995, Arain et al. 1996, Noilhan et al. 1997, McNaughton 1994, Raupach & Finnigan 1995, Lhomme et al. 1994, Chehbouni et al. 1995, 2000a). Nevertheless, the effectiveness of any of the above approaches cannot be fully assessed without a comparison between the model's fluxes simulations and the results from field measurements. This usually requires the deploy-ment of a network of point-sampling measurement devices such as Eddy correlation systems. However, due to their high price and the requirement for continuous availability of well-trained staff to operate and maintain them, these devices might not be the best choice for obtaining sur-face fluxes over large heterogeneous surfaces and it is worthwhile to explore other techniques which may be used to estimate directly the average flux over a complex, heterogeneous sur-face to validate the parameterization of surface heterogeneities in large-scale hydro-atmospheric models.

In this regard, the technology for scintillometry which has been developed over the last 25 years, initially at NOAAs Wave Propagation Laboratory in Boulder, CO (Clifford et al. 1974, Hill 1980, Hill 1989) represents an attractive alternative to traditional flux measurement systems since it can be potentially used to estimate surface fluxes over distances varying from a few metres up to a several kilometers. The technique consists of transmitting a beam of electromagnetic radiation and measuring the intensity variations of the received signal, which provides the structure parameter for refractive index (C_n^2 directly. This can be related to the structure parameters for temperature C_T^2 and humidity C_Q^2, which can be in turn converted into temperature and humidity scales using Monin Obukhov similarity theory (MOST) (Wesely 1976, Moene 2003). For visible and near infrared wavelengths, the signal is much more sensitive to C_T^2 and allows the estimation of the sensible heat flux. A direct measurement of water flux requires using scintillometers in the microwave region. As it is relatively cheap and robust, the large aperture scintillometer (LAS), which uses an incoherent beam in the near infrared region (as described by Ochs & Wilson 1993) is now becoming popular in hydrometeorological studies (de Bruin et al. 1995, Green et al. 1994, De Wekker 1995, McAneney et al. 1995; Lagouarde et al. 1996, 2000, 2005, Hartogensis et al. 2003, Chehbouni et al. 1999, 2000b; Watts et al. 2000, Meijninger et al. 2002a, Hoedjes et al. 2002, 2006, Ezzahar et al. 2006a,

Asanuma et al. 2006) and very attractive for the validation of satellite estimates of fluxes over heterogeneous surfaces.

Up to now most of studies have been devoted to homogeneous surfaces or a transect of two surfaces (Chehbouni et al. 1999, Lagouarde et al. 2002, Meijninger et al. 2002b). The objective of this study is to assess the potential and the limitations of the LAS in inferring path-average surface fluxes over a patch work of contrasting agricultural fields. The composite transect is was about 1.8 km long and made up of three contrasting agricultural fields in semi-arid region in northern Mexico. The following two sections provide a physical background and a description of the site and data used in this study. Then a comparison of eddy correlation based and LAS based estimates of path-averaged sensible heat flux is made. Finally the issues of the closure of the energy balance, the applicability of the Monin-Obukhov Similarity Theory (MOST) over heterogeneous surfaces are discussed.

2   PHYSICAL BACKGROUND

For a LAS, the variance of the natural logarithm of the irradiance I incident at the receiver is:

$$\sigma^2_{\ln(I)} = \overline{[\ln(I) - \overline{\ln(I)}]^2} = \int_0^1 C_n^2(ux)W(ux)du \tag{1}$$

where the overbar represents a spatial average, C_n^2 is defined by:

$$C_n^2 = \frac{\overline{(n(r_1) - n(r_2))^2}}{r_{1,2}^{2/3}} \tag{2}$$

where $n(r_1)$ is the refraction index at location r_1, and the distance $r_{1,2}$ lies in between the inner scale of turbulence (of the order of 5–10 mm) and the outer scale of turbulence (of the order of the measuring height), and $W(u_x)$ is a spatial weighing function, given by :

$$W(u_x) = 16\pi^2 k^2 L \int_0^\infty K\Phi_n(K)\sin^2(K^2 L u_x(1-u_x)/2k\lambda)\,[2J_1(x)/x]^4 dK \tag{3}$$

where $u_x = x/L$ is the normalized pathlength; L is the pathlength; $k\lambda = 2\pi/\lambda$ is the optical wave number; $x = KDu x/2$, where is D the receiver/transmitter aperture; K is the three-dimensional spatial wave number; J_1 is a Bessel function of the first kind of order one; and Φ_n, the three-dimensional Kolmogorov spectrum of the refractive index describes the turbulent medium in terms of its Fourier components K:

$$\Phi_n(K) = 0.033 C_n^2 K^{-11/3} \tag{4}$$

After integrating Eq. (3), and using Eqs. (1) and (4), C_n^2 can be obtained as a linear function of $\sigma^2_{\ln(I)}$ measured by the scintillometer as:

$$C_n^2 = C\sigma^2_{\ln(I)}D^{7/3}L^{-3} \tag{5}$$

where C = 1.12.

The spatially averaged refractive index structure parameter obtained by the scintillometer C_n^2 is related to the temperature structure parameter C_T^2 as:

$$C_T^2 = C_n^2 \left(\frac{T_a^2}{\gamma P}\right)^2 \left(1 + \frac{0.03}{\beta}\right)^{-2} \qquad (6)$$

where the Bowen ratio β is incorporated as a humidity correction. P is the atmospheric pressure (P_a), T_a the air temperature (K), and γ the refractive index for air ($\gamma = 7.9$ 10-7 K Pa-1). C_n^2 and C_T^2 are in m-2/3 and K2 m-2/3 respectively.

For a homogeneous surface the temperature scale T^* $(= -H/\rho\, c_p\, u^*)$ and C_T^2 are related by:

$$\frac{C_T^2(z - d)^{2/3}}{T^{*2}} = f\left(\frac{z - d}{L}\right) \qquad (7)$$

z is the measurement height and d the displacement height. L is the Monin-Obukhov length defined as:

$$L = \frac{Tu^{*2}}{kgT^*} \qquad (8)$$

For $f(\frac{z-d}{L})$ under unstable conditions, De Bruin et al. (1993) propose:

$$f\left(\frac{z - d}{L}\right) = 4.9\left(1 - 9\frac{z - d}{L}\right)^{-2/3} \qquad (9)$$

and under stable conditions, Wyngaard (1973) suggests:

$$f\left(\frac{z - d}{L}\right) = 4.9\left(1 + 2.4\left|\frac{z - d}{L}\right|^{2/3}\right) \qquad (10)$$

The friction velocity is given by:

$$u^* = \frac{k\,u}{\ln\left(\frac{z-d}{z_o}\right) - \psi_m\left(\frac{z-d}{L}\right) + \psi_m\left(\frac{z_o}{L}\right)} \qquad (11)$$

where ψm is the integrated stability function, z_o the roughness length.

Using T^* and u^* respectively provided by Eqs. (7) and (11), the sensible heat flux H can be computed as:

$$H = \rho\, c_p\, T^*\, u^* \qquad (12)$$

where ρ (kg m^{-3}) and cp (J kg^{-1} K^{-1}) are the air density and specific heat capacity at constant pressure, respectively. The sensible heat and momentum fluxes together determine atmospheric stability, and this in turn influences turbulent transport. An iterative procedure is thus needed to calculate sensible heat flux. Latent heat flux can then be obtained as the residual term of the energy balance equation providing measurement or estimate of available energy (Lagouarde et al. 1996; Meijninger et al. 2002; Hoedjes et al. 2002; Hemakumara et al. 2003; Ezzahar et al. 2006b).

Over homogeneous surfaces, d and z_o can be formulated as a fraction of vegetation height, while an aggregation scheme is required over heterogeneous surfaces. In this study effective displacement height and effective roughness length are obtained following Shuttleworth et al. (1997) and

Chehbouni et al. (1999 and 2000b) as:

$$d = \sum_i w_i d_i \quad \text{with } \ln^{-2}\left(\frac{z_b - d}{z_o}\right) = \sum_i w_i \ln^{-2}\left(\frac{z_b - d_i}{z_{0i}}\right) \tag{13}$$

where z_{o_i} and d_i are the patch scale roughness length and displacement height, respectively. w_i is the fraction of the surface covered by the patch i with obviously $\sum_i w_i = 1$.

The blending height z_b can be defined as the height above which the area-averaged wind speed follows a logarithmic shape and the flow is in equilibrium with the local surface and independent of horizontal position. For neutral conditions, Wieringa (1986) expressed it as a function of the friction velocity, wind speed, and the average horizontal length scale of the individual patches (Lpatch):

$$z_b \cong 2\left(\frac{u^*}{u}\right)^2 Lpatch \tag{14}$$

3 SITE AND DATA DESCRIPTIONS

The Yaqui Valley (27° N and 110° W) is the largest agricultural district in the state of Sonora, Northwest Mexico with an area of 220,000 hectares (Figure 1).

It is bordered on the west by the Gulf of California and on the east by the foothills of the Sierra Madre Occidental. The water for irrigation in the valley is provided by the Alvaro Obregon Reservoir situated on the Yaqui river, which has a capacity of approximately 3,000 hm^3. The climate of this region is very dry, i.e. the potential evaporation is very high, greatly exceeding the annual rainfall. The average yearly precipitation ranges from 200 mm at the coast to 400 mm at the foot of the Sierra Madre mountains. In this region, irrigation represents the largest water consumption (about 90%) in the valley. An important step towards sound management of the scarce water resources

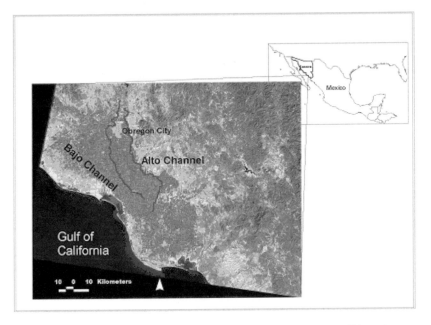

Figure 1. The Yaqui valley (Northwest Mexico) as described by Landsat ETM+ (false colour composite, 4, 3, 2 channels) February 26, 2000 (see colour plate page 375).

in Northern Mexico is providing accurate estimates of the spatial and temporal variability of the water lost to the atmosphere through evapotranspiration.

In this study, three fields of about 100 hectares each have been instrumented (cotton, wheat and chickpeas). The aerodynamic characteristics of the three vegetation types were very different during the study period. The height of the chickpeas, cotton and wheat were about 0.5 m, 0.25 m and 0.95 m, respectively. Similarly, the leaf-area-index (LAI) of the chickpeas, the cotton and the wheat were 0.5, 0.15 and 2.5, respectively.

A 9 m meteorological tower was equipped with a set of standard meteorological instruments to measure wind speed and direction (R.M. Young, MI, USA), air temperature and humidity and incoming (Vaisala, Sweden) short-wave radiation (Kipp, The Netherlands). Net radiation was measured using Q7.1 net radiometers (REBS Inc., WA, USA) over the chickpeas and the wheat and a CNR1 (Kipp & Zonen, the Netherlands) over the cotton. At each of the 3 sites, soil heat flux was measured using 3 HFT3 plates (REBS Inc., WA, USA) buried at 0.025 m depth. Soil moisture was measured at different depths (0.15, 0.15 and 0.30 m) using 3 CS600 TDR (Time Domain Reflectometer from Campbell Scientific Inc., Logan). All these meteorological measurements were sampled every ten seconds and an average was recorded every 30 minutes. Sensible and latent heat fluxes were measured at 2.9 height over each sites using 3D sonic anemometers and fast response hygrometers (Campbell Scientific Inc., USA), with a sampling frequency of 10 Hz for wheat and 8 Hz for Cotton and Chickpeas. The flux data were stored every 30 minutes using 21X datalogger (Campbell Scientific Inc., USA) in the wheat and CR10X (21X, Campbell Scientific Inc., USA). It is worthwhile to mention that an inter-comparison of the 3 eddy correlation systems has been performed at the beginning and at the end of the experiment. The agreement between the 3 systems was within the expected range.

The LAS used in this study were built at the Department of Meteorology of the Wageningen Agricultural University (WAU). They were derived from the basic design described in Ochs and Wilson (1993). The LAS has an aperture size of 0.15 m and the light source is a light emitting diode operating at a peak wavelength of 0.94 μm, which is placed at the focal point of a concave mirror. The receiver also employs an identical mirror to focus light on a photo diode detector. C_n^2 is given by $C_n^2 = 10^{(Vout-12)}$, where Vout is the output voltage of the scintillometer. Averaged values of Vout over 30 minute time steps were stored on a data logger. In this study, the transmitter and the receiver were installed perpendicularly to the dominant wind at a height of 6.4 m. The path-length was 1800 m spanning a transect made up of 48% of chickpeas, 26% wheat and 26% of cotton. However, according the scintillometer spatial weighting function (Eq. 3), the C_n^2 measured by the LAS corresponds to a spatial sampling of 41.5% from chickpea, 47.5% and 11% from wheat and cotton, respectively.

4   RESULTS AND DISCUSSION

Daytime data (from 10h to 18h L.T.) during a period of 7 days DOY 98 to DOY 104 in 1998 has been considered for this particular study. This period provided an ideal opportunity to test the performance of the LAS under complex situation. As a matter of fact, the chickpea field was very dry while the cotton and the wheat fields had been recently irrigated (DOY 96 for the cotton and DOY 90 for the wheat). Figure 2 presents the variation of measured daily averaged of available energy (2a), sensible heat flux (2b) and latent heat flux (2c) during the study period. These figures exhibit an expected behavior in the crops, knowing the moisture status of the individual fields. However, it is of interest to note that the sensible heat flux emanating from the cotton field was negative during the first two days. This creates an interesting situation with regard to the interpretation of the LAS measurements since a change of sign in the heat flux occurred along the path.

Path-averaged C_n^2 has been constructed using two methods. The first one consisted of using measurement of sensible (H_{EC}) and latent heat ($L_V E_{EC}$) fluxes, friction velocity (u_{EC}^*), and Monin-Obukhov length (L_{EC}) over each field to derive C_n^2 of each individual fields (using Eqs. (6)–(12)) and then use the scintillometer weighting function W(ux) to simulate the path-average C_n^2 (denoted

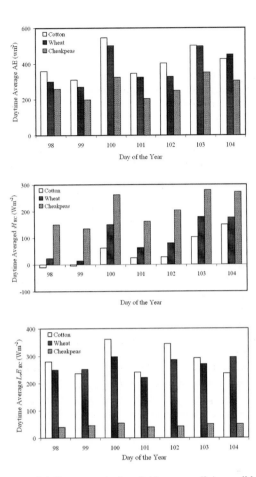

Figure 2. Variation of measured daily averaged of available energy (2a), sensible heat flux (2b) and latent heat flux (2c) during the study period.

$C_{n\ EC\_W}^2$). The second method consists of estimating C_n^2 (denoted $C_{n\ EC\_eff}^2$) from measured area-weighted averages of sensible (H_{EC_eff}) and latent heat ($L_V E_{EC\_eff}$) fluxes in conjunction with effective friction velocity ($u_{EC\_eff}^*$) and effective Monin-Obukhov length (L_{EC_eff}) (using Eqs. (6)–(12)). H_{EC_eff} and $L_V E_{EC\_eff}$ are obtained as a simple linear weighted average of the fluxes measured at three sites. Effective friction velocity is obtained by applying the matching rule for momentum fluxes (Chehbouni et al. 1999), which lead as:

$$u_{EC\_eff}^* = \left(\sum_i wi\ u_{iEC}^{*2} \right)^{0.5}$$

(15)

where u_{iEC}^* is the friction velocity associated with the patch i. Effective L_{EC_eff} is then derived from H_{EC_eff} and effective $u_{EC\_eff}^*$ as:

$$L_{EC\_eff} = \frac{-\rho c_p T_a u_{EC\_eff}^{*3}}{k\ g\ H_{EC\_eff}}$$

(16)

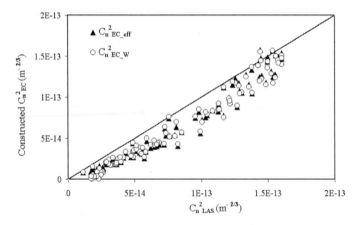

Figure 3. Comparison between measured path averaged values of $C^2_{n\,LAS}$ using the LAS and those derived from eddy correlation measurements, and $C^2_{n\,EC\_eff}$, using the average and effective methods. (see text).

A comparison between half-hourly averages of measured C^2_n (denoted $C^2_{n\,LAS}$) and those simulated using both methods is presented in Figure 3.

Both methods give very similar results and both significantly underestimate the observed $C^2_{n\,LAS}$. LAS-based estimates of area-averaged values of sensible, latent heat fluxes and friction velocity have been derived following the approach outlined in section 2. In Figure 4, the simulated sensible heat flux from the LAS (H_{LAS}) is compared with the H_{EC_eff}. Similarly, Figures 5 and 6 present the same comparison for friction velocity and latent heat flux, respectively.

It can be seen that the observed underestimation of C^2_n is translated into an overestimation of sensible heat flux and an underestimation of latent heat flux and friction velocity (see Table 1 for statistical results). Such behavior can be explained by several factors. First, water status differences among the 3 fields leads to a change of the sensible heat flux direction along the path (downward for cotton in stable conditions and upward for the wheat and chickpea both in unstable conditions). The LAS does not distinguish between upward and downward eddies, both contribute to the recorded signal. However, analysis of the daily course of measured heat flux (and L) showed that in most cases, the change in the sign occurred mostly during early morning and late afternoon, which only explain the overestimation for low values. Therefore, the change of sign along the pathway cannot explain the systematic overestimation of the LAS based estimates of the heat flux. The second explanation can be related to the fact that the contrast among the fields, i.e. dry-wet transition, creates extra turbulence (Brunet al.1994, Itier et al.1994, De bruin et al. 1991). This turbulence is not seen by the eddy correlation systems (point measurement devices) but causing an increase of the signal measured by the LAS. Both explanations might be plausible since according to Eq. (14) the blending height was about 10 m while the LAS measurements were made at 6.4 m. This may indicate that under such conditions, the basic requirement for the applicability of the Monin-Obukhov Similarity Theory (MOST) is not met. However, several authors (Shuttleworth 1988, De Bruin et al. 1989; Ronda and De Bruin 1999) have showed that for surfaces with disorganized heterogeneity there is a layer below the blending height where MOST still applies but contributions from separate fields can still be "seen". Using the same scaling reported by de Bruin et al. (1993), Meijninger et al. (2002) have found that a reasonable fluxes could be derived from path-averaged structure parameters measured with a LAS, installed well bellow the blending height. In the same vein, Kohsiek et al. (2002) reported that when deploying the XLAS (Extra Large Aperture Scintillometer, which can be used over pathlengths of up to 10 km) below the blending height, the violation of MOST is negligible. Recently, Ezzahar et al. (2006a) have confirmed these previous results by applying the MOST over heterogeneous terrain below the blending height which was at about 26m. Additionally, under heterogeneous conditions, the non-linearity of the flux- C^2_T relationship and the fact the sensitivity of the LAS along the optical path is not uniform but follows a bell-shaped

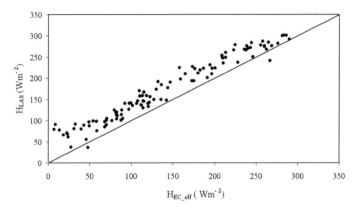

Figure 4.   Comparison between LAS-derived of area-averaged sensible heat flux H_{LAS} and H_{EC_eff} derived by weighing the sensible heat fluxes observed from the eddy correlation systems at three fields (cotton, wheat and chickpea).

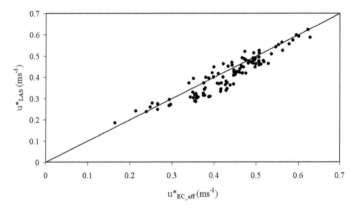

Figure 5.   Comparison between LAS-based of area-averaged friction velocity u^*_{LAS} and $u^*_{EC\_eff}$ obtained by applying the matching rule for momentum fluxes (Chehbouni et al. 1999).

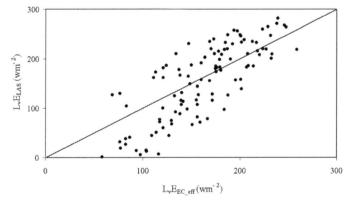

Figure 6.   Comparison between LAS-derived of area-averaged latent heat flux $L_V E_{LAS}$ and $L_V E_{EC\_eff}$ derived by weighing the latent heat fluxes observed from the eddy correlation systems at three fields (cotton, wheat and chickpea).

curve, may lead to an over-estimation by the LAS (Lagouarde et al. (2002)). To investigate the impact the LAS spatial response, we re-compute the average $C_n^2 (C_{n\ \text{EC\_w}}^2)$ by weighting individual C_n^2 not according to W(ux) function but to the area covered by each patch. The result showed that there is no significant difference compared to Figure 3. This is not surprising since in our case (spatial arrangement of the fields), the weighting methods affect mostly the relative contribution of the cotton versus that of the Wheat and both fields were wet during the study period. This result means that the reason of the discrepancy between LAS and eddy correlation based estimates of area-averages fluxes should be looked for elsewhere.

In this regard, based on the analysis of a large amount of data collected during the Southern Great Plains field campaign, Twine et al. (2000) have shown that the independent measurements of the major energy balance flux components are often not consistent with the principle of conservation of energy. The sum of latent and sensible heat flux measured by eddy correlation systems (from different manufacturers) is normally less than the available energy. They also reported that the problem of the energy balance closure can be explained neither by the mismatch between areas representative of the fluxes and available energy measurements, nor by uncertainties associated with measurement of soil heat flux and net radiation. This hypothesis was also supported by the analysis performed by Hoedjes et al. (2002) over a single wheat field in Mexico. The analysis of the energy balance closure of our data set confirmed the observations made above. Over all fields, the available energy was systematically higher that the sum of sensible and latent heat flux. The absolute value of average closure was about 13% of available energy over the cotton, 5% over the wheat and 10% over the chickpea.

Twine et al. (2000) suggested that eddy correlation systems underestimate sensible and latent heat flux but that their ratio (the Bowen ratio) is correctly measured. Based on this assumption, we re-computed sensible and latent heat flux over each individual field using measured Bowen ratio and available energy, thus forcing the closure of the energy balance. Figures 7, 8 and 9 display the same comparison as in Figures 3, 4 and 6, but using the recomputed fluxes.

A clear improvement for the correspondence between LAS and eddy correlation based estimates is observed. Table 2 presents the statistical results associated with each flux. Notwithstanding the improvements, some discrepancies still remain between observed and simulated quantities. First of all, this Bowen ratio closure method does not (and it is not meant to) solve the problem associated with the change in heat flux sign along the pathway. An overestimation (under-estimation) of the sensible heat flux (latent heat and C_n^2) for low values can still be seen in these figures. Second, the underestimation at higher heat flux values might be due the fact the Bowen ratio closure method distributes the closure error evenly between H and LvE without regard to the field conditions (soil

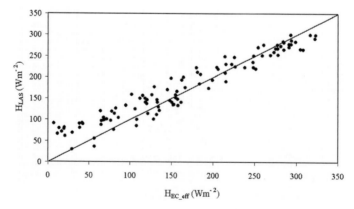

Figure 7. Comparison between measured path averaged values of $C_{n\ \text{LAS}}^2$ using the LAS and those constructed from eddy correlation measurements, $C_{n\ \text{EC\_w}}^2$ and $C_{n\ \text{EC\_eff}}^2$, using average and effective methods but after forcing the energy balance to close.

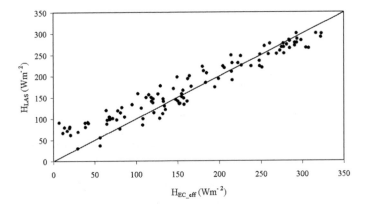

Figure 8. Comparison between LAS-based area-averaged sensible heat flux H_{LAS} and H_{EC_eff} derived by weighing the sensible heat fluxes observed from the eddy correlation systems at three fields (cotton, wheat and chickpea) after forcing the closure of the energy balance.

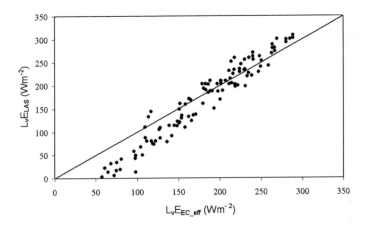

Figure 9. Comparison between LAS-based area-averaged latent heat flux $L_V E_{EC\_eff}$ and $L_V E_{EC\_eff}$ derived by weighing the latent heat fluxes observed from the eddy correlation systems at three fields (cotton, wheat and chickpea) after forcing the closure of the energy balance.

moisture state and vegetation cover) whereas the magnitude of the closure error depends on these conditions (Hoedjes et al. 2002).

5   CONCLUDING REMARKS

A combination of scintillometer measurements, ancillary meteorological data and an aggregation scheme has been used to derive area-averaged sensible and latent heat flux heat over a transect spanning three adjacent and contrasted vegetation fields: cotton, wheat and chickpea. Comparison between eddy correlation and scintillometer based estimates of area-averaged sensible heat flux and latent showed that the overall LAS performance is fairly good. However, the scintillometer systematically overestimates the heat flux.

During early morning and late afternoon, this behavior appeared to be associated with a change in the direction of the heat flux sign along the path (downward for cotton and upward for the wheat and chickpea). In fact, the LAS does not distinguish between upward and downward eddies and both contribute to the recorded signal. In this case however, all approaches based on MOST are expected to fail due to "asymmetric" behavior of "stable" and "unstable" turbulence (Ronda and De Bruin, 1999). To overcome this problem, the LAS should be installed high enough within the surface layer so that the PBL do not "see" the contribution of the individual fields. However, as the LAS moves higher, it may begin to capture contributions from fields outside the instrumented transect which will further complicate attempts to validate the method.

During the remaining part of the day, the overestimation can be partially explained by increased turbulence at the intersection of the fields with give an additional contribution to the observed Cn2. Another plausible explanation is associated with the fact that eddy correlation technique tends to provide underestimation of latent and sensible heat flux values leading to a non-closure of the energy balance, but allows a correct retrieval of the Bowen ratio. This hypothesis has been supported by the improved agreement between the LAS and the corrected eddy correlation fluxes. These first results confirm the potential of large aperture scintillometer for estimating surface fluxes over composite surfaces. Nevertheless a modeling effort remains to be done for the case where stable conditions are encountered on a portion of surface (i) to better evaluate the sensitivity of the scintillometer measurements to the characteristics of the patches, and (ii) to test the effect of the non linearity of the weighing function of C_n^2 on the accuracy of retrieved sensible heat flux, as done by Lagouarde et al. (2002) for a 2 patch-composite surface.

ACKNOWLEDGEMENTS

We are grateful for funding from CONACyT (project: 29340T), the PLEIADE project (CONTACT No. 037095) funded by European Union.

REFERENCES

Arain, A.M., Michaud, J.D., Shuttleworth, W.J. & Dolman, A.J. 1996. Testing of vegetation parameter aggregation rules applicable to the Biosphere-Atmosphere Transfer Scheme (BATS) at the Fife site. *Journal of Hydrology* 177:1–22.

Asanuma, J. & Lemoto, K. 2006. Measurements of regional sensible heat flux over Mongolian grassland using large aperture scintillometer. *Journal of Hydrology* (in press).

Avissar, R. 1992. Conceptual aspects of a statistical-dynamical approach to represent landscape subgrid-scale heterogeneities in atmospheric models. *Journal of Geophysical Research* 97: 2729–2742.

Avissar, R., 1998. Which type of SVAT is needed for GCMs. *Journal of Hydrology* 50: 3751–3774.

Blyth, E.M., Dolman, A.J. & Wood, N. 1993. Effective resistance to sensible and latent heat flux in heterogeneous terrain. *The Quarterly Journal of the Royal Meteorological Society* 119: 423–442.

Boulet, G., Kalma, J.D., Braud, I. & Vauclin, M. 1999. An assessment of effective land-surface parameterisation in regional-scale water balance studies. *Journal of Hydrology* 217: 225–238.

Braden, H. 1994. Energy fluxes from heterogeneous terrain: averaging input parameters of the Penman-Monteith formula. *Agricultural and Forest Meteorology* 75: 121–133.

Brunet, Y., Itier B., McAneney K.J. & Lagouarde J.P. 1994. Downwind evolution of scalar fluxes and surface resistance under conditions of local advection. Part II: measurements over barley. *Agricultural and Forest Meteorology* 71 (3–4): 227–245.

Chehbouni, A., Watts, C., Kerr, Y.H., Dedieu, G., Rodriguez, J.-C., Santiago, F., Cayrol, P., Boulet, G. & Goodrich, D.C. 2000a. Methods to aggregate turbulent fluxes over heterogeneous surfaces: application to SALSA data set in Mexico. *Agricultural and Forest Meteorology* 105: 133–144.

Chehbouni, A., Watts, C., Lagouarde, J.-P., Kerr, Y.H., Rodriguez, J.-C., Bonnefond, J.-M., Santiago, F., Dedieu, G., Goodrich, D.C. & Unkrich, C. 2000b. Estimation of heat and momentum fluxes over complex terrain using a large aperture scintillometer. *Agricultural and Forest Meteorology* 105: 215–226.

Chehbouni, A., Kerr, Y.H., Watts, C., Hartogensis, O., Goodrich, D.C., Scott, R., Schieldge, J., Lee, K., Shuttleworth, W.J., Dedieu, G. & De Bruin, H.A.R. 1999. Estimation of area- average sensible heat flux using a large aperture scintillometer. *Water Resources Research* 35: 2505–2512.

Chehbouni, A., Njoku, E.G., Lhomme, J.-P., & Kerr, Y.H. 1995. An approach for averaging surface temperature and surface fluxes over heterogeneous surfaces. *Journal of Climate* 5: 1386–1393.

Clifford, S.F., Ochs, G.R. & Lawrence, R.S. 1974. Saturation of optical scintillation of strong turbulence. *J. Opt Soc Amer.* 64: 148–154.

De Bruin, H.A.R., Nieveen, J.P., de Wekker, S.F.J. & Heusinkveld, B.G. 1996. Large aperture scintillometry over a 4.8 km path for measuring areally-average sensible heat flux. *Proc. 22nd AMS Symposium on Agricultural and Forest Meteorology, 28 January-2 February, Atlanta, Georgia, USA.*

De Bruin, H.A.R., van den Hurk, B.J.J.M. & Kohsiek, W. 1995. The scintillation method tested over a dry vineyard area. *Boundary-layer meteorol*ogy 76: 25–40.

De Bruin, H.A.R., Kohsiek, W. & van den Hurk, B.J.J.M. 1993. A verification of some methods to determine the fluxes of momentum, sensible heat, and water vapour using standard deviation and structure parameter of scalar meteorological quantities. *Boundary-layer meteorol*ogy 63: 231–257.

De Bruin, H.A.R., Bink, N.J. & Kroon, L.J.M. 1991. Fluxes in the surface layer under advective conditions. In: Schmugge, T.J. & André, J.C. (eds), Land surface evaporation: 157–171.

De Bruin, H.A.R. 1989. Physical aspects of the planetary boundary layer with special reference to regional evapotranspiration. *Proc. Workshop on the Estimation of Areal Evapotranspiration, Vancouver B.C., August 1987. IAHS Publ. 177*: 117–132.

De Wekker, S.F.J. 1995. The estimation of areally-averaged sensible heat fluxes over complex terrain with a large-aperture scintillometer. *M.S. thesis. Dept. of Meteorology, Wageningen Agricultural University, 42 p.*

Ezzahar, J., Chehbouni, A., Hoedjes, J.C.B. & Chehbouni, A. 2006a. On the application of scintillometry over heterogeneous grids. *Journal of Hydrology* (in press).

Ezzahar, J., Chehbouni, A., Hoedjes, J.C.B., Chehbouni, Ah., Er-Raki, S., Chehbouni, Ah, Boulet, G., Bonnefond, J-M. & De Bruin, H.A.R. 2006b. The use of the scintillation technique for monitoring seasonal water consumption of olive orchards in a semi-arid Region. *Agricultural Water Management* (accepted).

Green, A.E., McAneney, K.J. & Astill M.S. 1994. Surface-layer scintillation measurements of daytime sensible and momentum fluxes. *Boundary-layer meteorol*ogy 68: 357–373.

Hartogensis, O.K., Watts, C.J., Rodrigues, J.-C., and De Bruin, H.A.R. 2003. Derivation of an effective height for scintillometres: La Poza experiment in Northwest Mexico. *Journal of Hydrometeorology* 4: 915–928.

Hemakumara, H.M., Chandrapala, L. & Moene, A. F. 2003. Evapotranspiration fluxes over mixed vegetation areas measured from large aperture scintillometer. *Agricultural Water Management* 58: 109–122.

Hill, R.J., Clifford, S.F. & Lawrence R.S. 1980. Refractive-index and absorption fluctuations in the infrared caused by temperature, humidity, and pressure fluctuations. *J of Opt Soc Amer.* 70: 1195–1205.

Hill, R.J. 1998. Implications of Monin-Obukhov similarity theory for scalar quantities. *Journal of the Atmospheric Sciences* 6: 2236–2244.

Hoedjes, J.C.B., Chehbouni, A., Ezzahar, J., Escadafal, R. & De Bruin, H.A.R. 2006. Comparison of large aperture scintillometer and eddy covariance measurements: Can thermal infrared data be used to capture footprint induced differences? *Journal of hydrometeorology* (accepted).

Hoedjes, J.C.B., Zuurbier, R.M. & Watts, C.J. 2002. Large aperture scintillometer used over a Homogeneous Irrigated Area, Partly Affected by Regional Advection. *Boundary-layer meteorology* 105: 99–117.

Itier, B., Brunet, Y., McAneney, K.J. & Lagouarde, J.P. 1994. Downwind evolution of scalar fluxes and surface resistance under conditions of local advection. Part I: a reappraisal of boundary conditions. *Agricultural and Forest Meteorology* 71: 211–225.

Kohsiek, W., Meijninger, W.M.L., Moene, A.F., Heusinkveld, B.G., Hartogensis, O.K., Hillen, W.C.A.M. & de Bruin, H.A.R. 2002. An extra large aperture scintillometer (XLAS) with a 9.8 km path length. *Boundary-Layer Meteorol.* 105: 119–127.

Koster, R.D. & Suarez, M.J. 1992. Modeling the land surface boundary in climate models as a composite of independent vegetation stands. *J Geophys. Res.* 97: 2697–2715.

Lagouarde, J.-P., Irvine, M., Bonnefond, J.-M., Grimmond, C.S.B., Long, L., Oke, T.R., Salmond, J.A., & Offerle, B. 2005. Monitoring the sensible heat flux over urban areas using the large aperture scintillometer: case study of Marseille city during the escompte experiment. *Boundary-layer meteorology* DOI 10.1007/s10546-005-9001-0.

Lagouarde, J.-P., Bonnefond, J.-M., Kerr, Y.H., McAneney, K.J. & Irvine, M. 2002. Integrated sensible heat flux measurements of a two-surface composite landscape using scintillometry. *Boundary-layer meteorology* 105: 5–35.

Lagouarde, J.-P., Chehbouni, A., Bonnefond, J.-M., Rodriguez, J.-C., Kerr, Y.H., Watts, C. & Irvine, M. 2000. Analysis of the limits of the CT-profile method for sensible heat flux measurements in unstable conditions. *Agricultural and Forest Meteorology* 105: 195–214.

Lagouarde, J.-P., McAneney, K.J. & Green, E.F. 1996. Scintillometer measurements of sensible heat flux over heterogeneous surfaces. In: Stewart, J.B. Engman, E.T. Feddes, R.A. & Kerr, Y. (eds.), *Scaling up in hydrology using remote sensing.* Institute of Hydrology.

Lhomme, J.P., Chehbouni, A. & Monteny, B. 1994. Effective parameters of surface energy balance in heterogeneous landscape. *Boundary-layer meteorology* 71: 297–309.

Li, B. & Avissar, R. 1994. The impact of spatial variability of land-surface characteristics on land-surface heat fluxes. *J of Climate*. 7: 527–537.

McAneney, K.J., Green, A.E. & Astill, M.S. 1995. Large aperture scintillometry: The homogenous case. *Agricultural and Forest Meteorology* 76: 149–162.

McNaughton, K.G. 1994. Effective stomatal and boundary-layer resistances of heterogeneous surfaces. *Plant, Cell and Environment* 7: 1061–1068.

Meijninger, W.M.L., Hartogensis, O.K. Kohsiek, W. Hoedjes, J.C.B. Zuurbier, R.M. & De Bruin, H.A.R. 2002a. Determination of area-averaged sensible heat fluxes with a large aperture scintillometer over a heterogeneous surface – Flevoland field experiment. *Boundary-layer meteorology* 105: 37–62.

Meijninger, W.M.L., Green, A.E., Hartogensis, O.K., Kohsiek, W., Hoedjes, J.C.B., Zuurbier & De Bruin, H.A.R. 2002b. Determination of area-averaged water vapour fluxes with a large aperture and radio wave scintillometers over a heterogeneous surface-flevoland field experiment. *Boundary-layer meteorology* 105: 63–83.

Moene, A.F. 2003. Effects of water vapour on the structure parameter of the refractive index for near-infrared radiation. *Boundary-layer meteorology* 107: 635–653.

Noilhan, J., Lacarrere, P., Dolman, A.J. & Blyth, E.M. 1997. Defining area-average parameters in meteorological models for land surfaces with mesoscale heterogeneity. *Journal of Hydrology* 190: 302–316.

Noilhan, J. & Lacarrere, L. 1995. GCM grid scale evaporation from mesoscale modelling: a method based on parameter aggregation tested for clear days of Hapex-Mobilhy. *Journal of Climate* 8: 206–223.

Ochs, G.R. & Wilson, J.J. 1993. A second-generation large-aperture scintillometer. *NOAA Tech. Memo, ERL WPL-232, NOAA Environmental Research Laboratories, Boulder, Co. Publ*. 177: 117–132.

Raupach, M.R. & Finnigan, J.J. 1995. Scale issues in boundary-layer meteorology: surface energy balances in heterogeneous terrain. *Hydrological Processes*. 9: 589–612.

Raupach, M.R. 1991. Vegetation-atmosphere interaction in homogeneous and heterogeneous terrain: some implications of mixed-layer dynamics. *Vegetatio* 91: 105–120.

Ronda, R.J. & de Bruin, H.A.R. 1999. A note on the concept of 'effective' bulk exchange coefficients for determination of surface flux densities. *Boundary-layer meteorology* 93: 155–162.

Sellers, P.J., Heiser, M.D., Hall, F.G., Verma, S.B., Desjardins, R.L., Schuepp, P.M. & MacPherson, J.I. 1997. The impact of using area-averaged land surface properties—topography, vegetation condition, soil wetness—in calculations of intermediate scale (approximately 10 km2) surface-atmosphere heat and moisture fluxes. *Journal of Hydrology* 190: 269–301.

Shuttleworth, W.J., Yang, Z.-L. & Arain, M.A. 1997. Aggregation rules for surface parameters in global models. *Hydrology and Earth System Sciences*. 2: 217–226.

Shuttleworth, W.J. 1991. The modellion concept. *Reviews of Geophysics*. 29: 585–606.

Shuttleworth, W.J. 1988. Macrohydrology- the new challenge for process hydrology. *Journal of Hydrology* 100: 31–56.

Twine, T.E., Kustas, W.P., Norman, J.M., Cook, D.R., Houser, P.R., Meyers, T.P., Prueger, J.H., Starks, P.J. & Wesley, M.L. 2000. Correcting eddy-covariance flux underestimates over a grassland. *Agricultural and Forest Meteorology* 103: 279–300.

Watts, C.J., Chehbouni, A., Rodriguez, J.-C., Kerr, Y.H., Hartogensis, O. & de Bruin, H.A.R. 2000. Comparison of sensible heat flux estimates using AVHRR with scintillometer measurements over semi-arid grassland in northwest Mexico. *Agricultural and Forest Meteorology* 105: 81–89.

Wesely, M.L. 1976. The combined effect of temperature and humidity fluctuations on refractive index. *J. Applied Meteorology* 15: 43–49.

Wieringa, J. 1986. Roughness dependent geographical interpolation of surface wind speed averages. *Quarterly Journal of the Royal Meteorological Society* 112: 867–889.

Wyngaard, J.C. 1973. On surface-layer turbulence. *Workshop on Micrometeorology, Denver, Colorado. American Meteorological Society* 101–149.

Zhang, X., Friedl, M.A., Schaaf, C.B., Strahler, A.H., Hodges, J.C.F., Gao, F., Reed, B.C. & Huete, A. 2003. Monitoring vegetation phenology using MODIS. *Remote Sensing of Environment* 84: 471–475.

Part 2
The global perspective: strategies for large area mapping

Recent Advances in Remote Sensing and Geoinformation Processing
for Land Degradation Assessment – Röder & Hill (eds)
© 2009 Taylor & Francis Group, London, ISBN 978-0-415-39769-8

Potential of long time series of FAPAR products for assessing and monitoring land surface changes: Examples in Europe and the Sahel

N. Gobron, M.M. Verstraete, B. Pinty, M. Taberner & O. Aussedat
European Commission—Joint Research Centre, Institute for Environment and Sustainability Global Environment Monitoring Unit, Ispra (VA), Italy

ABSTRACT: Earth observation systems provide high-quality tools to review the state of terrestrial surfaces at the global scale over long periods. In the special context of monitoring land degradation and desertification, long time series of remote sensing products are needed to evaluate the changes in terrestrial surfaces. As an example, plant photosynthesis in terrestrial environments can be documented from spectral measurements made in space. Advances in the understanding of radiation transfer, and the availability of high performance instruments, have led to the development of a new generation of geophysical products providing reliable, accurate information on the state and evolution of terrestrial environments. Specifically, a series of optimized algorithms has been developed and used to estimate the Fraction of Absorbed Photosynthetically Active Radiation (FAPAR) for a suite of recent instruments. This paper summarizes the methodology and performance of these FAPAR algorithms and presents various examples of applications showing an analysis of seasonal cycles and maps of vegetation activity anomalies in Europe and the Sahel.

1 INTRODUCTION

The United Nations (UN) Conference on Desertification (UNCOD) in Nairobi in 1977 was held, in part, as a response to the severe environmental degradation that affected the Sudano-Sahelian region, itself the combined result of a prolonged drought and of extensive over-exploitation of that fragile environment throughout the 1960's. These events have been largely reported upon in the literature (e.g., United Nations Conference on Desertification 1977). At this conference a 'Plan of Action to Combat Desertification (PACD)' was adopted and was intended to propose a comprehensive set of practices to reduce the impact of the harmful processes and even reverse the degradation by rehabilitating affected lands. Lack of progress along these lines led the UN Conference on Environment and Development (UNCED), held in Rio de Janeiro, Brazil, in 1992, to stimulate a renewed interest and encourage greater international commitment towards this issue. This initiative culminated in the form of a Convention to Combat Desertification, which entered into force in 1996. The true scope and extent of desertification worldwide has been questioned (e.g., Thomas and Middleton 1994) opposing repeated claims of progressive and persistent devastation. Nevertheless, even using a somewhat restrictive definition of the process, many countries suffer from its consequences. Detecting and monitoring the causes and consequences of the relevant degradation processes remains an overarching goal in all projects dealing with land degradation and desertification being required before attempting to address the problem. The diverse environmental degradation processes labeled today as 'desertification' have been experienced throughout the history of humanity and known under a variety of names, including: drought; overgrazing; wind and water erosion; and salinization (e.g., United Nations Conference on Desertification 1977, Verstraete 1983, Verstraete and Dickinson 1986). In fact, significant efforts have been expended for long periods of time on the definition and generation of desertification indicators based on field or remote observations (e.g., Reining 1978, Tucker 1991, Pinty 1996, Anyamba and Tucker 2005). Reliable information on the causes and consequences of these processes is deemed essential when designing a rational plan to address the problem as well as to establish the effectiveness of measures and actions undertaken. Space-based remote sensing approaches have long been suggested as suitable solutions to the practical problem of repetitively surveying very large and often remote areas. In this context, the presence and productivity of vegetation plays a critical role, because: (1) a vigorous plant cover constitutes a common indicator of environmental health; (2) its removal or destruction (e.g., by grazing or through the extraction of fuel wood) constitutes a critical precursor for soil erosion by wind and water; and (3) the primary productivity of the vegetation essentially determines the carrying capacity of the land as far as sustainable cattle and human densities are concerned. For these reasons it is very useful to monitor the actual productivity of the plant cover in space and time. Clearly, the relevance and accuracy of the derived information hinges critically on the performance and reliability of the algorithms used to extract this information from the raw observations (e.g., Pinty et al. 1996). Earth observation systems have proved to be high-quality tools to review the state of terrestrial surfaces at the global scale (Eswaran et al. 2001). In the special context of monitoring land degradation and desertification, long time series of remote sensing products are needed to evaluate the changes in terrestrial surfaces. The time scale for studying modifications of landscapes may be longer than the typical sensor lifetime itself, thus continuity and synergy between the products derived from various sensors are essential. Long time series of biophysical indicators generated from different instruments are particularly interesting compared to other proxy records such as empirical spectral vegetation indices, which may depend on the spectral characteristics of instrument, the state of atmospheric conditions or the illumination and viewing geometries. To produce a long time series of relevant products a series of optimized algorithms has been developed to estimate the Fraction of Absorbed Photosynthetically Active Radiation (FAPAR) for various optical instruments (Gobron et al. 2000). This biophysical indicator has been chosen because it reveals the level of vegetation photosynthetic activity, which signifies the amount, state and health of vegetation canopies. This value is derived from the closure of the energy balance equation inside the canopy and corresponds to the normalized radiant flux in the Photosynthetic Active Radiation (PAR) region of the solar domain. The concept underpinning these algorithms for deriving optimized vegetation biophysical indicators was proposed by Verstraete and Pinty (1996) and then applied to multiple sensors including, amongst others, the Medium Resolution Imaging Spectrometer (MERIS) (Govaerts et al. 1999, Gobron et al. 1999), the Sea-viewing Wide Field-of-view Sensor (SeaWiFS) (Gobron et al. 2000) and the MODerate resolution Imaging Spectroradiometer (MODIS) (Gobron et al. 2006).

The MERIS FAPAR products have been produced operationally since the launch of the Envisat platform in March 2002 (see, for example[1]). The SeaWiFS FAPAR products are generated at the JRC thanks to a dedicated processing system that was developed to deliver daily, 10-day, and monthly composite products at the global scale with spatial resolutions ranging from about 2 km up to 0.5 degrees (Mélin et al. 2002). Seven years of SeaWiFS FAPAR products are available from September 1997 (beginning of the mission) to December 2004. Such a multi-annual global time series is quite suitable for conducting investigations aimed at assessing the performance of this category of FAPAR products over several geographical regions and thus for monitoring land degradation and desertification (Gobron et al. 2004, Knorr et al. 2005). The following section summarizes the methodology of optimizing the FAPAR algorithm and evaluates the performances of the FAPAR produced at the European Commission (EC) Joint Research Center (JRC). In this evaluation, SeaWiFS products are first analysed through a comparison with ground estimations over two semi-arid vegetation types. Seven years are then examined in time series of FAPAR over Spain, France and the Sahel through anomalies and seasonal cycles to illustrate and demonstrate the potential of geophysical products in monitoring land degradation and desertification.

---

[1] http://envisat.esa.int/

2 JRC-FAPAR PRODUCTS

The generic JRC algorithm produces a biophysical indicator from spectral data acquired in space without any a priori knowledge on the land cover. This indicator can be used to assess the state and health of vegetation surfaces. The algorithm is physically-based and has been developed for different optical sensors. This section summarizes the methodology used to derive the FAPAR values from the optical sensor data and describes the resulting JRC-FAPAR products. In fact, the FAPAR algorithm can be adapted to any sensor acquiring data separately in the blue, red and near-infrared regions of the solar spectrum. This algorithm exploits the physics of remote sensing measurements and its development accounts for the many operational constraints associated with the systematic processing and analysis of a large amount of data.

Basically, the desired information on the presence and state of vegetation is derived from the red and the near-infrared spectral band measurements. The information contained in the blue spectral band, which is very sensitive to the aerosol load, is used to account for the atmospheric effects on the red and near-infrared measurements. The original approach thus consists in analyzing the relationships between measurements in the blue spectral band and those available in the red and near-infrared regions (Govaerts et al. 1999, Gobron et al. 1999). Such relationships can be simulated for a variety of environmental conditions using radiation transfer models of the coupled vegetation-atmosphere system. The relationships can then be exploited with polynomial expressions optimized in such a way that top of atmosphere Bidirectional Reflectance Factor (BRF) measurements in the blue are related to those taken at other spectral bands at longer wavelengths, i.e., in the red and near-infrared regions. This approach aims at decontaminating the BRF from atmospheric effects without performing an explicit retrieval of the ambient atmospheric properties. The polynomial expressions are also designed to account for the bulk of the anisotropy effects simultaneously. The anisotropy effects are, themselves, approximated from an extensive set of radiation transfer simulations of the coupled surface-atmosphere system designed to mimic typical vegetation canopy conditions (Gobron et al. 2000). This same training data set is then used to relate the radiative measurements from each typical vegetation canopy condition to their corresponding FAPAR values.

In practice, the generic FAPAR algorithm is implemented in a two-step procedure where the spectral BRFs measured in the red and near-infrared bands are, first, rectified in order to ensure their optimal decontamination from atmospheric and angular effects and, second, combined together to estimate the instantaneous FAPAR value. The most recent versions of the appropriate formulae and coefficients derived from the mathematical optimization are given in Gobron et al. (2002) for SeaWiFS and Gobron et al. (2007) for MERIS, respectively. The time composite algorithm of Pinty et al. (2002) has been applied to sequences of daily FAPAR products to provide representative values for longer time periods. Over these selected pixels, time composite algorithms can be applied to eliminate outliers and to limit the impact of uncertainties inherent in the algorithm, remaining biases due to changing Sun and view geometries or unforeseen atmospheric conditions, intermittent presence of sub-pixel clouds, or any other undesired events such as occasional water or snow during the compositing period.

3 PERFORMANCE

The validation of geophysical products derived from remote sensing data, such as FAPAR, includes evaluating whether the quality of these products is in conformity with the needs of the users and intended applications. Space agencies have, therefore, developed their own calibration and validation projects on both atmospheric and land surfaces products (see, for example[2,3] and[4]). In the

---

[2] http://envisat.esa.int/
[3] http://www-misr.jpl.nasa.gov/
[4] http://landval.gsfc.nasa.gov/MODIS/

specific cases of the MODerate resolution Imaging Spectroradiometer (MODIS) and the Multi-angle Imaging SpectroRadiometer (MISR) sensors operating on-board the NASA Terra platform, significant efforts have been devoted to the validation of surface products such as the Leaf Area Index (LAI) and FAPAR generated from data acquired by these sensors (Huemmrich et al. 2005, Wang et al. 2004, Shabanov et al. 2003).

The scientific strategy adopted when designing the MODIS and MISR LAI/FAPAR algorithms is such that the LAI product values need to be carefully evaluated since it impacts the results of the FAPAR algorithm (Knyazikhin et al. 1998). Both LAI, a state variable of the radiation transfer problem, and FAPAR, a normalized radiant flux in the visible region of the solar domain, products correspond to physical quantities that can be measured in the field with varying levels of difficulty. Some of the issues associated with the generation of accurate ground-based estimations of FAPAR are discussed in Gobron et al. (2006).

The two panels of Figure 1 provide comparisons between the SeaWiFS FAPAR products and the ground-based estimations available from two experimental sites: Dahra [15° 24' N; 15° 26' W]

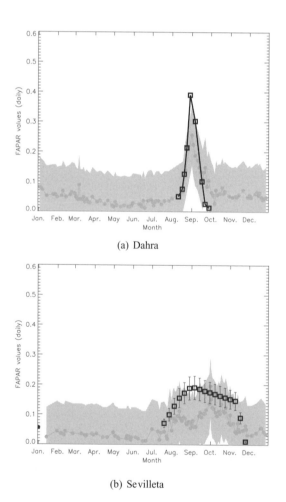

(a) Dahra

(b) Sevilleta

Figure 1. Comparisons of ground-based FAPAR profiles (empty square symbols) and instantaneous daily SeaWiFS FAPAR JRC products (full circle symbols) over the sites of (a) Dahra and (b) Sevilleta. The zone shaded in grey delineates the ±0.1 uncertainty range around the FAPAR JRC products. The vertical bars indicate the uncertainty range, when available, around the ground-based estimations.

in Senegal; and Sevilleta [34° 2′ N; 106° 42′ W] in the USA. The vegetation types at these sites include semi-arid grass savannah (Dahra) and desert grassland (Sevilleta) (Fensholt et al. 2004, Turner et al. 2005 and Gobron et al. 2006). The baseline FAPAR value for these two sites is close to zero and the signature of the different vegetation phenological cycles, both for the growing and decaying periods, are well identified by both remote sensing and ground-based estimations. Furthermore, the amplitudes, both maxima and minima, are in agreement with each other.

4 APPLICATIONS

4.1 *Drought in Europe*

Land degradation problems surrounded the Mediterranean basin, with, in Europe, Spain being the most severely affected. Extensive studies, both at the national and European level, have documented the severity and extent of the impact of these processes on the environment and the economy (Mediterranean Desertification and Land Use, MEDALUS[5] project financed by European Commission DG-Research) at the local and regional (sub-national) scales. According to a report published by the European Environmental Agency, EEA (European Environment Agency 2004), Spain and Portugal are the European countries that have been most affected by climate change in the last few years. Average temperatures in Europe have increased approximately 1° C more than the average in the rest of the world, and the warming in Spain and Portugal has been even higher. The report warns of: increased drought in southern Spain; more frequent forest fires; widespread heat waves; and associated risks for human health. The impact of drought on the state and health of land surfaces is illustrated in Figure 2 which shows anomalies between the monthly FAPAR products for Spain in April 1999 and France in August 2003 with the corresponding 6-year average from 1998–2003 (top panels). The regional pattern is clearly visible, with deficits in FAPAR values represented in warm colors, excesses in cold and normal conditions in grey. The chart (bottom panel) illustrates the seasonal cycle of monthly values of nationally averaged FAPAR for 1999 and 2003. Clearly, differences between 1999 and 2003 were less important at the national scale than locally: Spain suffered much more at the beginning of 1999 and France in summer 2003, as was shown in Gobron et al. (2005).

Figure 3 illustrates the impact of a fire event that occurred in August 2000 in Spain. The top panel shows Corine Land Cover (CLC)[6] map for northern Spain derived using data acquired in 2000. The seasonal cycle of FAPAR was extracted over one of the burned areas located at [2.65° W, 41.545° N] and the values plotted as function of time from 1999 to 2001 in the bottom panel. It can be seen that the photosynthetic activity decreased in September 2000. The low FAPAR values following this fire event show that vegetation photosynthesis did not return to the earlier level, at least until the end of 2004 (not shown here). This example and the one over France for 2003 (Gobron et al. 2005) show that vegetation stress can be assessed on the basis of long times series of biophysical indicators derived from space.

4.2 *Interannual change in the Sahel region*

During the last four decades or more, the Sahel has become the paradigm for regions subject to drought and desertification. This region has been the subject of numerous studies and intense discussions are still going on concerning the causes of the observed changes (Hein and de Riccer 2006). In this paper, FAPAR anomalies with respect to the long-term mean will be examined in order to detect possible land degradation. The time series covers January 1998 to December 2004 over the geographical region defined by [10° N, 20° N, 20° W, 10° E] (top panel, Figure 4).

---

[5] http://www.medalus.demon.co.uk/
[6] http://terrestrial.eionet.eu.int/CLC2000

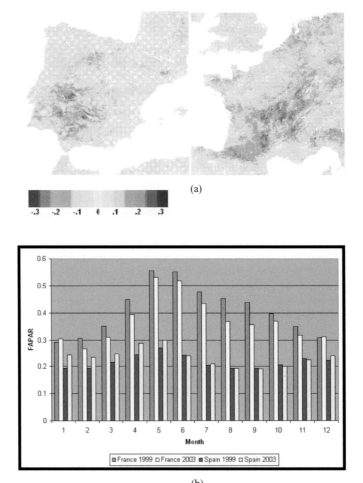

(a)

(b)

Figure 2.   (a) FAPAR anomalies over Spain, April 1999 (left), and France, August 2003 (right). (b) The seasonal cycle of nationally averaged FAPAR for France and Spain for 1999 and 2003 (see colour plate page 376).

The bottom two panels on Figure 4 show the time series of FAPAR (b) and the anomalies (c) spatially averaged over the entire region from January to December. The different colours indicate the year. The strong seasonal cycle is clearly visible and the high inter-annual variability is also noticeable. Two particular years may be used to underscore the changing environment: in 2000, the year started with the largest values of FAPAR of the period 1998 to 2004, but the positive anomaly decreased and became slightly negative by the end of the year. By contrast, 2003 started as the worst year within this record and maintained strongly negative deviations until July, when a large positive deviation was registered. The rest of the year was close to the expected average over the available record.

Figure 5 shows FAPAR time profiles since January 1998 averaged as a function of latitude and longitude. The top panel shows the high seasonality of the response of plants to climatic constraints in the region, as well as the well-known latitudinal dependency of the growing period. FAPAR was generally lower at all latitudes in 2002. The bottom panel shows a similar space-time diagram, but this time the FAPAR has been averaged along latitudinal bands. It can be seen that the highest FAPAR values are found between longitudes $-15°$ to $-5°$, and that FAPAR was lowest in year 2002 at all longitudes.

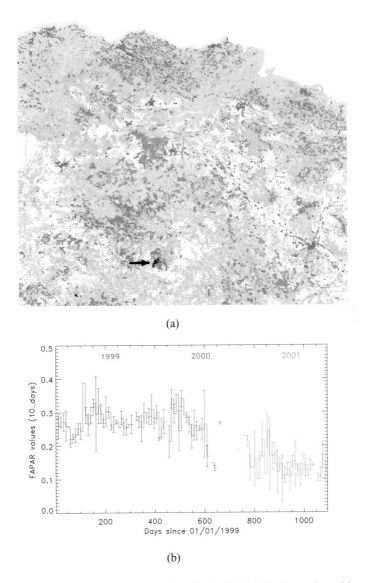

(a)

(b)

Figure 3. (a) Corine Land Cover 2000 over northern Spain. The full land cover legend is available from the source web site. In general, yellow, green and red colours stand for agricultural, forest and urban areas, respectively. (b) The seasonal cycle of monthly FAPAR products over a 'burnt area' (black spot on the map), located at [2.65° W, 41.545° N], is shown from January 1999 to December 2001. The FAPAR data are from the JRC-FAPAR product (see colour plate page 377).

Figure 6 exhibits the corresponding FAPAR anomalies with respect to the average over the entire record. The deficits in FAPAR values are reported in per cent and are represented in red colors, the excesses in blue, and normal conditions and very low FAPAR values (≤0.05) in grey. The relatively good conditions prevailing in 2000 and the extended drought that affected the region from September 2001 to May 2003 are clearly noticeable.

In Figure 7, the FAPAR monthly anomalies computed over the entire region (red line) are plotted along with the Sahel rainfall index (black line) from (Janowiak 1988) for the same time period (January 1998 to December 2004). The Sahel rainfall index is constructed from a rotated principal component analysis of 14 stations in the regions with extensive data records (typically from 1950

95

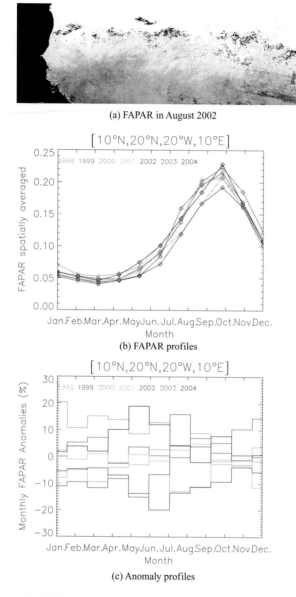

(a) FAPAR in August 2002

(b) FAPAR profiles

(c) Anomaly profiles

Figure 4. (a) Map of FAPAR for the Sahel region in August 2002. The colour scale is the same as for Figure 5. Black pixels correspond to missing values, either due to instrumental issues or the persistent presence of clouds. Time series of FAPAR (b) and FAPAR anomalies (c) averaged over the Sahel region for 1998 to 2004 (see colour plate page 378).

onwards). Precipitations are notoriously patchy in space and intermittent in time, especially in marginal regions such as the Sahel. Individual station records may not, therefore, be representative of a large area. Nevertheless, this index attempts to summarize the general trend (see[7] for the original data and more information on the index). Visual inspection of this diagram shows some correspondence between the two signals, indicating that FAPAR is somewhat responsive to precipitation,

[7] http://www.jisao.washington.edu/data_sets/sahel/

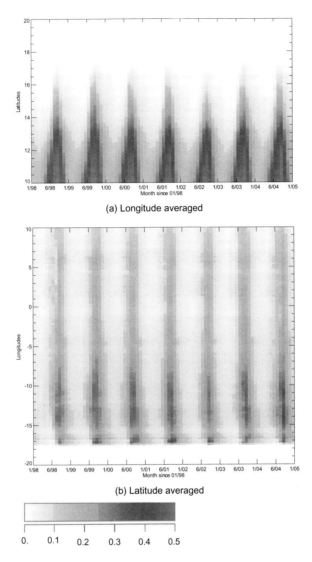

Figure 5. Time evolution of FAPAR with respect to latitude from 10° N to 20° N (a) and longitude from 20° W to 10° E (b) for the period 01/1998 to 12/2004 (see colour plate page 379).

but it is known that there are likely many other factors that affect the photosynthetic activity of plants. A more detailed analysis would require much better meteorological data than is currently available in the region. Nevertheless, the data convincingly show the strong seasonality and inter-annual variation of the vegetation cover. Not only, therefore, can this be detected and quantified from space but also this may prove useful to document, a posteriori, the net spatial and temporal response of vegetation to the combined effect of all stresses, whether climatic or otherwise.

5 SUMMARY AND CONCLUSIONS

In this paper the methodology for deriving FAPAR values from different optical sensors was out-lined. It was shown, using a SeaWiFS example, that a long time series of coherent products could

97

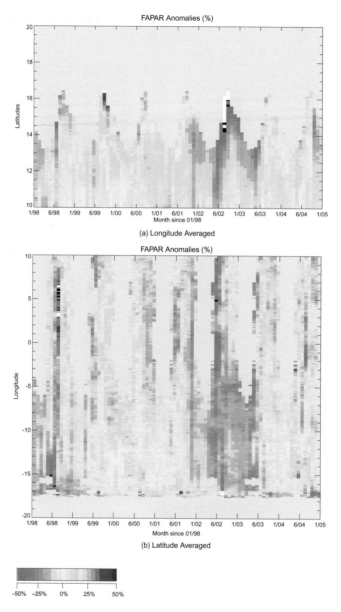

Figure 6.   Time evolution of FAPAR anomalies with respect to the average value during the record period, displayed as a function of latitude from 10° N to 20° N (a) and longitude from 20° W to 10° E (b) for the period from 01/1998 to 12/2004 (see colour plate page 380).

be produced. The results of two validation studies comparing directely space-derived products and ground-based estimations over semi-arid vegetation sites were presented.

The FAPAR product generated at the JRC (and available for the period from September 1997 up to December 2004 at the time of writing this contribution), was used to characterize the spatial and temporal evolution of plant productivity over (1) France and Spain, where drought and fire events were clearly detectable, and (2) a large area of the Sahel where both a strong seasonal cycle and substantial inter-annual variations in productivity were largely, but not uniquely, related to precipitation patterns.

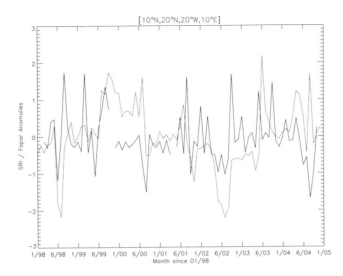

Figure 7. FAPAR anomalies (red line) and Sahel rainfall index (black line) over the Sahel region from 01/1998 until 12/2004 (see colour plate page 381).

In the case of the Spanish site the drastic change in the vegetation cover, lasting for an extended period, following a fire, was shown. This type of event, and its prolonged consequences, implies that land cover maps such as the CLC must be regularly updated, and that remote sensing biophysical products are useful for this purpose.

The Sahelian example suggests that the technique can be used to monitor large scale and long term evolution of substantial regions: clearly, the average variability for large regions is smaller than that for individual sites, but when small variations do occur in these aggregated statistics, they imply severe and widespread consequences for many individual locations.

Since land degradation issues occur over period spanning multiple decades, it is relevant to extend the availability of these FAPAR products both in the future and in the past. MERIS FAPAR products are available since June 2002 and work is on going to develop compatible algorithms for the Sentinel missions of European Space Agency (ESA). Separately, a complete reprocessing of Advanced Very High Resolution Radiometer (AVHRR) would be appropriate to extend the period of coverage in the past.

ACKNOWLEDGMENTS

The authors are grateful to the SeaWiFS Project (Code 970.2) and the Distributed Active Archive Center (Code 902) at the Goddard Space Flight Center, Greenbelt, MD 20771, for the production and distribution of the SeaWiFS data, respectively. Frédéric Mélin (EC-JRC) kindly performed the processing of original SeaWiFS data and Monica Robustelli (EC-JRC) developed the JRC-FAPAR data base from which analyses have been performed.

REFERENCES

Anyamba, A. and C.J. Tucker (2005). Analysis of Sahelian vegetation dynamics using NOAA AVHRR NDVI data from 1981–2003. *Journal of Arid Environments 63*, 596–614.
Eswaran, H., R. Lal, and P. Reich (2001). Land degradation: An overview. In *Proceedings of 2nd. International Conference on Land Degradation and Desertification, Khon Kaen, Thailand. Oxford Press, New Delhi, India*. Bridges, E.M., I.D. Hannam, L.R. Oldeman, F.W.T. Pening de Vries, S.J. Scherr, and S. Sompatpanit.

European Environment Agengy (2004). Impacts of Europe's changing climate, An indicator-based assessment. Report No. 2/2004, EEA.

Fensholt, R., I. Sandholt, and M.S. Rasmussen (2004). Evaluation of MODIS LAI, fAPAR and the relation between fAPAR and NDVI in a semi-arid environment using in situ measurements. *Remote Sensing of Environment 91*, 490–507.

Gobron, N., O. Aussedat, and P. Bernard (2006). MODerate Resolution Imaging Spectroradiometer, JRC-FAPAR Algorithm Theoretical Basis Document. EUR Report No. 22164 EN, Institute for Environment and Sustainability.

Gobron, N., F. Mélin, B. Pinty, M.M. Verstraete, J.-L. Widlowski, and G. Bucini (2001). A global vegetation index for SeaWiFS: Design and applications. In M. Beniston and M.M. Verstraete (Eds.), *Remote Sensing and Climate Modeling: Synergies and Limitations*, pp. 5–21. Dordrecht, The Netherlands: Kluwer Academic Publishers.

Gobron, N., B. Pinty, O. Aussedat, J.M. Chen, W.B. Cohen, R. Fensholt, V. Gond, K.F. Huemmrich, T. Lavergne, F. Mélin, J.L. Privette, I. Sandholt, M. Taberner, D.P. Turner, M. Verstraete, and J.-L. Widlowski (2006). Evaluation of FAPAR Products for Different Canopy Radiation Transfer Regimes: Methodology and Results using JRC Products Derived from SeaWiFS against ground-based estimations. *Journal of Geophysical Research 10.1029/2005JD006511*.

Gobron, N., B. Pinty, F. Mélin, M. Taberner, M. Verstraete, M. Robustelli, and J.-L. Widlowski (2007). Evaluation of the MERIS/ENVISAT FAPAR product. *Advances in Space Research 39*, 105–115.

Gobron, N., B. Pinty, F. Mélin, M. Taberner, and M.M. Verstraete (2002). Sea-viewing Wide Field-of-View Sensor (SeaWiFS) - Level 2 land surface products - Algorithm Theoretical Basis Document. EUR Report No. 20144 EN, Institute for Environment and Sustainability.

Gobron, N., B. Pinty, F. Mélin, M. Taberner, M.M. Verstraete, A. Belward, T. Lavergne, and J.-L. Widlowski (2005). The state of vegetation in Europe following the 2003 drought. *International Journal Remote Sensing Letters 26*, 2013–2020.

Gobron, N., B. Pinty, M.M. Verstraete, and Y. Govaerts (1999). The MERIS Global Vegetation Index (MGVI): Description and preliminary application. *International Journal of Remote Sensing 20*, 1917–1927.

Gobron, N., B. Pinty, M.M. Verstraete, and J.-L. Widlowski (2000). Advanced spectral algorithm and new vegetation indices optimized for up coming sensors: Development, accuracy and applications. *IEEE Transactions on Geoscience and Remote Sensing 38*, 2489–2505.

Govaerts, Y., M.M. Verstraete, B. Pinty, and N. Gobron (1999). Designing optimal spectral indices: A feasibility and proof of concept study. *International Journal of Remote Sensing 20*, 1853–1873.

Hein, L. and N. de Riccer (2006). Desertification in the sahel: a reinterpretation. *Global Change Biology 12*, 751–758.

Huemmrich, K.F., J.L. Privette, M. Mukelabai, R.B. Myneni, and Y. Knyazikhin (2005). Time-series validation of MODIS land biophysical products in a Kalahari woodland, Africa. *International Journal of Remote Sensing 26*, 4381–4398.

Janowiak, J.E. (1988). An investigation of interannual rainfall variability in Africa. *Journal of Climate 1*, 240–255.

Knorr, W., N. Gobron, M. Scholze, T. Kaminski, and B. Pinty (2005). Global drought conditions causing recent atmospheric carbon dioxide increase. *EOS Transactions, American Geophysical Union 86*, 178–181.

Knyazikhin, Y., J.V. Martonchik, D.J. Diner, R.B. Myneni, M.M. Verstraete, B. Pinty, and N. Gobron (1998). Estimation of vegetation canopy leaf area index and fraction of absorbed photosynthetically active radiation from atmosphere-corrected MISR data. *Journal of Geophysical Research 103*, 32, 239–32, 256.

Mélin, F., C. Steinich, N. Gobron, B. Pinty, and M. Verstraete (2002). Optimal merging of LAC andGAC data available from the SeaWiFS sensor. *International Journal of Remote Sensing 23*, 801–807.

Pinty, B., N. Gobron, F. Mélin, and M.M. Verstraete (2002). A Time Composite Algorithm Theoretical Basis Document. EUR Report No. 20150 EN, Joint Research Centre, Institute for Environment and Sustainability.

Pinty, B., M.M. Verstraete, J. Iaquinta, and N. Gobron (1996). Advanced modeling and inversion techniques for the quantitative characterization of desertification. In *The Use of Remote Sensing for Land Degradation and Desertification Monitoring in the Mediterranean Basin: State of the Art and Future Research*, pp. 79–93. European Commission.

Reining, P. (1978). *Handbook of Desertification Indicators*. Washington DC, USA: American Association for the Advancement of Science.

Shabanov, N.V., Y. Wang, W. Buermann, J. Dong, S. Hoffman, G.R. Smith, Y. Tian, Y. Knyazikhin, and R.B. Myneni (2003). Effect of foliage spatial heterogeneity in the MODIS LAI and FPAR algorithm over broadleaf forests. *Remote Sensing of Environment 85*, 410–423.

Thomas, D. and N. Middleton (1994). *Desertification: Exploding the Myth*. Chichester: John Wiley & Sons.

Tucker, C.J., H.E. Dregne, and W. Newcomb (1991). Expansion and contraction of the sahara desert from 1980–1990. *Science 253*, 299–301.

Turner, D.P., W.D. Ritts, W.B. Cohen, T. Maeirsperger, S.T. Gower, A. Kirschbaum, S.W. Running, M. Zhao, S. Wofsy, B. Dunn, A. Law, J. Campbell, W. Oechel, H.J. Kwon, T. Meyers, E. Small, S. Kurc, and J. Gamon (2005). Site-level evaluation of satellite-based global terrestrial gross primary production and net primary production monitoring. *Global Change Biology 11*, 666–684.

United Nations Conference on Desertification (1977). *Desertification: Its Causes and Consequences*. Oxford: Pergamon Press.

Verstraete, M.M. (1983). Another look at the concept of desertification. In S.G. Wells and D.R. Haragan (Eds.), *Origin and Evolution of Deserts*, pp. 213–228. University of New Mexico Press.

Verstraete, M.M. and R.E. Dickinson (1986). Modeling surface processes in atmospheric general circulation models. *Annales Geophysicae 4*, 357–364.

Verstraete, M.M. and B. Pinty (1996). Designing optimal spectral indices for remote sensing applications. *IEEE Transactions on Geoscience and Remote Sensing 34*, 1254–1265.

Wang, Y., C.E. Woodcock, W. Buermann, P. Stenberg, P. Voipio, H. Smolander, T. Hame, Y. Tian, J. Hu, Y. Knyazikhin, and R.B. Myneni (2004). Evaluation of the MODIS LAI algorithm at a coniferous forest site in Finland. *Remote Sensing of Environment 91*, 114–127.

Recent Advances in Remote Sensing and Geoinformation Processing
for Land Degradation Assessment – Röder & Hill (eds)
© 2009 Taylor & Francis Group, London, ISBN 978-0-415-39769-8

Inter-comparison of MEDOKADS and NOAA/NASA pathfinder AVHRR land NDVI time series

K. Friedrich & D. Koslowsky

Institut für Meteorologie, Freie Universität Berlin, Berlin, Germany

ABSTRACT: The Mediterranean Extended Daily One-km AVHRR Data Set (MEDOKADS) has been generated at the Free University of Berlin (FUB) since 1989, continues in near real-time and already covers a climatological relevant period. At a spatial resolution of about 1 km it extends over Central Europe and the Mediterranean basin. Calibrated reflectances and brightness temperatures in different spectral regions are included as well as derived quantities like albedo, vegetation index, land and sea surface temperatures, cloud cover and classification. National Oceanic and Atmospheric Administration (NOAA) and National Aeronautics and Space Administration (NASA) have released the Pathfinder AVHRR Land (PAL) as long-term data set that has been processed in a consistent manner at a spatial resolution of 8 km covering a period of more than 20 years for global change research. 10 days and monthly maximum NDVI composites have been generated from daily Global Area Coverage (GAC) data. Both data sets represent the primary remote sensing data input to the DeSurvey (A Surveillance System for Assessment and Monitoring of Desertification) 6th framework EU research project in order to detect and analyze hot spots of desertification risks as well as trends in vegetation cover in the Mediterranean basin.

1 INTRODUCTION

The Advanced Very High Resolution Radiometer (AVHRR) flown on board the NOAA satellites originally was designed for use in weather analysis and forecasting. At the time of its first operation on the NOAA 6 satellite in 1978 only a limited number of stations were capable to process and store the huge amount of data. In the meantime new storage media and high speed computer facilities became available at low price and enabled a larger number of institutions to handle and process AVHRR data. Simultaneously the successful compromise of temporal and spatial resolution, processing and storage requirements extended the application towards land surface investigations.

More than 25 years of operation without major changes now offer the possibility to use the data for studies of environmental changes. On the other hand this opportunity causes additional efforts to pre-processing, validation and harmonisation of these long-term data sets. Especially the detection of non-periodic changes of the earth's surface requires the removal of all possible contributions, of changing observation conditions and changes due to the instrument's characteristics. For users who are not interested in these processing steps, but want to retain as much information as possible information, an intermediate product has been compiled at the Free University of Berlin (FUB): the Mediterranean Extended Daily One-Km AVHRR Data Sets (MEDOKADS) (Koslowsky 2003).

Discussions are still going on whether AVHRR based time series are suited to give reliable results in near real-time monitoring and trend analysis of desertification risks in the Mediterranean basin. Additional efforts in calibration, normalisation and harmonisation are still required to fulfil this task not only as a prerequisite to contribute to an early warning system in the Mediterranean basin, but also to establish a long-term normalisation standard for inter-comparison with results of new sensors like MODIS, AATSR and SEVIRI. Inter-comparison with the global PAL data set provided by NOAA is ongoing to extrapolate results globally.

2   GENERAL DESCRIPTION OF THE MEDOKADS

2.1   *Research requirements for a remote sensing data set for studies of land surface processes in the Mediterranean basin*

In 1991 the ECHIVAL[1] Field Experiment in Desertification Threatened Areas (EFEDA) started as a multidisciplinary activity to study the full range of water transfer processes from sub-surface transports up into the free atmosphere including observations from space (Bolle et al. 2006). It was continued by a number of follow-on projects like Remote Sensing of the Mediterranean Desertification and Environmental Changes (RESMEDES), Synthesis of Change Detection Parameters Into A Land-surface Change Indicator for Long Term Desertification Studies in the Mediterranean Area (RESYSMED) and Land Degradation Assessment in Mediterranean Europe (LADAMER).

In-situ measurements, observations by air plane and upscaling of data gained by different satellite bond instruments were used to derive methods to calibrate and standardize AVHRR derived products like spectral reflectances, albedo, fluxes of sensible and latent heat and Normalized Difference Vegetation Index (NDVI) (Tucker 1979). While qualitative changes in time and regional gradients could be shown and described easily, the extraction of validated quantitative harmonized time series proved to be a great challenge (Koslowsky 1996, Bolle et al. 2006). Calibration problems (s. 2.3), orbit shift, illumination and observation conditions were found to have strong impacts on the derived products and have till now to be treated as preliminary and research is going on to overcome these limitations in order to make assured statements of environmental changes.

The necessity to establish a long term data set of AVHRR data and products that are easy to exchange and to handle by different disciplines (biology, ecology, geography, hydrology, meteorology) with a broad variety of processing skills was the cause to evoke the MEDOKADS. Special efforts were made to avoid irreversible corrections and processing steps offering the opportunity to apply later on sophisticated calibration and processing algorithms without the need to start on the raw data level.

2.2   *Design of the MEDOKADS*

Interactively supervised fine-navigated sub-areas from directly received and pre-processed High Resolution Picture Transmission (HRPT) data are used to create the MEDOKADS product. This data set consists of full resolution AVHRR channel data and collateral data in "geographic" projection, i.e. latitude-longitude presentation, with a resolution of 0.01° in both directions.

The area of interest for the studies in the Mediterranean is defined by its corner coordinates: upper left corner: 55°N/10°W, lower right corner: 27°N/42.36°E and amounts to 2800 lines by 5236 columns with 14,660,800 raster positions. Each of the four sub-regions marked in Figure 1 is stored separately covering the Iberian Peninsula (S), Italy (I), the Balkans (B), and the Eastern Mediterranean area (E) respectively. The stripes S and I have the same size of 2800 lines and 1536 columns, while B and E consist of 1900 lines by 1536 pixel. Smaller windows regardless to the stripe boundaries can be selected and are completely defined by the geographic coordinates of the upper left corner and the number of lines and columns. These data sets are easy to merge, if larger areas or the whole Mediterranean have to be considered. Areas for which data are not yet available in supervised form are kept blank. The coordinates of the upper left corners of the sub-areas are:

S:	Iberian Peninsula	55°N/10°W
I:	Italy	55°N/5°E
B:	Balkan Peninsula	46°N/15°E
E:	East Mediterranean	46°N/27°E

---

[1] European International Project on Climatic and Hydrological Interactions between Vegetation, Atmosphere, and Land-surfaces (ECHIVAL).

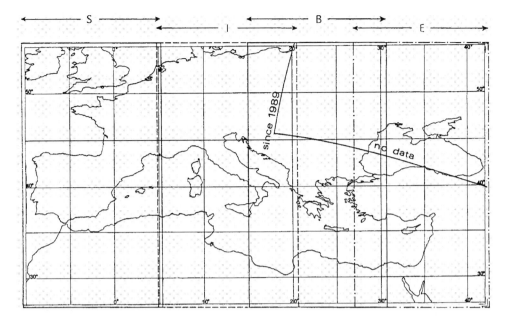

Figure 1.   Area covered by the MEDOKADS. S, I, B and E shows the four subset stripes for the Iberian Peninsula, Italy, Balkan, and Eastern Mediterranean. No data are processed so far for the north-east corner covering Eastern Europe and the northern Black Sea.

There is some overlapping between the stripes to cover the four main land surface sub-areas. To avoid problems of digital data representation, all values are converted to signed 16-bit integers.

The formal digital resolution is 0.01 of units (i.e. for reflectance, degree for temperatures (Celsius) and angles) multiplied by a factor of 100 without any offset. The calibrated data of the five AVHRR channels at top of atmosphere (TOA) form the first five files. Channel 1 and 2 data are corrected for sun height and sun-earth distance. A number of pseudo-channels are added containing collateral data or products coincident pixel by pixel: local satellite and sun zenith distance and azimuth, local scattering angle, broad-band albedo, NDVI, SST/LST, local time and a bitmap. The lower byte of the bitmap channel represents the origin indicator, a 6 bit bitmap inserted into the spare bits of each AVHRR super-pixel at data reception time, which allows to identify the date and the orbit number of each individual pixel, even in the case of multi-day composites like the maximum NDVI. It allows to recover the original digital counts and to apply sophisticated calibration and correction algorithms later on to each individual pixel without a loss in digital resolution. The other byte contains a land/water bit and a cloud mask and allows to switch between different evaluation algorithms for land and water surfaces. Thus the MEDOKADS builds up a 15 channel pixel-congruent data set (Table 1). Daily as well as ten day composite data sets according to the maximum NDVI are generated. As a sample the 16 year mean NDVI (January to December) is shown in Figure 2.

The west to east extent of the area is 52.36° corresponding to about 4400 km. Thus 2 to 3 adjacent satellite passes are necessary to cover the whole region resulting in an absolute time difference up to 31/2 hours between the western and eastern part. The total amount of data for a complete set with 16 pseudo-channels for one day of the whole basin is about 580 Mbyte. Thus up to now the MEDOKADS of more than 16 years from 1989 to 2005 amounts to about 3 Tbyte.

2.3   *Calibration of the short-wave channels and changes in spectral response*

While internal calibration targets can be used to calibrate the thermal channels of the AVHRR instrument (Kidwell, 1991, Goodrum et al. 2000) external targets have to be used to monitor the gain

Table 1.   Channels and pseudo-channels of the MEDOKADS.

No.	Content	No.	Content
1	Calibrated AVHRR CH1	9	Local sun azimuth
2	Calibrated AVHRR CH2	10	Local scattering angle sun to satellite
3	Calibrated AVHRR CH3	11	"Origin indicator" and bitmap
4	Calibrated AVHRR CH4	12	TOA broadband reflectance
5	Calibrated AVHRR CH5	13	NDVI
6	Local satellite zenith distance	14	Sea/land surface temperature
7	Local sun zenith distance	15	local time since ascending node
8	Local satellite azimuth		

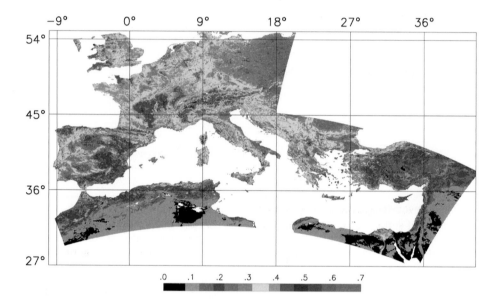

Figure 2.   Sample of a MEDOKADS derived product. Annual mean of the period 1989 to 2004 showing the region covered with data (see colour plate page 381).

of the shortwave channels. Starting with carefully evaluated post-launch calibration coefficients for the NOAA 11 satellite (Rao and Chen 1994) a test site in the Western Great Erg was monitored by FUB to derive degradation coefficients for the AVHRR instrument using time series corrected for BRDF (Bidirectional Reflectance Distribution Function) and inter-calibration of the instruments of the successive satellites NOAA 11, NOAA 14 and NOAA 16 (Koslowsky et al. 2001). This work resulted in somewhat different degradation and calibration coefficients than used for the PAL data set. As an example of different approaches for deriving degradation coefficient for NOAA 14 refer to Figure 3. A broad spread can be found with the curves of FUB in the middle for the early years of operation. The slope of the degradation functions of FUB for the years 1998 to 2000 in comparison to others decreases as in 2000 nearly no degradation could be found. Figure 4 shows a plot of the reflectance of the calibration target without (upper panel) and with inter-calibration for the 16 year period from 1989 to 2004.

Starting with NOAA 15 the AVHRR/3 instrument is flown on the polar orbiting satellites, which shows a sharper separation between the red and near infrared channel resulting in higher NDVI data (Figure 8). The correction functions presented by Trishchenko (2002) have been used to normalize the data series to NOAA 9 standard. But it is felt, that the corrections for the AVHRR/3 instrument are not sufficient for most part of the MEDOKADS area.

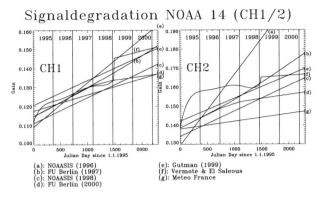

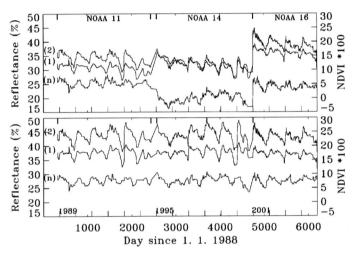

Figure 3.   Signal degradation of channel 1 and 2 of the AVHRR of NOAA 14 by different authors.

Figure 4.   Time series of daily reflectance of AVHRR channel 1 (1), channel 2 (2) and derived NDVI (n) of the satellites NOAA 11 to NOAA 16 of a calibration test site in the Great Erg Occidental in Algeria.

As shown in Figure 7 the NDVI for SPOT-VEGETATION decreases slightly from year 1999 to year 2001 by about 2%, while the NDVI of MEDOKADS of the year 2001 is 14.3% higher than for 1999. Even after normalization with the correction functions presented by (Trishchenko et al. 2002) an increase of the NDVI by 7.7% for the MEDOKADS area is to be detected that has to be attributed to change in spectral characteristics of the AVHRR instrument.

Still existing uncertainties in calibration and incomplete corrections for changes to the AVHRR instrument's characteristics flown on successive satellites that have strong impact to derived NDVI values require special caution to statements to changes of real vegetation as made by Myneni et al. (1997).

3   GENERAL DESCRIPTION OF THE PATHFINDER AVHRR LAND (PAL) DATA SET

The PAL data set has been released by NOAA and NASA as a long-term data set. It has been processed in a consistent manner for global change research as 10 day composite synchronised at the first of each month. This data set is generated from daily data covering now a period of more than 20 years. The PAL is based on the Global Area Coverage (GAC) data. The processor onboard the satellite samples the real-time AVHRR data to produce reduced resolution GAC data. Four out of every five samples along the scan line are used to compute one average value, and the data from only every third scan line are processed. As a result, the spatial resolution of GAC data near the nadir is actually 1.1 km by 4 km with a 3 km gap between pixels across the scan line, although generally treated as 4 km resolution. All of the GAC data computed during a complete pass are recorded onboard the satellite for transmission to the command and data acquisition stations (Kidwell 1991). The 10-bit precision of the AVHRR data is retained. The Re-sampling and averaging processes used to produce GAC data are important for the inter-comparison of PAL and MEDOKADS.

According to the maximum NDVI of periods of 10 day's GAC AVHRR data and all related collateral parameters are re-sampled to 8 km resolution.

4   DIFFERENCES IN PROCESSING OF THE MEDOKADS AND PAL DATA SET

Both data sets are derived from the Advanced Very High Resolution Radiometers (AVHRR) of the "afternoon" NOAA operational meteorological satellites (NOAA7, 9, 11, 14, 16 and 18). Differences in the collected data, navigation process, calibration, atmospheric correction and restriction in compiling NDVI composites leads to differences in the data sets. Table 2 summarizes these differences, which are discussed in detail in the following subchapters.

4.1   *Data Source and Resolution*

The PAL daily data are derived from GAC through averaging and re-sampling of original data onboard the NOAA satellites. PAL has a resolution of 8 km which means that re-sampling and averaging has been applied to get the data in the Interrupted Goode Homolosine global projection. MEDOKADS is remapped to geographical projection with a nominal resolution of 0.01° to retain the full resolution of the AVHRR instrument.

It is well known that a reduction in spatial resolution of a scene by averaging as well as re-sampling diminishes the range of values because at least some of the minima and maxima of the data set vanish. Thus the slope of a linear regression line between original and reduced resolution data will be clearly less than 1 and the histogram of the data gets narrower. This effect is demonstrated in Figure 5 for a sample area in Spain and is representative for the sampling and averaging differences between the PAL and MEDOKADS data sets.

Table 2.   Differences in processing MEDOKADS and PAL data set.

	MEDOKADS	PAL data set
Data source	HRPT	GAC
Resolution	1 km	8 km
Calibration	Updated during the whole operation time	Data from the beginning of each mission
Atmospheric correction	No atmospheric correction	Atmospheric correction only for ozon und rayleigh scattering
Navigation	Orbital model with supervised fine-navigation	Orbital model
Restrictions in observation geometry to compile NDVI Composites	No restrictions	<80° sun-zenith <42° viewing angle

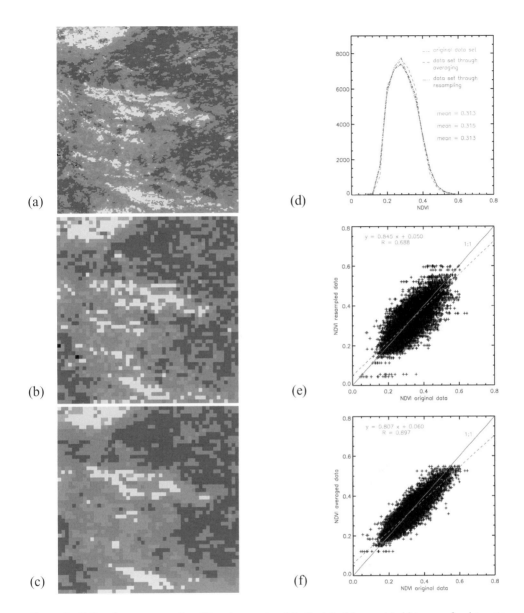

Figure 5. Effect due to re-sampling (b) and averaging (c) of original image (a), histogram for data sets (d), linear regression between original and re-sampled data (e) and averaged data (f) for an NDVI image (see colour plate page 382).

4.2 *Atmospheric correction, Recalibration*

The radiance which is measured at the satellite is attenuated by the atmosphere on its path from sun to earth and earth to satellite. Different processes have to be considered for correcting this atmospheric influence. Radiant energy interacts with the atmosphere by scattering, absorption and emission. The emission for the visible and near infrared channels of AVHRR instrument is negligible. The absorption by ozone affects channel 1 and by water vapour channel 2. Mie and Rayleigh scattering are stronger for channel 1 than for channel 2. Dark surfaces appear brighter when observed from space and bright surfaces darker. There exists a "neutral point" at which the atmosphere has no effect, the atmospheric backscatter just compensates the loss of radiation

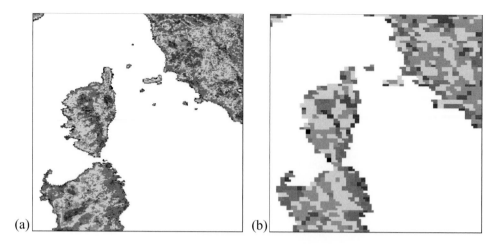

(a)

(b)

Figure 6.   Example for MEDOKADS (a) and PAL (b) data (last decade of month May 1995, region around Corsica) (see colour plate page 383).

from surface. For channel 1 the "neutral point" is nearly at a reflectivity of 0.15 and for channel 2 at 0.06 and varies slightly with the aerosol and water vapour content of the atmosphere (Koslowsky 1996).

A complete correction for atmospheric effects for large areas is difficult because of the high variability of water vapour and aerosol load of the atmosphere in space and time. The PAL data set is only corrected for Rayleigh scattering and ozone absorption. The MEDOKADS is not atmospheric corrected.

For the purpose of comparison of the MEDOKADS and the PAL data set following procedure has been applied to the MEDOKADS: geographical resolution was reduced to 0.04°, recalibration by using coefficients of Rao and Chen (1995; 1999), and Koslowsky et al. (2001) was performed. An atmospheric correction was added, using the SMAC algorithm (Simplified Method for Atmospheric Corrections) which is described in Rahman and Dedieu (1994). This algorithm is a semi-empirical approximation of the radiative transfer in the atmosphere. Surface pressure, ozone and water vapour content as well as aerosol type and distribution are input parameters. If one or more of these parameters are missing the corresponding part of atmospheric correction is omitted.

4.3   *Navigation*

The PAL data set is geo-coded and re-mapped using an orbital model and predicted orbital parameter. MEDOKADS is navigated using an orbital model and additionally a supervised fine-navigation with a final accuracy of about 1 km. Thus, smaller structures, as small forests and valleys are still existent in MEDOKADS as well as islands and lakes. Figure 6 shows clearly the different resolution and effects of navigation.

4.4   *Restrictions to compile NDVI composites*

To generate the maximum *NDVI* for 10 day periods the PAL data set is restricted to pixels of scan angles <42°, or solar zenith angles <80°. If no appropriate pixel exists even pixel beyond this restrictions are selected. In the case of MEDOKADS no angular restrictions are applied. The shift of observation time towards late afternoon at the end of each mission of the NOAA satellites results in extreme illumination conditions, shadowing and observation geometry, where atmospheric corrections for MEDOKADS overestimates the atmospherical attenuation and the calculated *NDVI* gets unrealistic.

5 INTER-COMPARISON OF PAL DATA SET AND MEDOKADS NDVI TIME SERIES

The PAL data set and MEDOKADS overlap for about 12 years. In this period the mean of PAL NDVI is more than 35% higher for parts of the Mediterranean than the mean of MEDOKADS NDVI without atmospheric correction (Figure 7). The variations are the same for both data sets over the displayed period. Due to extreme observation conditions and problems to get valid calibration coefficients in 2000 the NDVI declines at the end of the NOAA 14 mission. It is obvious that the year 2000 needs a further recalibration in order to remove the dip in the curves. This is confirmed by the NDVI time series of the Spot Vegetation instrument which has been added to Figure 7.

In contrast to PAL a full atmospheric correction is applied to the Spot Vegetation data. Additionally slight differences in the response function of the instrument result in higher absolute NDVI values, but confirm that the dip in the MEDOKADS and the PAL data set time series in 2000 is an artificial one. Beginning with NOAA 15 AVHRR/3 is flown with changes in the filter response functions of channel 1 and 2 (Figure 8) causing an increase in NDVI at a magnitude

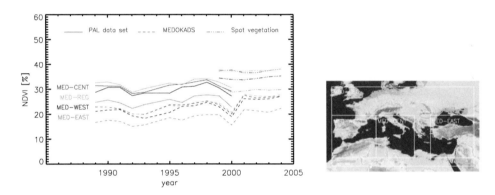

Figure 7.   Annual mean of PAL, MEDOKADS (without atmospheric correction) and Spot Vegetation NDVI for different regions.

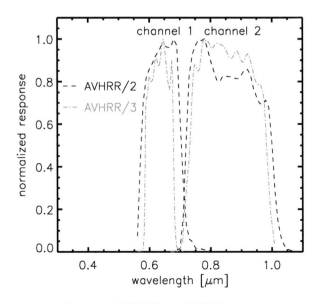

Figure 8.   Instrument response function of AVHRR/2 and AVHRR/3.

111

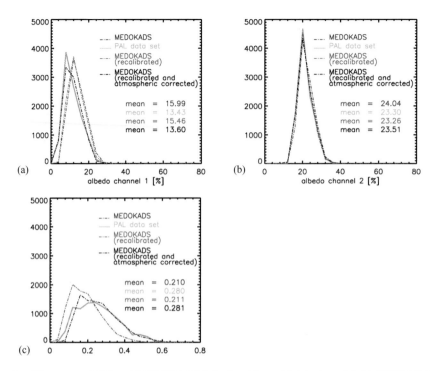

Figure 9. Histograms of PAL and MEDOKADS daily data (channel 1 (a), channel 2 (b), NDVI (c)) before, after recalibration and after recalibration and atmospheric correction (June 1st, 1995).

of 20%. This is only partly corrected by the formulas given by Trishchenko et al. (2002) in the MEDOKADS from 2001 onward.

Means, histograms and linear correlations over small and large areas are useful statistical parameters to compare different data sets. Monthly and yearly time periods have been chosen for averaging to identify short-term and long-term phenomena. Direct comparison of channel 1 and 2 is only possible for the first day of each month because other days are missing in the PAL data set.

Figure 9 shows the inter-comparison of the PAL data set and MEDOKADS in different processing steps. As shown in Figure 9c the estimated NDVI from PAL is higher than the NDVI from MEDOKADS, mainly caused by the lower mean of channel 1 for the PAL data set (Figure 9a). The shape of the histogram and the mean for channel 1 (Figure 9a) agrees very well after recalibration and atmospheric correction for MEDOKADS. The influence of atmospheric correction for channel 2 is not as pronounced (Figure 9b).

Figure 9 shows that the effect of recalibration has less importance than the atmospheric correction. The means of NDVI for the PAL data set and MEDOKADS agree good after recalibration and atmospheric correction but the shape of the histogram shows some differences (Figure 9c) that have to be attributed to different re-sampling methods during data treatment.

6 DEVIATIONS BETWEEN MEDOKADS AND PAL DATA SET

It has been investigated if a procedure could be established to fit the two data sets together, since the MEDOKADS seems to contain more information due to the higher resolution and better navigation.

Figure 10 shows the regression between the PAL data set and MEDOKADS as they are generated without any supplementary correction. The dashed gray line shows the linear regression between both data sets and the solid gray line the second degree polynomial. These regression functions can

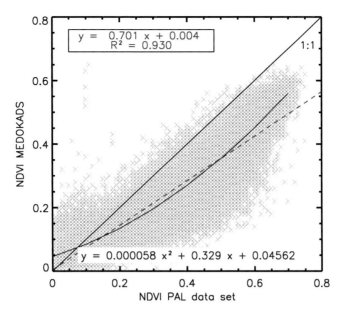

Figure 10. Regression between PAL data set and MEDOKADS for all pixels which are defined in MEDOKADS.

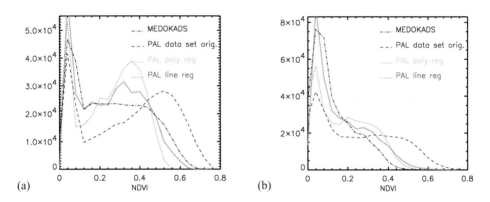

Figure 11. Histogram of NDVI for MEDOKADS and PAL data set before and after linear and polynomial correction area of Western Europe, (a) first decade of June 1989, (b) first decade of December 1996.

be used to convert the data sets even for regions outside of the Mediterranean area. The regression function has been calculated through averaging all pixels which are defined for MEDOKADS and the PAL data set in a 1° raster for each month of each year.

The calculated regression functions (Figure 10) have been used to transform the PAL data set to MEDOKADS window S (Iberian Peninsula). Figure 11 shows the histograms for two examples of this calculation. Figure 11a shows data from the beginning of MEDOKADS for the first decade of June 1989 and Figure 11b shows data from the first decade of December 1996. In Figure 11a histograms of the PAL data set and MEDOKADS agree well in the range of NDVI 0.1–0.25 and 0.5–0.7, the polynomial regression reduces the big difference in the range between. But there still exists a remarkable difference. For the second case (first decade of December 1996) the agreement is quite better. The high values of the PAL data set are minimized for both examples.

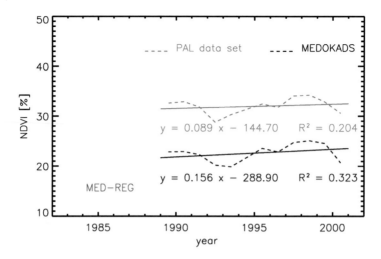

Figure 12.   Trends in annual mean of MEDOKADS and PAL data set.

Both data sets are used for trend analysis within the DeSurvey project (A Surveillance System for Assessment and Monitoring of Desertification). It was checked whether differences in data treatment lead to different results in trend analysis.

Figure 12 shows annual means of the whole MEDOKADS area of both data sets for the time where both data sets are defined. The linear regression for the PAL data set gives a slope of 0.089 and for MEDOKADS a slope of 0.156. The absolute differences in NDVI increase over the period of 12 years is only 0.8 % and due to the small correlation coefficients negligible.

SUMMARY

Reasons for differences between the PAL data set and MEDOKADS could be identified and partly quantified: different resolution, navigation accuracy, calibration, atmospheric correction and selection of different pixels, mainly in wintertime. The bias between the PAL NDVI values and the top of atmosphere (TOA) NDVI values used in MEDOKADS is explained by the simplified atmospheric correction applied to PAL data. Atmospheric correction mainly results in lower AVHRR channel 1 data causing considerable higher NDVI values. This difference in data treatment can be removed easily.

Long-term analysis of surface characteristics in the framework of DeSurvey is sensible to artificial trends due to calibration, changing instrument characteristics and data treatment. No significant differences in long-term trends between the PAL and the MEDOKADS NDVI time series (TOA) could be detected. Thus trends found in either PAL or MEDOKADS are expected to be not only qualitatively but also quantitatively comparable.

The broad scatter in the regression of PAL and MEDOKADS NDVI values has to be attributed mainly to the re-sampling and averaging during the PAL data processing going along not only with a considerable loss in resolution but also in a bias of the derived area and temporal means. Mean effects can be removed by regression formulas. But effects of navigation errors, re-sampling and averaging cannot be removed satisfactorily. Thus regional and local surface characteristics may show differences which have to be attributed to data treatment. The extrapolation to the PAL data set of the additional information given by the MEDOKADS with higher resolution and more precise geo-location will show distinct limitations and need special attention.

Some differences in the applied calibration coefficients show no significant influence in long-term trends but may be more pronounced if shorter periods are considered. Attention has to be paid to the life cycle of the operational instruments of about 6 years with increasing uncertainties at the end of the operation.

The PAL data set is a closed archive which covers 20 years of data and ends in 2001. The MEDOKADS is still prolonged in near real time retains the full dynamic range of possible AVHRR data. It is hence not only better suited for desertification assessment, but also for near real time monitoring in the framework of the DeSurvey project. The severe increase in NDVI due to the change to the AVHRR/3 instrument onboard starting with the NOAA 15 satellite is only partly corrected by the formulas given by (Trishchenko et al. 2002) and requires further consideration in a forthcoming study.

REFERENCES

Bolle, H.-J., Eckardt, M., Koslowsky, D., Maselli, F., Meliá-Miralles, J., Menenti, M., Olesen, F.-S., Petkov, L., Rasool, I. & Van de Griend, A. (2006). *Mediterranean Land-surface Processes Assessed from Space*. Berlin, Heidelberg, New York: Springer.

Goodrum, G., Kidwell, K.B. & Winston, W. (2000). *NOOA KLM user's guide*. Washington, DC: U.S. Department of Commerce National Oceanic and Atmospheric Administration; National Environmental Satellite, Data and Information Service.

Kidwell, K. (1991). NOAA Polar Orbiter Data User's Guide. In, *NCDC/SDSD. National Climatic Data Center, Washington, DC*. Washington, DC.

Koslowsky, D. (1996). *Mehrjährige validierte und homogenisierte Reihen des Reflexionsgrades und des Vegetationsindexes von Landoberflächen aus täglichen AVHRR-Daten hoher Auflösung*. Berlin: Free University.

Koslowsky, D. (2003). The MEDOKADS data set as a substantial part of a remote sensing data network for a Mediterranean research and application network. In H.-J. Bolle (Ed.), *Mediterranean Climate* (pp. 165–177). Berlin Heidelberg New York: Springer-Verlag.

Koslowsky, D., Billing, H., & Eckardt, M. (2001). Sensor degradation and inter-calibration of the shortwave channels of the AVHRR - NOAA 11/14/16 satellites In, *Proceedings of the 2001 EUMETSAT Meteorological Satellite Data Users' Conference* (pp. 107–113). Antalya Turkey.

Myneni, R.B., Keeling, C.D., Tucker, C.J., Asrar, G. & Nemani, R.R. (1997). Increased plant growth in the northern high latitudes from 1981 to 1991. *Nature, 698*, 698–702.

Rahman, H., & Dedieu, G. (1994). SMAC: A simplified method for atmospheric correction of satellite measurements in the solar spectrum. *International Journal of Remote Sensing, 15*, 123–143.

Rao, C.R.N. & Chen, J. (1994). Post-Launch Calibration of the Visible and Near-Infrared channels of the Advanced Very High Resolution Radiometer on NOAA-7, -9 and -11 Spacecraft. In (p. 22): NOAA Technical Report NESDIS 78.

Rao, C.R.N. & Chen, J. (1995). Inter-satellite calibration linkages for the visible and near-infrared channels of the Advanced Very High Resolution Radiometer on the NOAA-7, -9, and -11 spacecraft. *International Journal of Remote Sensing, 16*: 1931–1942.

Rao, C.R.N. & Chen, J. (1999). Revised post-launch calibration of the visible and near-infrared channels of the Advanced Very High Resolution Radiometer (AVHRR) on the NOAA-14 spacecraft. *International Journal of Remote Sensing, 20*: 3485–3491.

Trishchenko, A.P., Cihlar, J. & Li, Z. (2002). Effects of spectral response function on surface reflectance and NDVI measured with moderate resolution satellite sensors. *Remote Sensing of Environment, 81*: 1–18.

Tucker, C.J. (1979). Red and photographic infrared linear combinations for monitoring vegetation. *Remote Sensing of Environment, 8*: 127–150.

Recent Advances in Remote Sensing and Geoinformation Processing
for Land Degradation Assessment – Röder & Hill (eds)
© 2009 Taylor & Francis Group, London, ISBN 978-0-415-39769-8

Change detection in Syria's rangelands using long-term AVHRR data (1982–2004)

Th. Udelhoven
Centre Recherche Gabriel Lippmann, Esch/Alzette, Luxembourg

J. Hill
Remote Sensing Department, FB VI Geography/Geosciences, University of Trier, Trier, Germany

ABSTRACT: The present study focuses on long-term variations (1982 to 2004) in the Syrian rangelands using the "Mediterranean Extended Daily One Km AVHRR Data Set" (MEDOKADS) and the 8 km Global Inventory Modeling and Mapping Studies (GIMMS) data set. The major drivers that are putting pressure on Syria's rangeland systems are climate variability and human activities with ecological and social impacts at various temporal and spatial scales. In agreement with other studies it was found that the Normalized Difference Vegetation Index is a suitable proxy for land surface response to precipitation variability even at low vegetation coverage. Although it is well known that the rangelands are progressively vulnerable to dry spells and recent episodes of droughts, changes in productivity cannot be understood without reference to additional factors such as food security, poverty and man induced land-use changes. To decide whether detected changes in surface reflectance are related to improvement (*bright-spot*) or reduction (*hot-spot*) in resource productivity a synoptical view of time-series analysis results integrated with physio-geographical and socio-economical information is required. *Bright-spot* areas could be linked to modifications in land management schemes, expansion of irrigation systems, and land reclamation. Main explanations for *hot-spots* were the expansion of irrigation beyond mid- and long-term sustainability, water erosion, crusting and salinization, the depletion of biomass and vegetation cover, and the explosive expansion of urban agglomerations.

1 INTRODUCTION

The objective of the present study is to investigate surface conditions in Syria within a 22-year period using data from the Advanced Very High Resolution Radiometer (AVHRR) and to discuss the impact of local and national policies on the observed results. The status of Syria's drylands is controlled by a complex interaction of climate, soil, vegetation, water resources, stocking rates and human impacts. As in other Eastern Mediterranean countries, a fast growing population, which is estimated to increase currently at a rate of 2.3% in Syria (CIA 2007), triggers land use changes and leads to conflicts between ecological and economical priorities. Syrian's rangelands are considered to be widely degraded due to deforestation, forest fires, urbanization, inappropriate agricultural practices, overgrazing and uprooting of woody species (Nahal 1995, Rae et al. 2001, Hirata et al. 2001, Geerken and Evans 2004).

The problem of degradation received worldwide attention with the United Nation Convention to Combat Desertification (UNCCD). Using their definition "Land degradation means a reduction or loss, in arid, semi-arid and dry sub-humid areas, of the biological or economic productivity [...] resulting from land uses or from a process or combination of processes [...], such as: (1) soil erosion caused by wind and/or water; (2) deterioration of the physical, chemical and biological or economic properties of soil; and (3) long-term loss of natural vegetation." (UNCCD 2003, Article 1). For change detection on country to global levels archives of satellite data with coarse geometric but high temporal resolution are the preferable choices. Since most of the fluctuation in the signal from optical sensors is due to the vegetation, it is essential that the time series cover complete phenological periods and have as long a duration as possible, so as to contain data for a variety of meteorological conditions. Only this way it can be discriminated between short-lived fluctuations in surface reflectance and long-term trends as indicators of enduring environmental pressure. The Advanced Very High Resolution Radiometer (AVHRR) provides the longest and most comprehensive record for monitoring the earth terrestrial biosphere that is currently available (Gutman & Ignatov 1995, Tucker et al. 2005). Consequently it is the best source for distinguishing between long-term changes in the biosphere and short-term impacts caused by climatic variability.

Referring to the UNCCD definition the term *bright-spot* is used in this study to indicate areas which have or are taking up positive development incentives leading to improvements in the health of ecosystems or economic improvements in compliance with criteria for sustainable development. *Hot-spots* in turn flag areas suffering from various types of degradation processes; these might be triggered from changes in natural factors and/or socio-economic conditions.

As a tool to identity potential *hot-* and *bright-spots* long-term trends in Normalized Difference Vegetation Index (NDVI) data were investigated. Although this index is often used as surrogate for ground vegetation cover (Tucker et al. 2005), NDVI variations cannot be linked directly to the health of dryland ecosystems. Thus, an assessment of suspicious areas requires the consideration of auxiliary climatic and socioeconomic information.

2 DATA AND METHODS

2.1 *Data*

The AVHRR data used in this study are the "Mediterranean Extended Daily One Km AVHRR Data Set" (MEDOKADS) and the 8 km Global Inventory Modeling and Mapping Studies (GIMMS) data set. The MEDOKADS are compiled from the NOAA-11, -14- and 16 sensors and is distributed by the Institute of Meteorology, Free University of Berlin. The archive covers the time period from 1989 until present. The MEDOKADS are corrected for sensor degradation and orbital drift effects that cause non-linear changes in the measured signal (Koslowsky 1998). The global 8 km GIMMS data set is constituted from NOAA 7, 9, 11, 14 and 16 and comprises bimonthly NDVI composites for the period from (1981–2003) (Tucker et al. 2005). For this study monthly maximum value composites from both data sets were used.

Monthly precipitation data from six selected stations for the time period 1982 to 2003 in Syria were provided by the Arabic Centre for Semi-arid and Dry Lands in Damascus (ACSAD).

Annual rainfall and NDVI data were computed from monthly data for hydrological years (October–September).

2.2 *Methods*

All time-series calculations were carried out using the TimeStats software package (Udelhoven 2006). To assess the variability of the NDVI and data from selected rainfall stations the coefficient of variability (V_{rain} and V_{NDVI}) was used:

$$V = \frac{s}{\bar{x}} \qquad (1)$$

where s is the standard deviation and $\bar{x}$ the mean of the rainfall or NDVI series. $V_{rain/NDVI}$ can be understood as a simple concentration measure of a series.

Long-term variations in the data and their significances were assessed using the non-parametric Modified Seasonal Mann-Kendall (MSK) test and Seasonal Kendall slopes (SKS) (Schönwiese 2000). The Null-hypothesis of the MSK test assumes all observations in each season are randomly ordered, while the alternative hypothesis expects a monotone trend (increasing or decreasing) in at least one season (Hirsch & Slack 1984). The SKS represents a non-parametric alternative for the

118

slope coefficient in a regression analysis (EPA 2000). Areas characterized by significant long-term variations were considered as candidates for *hot-spots* and *bright-spots*, respectively.

To characterize the phenology in potential *hot-* and *bright spots* areas Discrete Fourier transform (DFT) was applied to compute the magnitude $I(\lambda)$ information from the frequency λ related to the annual vegetation growth cycle:

$$\mathrm{NDVI}_{I(\lambda)} = \frac{\sqrt{\Re(\mathrm{DFT}(\lambda))^2 + \Im(\mathrm{DFT}(\lambda))^2}}{N} \tag{2}$$

The phase term for λ was recovered from the real $\Re$ and imaginary $\Im$ part of $F(\lambda)$ to assess the annual peaking times of the vegetation:

$$NDVI_{P(\lambda)} = \tan^{-1}[\Im\{DFT(\lambda)\}/\Re\{DFT(\lambda)\}] \tag{3}$$

Magnitudes and phases in NDVI data are closely linked to agro-biological phenomena, such as land cover conditions in response to seasonal pattern of rainfall and temperatures (Azzali and Menenti 1999).

A crucial restriction of the Fourier spectrum is that it loses track of locations for temporal events. Continuous Wavelet Transform (CWT) overcomes this problem, since this technique decomposes a series such that both time and frequency information remains assessable. CWT has been proven to be very useful in analyzing data with gradual frequency changes, which are hardly accessible by DFT (Torrence & Compo 1998, Anctil & Tape 2004). The CWT describes a time series by sequences of wavelet coefficients $W_n^2(s)$, each of which representing the amplitude of the wavelet function at a particular location within the data vector and at different wavelet scales s. This results in a scalogram that describes the wavelet power of Fourier frequencies in dependence of the time and s:

$$W_n^2(s) = FFT^{-1}\left[\sum_{k=0}^{N-1} \widehat{x} \left(\sqrt{\frac{2\pi s}{\delta t}} \, \widehat{\psi_0}^*(s\omega_k) e^{i\omega_k n\delta t} \right) \right] \tag{4}$$

where $\widehat{x}$ and $\widehat{\psi_0}$ denote the Fourier transforms of the series x and the mother wavelet function ψ, $*$ symbolizes the complex conjugate, N is the series length, the wavelet scale s ($s > 0$) indicates the wavelet's width, and the translation parameter n defines its position in time t and ω_k is the angular frequency.

The average of all the local wavelet spectra for over time is the global wavelet spectrum:

$$\overline{W}_n^2(s) = \frac{1}{n_a} \sum_{n=0}^{N-1} |W_n(s)|^2 \tag{5}$$

The global wavelet spectrum provides an unbiased and consistent estimation of the true power spectrum of a time-series (Torrence & Compo 1998).

2.3 *Study area*

The Syrian climate is typical of the Mediterranean Climatic Regime, with rainy winters and hot dry summers and short transitional seasons in between. Considering the annual rainfall amount Syria can be divided into several zones, which are illustrated in figure 1. The dominant climatic zone in the country, representing about 45% of the total area has less than 200 mm of rain per year and is designated as 'badiah' or steppe (stabilization zone 5). Average NDVI values within the central steppe are rather low (0.05–0.25). Rainfed agriculture is historically practiced in northern and western plateaus and plains in the stabilization zones 1–3. Irrigation is common in the large river valleys of Euphrates, Khabour and Orontes (Wirth 1971, Republic of Syria 2003).

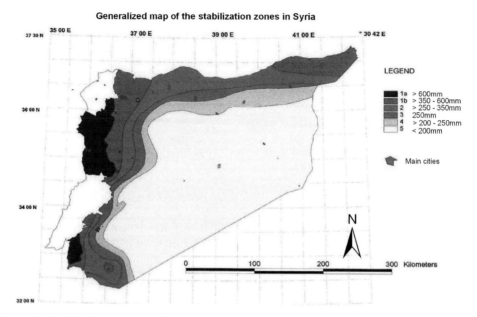

Figure 1.   Stabilization zones of Syria (Republic of Syria, Ministry of defence, 1977).

Syria's most important agricultural products include grain crops, cotton, sugar beet and olives. In dryland agricultural areas the rotation of winter crops with fallow is widespread, whereas permanent irrigation allows a rotation of winter and summer crops.

The dry central steppe area is dominated by Aridisols while the surrounding Inceptisols indicate areas outside the central steppe with rainfalls amounts between 250 and 600 mm. Entisols cover the coastal area and along the terraces of Syria's rivers and Vertisols can be found in the South of the country.

Depending on the climate conditions, several zones of natural vegetation can be distinguished. The forested areas in sub-humid or humid climates are dominated by *Quercus calliprinos* and *Quercus infectoria*, *Cedrus libani*, and *Juniperus excelsa*. At least 400 mm annual rainfall is required for evergreen forests and maquis, while dryer conditions favour evergreen shrub or dwarf shrub formations (Conacher & Sala 1998, Ilaiwi 1999). Relevant trees and shrubby pseudo-steppe species in semi-arid to arid climates are *Pistacia atlantica* and *Rhamnus palestina*. Pseudo-steppe and sub-desert formation (*Artemisia herba-alba*, *Haloxylon salicornicum* and *Anabasis setifera*) are widespread in the arid climate (Ibrahim 2002).

3   RESULTS

3.1   *Sensitivity of the NDVI toward rainfall conditions*

To assess the impact of climatic conditions on surface reflectance, rainfall and NDVI series at selected stations in the rangelands were compared. Intra- and inter-annual climatic variability profoundly influences plant phenological status such as germination, greening-up, the rate of biomass accumulation, and the onset and rate of vegetation senescence and thus significantly affects NDVI time-series properties (Lee et al. 2002, Erian et al 2006). Studies carried out by DuPlessis (1999), Hielkema et al. (1986), Nicholson et al. (1990), Wang et al. (2001), Anyamba & Tucker (2005) and others have demonstrated that the NDVI is a suitable proxy for land surface response to precipitation variability in drylands. The correlation between climatic parameters and

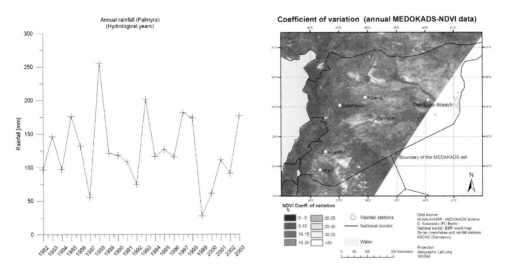

Figure 2. Annual rainfall amount in Palmyra (left) and coefficients of variability (V_{NDVI}) of annual NDVI data (MEDOKADS, 1989–2004, right).

NDVI is strongly determined by the degree of aggregation of the variables in the time dimension. Furthermore the NDVI lags behind rainfall by several weeks or months (Evans & Geerken 2004).

As an example the annual rainfall data for Palmyra that is located in the central steppe is shown in figure 2 (left). The variability in this series suggests that Syrian's rangelands are vulnerable to recurrent droughts. Palmyra experiences in the average a long-term average precipitation amount of approximately 120 mm, but it received less than 50 mm during the growing season 1987 and only 25 mm in 1999. The steppe and its fringe (stabilization zones 3–5) are characterized by higher NDVI variability compared to the regions that belong to stabilization zones 1–2 (Fig. 2, right). The largest coefficients of variations (>35%) in the Southwest of Syria coincide with areas receiving less than 100 mm annual rainfall (Erian et al. 2006).

Table 1 shows the bivariate correlation coefficients (Spearman, r_s) for six rainfall stations in the steppe and the surrounding fringe (compare figure 2, right) among rainfall and NDVI characteristics. The average annual rainfall amounts vary between 125 mm (Palmyra) and 284 mm (Izrah); the related average NDVI values (GIMMS), which represent a buffered area of 10×10 pixels around each station, are 0.1 and 0.24, respectively.

The correlation between annual rainfall amount and V_{rain} is negative for all rainfall stations in stabilization zone 5. Thus, low average rainfall amounts coincide with increased inter-annual rainfall variability. This explains the high degree of variability of the NDVI in the steppe in figure 2 (right). The broad range of the r_S values between NDVI and Rf (0.36 in Atheria to 0.8 in Itzah) suggests that the rainfall-NDVI relation is overlaid by individual response characteristics of different vegetation communities toward climatic variability (Evans & Geerken 2004).

The data in table 1 further indicate that high NDVI values and rainfall amounts go along with a strong annual vegetation growth cycle (I_{NDVI}). In Salamiyeh and Izrah, which both belong to stabilization zone 2, the significant correlation between P_{NDVI} and rainfall amount indicates forward shifts of the greenpeak in those years with above average annual rainfall amounts. In this zone rainfed agriculture of wheat, barley and summer crops is common.

To visualize the inter-annual variability of the annual vegetation growth cycle P_{NDVI}-values were calculated for each year using a windowed DFT (Fig. 3). On average the annual green-peak in the central steppe occurs early after the winter rainfalls in February, while the peaking time is shifted toward March to May in the surrounding rainfed areas. Along the Euphrates River, where cotton is grown, the green-peak is postponed towards July and September. The individual maps in figure 3 impressively demonstrate the high variability embodied in the steppe region and its fringe

121

Table 1. Correlation coefficient (Spearman, r_s) between annual rainfall amounts, V_{rain}, NDVI, P_{NDVI}, and I_{NDVI} at six climatic stations (only $r_s > 0.3$ are shown).

		Rf [mm]	Vrain	NDVI	PNDVI	INDVI
Salamiyeh	Rainfall [mm]	1.000	−.688	.621	.419	.582
Mean NDVI: 0.18	Rf. coeff. of var.	−.688	1.000	−.477	−.375	−.402
mean Rf: 277 mm	NDVI	.621	−.477	1.000		.921
(Stabil. Zone 2)	NDVIP e	.419	−.375		1.000	
	NDVII p	.582	−.402	.921		1.000
Palmyra	Rainfall [mm]	1.000	−.443	.563		.537
Mean NDVI: 0.10	Rf. coeff. of var.	−.443	1.000	−.525	.385	
Mean Rf: 125 mm	NDVI	.563	−.525	1.000		.860
(Stabil. Zone 5)	NDVI Phase		.385		1.000	
	NDVI Amp	.537		.860		1.000
Izrah	Rainfall [mm]	1.000	.762	.803	.508	.775
Mean NDVI: 0.24	Rf. coeff. of var.	.762	1.000	.605	.552	.633
Mean Rf: 284 mm	NDVI	.803	.605	1.000		.839
(Stabil. Zone 2)	NDVI Phase	.508	.552		1.000	
	NDVI Amp	.775	.633	.839		1.000
Atheria	Rainfall [mm]	1.000	−.482	.358		
Mean NDVI: 0.11	Rf. coeff. of var.	−.482	1.000			
mean Rf: 181 mm	NDVI	.358		1.000		.533
(Stabil. Zone 5)	NDVI Phase				1.000	
	NDVI Amp			.533		1.000
Deir-Ezzor-Mreayh	Rainfall [mm]	1.000	−.635	.667		
Mean NDVI: 0.16	Rf. coeff. of var.	−.635	1.000			
mean Rf: 147 mm	NDVI	.667		1.000		
mean(Stabil. Zone 5)	NDVI Phase				1.000	.483
	NDVI Amp				.483	1.000
Kharabo	Rainfall [mm]	1.000	−.591	.562		.377
Mean NDVI:0.17	Rf. coeff. of var.	−.591	1.000	−.625		
mean Rf: 145 mm	NDVI	.562	−.625	1.000		.512
(Stabil. Zone 5)	NDVI Phase				1.000	
	NDVI Amp	.377		.512		1.000

that corresponds well with the V_{NDVI}- map shown in figure 2 (right). Recurrent drought spells postpone the annual phases in the fringe of the central steppe area toward February and January. In 1999, 2002, 2003 and 2004 the central rangelands seem to expand since land remained fallow during the growth period. In contrast, 1998 was an above average wet year (compare fig. 3, left) in which the maximum of the NDVI has been shifted toward the second half of May in most parts of the central steppe. The maps also illustrate the risk for rainfed farming system in Syria, especially in the South of the 36th latitude that corresponds to the 250 mm isohyet.

3.2 *Assessment of long-term changes in surface reflectance*

To describe long-term variations in biomass production trend analysis was accomplished based on the monthly NDVI values.

The maps in figure 4 illustrate the results from the MSK-test based on 1km Medokads (1989 to 2004) and GIMMS data (1982 to 2003). Spots characterized by significant negative and positive developments are considered as candidates for *hot-spots* and *bright-spots*, but require further evaluation.

3.2.1 *Hot-spot areas*

Both trend maps illustrate that negative NDVI trends are not the overall impression in Syria. In contrast, most parts of the central rangelands, the fringe around the central steppe, the coastal areas and the alluvial plains along the Euphrates River are characterized by neutral or positive trend

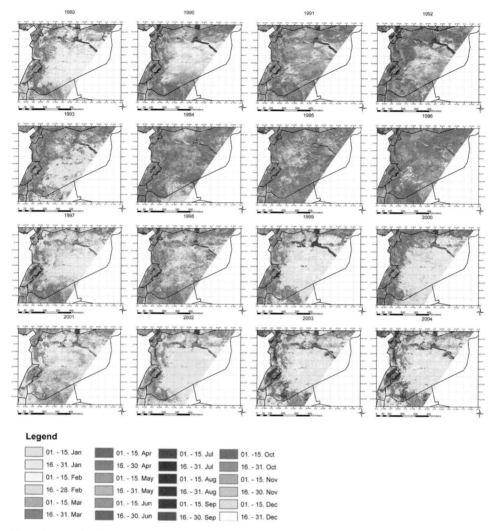

Figure 3. Peaking times of the NDVI, calculated for each year of the observation period (see colour plate page 384).

inclinations. Minor differences in the MSK maps are rather related to the different observation periods of the two AVHRR data sets than to the different at-ground resolutions.

The MSK and the Kendall slopes maps show negative NDVI trends along the Euphrates river upstream the Al-Assad lake and along the Khabour river which are related to recent dam constructions. The critical limit for dryland cultivation is the 200 mm isohyet (Arabic Centre for Semi-arid areas and Dry lands (ACSAD), personal communication). Thus, Syria's government is pushing to further development of the irrigation sector in order to maintain crop production during drought years. Together with its two major tributaries, the Balikh and Khabur (Nahr al Khabur) rivers, the Euphrates is the major irrigation source. The declining NDVI trends along the Euphrates are caused by flooding the lower river terraces by a new dam between Ash Shajarah, situated on the northern edge of Lake Assad, and Jarabulus, located near the Turkish border.

At the Khabour River a new dam was finished in the middle of the 1990s (Fig. 7). Negative trends along that river suggest that the dam construction had also downstream impacts. Zaitchik et al. (2002) investigated the effects of the Khabour dam on the lower basin, where after closing

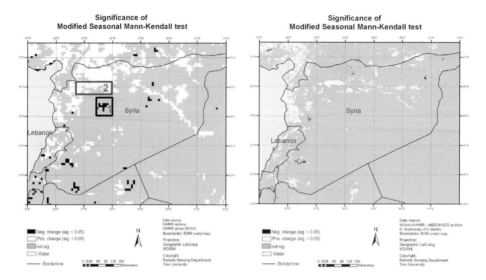

Figure 4. Significance for the MSK test for trend analysis from GIMMS (1982 to 2003, left) and MEDO-KADS data (1989 to 2004, right). Black and white colors indicate significant negative and positive trends, respectively.

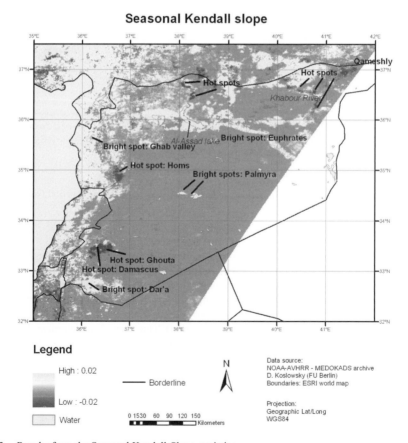

Figure 5. Results from the Seasonal Kendall-Slope statistics.

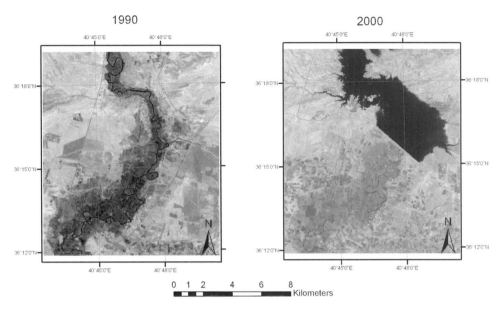

Figure 6.   Landsat scenes in 1990 and 2000 showing the Khabour river dam project.

the dam floodplain irrigation was replaced by groundwater extraction. The authors report that in the lower Khabour basin groundwater irrigators have encountered problems with water quantity and quality and secondary salinization of marginal agricultural soils. Furthermore, groundwater withdrawals lowered the regional water table to a point that the lower Khabour dries up every summer.

Two other spots with significantly declining NDVI are located in the western part of Syria around Damascus and Homs cities. The first area around Damascus can be separated into two parts. The first area is attributable to urban encroachment of Damascus and its suburbs during the last two decades. The second spot in the eastern part corresponds to the Ghouta region of Damascus where orchards and horticultures are grown. Food production in Damacus' Ghouta is traditionally supported by wells and surface irrigation, mainly from the Barada River, which forms two table lakes in the East, Bahr el Aateibe and Bahret Hijjane. By excessive withdrawal of water, these lakes regularly run dry during the summer season. Meanwhile, the exploitation of the water resources has exceeded mid- and long-term sustainability and groundwater pumping has led to salinity build-up in the soil matrix (personal communication, ACSAD). A comparison of the two Landsat TM scenes from 1987 and 2000 in figure 7 suggests a remarkable land cover change, although the lower overall greenness in the 2000 image can be partly explained by the severe drought in 1999 (compare the rainfall graph for Palmyra in figure 3). Figure 4 shows that in the dry areas in 1999 and 2000 the peaking time of the NDVI has been shifted backward. However, the graph in Figure 9 (left) that illustrates the individual PNDVI values in the total observation period for the Ghouta area also point on a land-cover change. In the first years of observation time the maximum of the green peak appeared between April and June, while the peaking time had been shifted to January to March just after winter rainfall at the end. A similar development suggests the patterns in the scalogram in figure 8 (right), which shows a wavelet power at the period of one year that was fading out over time.

The situation in the conurbation of Homs is similar to that in the vicinity of Damascus. The southern area is suffering from urban encroachment and the horticulture region in the North from shortage of water. Both processes result in negative long-term NDVI trend. Due to excessive irrigation in the past the water tables of the Orantes River and the groundwater have been significantly lowered (Arabic Centre for Semi-arid areas and Dry lands (ACSAD), personal communication).

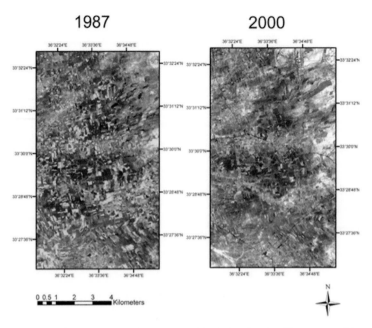

Figure 7.   Landsat TM sub-scenes (26.5.1987 and 21.5.2000) showing the Ghouta area of Damascus (RGB channels: 5-4-3) (see colour plate page 385).

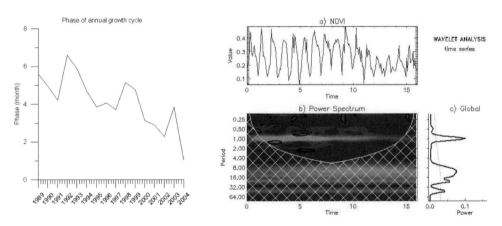

Figure 8.   Development of the annual growth cycle's phase in the Ghouta region from a windowed FFT (left) and results from a CWT using MEDOKADS data (1989–2004, Morlet wavelet, x-/y-axis: annual scale). Cross-hatched regions in the scalogram indicate the cone of influence (COI), in which the wavelet spectrum is lowered by edge effects and, therefore, unreliable results might be expected.

The Kendall slope map shows one other potential hot-spot area that is located in the Northeast of Syria around Qameshly city.

3.2.2   Bright-spot areas

Significant positive trend inclinations in figure 5 are confined to areas outside the steppe with above average NDVI levels between 0.30 and 0.55 and to irrigated places. The lower Euphrates terraces downstream the Al-Assad Lake with their fertile soils (Torrifluvents, a suborder of Entisols) is an important cropping region in Syria.

126

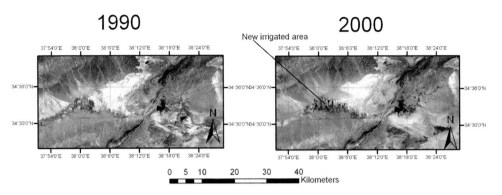

Figure 9.   Landsat TM sub-scenes (26.5.1987 and 21.5.2000) showing the Palmyra area.

Figure 10.   Stony soils area in the Dar'a province (South Syria).

In western Syria positive trends correspond to the coastal area and the ridges of the coastal mountains. Belonging to stabilization zone 1 this region receives more than 600 mm annual precipitation. One bright-spot area in that region is the Al Ghab Valley East of Homs. Since 1954 the government had reclaimed land from this swampy area, where periodically the Orontes River overflowed its banks in destructive floods (Wirth 1971).

In the central part of the steppe area there are two small neighboured bright-spots visible from the MSK map. The spot in the East marks the Palmyra oasis; the second spot corresponds to an irrigated area with a size of 5650 ha that has grown during the 1990s along the road between Tiyas and Tadmur (Palmyra). The two Landsat images in figure 9 shows this location in 1990 and 2000.

One further bright-spot area is located in the South of Syria, in the Dar'a province. Belonging to stability zone two, rainfall amount is not the most limiting factor for rainfed agriculture in this region, but a high density of stones in the topsoil. In 1994, the Syrian Government initiated a project to clean the topsoils from stones (Fig. 10). The impact of this action is enhanced agricultural productivity that is reflected in the NDVI trend map.

Two other examples for land reclamation are highlighted in the GIMMS MSK trend map. The spots, which are marked as 1 and 2, are located in the Hamah-province and belong to stabilization zones 3–4 that enable marginal dryland agriculture. Decreasing trends characterize the first, and increasing trends the second spot, respectively. Figure 11 shows the results from CWT-analysis of a NDVI series (GIMMS) from the first place. In 1989, seven years after the beginning of the observation period, both the decreasing NDVI curve and the temporal discontinuity of the annual vegetation growth cycle in the scalogram suggest a land-cover change. During the first years an annual cycle and in addition a three-year period is visible in the scalogram that documents temporally fallow periods. After 1989, both cycles disappear and the NDVI sharply drops below 0.10. The fact that formerly cultivated areas typically display lower NDVI values in the rangelands in

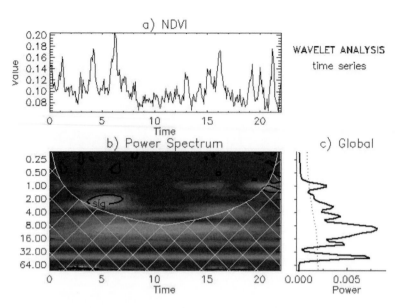

Figure 11. CWT result for the Hamah-province using GIMMS data (1982–2003, Morlet wavelet, x-/y-axis: annual scale).

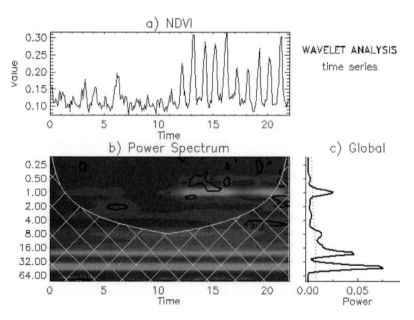

Figure 12. CWT result for a NDVI series from the Hamah province using GIMMS data (1982–2003, Morlet wavelet, x-/y-axis: annual scale).

subsequent years is explained by the destruction of seed banks and uprooting the natural shrub cover through plowing (Geerken and Hansmann 2000, Evans & Geerken 2004). The period from 1989 to 1991 was exceptionally dry and Syrian's Steppe Directorate, founded in 1961 for range management, prohibited further cropping in that area (ACSAD, personal communication). The temporally increasing NDVI trend and the growing magnitude of the annual vegetation growth

128

cycle after abandonment of the place is related to a recovery of natural vegetation and not to cropping activities. In 1999 and 2000 another drought explains again a temporally decrease of the NDVI and the weaker annual growth cycle.

The second location that is highlighted in the GIMMS-MSK-map indicates an area that has been taken under cultivation by irrigation in 1992. The CWT analysis (Fig. 12) illustrates the reinforcing of the annual vegetation growth cycle in the 12th year (corresponding to the year 2000) of the observation period. The missing significant two- or three-year component in the scalogram documents that cultivation is not interrupted by fallow periods due to permanent irrigation.

4   DISCUSSION

The rangelands in Syria are widely considered to be degraded. Nevertheless, results from this study demonstrate that ongoing degradation tendencies are not the overall impression in the rangelands, but that negative NDVI trends are rather restricted to local areas. The NDVI is often assumed to be insensitive toward subtle environmental changes in semi-arid and arid areas because the index suffers from several limitations, such as the influence from soil background. Indeed, in a monotemporal context comparable vegetation cover might correspond to different NDVI levels due to different soil types. Nevertheless, in a temporal context the NDVI is a sensitive measure for climatic and long-term environmental conditions even at low biomass coverage. A restriction of the method is that the usage of low-spatial-resolution imagery complicates the detection of small land-cover features and related processes. The results suggest that time-series analysis products cannot be apriori assigned to certain causative environmental processes; they rather require a careful check of plausibility of identified patterns and a second assessment state. Once significant long-term tendencies are identified on the coarse scale, spatially and spectrally higher remote sensing systems or ground-based observation networks might come into play to study such phenomenon rather on a regional to local scale to verify what is happening on the ground (Tucker et al 2006). From this perspective multi-/hyperspectral and hypertemporal satellite systems provide complementary information for the assessment of surface reflectance.

The results further demonstrate that the drylands are vulnerable towards dry spells and recent episodes of droughts; however short-term climate impacts are not accountable for degradation processes in Syria's rangelands. The main reasons for the identified *hot-spots* are the expansion of the irrigation system, soil erosion, crusting and salinization, the depletion of biomass and vegetation cover, the destruction of wetlands and explosive expansion of urban agglomerates. In contrast, positive NDVI trends could be related to improved management systems in grazing and irrigation, and by land reclamation schemes. Modern agricultural trends in drylands include expanding greenhouse and plastic cultures (which is not visible by the NDVI), less fallow, greater use of herbicides, deeper tillage and the usage of legumes as self-regenerating pasture plants in rotation systems with cereals (Cocks & Ehrman 1987). Crop production is further increased by the development of drought resistance breeding lines and by the use of early-maturing genotypes as a way of withstanding drought and high temperatures during the grain-filling and cropping period. Additionally, different phenologies are mixed to maximize grain yield in wet and dry environments (Shakhatreh et al. 2001).

Syria's increasing population reinforces the expansion of agriculture into the fringe of the steppe, into a landscape where agricultural activity is risky without irrigation. This bears the risk that even initially efficient reclamation schemes may later enter a phase where undesired effects become important, due to endured exploration of the natural resources beyond mid- and long term sustainability or by imprudent land management developments. Cultivating the fringe of the steppe must be reconceived within this context. Eliminating the shrubby vegetation cover leaves the soil particles vulnerable to the action of the wind. Weak structural stability and light texture that mainly characterize the aridisols in the steppe support the vulnerability towards wind erosion (Ibrahim 2002). One other critical example is Ghab valley: Although problems are yet not visible in the small scale, parts of the valley in the meantime suffer from salinization problems due to a high ground water table and the usage of brackish water in irrigation (ACSAD, personal communication). In other regions enduring exploration of the natural resources has already generated *hot-spots* such as in the Ghouta area of Damascus. On the other hand land abandonment in the fringe of the central steppe, such as demonstrated in the Hamah province, is actually a positive land-use development in those marginal landscapes despite documented negative long-term NDVI trends.

5   CONCLUSION

Data of high temporal resolution remote sensing systems combined with appropriate TSA-methods are extremely valuable base for the extensive monitoring of land surface properties, providing spatially distributed information about environmental processes in their temporal contexts, despite the moderate spatial resolution of those systems. The availability of high-frequency data thus opens an information level which is not tangible in that detail using data from systems providing higher spatial and radiometrical resolution at the expense of temporal resolution. The intercalibration of existing AVHRR data with new data from the recent generation of moderate resolution narrow band satellite systems such as Spot VEGETATION or MODIS is a promising source of data for future land-use/land-cover change monitoring that was already realized in the GIMMS data set (Tucker et al 2005). Nevertheless, the results demonstrate that from time-series analysis alone the assignment of areas into the categories bright and hot spots is not feasible. Remote Sensing systems cannot be used to detect directly whether changing surface reflectance are related to the health of dryland ecosystems. Especially the exclusive focus on trend inclination describing changes in biomass production is no adequate criterion to assess the condition of related patches. Rather a careful assessment of time-series analysis results, additional environmental data sets and substantial ancillary information seem more suitable.

ACKNOWLEDGMENTS

We would like to thank Dirk Koslowksy (Free University Berlin) and the GIMMS groups (NASA) who provided the AVHRR Medokads and GIMMS data sets. Rainfall data were used by courtesy of the Arabic Centre for Semi-arid and Dry Lands (ACSAD). Furthermore we are very grateful to C. Torrence and G. P. Compo for their CWT tool that is used in the TimeStats software. Their source code is available at http://paos.colorado.edu/research/wavelets.

REFERENCES

Ancil, F. & Tape, D.G. 2004. An exploration of artificial neural network rainfall-runoff forecasting combined with wavelet decomposition. *Journal of Environmental Engineering and Science* 3: 121–S128.

Anyamba, A. & Tucker, C.J. 2005. Analysis of Sahelian vegetation dynamics using NOAA-AVHRR NDVI data from 1981–2003. *Journal of Arid Environments* 63: 596–614.

Azzali, S. & Menenti, M. 1999. Mapping isogrowth zones on continental scale using temporal Fourier analysis of AVHRR-NDVI data. *International Journal of Applied Geosciences* 1: 9–20.

Barkoudah, Y., Darwish, A.I. & Antoun, M.A. 2000. *Biological Diversity National Report, Biodiversity Strategy and Action Plan and Report to the Conference of the Parties*, NBSAP Project SY/97/G31, Ministry of Environment, Syria, United Nation Development Program. 34 p. http://www.biodiv.org/doc/world/sy/sy-nr-01-en.pdf

DuPlessis, W.P. 1999. Linear Regression Relationships Between NDVI, Vegetation and Rainfall in Etosha National Park, Namibia, *Journal of Arid Environments* 42: 235–260.

EPA, 2000. *Guidance for Data Quality Assessment—Practical Methods for Data Analysis*. United States Environmental Protection Agency (EPA), Washington, DC 20460. EPA/600/R-96/084.

Erian, W.F., Fares, F.S., Udelhoven, T. & Katlan, B. 2006. Coupling long-term "NDVI" for monitoring droughts in Syrian rangelands. *The Arab Journal for Arid Environment* 1: 77–87.

Evans, J. & Geerken, R. 2004. Discrimination between climate and human-induced dryland degradation. *Journal of Arid Environments* 57: 535–554.

Geerken, R. & Hansmann, B., 2000. Combating desertification in the near east—identification of rehabilitation measures and impact monitoring. In: Hongbo, J. (ed.), *Workshop of the Asian Regional Thematic Programme Network on Desertification Monitoring and Assessment. UNCCD, Tokyo*: 56–69.

Gutman, G. & Ignatov, A. 1995. Global land monitoring from AVHRR: potential and limitations. *International Journal of Remote Sensing* 16: 2301–2309.

Hielkema, J.U., Prince, S.D. & Astle, W.L. 1986. Rainfall and vegetation monitoring in the Savanna zone of democratic Republic of Sudan Using the NOAA advanced Very High-Resolution Radiometer. *International Journal of Remote Sensing* 7: 1499–1513.

Hill, J. 2000. Semiarid Land Assessment: Monitoring Dry Ecosystems with Remote Sensing: In: Meyers, R.A. (ed.): *Encyclopedia of Analytical Chemistry*: 8769–8794, John Wiley & Sons, Chichester.

Hirsch, R. & Slack, J. 1984. A nonparametric trend test for seasonal data with serial dependance. *Water Resources Research* 20: 727–732.

Ibrahim, W.Y. 2002. http://www.fao.org/ag/agl/swlwpnr/reports/y_nr/z_sy/sy.htm#hla.

Ilaiwi, M. 1999. Land Degradation in Syria (Case Study Jabel Al Bishri). *Proceedings of the conference on the use of remote sensing techniques and GIS in monitoring and combating desertification*. ACSAD, Damascus, Syria.

Koslowsky, D. 1998, Daily extended 1-km AVHRR data sets of the Mediterranean, *Proceedings 9th Conf. Sat. Meteor. and Oceanogr. UNESCO, Paris; France, 25–29 May, AMS*, Boston, MA: 38–41.

Lee, R., Yu, F. & Price, K.P. 2002. Evaluating vegetation phonological patterns in inner Mongolia using NDVI time series analysis. *International Journal of Remote Sensing* 23: 2505–2512.

Mimi, Z.A. & Sawalhi, B.I. 2003. A decision tool for allocating the waters of the Jordan River basin between all riparian parties. *Water Resources Management* 17: 447–461.

Nicholson, S.E., Davenport, M.L. & Malo, A.R. 1990. A Comparison of Vegetation Response to Rainfall in the Sahel and East Africa, Using Normalized difference Vegetation Index from NOAA AVHRR. *Climate Change* 17: 209–241.

Rae, J, Arab, G., Nordblom, T., Jani, K. & Gintzburger, G. 2001. Tribes, state, and technology adoption in arid land management, Syria. *Capri working paper*, 15, International Food Policy Research Institute 2033 K Street, N.W. Washington, D.C. 20006 U.S.A.

Republic of Syria, Ministry of defence, 1977. *Climate Atlas of Syria*. Damascus.

Röder, A. 2005. *A Remote Sensing Based Framework for Monitoring and Assessing Mediterranean Rangelands. Case Studies from Two Test Sites in Spain and Greece*. PhD Thesis, Trier.

Schönwiese, C.-D. 2000. *Praktische Statistik für Meteorologen und Geowissenschaftler*, Berlin: Gebrüder Bornträger.

Shakhatreh, Y., Kafawin, O., Ceccarelli, S. & Saoub, H. 2001. Selection of Barley Lines for Drought Tolerance in Low-Rainfall Areas. *J. Agronomy & Crop Science* 186: 119–127.

Torrence, C. & Compo, G. 1998. A Practical Guide to Wavelet Analysis. *Bulletin of the American Meteorological Society* 79: 61–78.

Tucker, C.J., Pinzon, J.E., Brown, M.E., Slayback, D., Pak, E.W., Mahoney, R., Vermote, E. & El Saleous, N. 2005. An Extended AVHRR 8-km NDVI Data Set Compatible with MODIS and SPOT Vegetation NDVI Data. *International Journal of Remote Sensing* 26: 4485–4498.

Tucker, C.J., Anyamba, A. & Gonzales, P. 2006. Monitoring of land degradation and desertification dynamics using coarse-scale satellite imagery. In: Röder, A. & Hill, J.. (eds), Proceedings of the *1st Conference on Remote Sensing and Geoinformation Processing in the Assessment and Monitoring of Land Degradation and Desertification, 7th–9th September 2005, Trier*: 120–127.

Udelhoven, T. 2006. TimeStats: a software tool for analyzing spatial-temporal raster data archives: In: Röder, A. & Hill, J. (eds), Proceedings of the *1st Conference on Remote Sensing and Geoinformation Processing in the Assessment and Monitoring of Land Degradation and Desertification, 7th–9th September 2005, Trier*: 247–255.

UNCCD (2003): United Nations Concention to Combat Desertification. http://www.unccd.int/convention/convention.php.

Wang, J., Price, K.P. & Rich, P.M. 2001. Spatial Patterns of NDVI in Response to Precipitation and Temperature in the central Great Plains. *International Journal of Remote Sensing* 22: 1005–1027.

Wirth, E. 1971. *Syrien. Eine geographische Landeskunde*, Darmstadt: Wissenschaftliche Buchgesellschaft.

Zaitchik, B., Smith, R. & Hole, F. 2002. Spatial analysis of agricultural land use changes in the Khabour River basin in Northeaster Syria. *Pecora 15/Land Satellite Information IV/ISPRS Commission I/FIEOS 2002 Conference Proceedings*.

*Recent Advances in Remote Sensing and Geoinformation Processing
for Land Degradation Assessment – Röder & Hill (eds)
© 2009 Taylor & Francis Group, London, ISBN 978-0-415-39769-8*

'Hot spot' assessment of land cover change in the CWANA region using AVHRR satellite imagery

D. Celis & E. De Pauw

International Center for Agricultural Research in the Dry Areas, Aleppo, Syria

ABSTRACT: Remote sensing presents a valuable tool to address the highly complex issue of land degradation in dryland areas. Ten-daily composites of 8 km-AVHRR reflectance data, for band 1 (0.58–0.68 μm) and band 2 (0.725–1.1 μm), were merged to form a complete coverage of North Africa, the Horn of Africa, West and Central Asia regions (the 'CWANA' region), and transformed into monthly NDVI composites. This temporal NDVI dataset was converted into a land use/land cover classification, using a hierarchical decision-tree, based on the average values of the mean and maximum NDVI. In a second stage the NDVI thresholds for different agroclimatic zones were adjusted to account for weather variability. Using these procedures 17 annual land cover maps were produced for each year of the period and further condensed into 'change' maps, using a procedure of identifying hierarchical levels of change patterns.

The analysis of the time series shows that most of the land in the CWANA region has remained stable during the period 1982–1999. Nevertheless, major changes in land cover occurred, especially in the Sahel, the Near East and in North Central Asia. Intensification of agriculture, mainly by the conversion of rainfed into irrigated croplands, retrenchment of agriculture and natural vegetation, which are potential indicators of land degradation, but also intensification of natural vegetation, are the main trends throughout CWANA, with significant differences between sub-regions.

1 BACKGROUND

Land degradation is a serious threat to the prosperity of rural populations in dryland areas. Its assessment is hampered by both fundamental and operational problems. First, in most cases land degradation is a creeping phenomenon: it takes time to reach a degraded state. As a result, most of the degradation has already often occurred before the start of the observation period. In addition, it is not always easy to distinguish processes related to aridity and natural climatic fluctuations from human-induced negative trends. For example, changes in biomass, salinity and sand movement may be associated with wet and dry periods rather than human influence. A second problem is one of shortage of reliable basic data on the extent and severity of land degradation, and a third one is the difficulty of harmonizing land degradation information across national and even local boundaries.

The International Center for Agricultural Research in the Dry Areas (ICARDA) deals with germplasm enhancement and natural resource management issues, including land degradation in drylands. Its current mandate region includes North Africa, West Asia, Central and South Asia, and the Horn of Africa. For convenience this area is abbreviated as CWANA (= Central Asia + West Asia + North Africa). In a dryland region, as huge and diverse as CWANA, with limited reliable ground-based resource inventories and monitoring systems, remote sensing presents a highly valuable tool deal with the highly complex issue of land degradation. The main goal of this study is to use remote sensing for the detection of large-scale (regional to sub-continental) land use change trends and to assess whether any such trends can be associated with degradation. To achieve regional coverage for a substantial number of years, the system of the NOAA Advanced Very High Resolution Radiometer (AVHRR) satellites has been selected.

2 METHODOLOGY

Six hundred and twelve 10-daily composites of 8 km-AVHRR reflectance data, covering the period from January 1982 until December 2000, were downloaded from the NASA Web site (http://daac.gsfc.nasa.gov/data/dataset/AVHRR/01_Data_Products/04_FTP_Products/01_ftp_subsets/index.html) for band 1 (0.58–0.68 μm) and band 2 (0.725–1.1 μm). No complete time series was available for the year 1994. These data consisted of separate subsets for Africa (top left: 37.9 N, −20 W; lower right −2 S, 59.9 E) and Asia (top left 59 N, 26.5 E; lower right 4.5 S, 91 E), in 16-bit unsigned format and in Goode's Homolosine Interrupted Space projection. The data were imported as layer stacks and both subsets were mosaiced to form a complete coverage of the CWANA-region (Fig. 1).

The Normalized Difference Vegetation Index (NDVI) was calculated and aggregated into monthly NDVI composites, by retaining the highest NDVI value of three consecutive 10-daily composites, in order to reduce the effects of cloud cover. Additional corrections for noise and sensor drift were made, as well as conversion to geographic projection and merger into a single CWANA-dataset.

NDVI can be considered a proxy but integrated indicator for major phenological vegetation characteristics (green biomass, ground cover, leave area) and photosynthetic activity, and its correlation with vegetation productivity has been well established (Myneni et al. 1997).

2.1 Classification tree

In order to convert this temporal NDVI dataset into a land use/land cover classification, two procedures were developed. The first procedure consisted of developing an empirical hierarchical decision-tree for average weather conditions (Celis et al. 2007).

In contrast with unsupervised or supervised statistical classification procedures, a decision-tree algorithm does not assign all classes in one step, but groups pixels together based on stepwise conditions. This way the data are split up into smaller subgroups, which allow better control of the classification process. The classes that are easier to identify, are selected first.

We used as differentiating criteria the maximum and mean NDVI, the agroclimatic zone, the period of the dry season, and the latitude zone (Fig. 2).

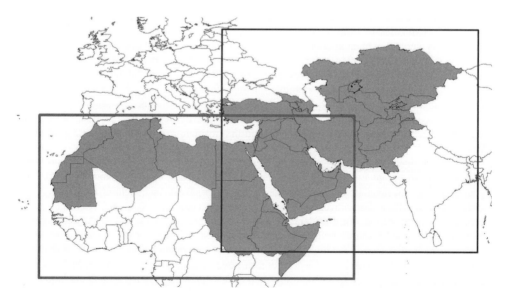

Figure 1.   Location of AVHRR subsets.

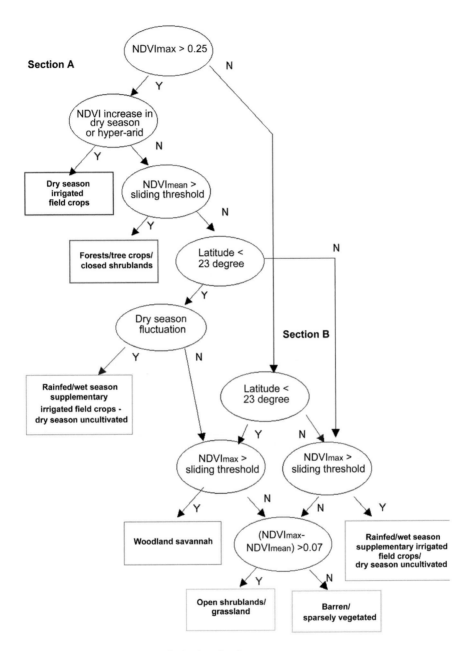

Figure 2. Hierarchical decision-tree for land use/land cover assessment.

The maximum and mean NDVI proved to be the best criteria for differentiating the land use/land cover classes. Instead of fixed thresholds, the decision-tree uses 'sliding thresholds', which vary according to the agroclimatic zone. The fine-tuning of the NDVI thresholds was based on a careful analysis of the available CWANA climate station data and land use maps in different agroclimatic zones, as defined and mapped by UNESCO in its study of the world aridity regimes (UNESCO 1979). To separate croplands the NDVI pattern during the dry season was an important indicator. The addition of latitude as a criterion in the decision-tree enabled to include the typical vegetation patterns of each latitudinal zone and to set different values on the timing of the dry season.

Table 1. Existing land cover/land use maps based on AVHRR information.

Land cover/land use map	Geographical scope	Derived from	Method for classification	KHAT
Asian Association for Remote Sensing (Tateishi 1999)	Asia	NDVI, 4, 5	Unsupervised statistical + decision tree	0.2667
University of Maryland (Hansen et al. 2000)	Global	All 5 AVHRR	Supervised statistical + decision tree	0.2373
International Global Biosphere Program (Belward et al. 1999)	Global	NDVI	Unsupervised statistical + post-classification refinement	0.1976
ICARDA	CWANA		Decision tree	0.5533

The performance of this decision tree-based classification, applied to AVHRR-NDVI imagery with 1-km resolution, covering the period 1 April 1992 to 31 March 1993, was compared to three regional or global land classifications based on statistical procedures (Table 1).

The accuracy of the four classifications was assessed by comparison with a higher-resolution land use/land cover map for Syria (De Pauw et al. 2004) using the KHAT accuracy index (Lillesand & Kiefer 1994). The land use/land cover map of Syria was obtained from visual interpretation of 1:200,000 scale false-color prints of Landsat 5 scenes from the period 1989/1990, followed by digitizing and integration in a GIS, and verified through field work. Additional accuracy tests were undertaken using ground truth samples collected by Tateishi (1999) in Central Asia during the summers of 1996 and 1997 and the spring of 1998.

The second procedure consisted of an adjustment of the NDVI thresholds in the decision tree of Fig. 2 to year-by-year variations in climatic conditions. As these NDVI thresholds apply to average climatic conditions, there is the risk that a given land use/land cover type might be misclassified, if weather conditions at a given place deviated substantially from normal. In fact, it has been well established that the NDVI is particularly sensitive to variations in weather. Strong correlations between NDVI and precipitation, particularly in tropical areas, have been much reported in the literature (e.g. Hielkema et al. 1987, Nicholson et al. 1990, Lotsch et al. 2003). Schutz & Halpert (1993) highlighted the influence of both precipitation and temperature on NDVI in temperate regions. This is confirmed by our own results in a climatologically homogeneous region, covering southern Anatolia, northern and western Syria. Figure 3 shows very high correlation ($r^2 = 0.88$) between the annual maximum NDVI (NDVImax), the annual precipitation and the absolute minimum temperature for the 6 stations in the hatched agroclimatic zone of Figure 4. The individual correlations in the same area between NDVImax and absolute minimum temperature ($r^2 = 0.61$) and between NDVImax and precipitation ($r^2 = 0.53$) are also high.

To account for the fact that in any particular year the actual weather can differ substantially from the average, a procedure was developed to adjust the NDVI thresholds in the decision tree. For this purpose, using the same UNESCO classification for Arid Zones (UNESCO 1979), the entire region was subdivided into 152 zones with homogeneous climatic conditions (Fig. 4). For each climatic zone the average NDVImax for the time series was determined, as well as the average threshold based on the classification tree in Fig. 2. On this basis a threshold adjustment factor was established as follows:

$$[\text{Threshold}]_{i,j} = \text{Threshold}_{av\_j} * \frac{\text{NDVI\_max}_{i,j}}{\text{NDVI\_max}_{av\_j}} \qquad (1)$$

where i = year; j = zone; av_j = average for zone j

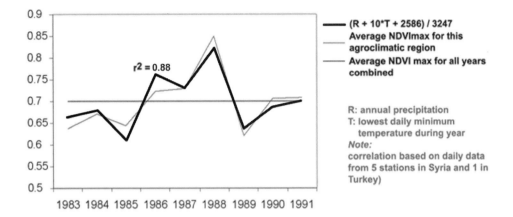

Figure 3. Relationship between average NDVImax and a linear combination of average annual rainfall (R) and minimum temperature of the coldest month (T) for the 6 stations shown in the hatched agroclimatic zone of Figure 4.

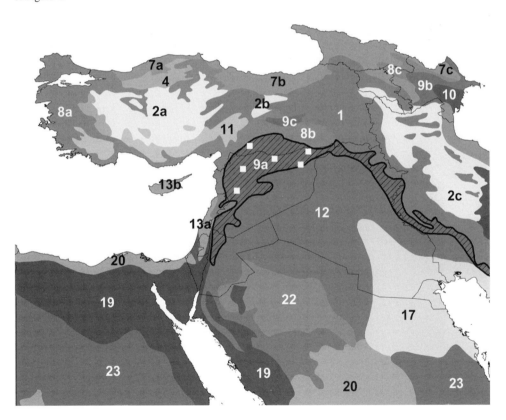

Figure 4. Local zones with homogeneous agroclimatic conditions in West Asia. White squares are climatic stations located in agroclimatic zone 9a.

Example: Zone j has an average NDVImean threshold of 0.8 for the class 'Forests/tree crops/closed shrublands'. The average NDVImax in zone j is 0.66 and the NDVImax for 1982 is 0.74, which indicates better than average growing conditions. So for a correct classification the NDVImean threshold also has to be higher. Following equation (1) it will be equal to 0.897.

137

Using the above algorithm for land use/land cover classification, 17 annual land cover maps were produced by running the decision-tree algorithm with the annually adjusted NDVI thresholds for each year in the period 1982–1999. This dataset was further condensed into four maps, showing the majority land cover classes for the following three-year key periods: 1982–1984, 1987–1989, 1992–1994, and 1997–1999. For each of the sampled three-year periods the majority land cover type in each pixel was retained. If the three years had three different classes, the pixel was classified as 'noise'. Figure 5 illustrates this process for the region around Syria's northwestern border with Turkey, using the period 1987–1989.

2.2 Change analysis

Depending on the value and sequence of this majority land cover type, the following kinds of change were allocated to each pixel: *noise, stable land use/land cover, stable land use/land cover mosaic and change pattern* (Table 2).

Seventeen stable classes were recognized, as well as 66 change combinations, which were regrouped into 22 change classes and four change trends. The 17 stable classes are shown for parts of West Asia in Figure 6.

The 66 change combinations refer to the transition patterns from one land use/land cover type to another between the four key periods. For example, if the land use/land cover type "irrigated croplands" is coded "1", "Forest/Tree crops" is coded "2", and "Rainfed croplands" is coded "3", the change combination "1113" stands for "transition from irrigated to rainfed crops in the period 1992–94" and the change combination "2111" stands for "transition from forest/tree crops to rainfed crops in the period 1987–89".

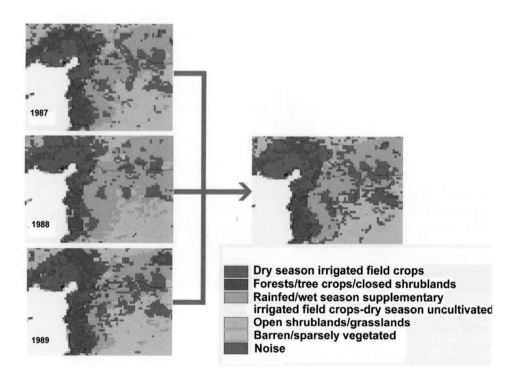

Dry season irrigated field crops
Forests/tree crops/closed shrublands
Rainfed/wet season supplementary irrigated field crops-dry season uncultivated
Open shrublands/grasslands
Barren/sparsely vegetated
Noise

Figure 5. Majority land cover classes for different years, subset northwest Syria (on the left the land cover classes for the years 1987, 1988, 1989; on the right the majority land cover class) (see colour plate page 385).

Table 2. Decision rules for change detection.

1. If all four classes are equal, the land cover remains stable.
   e.g. A A A A; B B B B

2. If a permanent change occurs from one class to another, a change class results;
   e.g. A B B B; A A B B; A A A B; B A A A; B B A A; B B B A

3. If the change from one class to another is not permanent, the combination is interpreted as a stable mosaic of both classes.
   e.g. A A B A; A B A A; A B A B; A B B A; B B A B; B A B B; B A B A; B A A B

4. If there are three or four different classes or a 'noise' class (E), the classification is 'noise'.
   e.g. A B C A; A B C D; A E A A; A A B E

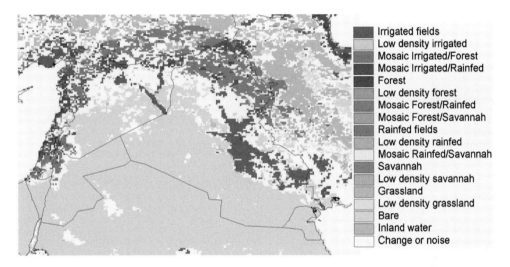

Figure 6. Stable classes for parts of West Asia (see colour plate page 386).

At a higher level, the change combination 1113 also corresponds with the change class "Irrigated > Rainfed cropland", and the change combination "2111" with the change class "Forest > irrigated cropland".

The four change trends differentiated are: *'intensification of agriculture', 'intensification of natural vegetation', 'retrenchment of agriculture',* and *'retrenchment of natural vegetation'*. Examples of 'intensification of agriculture' are the transitions from *Barren* to *Irrigated*, from *Forest* to *Rainfed*, or from *Rainfed* to *Irrigated*. Examples of 'intensification of natural vegetation' are the transitions from *Barren* to *Grassland*, or from *Savannah* to *Forest*. Examples of 'retrenchment of agriculture' are the transitions from *Irrigated* to *Barren*, from *Rainfed* to *Grassland*, or from *Irrigated* to *Rainfed*. Examples of 'retrenchment of natural vegetation' are the transitions from *Forest* to *Grassland*, or from *Grassland* to *Barren*.

Both *Intensification* and *Retrenchment* change trends may be associated with particular forms of land degradation. Intensification could, for example, be associated with deforestation, or be supported by depleting non-renewable water resources. On the other hand, the retrenchment trend could be associated with depleted water resources, salinization, loss of fertility, conversion of forest to grasslands etc. The large-scale change mapping will not provide specific answers, only point to areas where substantial change in the rural environment has taken place during the period 1982–1999.

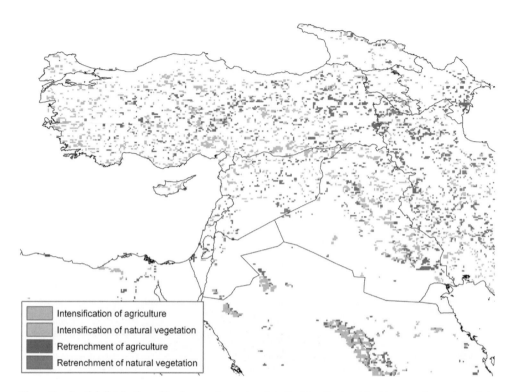

	Intensification of agriculture
	Intensification of natural vegetation
	Retrenchment of agriculture
	Retrenchment of natural vegetation

Figure 7.   Spatial distribution of land cover change trends in the Near East and the Caucasus (see colour plate page 386).

3   RESULTS

On the basis of this hierarchical classification 'change maps' were prepared for eight sub-regions in CWANA (the Maghreb, Northeast Africa, the Sahel and Horn of Africa, the Arabian Peninsula, West Asia and the Caucasus, Southwest Asia, South Asia and Central Asia), and areas belonging to individual change combinations, change classes and change trends were calculated. The results are summarized in Table 3. An example of a 'change map' is shown for the West Asia and Caucasus region in Figure 7.

3.1   *Regional trends*

In terms of total land area, the most dramatic changes in land cover have occurred in the Sahel, followed by North Central Asia. In the former region approximately 750,000 km^2 changed from one land cover to another, in the latter 430,000 km^2. In relative terms, the regions where the most change occurred are the Middle East and the Sahel, where about 14% of the land cover changed. Despite these significant changes, even in the regions with the highest change in land cover, most of the land has remained stable.

3.2   *Sub-regional trends*

At the level of the sub-regions the nature and extent of the changes can be quite different.

3.2.1   *North-West Africa*

The main trend was *intensification of agriculture* (50,800 km^2). This occurred mostly through deforestation to develop irrigated (12,900 km^2) and rainfed cropland (9600 km^2) in the coastal

Table 3. Summary of land cover/land use stability and change trends 1982–1999 in CWANA by sub-region (areas in km^2).

Sub-region	Stability/Change			Change trends			
	Stable	Change	Noise	IA*	RA	INV	RNV
North-West Africa	2,907,004	83,712	203,328	50,816	22,400	7040	3456
North-East Africa	2,554,686	19,392	21,632	6464	11,840	256	832
Africa Sahel	4,232,252	749,824	458,368	125,504	80,768	422,592	120,960
Arabian Peninsula	2,636,992	105,408	34,240	52,928	34,048	15,360	3072
West Asia	1,024,512	218,112	296,512	137,088	63,488	11,392	6144
Caucasus	114,688	19,520	52,224	11,840	6,976	64	640
South-West Asia	2,810,688	368,576	444,352	116,032	159,680	56,448	36,416
Central Asia	2,754,040	431,872	272,320	147,520	125,248	111,040	48,064
TOTAL CWANA	19,034,862	1,996,416	1,782,976	648,192	504,448	624,192	219,584

*IA: intensification of agriculture; RA: retrenchment of agriculture; INV: intensification of natural vegetation; RNV: retrenchment of natural vegetation.

zone, and by converting bare land into irrigated croplands inland (10,900 km^2). A second trend involved a *retrenchment of agriculture* (22,400 km^2), which occurred inland, where both rainfed (7000 km^2) and irrigated fields (12,600 km^2) were taken out of cultivation to become barren/sparsely vegetated areas. A third, less important, trend is one of intensification of the natural vegetation. This occurred mostly as a change from bare areas to grasslands/ open shrublands in the Atlas mountains of Algeria (4800 km^2).

3.2.2 North-East Africa
The only noticeably trends are *retrenchment of agriculture* on the right bank of the Nile (11,800 km^2, of which 7000 km^2 changed from irrigated cropland to bare land) and *intensification of agriculture* (6500 km^2, of which 4700 km^2 changed from bare land to irrigated cropland) on the western part of the Nile delta.

3.2.3 Sahel
The major trend detected was an *intensification of natural vegetation* (422,600 km^2), mainly in the form of a change from barren/sparsely vegetated areas to grasslands/open shrublands (333,100 km^2), but also the transition from grasslands/open shrublands into woodland savannah (82,900 km^2). *Intensification of agriculture* occurred in Southwest Sudan and West Ethiopia (125,500 km^2), in the form of a change from rainfed field crops to tree crops (49,700 km^2) or from woodland savannah to rainfed field crops (59,300 km^2). In the center of both Sudan and Ethiopia, a *retrenchment of agriculture* occurred (80,800 km^2), mainly in the form of a change from rainfed field crops to woodland savannah (58,800 km^2).

3.2.4 Arabian Peninsula
During the period 1982–1999 two contrasting trends of land cover change are observed. *Intensification of agriculture* occurred on 48,800 km^2 whereas *retrenchment of agriculture* occurred on 27,300 km^2. *Intensification of natural vegetation* was observed at the edges of the Asir and Yemen Highlands, in the form of a change from barren/sparsely vegetated areas to grassland/open shrublands (15,400 km^2).

3.2.5 West Asia and Caucasus
In relation to its size, West Asia is the sub-region with the highest degree of land cover change in the period 1982–1999. Of the different change trends, *intensification of agriculture* is the most important one. Intensification occurred both towards irrigated (83,300 km^2) and rainfed (53,800 km^2) cropland. At the coasts of Turkey and Syria deforestation occurred to establish

irrigated fields (36,000 km^2), in North-East Syria rainfed cultivation intensified to irrigated cultivation (18,700 km^2), and in Iraq barren/sparsely vegetated areas were taken into irrigation (25,600 km^2). In central Turkey, grasslands/open shrublands were taken into rainfed cultivation (27,600 km^2). *Retrenchment of agriculture* (63,500 km^2) occurred mostly in inland Turkey, in the form of a change from rainfed cultivation to grasslands/open shrublands (34,000 km^2), and in Iraq in the form of irrigated fields taken out of cultivation by change to barren/sparsely vegetated areas (14,800 km^2).

In the Caucasus *intensification of agriculture* (11,800 km^2) occurred mostly in the form of a change from rainfed fields to either tree crops (3900 km^2) or irrigated fields (3300 km^2). A *retrenchment of agriculture* (7000 km^2) occurred mainly in the form of a change from irrigated fields to rainfed fields (3300 km^2) or from rainfed fields to grasslands/open shrublands (2700 km^2).

3.2.6 *South-West Asia*

The main trend in Southwest Asia is one of *retrenchment of agriculture* (159,700 km^2) and is particularly noticeable in Pakistan and Afghanistan. Mostly in Pakistan irrigated croplands reverted to barren/sparsely vegetated areas (81,100 km^2), whereas in Afghanistan the conversion was mostly into grasslands/open shrublands (41,500 km^2).

3.2.7 *Central Asia*

In Central Asia *intensification of agriculture* (147,500 km^2) occurred mostly in the form of a change from grasslands/open shrublands into rainfed cultivation (84,200 km^2). The inverse trend, a change from rainfed cultivation to grasslands/open shrublands (81,000 km^2), is the major component of the retrenchment of agriculture (125,200 km^2). Intensification of natural vegetation is another important trend (111,000 km^2), almost entirely (110,000 km^2) due to the change from sparsely vegetated areas to grasslands/open shrublands. This intensification trend is very noticeable in South-West Kazakhstan near the Caspian Sea. The inverse trend, a change from grasslands/open shrublands into barren/sparsely vegetated areas (46,300 km^2) is particularly clear in Central Kazakhstan.

4   INTERPRETATION

A first issue to consider in the interpretation of the results is the resolution of the imagery used. The 8-km grid cell resolution of the Pathfinder dataset entails that in many situations the pixels are not internally homogeneous in their land use/land cover categories. This would certainly be the case in agricultural areas, where rotation practices ensure proximity between fields under crops and fallow areas. For such mixed pixels the total reflectance is the sum of the relative reflectance contributions by all elements in the pixel. For this reason the area estimates of land use/land cover changes provided in the previous section are only indicative. Since the majority land cover class in each 3-year period takes all the area of the pixel, even if it only occurs on a fraction of it, it may be overestimated. On the other hand the areas of minority classes are not taken into account and may thus be underestimated.

Another concern is the interpretation of the observed land use/land cover changes in terms of causes, triggers and degradation trends in different parts of the CWANA region.

In the Sahel the likely reason for the trend towards intensification of natural vegetation is a recovery of the natural vegetation, after the drought, experienced in the Sahel between 1961 until 1984, was followed by a period with normal rainfall. This 'greening of the Sahel' has been described by many authors (e.g. Hellden 1988, Tucker et al. 1991, Olsson et al. 2005). The northward shift of the desert boundary after 1984, also described by Tucker (Tucker et al. 1991) can therefore be interpreted as a regeneration of the grasslands/open shrublands when rainfall returned to normal after the long drought cycle. Anyamba & Tucker (2005) associate the recovery of the Sahelian vegetation with the transition from a high-deficit precipitation regime in the early part of the time series (1982–85) to a slight surplus pattern at the end (1994–99). However, it is not sure that a decadal-scale precipitation cycle is the only factor involved. Herrmann et al. (2005) hypothesize a human-induced trend on top of the climatic trend, whereas Olsson et al. (2005) consider the possibility of a contribution from migration and land use changes.

As for the trends in agriculture, intensification in the southwest of Sudan and the west of Ethiopia, retrenchment in the center of both Sudan and Ethiopia, no clear explanation suggests itself. In Ethiopia the areas with retrenchment of agriculture coincide with areas affected by severe drought in the early 1980s. It is therefore possible that the intensification of agriculture in western Ethiopia is due to settlement and agricultural development in parts of the country that were less affected by the drought.

In the Arabian Peninsula the area under irrigation more than doubled between 1980 and 1996 (FAO 2001), aided by the use of modern irrigation technology, such as center-pivot and drip irrigation. Spectacular growth in irrigated agriculture occurred mostly in the center of Saudi Arabia, with barely any irrigated areas in 1983 and ten years later reaching their maximum extent (De Pauw 2002). This trend towards intensification of agriculture was driven by pumping fossil water, itself made possible by generous government subsidies on fuel for well drilling (FAO 1997).

The reverse trend, retrenchment of agriculture, is the mirror image of the intensification trend. It was caused by the fact that large irrigated areas were also taken out of cultivation in Saudi Arabia, initially due to the depletion of the fossil aquifers and, as the unsustainability of the practices became clear to the government, by policy changes in respect of groundwater use. From the change analysis it appears about 400 km^2 was taken out of production between 1984 and 1987, 8300 km^2 between 1989 and 1992, and the largest contingent (18,600 km^2) between 1995 and 1997. The policy reversal, stopping fuel subsidies to save water and conserve the environment, came in 1994 (O'Sullivan, 1994). This is another example, in addition to the case studies reported by Nielsen & Adriansen (2005), of how government policies can impact on the spread or containment of land degradation.

The observed intensification of natural vegetation at the edges of the Asir and Yemen Highlands is probably related to increased rainfall after the Sahelian drought.

With some exceptions, the change patterns in other sub-regions are more fragmented. A first exception is a clear linear intensification trend of the natural vegetation in Central Asia, resembling in pattern, but at smaller scale, the Sahelian greening. With focus north of the Caspian Sea and extending along parallel 48°N to the west of the former Aral Sea, there is no clear explanation for this trend in the literature. It is possible that the trend is related to the decline of livestock herds and corresponding decrease in grazing intensity after the collapse of the Soviet Union in the early 1990s, a cause proposed for the NDVI patterns in other parts of Kazakhstan (de Beurs & Henebry 2004).

A second exception are the trends of intensification and retrenchment of agriculture in the Nile Valley in Egypt, with a clear pattern of retrenchment along the right bank of the Nile, mirrored to a lesser extent on the left bank of the Nile, and an intensification trend west of the Nile delta. This area coincides with the 'Mubarak' land reclamation project, which was initiated in 1987, as reported by Nielsen et al. (2005). The areas with retrenchment of agriculture along both banks of the Nile are probably related to urban encroachment and expansion of public infrastructure. Urban expansion at the expense of agricultural land is a long-term trend in Egypt, comprehensively documented by El Hefnawi (2005), and its extent has been quantified in several case studies using remote sensing with higher-resolution data (e.g. Lenney et al. 1996, Ghar et al. 2005).

In Pakistan a clear trend of retrenchment of agriculture is evidenced in two distinct strands, one running along the Kirthar, central Brahui and Sulaiman ranges in Balochistan Province, the other one bordering the Thar desert in Sindh Province. These arid to semi-arid areas are well known for resource and degradation problems, such as deforestation, sand encroachment, falling water tables, and soil erosion (e.g. Government of Pakistan et al. 2000, International Union for Conservation of Nature and Natural Resources et al. 2000). However, given the general lack of resource inventories there is no direct evidence linking these problems to the observed retrenchment trend.

5 CONCLUSIONS

After the year-to-year weather variations were compensated for and taking into account the local knowledge about some large-scale change trends, it is still a challenge to understand the causes of the land use/land cover changes, and even more difficult to interpret them in terms of degradation trends. An intensification of agriculture is the easiest to interpret because it is always human-induced, whereas a retrenchment of agriculture can have three possible causes: a long term weather change for the worse, a lack of inputs to maintain the cultivation or a degradation of the natural resource base (too intensive use of natural resources). Both the intensification and the retrenchment of natural vegetation could be either human-induced or result from a longer-term change in weather. The identification of a trend towards denser natural vegetation proves the possibility of natural regeneration and puts the issue of land degradation in a more cyclical perspective. In fact, in all degradation studies the time frame is paramount: length and start of the time series determine to a large extent whether we see deterioration, improvement, cycles or stable conditions.

The study evidences the main advantage of the 'hot spots' approach: its ability to detect macro-scale patterns of change with considerable savings in time and financial resources. It also underscores some of its (unsurprising) limitations. The issue of scale is probably the most important. Given the resolution of the AVHRR imagery, with one pixel covering approximately 64 km^2, it is obvious that only large-scale change patterns can be detected. This implies that changes occurring at smaller-scale, and associated degradation trends, can be masked by normal land use dynamics (e.g. fluctuations in the land use patterns, particularly in response to rainfall fluctuations). This inability to detect small-scale changes is bound to either under-estimate or over-estimate land use change and associated degradation trends.

In short, due to the low resolution of the imagery, there are considerable limitations to what can be seen. For this reason, a second assessment stage is necessary, in which the 'hot spots' are further characterized using ground-based observation networks, including biophysical and socioeconomic characterization components, complemented with high-resolution satellite imagery, such as Landsat or SPOT.

REFERENCES

Anyamba, A. & Tucker, C.J. 2005. Analysis of Sahelian vegetation dynamics using NOAA-AVHRR NDVI data from 1981 to 2003. *Journal of Arid Environments* 63: 596–614.

Belward, A., Estes, J. & Kline, K. 1999. The IGBP-DIS Global 1-km Land-Cover Data Set Discover: A project overview. *Photogrammetric Engineering & Remote Sensing* 65 (9): 1013–1020.

Celis D., De Pauw, E. & Geerken, R. 2007. *Assessment of land cover and land use in Central and West Asia and North Africa.* Part 2. Hot spots of land cover change and drought vulnerability. International Center for Agricultural Research in the Dry Areas (ICARDA), Aleppo, Syria, x + 69 p.

de Beurs, K.M. & Henebry, G.M. 2004. Land surface phenology, climatic variation, and institutional change: analyzing agricultural land cover change in Kazakhstan. *Remote Sensing of Environment* 89: 497–509.

De Pauw, E. 2002. *An agroecological exploration of the Arabian Peninsula.* ICARDA, Aleppo, Syria, 77 p. ISBN 92-9127-119-5.

De Pauw, E., Oberle, A. & Zöbisch, M. 2004. *Land cover and land use in Syria—An overview.* Jointly published by Asian Institute of Technology (AIT), International Center for Agricultural Research in the Dry Areas (ICARDA) and the World Association of Soil and Water Conservation (WASWC). 47 p + 1 map A3 + 1 CD.

El Hefnawi, A.I.K. 2005. *"Protecting" agricultural land from urbanization or "Managing" the conflict between informal urban growth while meeting the demands of the communities.* In Third Urban Research Symposium on "Land Development, Urban Policy and Poverty Reduction", Brasilia, 4–5 April 2005. URL: http://worldbank.org/urban/symposium2005/pp.html).

FAO (Food and Agriculture Organization of the United Nations) 1997. *AQUASTAT. Country Profile: Saudi Arabia.* URL: http://www.fao.org/AG/AGL/aglw/aquastat/countries/saudi_arabia/index.stm.

FAO (Food and Agriculture Organization of the United Nations) 2001. *FAOSTAT Agriculture Data.* FAO Statistical Databases.

Ghar, M.A., Shalaby A. & Tateishi, R. 2005. Agricultural land monitoring in the Egyptian Nile delta using Landsat data. *International Journal of Environmental Studies* 61 (6): 651–657.

Government of Pakistan, World Wide Fund for Nature, Pakistan and International Union for Conservation of Nature and Natural Resources. 2000. *Biodiversity Action Plan for Pakistan*. 86 p, ISBN 969-8141-35-9.

Hansen, M., Defries, R., Townshend, J. & Sohlberg, R. 2000. Global land cover classification at 1 km spatial resolution using a classification tree approach. *International Journal of Remote Sensing* 21 (6): 1331–1364.

Hellden, U. 1988. Desertification monitoring: Is the desert encroaching? *Desertification Control Bulletin* 17: 8–12.

Hermann, S.M., Anyamba, A. & Tucker, C.J. 2005. Recent trends in vegetation dynamics in the African Sahel and their relationship to climate. *Global Environmental Change* 15: 394–404.

Hielkema, J.U., Prince, S.D. & Astle, W.L. 1987. Monitoring of global vegetation dynamics for assessment of primary productivity using NOAA advanced very high resolution radiometer. *Advances in Space Research* 7 (11): 81–88.

International Union for Conservation of Nature and Natural Resources, Pakistan and Government of Balochistan 2000. *Balochistan Conservation Strategy*. IUCN. URL: http://www.bcs.iucnp.org/

Lenney, M.P., Woodcock, C.E., Collins, J.E. & Hamdi, H. 1996. The status of agricultural lands in Egypt: the use of multitemporal NDVI features derived from Landsat TM. *Remote Sensing of Environment* 56 (1): 8–20.

Lillesand, T.M. & Kiefer, R.W. 1994. *Remote Sensing and Image Interpretation*. 3rd ed. New York, USA.

Lotsch, A., Friedl, M.A., Anderson, B.T. & Tucker, C.J. 2003. Coupled vegetation-precipitation variability observed from satellite and climate records. *Geophysical Research Letters* 30 (14): 1774.

Myneni, R.B., Keeling, C.D., Tucker, C.J., Asrar, G. & Nemani, R.R. 1997. Increased plant growth in the northern high latitudes from 1981 to 1991. *Nature* 386: 698–702.

Nicholson, S.E., Davenport, M.L. & Malo, A.R. 1990. A comparison of the vegetation response to rainfall in the Sahel and East Africa, using normalized difference vegetation index from NOAA AVHRR. *Climatic Change* 17: 209–241.

Nielsen, T.T. & Adriansen, H.K. 2005. Government policies and land degradation in the Middle East. *Land Degradation and Development* 16: 151–161.

Olsson, L., Eklundh, L. & Ardoe, J. 2005. A recent greening of the Sahel—trends, patterns, and potential causes. *Journal of Arid Environments* 63: 556–566.

O'Sullivan, E. 1994. Saudi Arabia: Wheat harvest drop forecast. *Middle East Economic Digest*, May 2, 1994.

Schultz, P.A. & Halpert. M.S. 1993. Global correlation of temperature, NDVI and precipitation. *Advances in Space Research* 13 (5): 277–280.

Tateishi, R. 1999. AARS Asia 30" Land Cover Data Set with Ground Truth Information by the Land Cover Working Group of the Asian Association on Remote Sensing and CEReS, the Center for Environmental Remote Sensing of Chiba University, Japan.

Tucker, C.J., Dregne, H.E. & Newcomb, W.W. 1991. Expansion and contraction of the Sahara Desert between 1980 and 1990. *Science* 253: 299–301.

UNESCO 1979. *Map of the world distribution of arid regions*. Map at scale 1:25,000,000 with explanatory note. UNESCO, Paris, 54 p. ISBN 92-3-101484-6.

Recent Advances in Remote Sensing and Geoinformation Processing
for Land Degradation Assessment – Röder & Hill (eds)
© *2009 Taylor & Francis Group, London, ISBN 978-0-415-39769-8*

Fuzzy integration of satellite data for detecting environmental anomalies across Africa

P.A. Brivio, M. Boschetti, P. Carrara & D. Stroppiana
IREA CNR, Institute for Electromagnetic Sensing of the Environment, Milano, Italy

G. Bordogna
IDPA CNR, Institute for the Dynamics of Environmental Processes, Bergamo, Italy

ABSTRACT: New opportunities for the detection of anomalous environmental conditions at continental and global scale are offered by the exploitation of time series of satellite remotely sensed data coupled with appropriate geoinformation processing. The assessment of environmental conditions is generally based on complex models that require large dataset and whose performance depends on expert knowledge and specific tuning. In this paper we propose a synthetic indicator which is obtained by aggregating the scores of diverse observable factors that reinforce the convergence of anomaly evidence. The factors evaluation and aggregation are framed within the fuzzy set theory and approximate reasoning methods so as to take into account the uncertainty and incompleteness affecting the collection of factors, the estimation of their importance and the complexity of their interrelationships. The methodology based on parameters derived from the analysis of time series of low resolution satellite data concerning water availability and vegetation phenology is described and the results obtained from its application in the detection of anomalous environmental conditions over the African continent are presented and discussed.

1 INTRODUCTION

Changes and degradation in land surface attributes and vegetation cover status have considerable implications for a variety of ecosystem processes and profound impact on the societies exploiting these ecosystems (Lambin et al. 2003). These conditions can be particularly critical in large unmanaged areas, such as many regions of the African continent (Galvin et al. 2001, Linderman et al. 2005).

Operational monitoring of environmental conditions over large areas, from regional to continental scales, aims to provide reliable and understandable information to help decision-makers in destining resources where alarming situations are likely to occur. In such a wide context, several initiatives have been promoted and methodologies have been developed with the objective of depicting the status of the environment. The choice of the most suited approach is a trade off between the type of information that are to be provided, the users and project's requirements, and the constraints given by the availability of data and implementation tools.

At continental and global scales, data availability is particularly critical due to the need of consistent datasets over a sufficiently long period of time. Satellite-based Earth Observation (EO) systems constitute a valuable and often unique source of information for environmental monitoring: they provide consistent and continuous data, both in time and space, over large and remote areas; a necessary condition in operational monitoring (Loveland et al. 2000, Friedl et al. 2002, Bartholomé & Belward 2005). These data can be integrated with information derived from other sources, e.g. ground data, socio-economic statistics, etc. Multi-source spatial data fusion and interpretation models and systems are therefore required for environmental status assessment (Wickham et al. 1999, Pykh et al. 2000).

The understanding and quantification of the nature of land-use/cover change at global and continental scales and the assessment of environmental conditions can be formalised by complex models whose performance, due to the interdisciplinary character of the dynamics involved, require a large set of data, expert knowledge and specific tuning. In most cases data are lacking or incomplete and the knowledge of the phenomena and their inter-relationships is uncertain and approximate. For these reasons the development and implementation of continental and global models is a challenge.

Environmental indicators can help in overcoming these limitations by providing a synthetic picture of the phenomena under analysis (Lenz et al. 2000, Tran et al. 2002). An environmental indicator can be defined as a means devised to reduce a large quantity of data to its simplest form, retaining essential meaning of the questions that are being asked for the data, and supplying simplified information about complex system, or not easily measurable criteria (Ott 1978). However, considerable difficulties arise when trying to assess the environmental status by adopting classical mathematical approaches that cannot deal with the uncertainty and the imprecision/vagueness involved at distinct levels.

This paper proposes a new approach based on fuzzy logic and approximate reasoning techniques to integrate a set of contributing factors, e.g. vegetation conditions and phenology, climate and, eventually, socio-economic data, into a synthetic indicator. The contribution of each factor is evaluated through the use of fuzzy set theory (Zadeh 1965) and the aggregation is performed by soft integration techniques (Yager 1988) to derive an anomaly indicator. The Anomaly Indicator (AI) identifies those situations that can be interpreted as signals of changes occurring on the vegetated land surface. Although commonly used with a negative meaning, a change is not necessarily an adverse effect on the vegetation cover (Lambin et al. 2003).

Fuzzy set theory allows to flexibly model the subjective nature of uncertainty and imprecision/vagueness affecting the knowledge of environmental phenomena. Soft integration techniques can help in creating realistic, human-centred representations of environmental phenomena, reflecting the more or less conservative attitude of the human expert, thus avoiding oversimplifications and crisp decisions.

The AI here proposed was developed in the framework of the Observatory on Land Cover and Forest Change (OLF) of the GeoLand project (http://www.gmes-geoland.info) with the objective of performing periodic monitoring at continental scale through the provision of a synthetic picture of the environmental status. The AI highlights alarming conditions to enhance those areas where the occurrence of events, given the ambient conditions, may determine situations where changes are presently undergoing or are likely to occur in the near future.

2   MATERIALS AND METHODS

2.1   Conceptual model: anomaly indicator

Environmental systems present considerable difficulties to experts and modellers. Compared with mathematical and physical models applied to engineered systems, building an adequate model of the ecosystems is severely limited by the incomplete and inaccurate knowledge of the basic components of the ecosystem and their complex interrelationships. Moreover, it is almost impossible to make real experiments where we perturb system inputs to observe and understand the resulting behaviour and effects (Lenz et al. 2000).

Alternative approaches are based on the integrated modelling, where a large quantity of data and prior information are conveyed into a synthetic environmental indicator that captures as much as possible of the cause-effect relationships (Lenz et al. 2000).

The conceptual model behind the approach proposed here is that, whereas a single contributing factor is not sufficient to assess the status of a phenomenon, a synthetic indicator could be generated from several factors combined together to reinforce/weaken this assessment (Brivio et al. 2006). The detection of anomalous environmental conditions at continental scale is therefore based on a multi-criteria soft integration where individual factors contribute to the synthetic indicator of anomaly through the concept of reinforcement of evidence: the contribution relies on the anomaly of each factor defined as the departure from some reference condition, namely some ideal state of the factor. In this study, the ideal state for a particular factor was simply built by using its long-term average computed from a time series of available observations. Conceptually the anomaly indicator represents the distance of the ecosystem to an arbitrarily ideal ecosystem that has the ideal states for the contributing factor (Tran et al. 2002).

The *AI* is designed to highlight as anomalies departures from reference conditions: the greater the departure the greater the anomaly. The *AI* proposed here does not discriminate between positive (above long term average) or negative (below long term average) anomalies. Although this is highly desirable, it requires the identification of the type of contribution (positive or negative) for each factor and that this association is the same the whole area under analysis (in this case: the continent). However, we found that both these conditions are rarely satisfied: in most of the cases no "a priori" experience is available and nobody is able to uniquely identify the factors that lead to either positive or negative anomalies; in other cases experts have conflicting ideas and they disagree on the positive/negative contribution of the factors to the anomaly. A rainfall increase, for example, can contribute either to favourable conditions, when its contribution mitigates vegetation stress, or negatively when its excess can enhance flood risk. Similarly, a reduction of fire activity is not always a positive phenomenon; it can in fact produce negative effects altering the complex equilibrium between herbaceous and woody components of a specific ecosystem and enhancing plant encroachment risk (Palumbo et al. 2003). The lack of a consensus on the 'sign' of the contribution of each factor at continental scale led us to assume no distinction between positive and negative anomalies. However, the formal framework under which the proposed approach is rooted is flexible enough to manage the association "factors/sign of anomaly" if it can be determined by expert knowledge.

Figure 1 summarizes the main steps of the proposed methodology, which is described in detail in the following sections.

The multi-criteria integration approach is formally represented in the following expression:

$$AI = F(f_1, \ldots, f_n, I_1, \ldots, I_n) \tag{1}$$

where

- *AI* is the anomaly indicator, intended as a measure of the distance between the actual eco-system situation and the ideal reference state, and it is the higher-level information derived from multi-criteria integration. The integration of the contributing factors, i.e. the available parameters that interact with the process, is based on the concept of reinforcement of evidence.
- *F* is the operator that realises the integration of the contributions of each factor, by means of an Ordered Weighted Averaging (OWA) aggregation operator which makes it possible to take into account flexible compensations of the judgements (Yager 1988).
- *fj* is a contributing factor, i.e. a variable/parameter, which is relevant for the computation of the synthetic indicator. The computation of its contribution is framed in the fuzzy sets theory

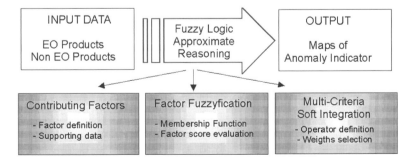

Figure 1.   Block-diagram of the main steps in the proposed methodology for the detection of anomalous environmental conditions.

(Zadeh 1965, 1975): it results in a score, i.e. a real number in the [0,1] range associated to each elementary unit (usually the pixel) of the area under investigation. This value is not intended as a quantitative measure of a process under analysis, instead it is an evaluation of the factor contribution in term of anomaly.
− Ij is the importance associated to the contributing factor fj.

2.2 Contributing factors

Factors contributing to the environmental assessment may include different type of information concerning both physical parameters and socio-economic information of the region. Physical parameters of the ecosystem related to the vegetation cover, water resources, fire activity, etc. can be derived from satellite observations (Tucker & Choudury 1987, Levine 1996, Tansey et al. 2004). Presently several coarse spatial resolution sensors, such as the MODerate Imaging Spectrometer (MODIS), SPOT-Vegetation, NOAA Advanced Very High Resolution Radiometer (AVHRR), are available to monitor the status and the changes of the ecosystems at continental scale trough the availability of time series of observations.

Since the approach has been simplified into the reinforcement of evidence, the selection of the input data can be merely based on the assumption of the presence of a relationship between the factor and the indicator without further formalisation.

In this analysis, the AI is based on the integration of contributing factors that describe the anomaly of both vegetation phenology and rainfall occurred during the season. The anomaly of each contributing factor is derived as the difference between the current observation and the long-term average (LTA) assumed as the reference/ideal condition: these distances from the reference are the actual factors to be integrated into the AI. The LTA is computed from the available historical datasets.

For water availability, ten-day 8 km Rainfall Estimates (RFE version 2.0) of the Famine Early Warning System Network (FEWS-NET) were available since 1995 over the African con-tinent (http://www.fews.net/). The RFE algorithm integrates data from the geostationary METEOSAT-7 satellite, the Special Sensor Microwave/Imager (SSM/I) and the Advanced Microwave Sounding Unit (AMSU) with rain gauge data from World Meteorological Organization (WMO) Global Telecommunication System (GTS). These estimates have been used to derive the contributing factor related to water availability (frain) as 30-day cumulated rainfall for each decade of the period 1996 to 2002.

For vegetation phenology, we used 10-day Normalized Difference Vegetation Index (NDVI) obtained via a Maximum Value Composite (MVC) of the 8 km NOAA-AVHRR PAL Pathfinder data, covering the entire African continent. Three contributing factors related to vegetation phenology were derived from the analysis of the NDVI temporal profiles as follows:

− *fstart*: start of apparent growing season, the inflexion point of the portion of the NDVI temporal profile that shows an increasing trend from the minimum value. Mathematically the inflexion point is derived as the maximum of the first derivative of the interpolated (piece-wise logistic) NDVI curve (Zhang et al. 2003).
− *fpeak*: peak of the growing season, i.e. the maximum NDVI value observed during the season.
− *flength*: duration of the growing season, the time difference between the end and the start of the growing season. The end of the growing season is identified as the time, occurring after the season's peak, when NDVI value reaches a threshold established as a fraction of the difference between the minimum and the maximum values.

Although NDVI of a month is highly correlated with cumulated rainfall the preceding 30-day, phenology may depend on multiple environmental factors (Cleland et al. 2006) among which precipitation is one of the important parameters; in some cases it is considered the main constraint for the "stability" of an ecosystem (Sankaran et al. 2005). In order to provide a synthetic indicator valid at continental scale both phenology and rainfall must be considered at the same time.

The long-term average was computed over the 1990–2002 period for NDVI-related parameters, while for rainfall LTA was computed for the 1996–2002 period.

Long time series of remotely sensed data, gathered by operational low-resolution EO systems are one of the main sources of information for long-term environmental monitoring. However, it has been widely acknowledged by the previous literature that those datasets have several caveats. Differences in design of different sensors of the same series, variation of the calibration coefficients over time, changes in the ground segment of the EO missions result in variations of the satellite data, which may not reflect actual changes occurred on the Earth surface (de Beurs & Henerbry 2004, Boschetti et al. 2006). The selection of contributing factors related to vegetation phenology instead of the direct sensor measurements partially reduces the influence of mission and sensor changes involved in the long time series acquisition.

2.3 *Fuzzy membership and factor score*

The factor values estimated from the available datasets do not contribute directly to the *AI*. Instead, the possible influence of a factor on the synthetic indicator is represented by an elastic constraint on the values of the contributing factor: only those factor values satisfying the constraint can be interpreted as partial evidences of the *AI* status and thus contribute to determine the *AI* value (Carrara et al. 2006).

The contribution of each factor f_i to the anomaly indicator is described by a fuzzy membership function

$$\mu(f_i) : D(f_i) \to [0,1] \tag{2}$$

The choice of a method based on the fuzzy sets theory is driven by its ability to deal with vague and incomplete information as well as the possibility of applying a gradual degree of membership for score extraction instead of crisp value. Indeed in environmental assessment it is more likely to deal with a smooth change between two different conditions rather than an abrupt change (Burrough 1996).

Moreover, when multi source data are to be integrated, it is necessary to normalize them to the same domain so as to achieve consistency and comparability; furthermore, in some cases it can be useful to perform a non-linear normalization. Since the contributing factors have distinct domains D(fi), the computation of their degrees of satisfaction of the elastic constraint serves also the purpose of non-linear normalization of each factor contribution.

The specification of grades of membership, i.e. of factor scores, is one of the key problems when applying fuzzy sets to spatial data. Major approaches that are used to assign grades of membership can be grouped in the following types: completely data-driven, partially data-driven, user-driven (Robinson 2003). In the last two approaches domain experts have a role in defining the membership functions.

In the case of the completely data-driven approach, the factor score functions $\mu(f_i)$ are defined purely on the basis of the statistical analysis of a set of values assumed by a specific contributing factor. For example, if the contributing factor is the difference between the present value and its LTA, its statistical distribution (frequency histogram) shows how the factor behaves on average and which range of values can be assumed as normal (i.e. the most frequent ones).

Most standard membership functions can be categorized as open form or closed form, and as linear or S-shaped membership function (Burrough & McDonnel 1998). Despite the simplicity and elegance of the triangular and trapezoidal membership functions, the Gaussian function $G : X \to [0,1]$ found large use in GIS applications related to the environmental science due to its smooth shape. It is common to define the Gaussian function as:

$$G(f_i; \alpha, \sigma) = \mu_G(f_i) = e^{-\left(\frac{f_i - \alpha}{2\sigma^2}\right)} \tag{3}$$

where the parameters α and σ can be easily related to familiar concepts like mean and variance.

In this study the data-driven approach was used to assign grades of membership and the function $\mu(f_i)$ was modelled as the complement of the Gaussian function interpolating the frequency histogram.

This step relies on a spatial stratification necessary to identify areas where the factors are assumed to have, on average, the same behaviour. Over these homogeneous areas the factor's variability is assumed to be solely due to changes in the status of the environment and not to be related to the spatial variability of the ecosystem characteristics. The choice of the most suited stratification scheme is therefore an important step since it influences the shape of the factor score function. An optimal solution should foresee stratification layers specific for each contributing factor: ecosystems maps could be more appropriate for rainfall and land cover maps may be more suitable for phenological parameters. A possible stratification criterion could be derived from the intersection of more information layers such as the eco-regions and the land cover map, e.g. the Global 200 (Olson & Dinerstein 1998) and the GLC2000 (Bartholomè & Belward 2005). However, this solution may produce an over-fragmentated stratification that is not easily applicable and practically usable. Our choice was to adopt, as suitable stratification of the African continent, the GLC2000 land cover map of 27 classes derived from SPOT-Vegetation data. The 1 km GLC2000 land cover map was resampled to 8 km spatial resolution using a majority filter and non vegetated classes, such as deserts, were masked out. For a particular factor fi a unique Gaussian membership function $\mu_G(f_i)$ is defined for each homogeneous area and for each ten-day of the year (and the complement to 1 used as anomaly function). The fitting accuracy, which varies with both the land cover class and the factor, was quantified using the Root Mean Squared Error (RMSE) and used in the integration step. The scores of individual contributing factor were actually computed for each ten-day period of each year and then averaged monthly.

Figure 2 shows the histogram (circle markers), the interpolation curve $\mu_G(f_i)$ (dashed line) and the anomaly score function membership function $\mu_G(f_i)$ (continuous line) for the contributing factor "season length" in two GLC2000 vegetation classes. Before interpolation, the frequency values were rescaled between 0 and 1 so that the maximum frequency corresponds to 1 (normality) and the least frequent values to 0 (least frequent cases far from the ideal state). The interpolated Gaussian function represents the normality where the most frequent cases are the normal ones.

The membership function $\mu_G(f_i)$ indicates that a greater distance from the ideal condition (long-term average) implies a higher factor score, i.e. more reinforcement to an anomaly situation can be brought by the present observation of the contributing factor. A narrow membership

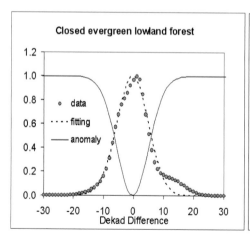
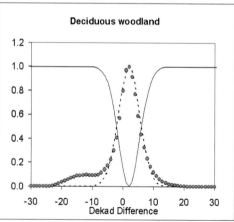

Figure 2. The frequency histograms (circle markers) of the historical dataset, the fitting Gaussian functions (dashed line) and the anomaly membership functions (continuous line) for the 10-day period April 1–10 for the contributing factor f_{lenght}. The functions were derived for the land cover classes Closed Ever-green Lowland Forest and Deciduous Woodland of GLC2000 map.

function implies that even a small departure from the average value can be labelled as highly anomalous. For instance, the class Deciduous Woodland is characterised by a more homogeneous season, in term of length, with respect to Closed Evergreen Lowland Forest class: the same deviation brings therefore larger anomaly contribution than for Evergreen forest.

2.4 Soft integration of contributing factors

The proposed strategy for soft integration allows coping with the imperfect knowledge of the relationships between the contributing factors and the modelled synthetic indicator *AI*.

This piece of knowledge concerns the representation of the semantics of the integration operation resembling the expert's attitude and expertise. Soft integration is able to formalise the expert's attitude in a context of imprecise knowledge that can be expressed by fuzzy linguistic majority quantifiers such as most of . . . (e.g. most of the criteria are to be met). The integration is modelled as a compensatory function implementing the mutual reinforcement of partial complementary and redundant pieces of evidence of the phenomenon.

The integration of multiple contributing factors into a synthetic indicator score is conceived as a quantifier guided aggregation of fuzzy factor scores. Aggregation operations on fuzzy sets are operations that combine several fuzzy sets in a desirable manner to produce a single fuzzy set (Klir & Yuan 1995). Moreover, the integration has been conceived so as to allow the selection of the way in which factor scores are aggregated and possibly weighted by their importance.

Within this formal framework the Ordered Weighted Averaging (OWA) operators were introduced to offer a flexible and comprehensive way to define a complete family of soft integration operators reflecting different attitudes in combining a set of contributing factors (Yager 1988, 1996).

An OWA operator of dimension n is a function $OWA: R^n \rightarrow R$, that has associated a set of weights $(w_1, \ldots, w_n)$, so that $w_i \in [0, 1]$ and $\sum_{i=1}^{n} w_i = 1$, and is defined to aggregate a list of values $(p_1, \ldots, p_n)$ according to the following expression:

$$OWA(p_1, \ldots, p_n) = \sum_{i=1}^{n} w_i p_{\sigma(i)} \qquad (4)$$

where $\sigma : \{1, \ldots, n\} \rightarrow \{1, \ldots, n\}$ is a permutation such that $p_{\sigma(i)} \geq p_{\sigma(i+1)}, \forall i = 1, \ldots, (n-1)$, i.e. $p_{\sigma(i)}$ is the i-th highest value in the set $\{p_1, \ldots, p_n\}$. The permutation σ sorts the elements $\{p_1, \ldots, p_n\}$ in decreasing order and hence each weight w_i of the OWA is associated to an ordered position rather than to a specific factor.

The weights vector uniquely identifies an OWA operator. By appropriate selection of the weights, distinct semantics of the aggregation can be implemented so that exact approaches are limiting cases of fuzzy approaches (Yager 1996). The arithmetic mean is modelled with $w = (1/k, \ldots, 1/k)$. The weights vector $W* = (1, 0, \ldots, 0)$ identifies an OWA that selects the maximum of the factor scores, while the vector $W* = (0, 0, \ldots, 1)$ selects the minimum.

The definition of the weighting vector can be accomplished by either exploiting a learning mechanism or trying to assign some semantics or meaning to the weights, such as in the area of quantifier guided applications. In fact, the OWA operators have been used to implement the concept of fuzzy majority in the aggregation phase of fuzzy criteria by means of a fuzzy linguistic quantifier (Zadeh 1983), which indicates the proportion of criteria 'necessary for a good solution' (Yager 1996).

Furthermore, the calculation of the weighting vector can also take into account the importance I_i attributed to each considered factor, that shouldn't be confused with the weights w_i of an OWA operator. While it is often difficult to specify the values of importance by exact score, it would be more natural to specify them either in a linguistic form with the use of qualifiers such as important, very important, fairly important, etc. or by the aid of a graphic representation such as by moving a cursor on a bar or by selecting a grey level on a scale. The adoption of the above described formal framework allows implementing a software environment in which it is possible to select different aggregation operators and obtain therefore different results, which can be compared and validated, and to attribute relative importance to the contributing factors.

The synthetic *AI* for the African continent was computed by integrating the monthly single factor scores using an OWA operator with the semantics of a weighted average. We defined the relative importance weights of the factors at continental scale inversely proportional to the average RMSE of their related fitting functions for each contributing factor (Stroppiana et al. 2006). The idea is that the higher the RMSE, the more critical is the quality of the data or the validity of the hypothesis of normal distribution and then the lower the importance of the factor in the integration.

Therefore the OWA operator has been defined as a weighted average of the scores of the contributing factors $\{f_{rain}, f_{start}, f_{peak}, f_{length}\}$ with importance vector $I = (0.11, 0.22, 0.37, 0.30)$.

3 RESULTS AND DISCUSSION

Monthly maps of *AI* for the African continent were produced for the period 1996 to 2002 with the soft integration represented by the above defined OWA operator.

Figure 3 shows the *AI* maps for the month of October of seven years, from 1996 to 2002. The oceans, the inland water basins and deserts have been masked out and therefore appear as white regions. Green areas indicate regions with low anomaly values, where environmental conditions are not far from their reference state. The orange and red areas highlight the most anomalous regions where changes are likely occurring.

The *AI* maps show a high inter-annual variability; although green areas of low anomaly are predominant, some anomalous pattern can be clearly observed. In particular, our results show high anomalies in 1997 and 1998 in the Eastern Africa regions, that are partially recovered in subsequent years, and in Southern Africa in 1999 and 2000. The anomaly in the Horn of Africa is in agreement with the occurrence of several years of below-normal rainfalls that affected this region at the end of the last century: some regions, in fact, experienced prolonged drought beginning in 1998 and beyond (WMO 2001).

The indicator proposed in this work resulted also correlated with El Niño Southern Oscillation (ENSO) that is a quasi-periodical fluctuation between El Niño and La Niña states of the Pacific sea surface that takes place typically every 2 to 7 years (Philander 1990). The connections between

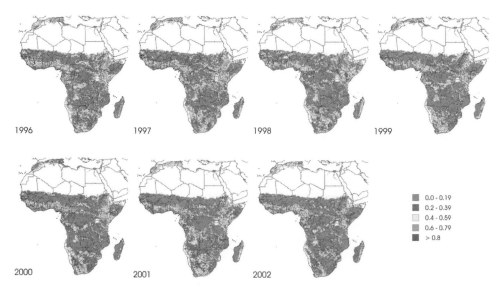

Figure 3. Maps of *AI* showing inter-annual variation for the month of October during the seven years period 1996–2002 (see colour plate page 387).

ENSO events and the inter-annual variability of tropical regions have been widely studied (Kogan 2000, Plisnier et al. 2000) and past studies have shown a vegetation response to these climatic anomalies (Anyamba et al. 2002). However, it is difficult to consistently indicate a cause-effect relation between ENSO and the vegetation conditions (Thomson et al. 2003).

These results show that 1999 and 2000 were the most anomalous years for Southern Africa (Fig. 3) despite the 1997–98 El-Niño event is recognized as the strongest on record. A shift from cold to warm ENSO was reported beginning in 2000, typically resulting in a shift from wetter to drier conditions in Southern Africa (Lindermann et al. 2005). The opposite situation, i.e. a shift from drier to wetter conditions in Northern Kenya, Ethiopia and Somalia (Lindermann et al. 2005), could confirm the partial recovery in the Eastern Africa as shown in Figure 3 for the years 2001 and 2002.

In order to characterize ENSO intensity and development, climatologists observe Sea Surface Temperature (SST) in the tropical Pacific between 5° North and 5° South. Based on the SST a set of indices have been developed to highlight the occurrence of ENSO events; in this study we used SSTs for the Niño 3.4 region in the central Pacific (NOAA 1999).

The mean monthly AI was computed over a 500 pixel window located in the Eastern Africa at the border between Kenya, Ethiopia and Somalia (Fig. 4) for the period 1996–2002; the AI monthly values range between 0.2 and 0.6 and the highest peaks highlight the most anomalous conditions. In the same figure, the temporal variation of AI is compared to the SST of the Niño 3.4 region for the seven years analyzed. The two time series of data show a general agreement with a time lag of some months; moreover, AI values higher than 0.5 can be observed in correspondence with the well known El-Niño episode of 1997–98.

Although a comparison with independent datasets is necessary, the continental scale of the analysis carried out and the synthetic character of the AI make this exercise hardly feasible. In fact, most of the analysis of anomalous behaviour performed at continental and/or global scales concerns the quantification of the anomaly of a single meteorological parameter, considered as driving force for vegetation condition, such as sea surface temperature or rainfall.

Although comparison with independent datasets is necessary, the continental scale of analysis and the synthetic character of the Anomaly Indicator make this exercise hardly feasible. In fact most of the analysis of anomalous behaviour performed at continental and/or global scales concerns the quantification of the anomaly of a single meteorological parameter, considered as driving force for vegetation condition, such as sea surface temperature or rainfall. It is therefore difficult to compare a synthetic (i.e. integrated) indicator with a single parameter.

Moreover, no consistent and regular information exist on ecosystem's response to these meteorological anomalies at continental scale.

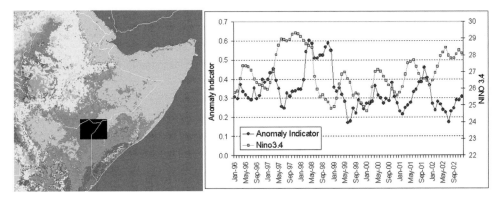

Figure 4. Comparison between the sea surface temperature of Niño 3.4 and the Anomaly Indicator averaged over a region of interest (500 pixels) in Eastern Africa. High anomaly values are evident in correspondence with the 1997–98 El Niño event (see colour plate page 387).

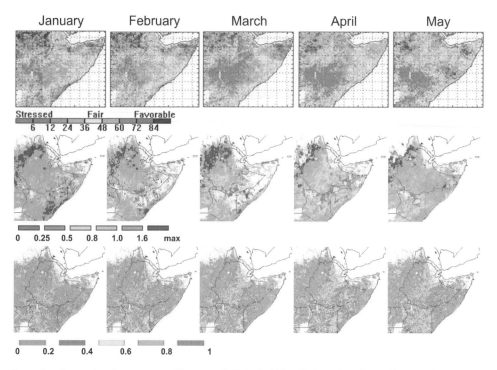

Figure 5.   Comparison between monthly maps of *AI* (top), AN-NPP (centre) and VHI (bottom) for the Horn of Africa during the first five months of the year 2000 (see colour plate page 388).

For these reasons, a qualitative comparison was attempted with two reference products: the Vegetation Health Index (Kogan 2000) and the Net Primary Production (NPP) derived from Global Production Efficiency Model (GloPEM) (Prince & Goward 1995).

The maps of Vegetation Health Index (VHI), which describes overall vegetation conditions, are produced as a weighted average between VCI (Vegetation Moisture Index) and TCI (Thermal Condition Index) derived by NOOA-AVHRR and were available directly on the NOAA-NESDIS web site (http://www.orbit.nesdis.noaa.gov/smcd/emb/vci/VH/index.html). Monthly VHI maps were downloaded from the web site for a qualitative comparison with our results.

Decadal NPP data were downloaded from ftp://ftp.glcf.umiacs.umd.edu/glcf/GLOPEM/ and NPP-LTA was computed by exploiting the entire 20 years of archived NPP data 1981–2000 derived from AVHRR. Using NPP product we calculated NPP anomaly (AN-NPP) for each decade as the ratio between actual decadal NPP data and the correspondent long-term average (NPP-LTA). Monthly anomaly values were obtained by averaging decadal NPP anomalies, where missing values or no data were omitted from the calculation.

Figure 5 shows a zoom over the Horn of Africa of the *AI* maps for the first five months of the year 2000 and it compares this product with monthly VHI images and with the NPP anomaly. In this region of Africa, the first five months of the year 2000 were extremely anomalous, according to the agency's NOAA-NESDIS, due to drought conditions in some parts of Kenya, Ethiopia and Somalia. Although general anomalous conditions are highlighted by all the three products in this region from January to May, some differences are present in the spatial patterns.

VHI data (Fig. 5 bottom panel) show a persistent vegetation stress in the region around Lake Turkana at the border between Kenya, Ethiopia and Sudan. Favourable conditions were high-lighted in the Northern part of Ethiopia and Sudan in agreement with both AN-NPP and *AI* maps: NPP values are, in fact, higher than two times the NPP-LTA and the *AI* values are lower than 0.2.

Intensive negative conditions, identified by the red-orange colours, are shown by AN-NPP maps (Fig. 5 middle panel) starting in January in the central part of Somalia and spreading in a wider area during the season; the same maps show a full recovery (favourable conditions) for Somalia in May, which is also confirmed also by the VCI and *AI* data.

Strong environmental anomalies are detected by *AI* maps (Fig. 5 top panel) at the beginning of the year in the region around Lake Turkana, North of Kenya, close to the border with Sudan and Ethiopia and between Kenya and Tanzania. In particular, a hot spot of *AI* (AI > 0.8) in southern east Sudan corresponds to a persistent unfavourable value for both VHI and NPP. Both *AI* and VHI products show anomaly patterns in the region of Ethiopia close to the Red and Arabic Sea beginning in January, while AN-NPP highlights this phenomenon later in the year, beginning in March–April. Finally, the month of May shows a better agreement between *AI* and VHI with respect to the AN-NPP product in Ethiopia, while all of the three products underline an increase of critical conditions for the entire Kenya.

A more exhaustive comparison between the three products is difficult because they are built on different concepts and techniques. VHI vegetation condition and health are estimated relative to the max-min interval of the NDVI and brightness temperature values. NPP is the output of a production efficiency model, based on physiological principles, in which the fraction of canopy absorption of photosynthetic active radiation (fAPAR) is derived from spectral vegetation index (SVI). The *AI* is based on a fuzzy integration of monthly cumulated rainfall and phenological anomalies which are estimated relative to the departure from LTA after a stratification procedure.

4 CONCLUSIONS

An innovative approach is proposed here to derive an indicator for the detection of anomalous environmental conditions at continental scale, i.e. for the identification of those areas where changes are presently undergoing or are likely to occur in the near future.

The methodology is based on the concept that environmental status assessment requires the integration of multiple and different factors that can be achieved by combining them for reinforcing-weakening the evidence of anomaly. The contribution brought by each factor (factor anomaly score) is quantified by a fuzzy membership function that represents the factor anomaly relative to an ideal reference state. The reference conditions are derived from the analysis of a time series of coarse resolution satellite data as the long-term average of vegetation phenology parameters (date of the start, peak and duration of growing season) and cumulated rainfall. The integration of the factor scores can be performed using the OWA operator by exploiting the advantages offered by approximate reasoning when dealing with incomplete knowledge of the phenomena under investigation.

The proposed methodology has been applied over the African continent for the period 1996–2002 to derive monthly maps of Anomaly Indicator. The comparison between the *AI* maps and the NPP anomaly and vegetation condition (VHI) maps, supports the validity of the obtained results. Although a validation exercise becomes hardly feasible due to the lack of independent datasets at the appropriate scale, good agreement was found between the temporal profiles of *AI* and the Niño 3.4 SST index.

The Anomaly Indicator approach has been designed to perform periodic monitoring at continental scale through the provision of a synthetic picture of the environmental status in support of expert analysis. For these reasons the approach, which relies on a very simple (reinforcement of evidence) and intuitive model (anomaly is proportional to the distance from the more frequent event), is flexible to manage uncertain/imprecise knowledge of the contributing factors and their inter-relationships and it is robust to provide results when one or more contributions are lacking.

The *AI* is definitely an improvement over previous approaches because it faces the uncertainty involved in environmental status assessment by exploiting the fuzzy set theory and the general scheme of the proposed methodology could easily accommodate datasets derived from different sources, such as socio-economic data.

ACKNOWLEDGEMENTS

This work has been carried out within the Observatory for Land cover and Forest change (OLF) of the GeoLand project (2004–2006). GeoLand is an Integrated Project of European Commission FP-6 and it is carried out in the context of GMES (Global Monitoring for Environment and Security), a joint initiative of European Commission and European Space Agency. Bruno Combal (JRC-EC) provided time series of data on vegetation phenology in the frame of Geo-Land-OLF activities.

REFERENCES

Anyamba, A., Tucker, C.J. & Mahoney, R. 2002. From El Niño to La Niña: Vegetation response pattern over East and Southern Africa during the 1997–2000 period. *Journal of Climate* 15: 3096–3103.

Bartholomé, E. & Belward, A.S. 2005. GLC2000: a new approach to global land cover mapping from Earth observation data. *International Journal of Remote Sensing* 26 (9): 1959–1977.

Boschetti, L., Brivio, P.A., Kunzle, A. & Mussio, L. 2006. Non parametric statistical tests for the analysis of multiple-sensor time series of remotely sensed data. *IGARSS 2006* Boulder (Co-USA), 31 July–4 August 2006 (in press).

Brivio, P.A., Boschetti, M., Carrara, P., Stroppiana, D. & Bordogna, G. 2006. A fuzzy anomaly indicator for environmental status assessment based on EO data: preliminary results for Africa. In Röder, A. & Hill, J. (eds), *Proceedings of the 1st Int. Conf. on Remote sensing and geoinformation processing in the assessment and monitoring of land degradation and desertification, Trier, Germany, 7–9 September 2005*: 383–390. Trier, Germany.

Burrough, P.A. 1996. Natural objects with indeterminate boundaries. In Burrough P.A. & Frank A.U. (eds.): *Geographic objects with indeterminate boundaries*: 3–28. London: Taylor & Francis.

Burrough, P.A. & McDonnel, R.A. 1998. *Principles of Geographical Information Systems*. Oxford: Oxford University Press.

Carrara, P., Bordogna, G., Boschetti, M., Stroppiana, D., Nelson, A. & Brivio, P.A. 2006. A flexible multi-source spatial da fusion system for environmental status assessment at continental scale. *International Journal of Geographical Information Science* (submitted).

Cleland, E.E., Chiariello, N.R., Loarie, S.R., Mooney, H.A. & Field, C.B. 2006. Diverse responses of phenology to global changes in a grassland ecosystem. *Proceedings of the National Academy of Sciences* 103 (37): 13740–13744.

De Beurs, K.M. & Henerbry, G.M. 2004. Trend analysis of the Pathfinder AVHRR Land (PAL) NDVI data for the desert of central Asia. *IEEE Geoscience and Remote Sensing Letters* 1(4): 282–286.

Friedl, M.A., McIver, D.K., Hodges, J.C.F., Zhang, X.Y., Muchoney, D., Strahler, A.H., Woodcock, C.E., Gopal, S., Schneider, A., Cooper, A., Baccini, A., Gao, F. & Schaaf, C. 2002. Global land cover mapping from MODIS: algorithms and early results. *Remote Sensing of Environment* 83: 287–302.

Galvin, K.A., Boone, R.B., Smith, N.M. & Lynn, S.J. 2001. Impacts of climate variability on East African pastoralists: linking social science and remote sensing. *Climate Research* 19: 161–172.

Klir, G.J. & Yuan, B. 1995. *Fuzzy Sets and Fuzzy Logic: Theory and Applications*. Upper Saddle River: Prentice-Hall.

Kogan, F.N. 2000. Satellite-Observed Sensitivity of World Land Ecosystems to El Niño/La Niña. *Remote Sensing of Environment* 74: 445–462.

Lambin, E.R., Geist, H.J. & Lepers, E. 2003. Dynamics of Land-Use and Land-Cover Change in Tropical Regions. *Annual Review of Environment and Resouces* 28: 205–241.

Lenz, R., Malkina-Pykh, I.G. & Pykh, Y. 2000. Introduction and overview. *Ecological Modelling* 130: 1–11.

Levine, J.S. 1996. *Biomass burning and global change*. Cambridge: MIT Press.

Linderman, M., Rowhani, P., Benz, D., Serneels, S. & Lambin, E. 2005. Land-cover change and vegetation dynamics across Africa. *Journal of Geophysical Research* 110: D12104 doi:10.1029/ 2004JD 005521.

Loveland, T.R., Reed, B.C., Brown, J.F., Ohlen, D.O., Zhu, Z., Yang, L. & Merchant, J.W. 2000. Development of a global land cover characteristics database and IGBP DISCover from 1 km AVHRR data. *International Journal of Remote Sensing* 21 (6–7): 1303–1330.

NOAA (National Oceanic and Atmospheric Administration) 1999. ENSO Advisory, El Niño Southern Oscillation Cold Phase. January, NOAA/National Weather Service, No.99/1.

Olson, D.M. & Dinerstein, E. 1998. The Global 200: a representation approach to conserving the Earth's most biologically valuable ecoregions. *Conservation Biology* 16: 502–515.

Ott, W.R. 1978. *Environmental indices: theory and practice.* Ann Arbor.

Palumbo, I., Grégoire, J.-M., Boschetti, L. & Eva, H. 2003. Fire regimes in protected areas of Sub-Saharan Africa, derived from the GBA2000 dataset. 4th EARSeL Int. Workshop on *Remote Sensing and GIS applications to Forest Fire Management: Innovative concepts and methods in fire danger estimation*: 139–149. Ghent, Belgium.

Philander, G.S. 1990. *El Niño, La Niña and the Southern Oscillation.* New York: Academic Press.

Plisnier, P.D., Serneels, S. & Lambin, E. 2000. Impact of ENSO on East African ecosystems: A multivariate analysis based on climate and remote sensing data. *Global Ecology and Biogeography* 9: 481–497.

Pykh, Y.A, Kennedy, E.T. & Grant, W.E. 2000. An overview of systems analysis methods in delineating environmental quality indices. *Ecological Modelling* 130: 25–38.

Prince, S.D. & Goward, S.N. 1995. Global primary production: a remote sensing approach. *Journal of Biogeography* 22: 815–835.

Robinson, P.B. 2003. A perspective on the fundamentals of Fuzzy Sets and their use in Geographic Information Systems. *Transactions in GIS* 7(1): 3–30.

Sankaran, M., Hanan, N.P., Scholes, R. et al. 2005. Determinants of woody cover in African savannas. *Nature* 438: 846–849.

Stroppiana, D., Boschetti, M., Carrara, P., Bordogna, G., & Brivio, P.A. 2006. Continental monitoring of vegetation cover status with a fuzzy anomaly indicator: an example for Africa. *Remote Sensing of Environment* (submitted).

Tansey, K., Grégoire, J.M., Binaghi, E., Boschetti, L., Brivio, P.A., Ershov, D., Flasse, S., Fraser, R., Graetz, D., Maggi, M., Peduzzi, P., Pereira, J.M., Silva, J., Sousa, A. & Stroppiana, D. 2004. A global inventory of burned areas at 1km resolution for the year 2000 derived from SPOT Vegetation data. *Climatic Change* 67 (2): 345–377.

Tran, L.T., Knight, C.G., O'Neill, R.V., Smith E.R., Riitters, K.H. & Wickham, J. 2002. Environmental Assessment Fuzzy Decision Analysis for Integrated Environmental Vulnerability Assessment of the Mid-Atlantic Region. *Environmental Management* 29: 845–859.

Thomson, M., Abayomi, K., Barnston, A., Levy, M. & Dilley, M. 2003. El Niño and drought in southern Africa. *The Lancet* 361: 437–438.

Tucker, C.J. & Choudury, B.J. 1987. Satellite remote sensing of drought conditions. *Remote Sensing of Environment* 23: 243–251.

Yager, R.R. 1988. On ordered weighted averaging aggregation operators in multi-criteria decision making. *IEEE Trans. Systems, Man and Cybernetics* 18: 183–190.

Yager, R.R. 1996. Quantifier guided aggregation using OWA operators. *International Journal of Intelligent Systems* 11: 49–73.

Wickham, J.D., Jones, K.B., Riitters, K.H., O'Neill, R.V., Tankersley, R.D., Smith, E.R., Neale, A.C. & Chaloud, D.J. 1999. Environmental auditing, an integrated environmental assessment of the US Mid-Atlantic region. *Environmental Management* 24 (4): 553–560.

WMO, 2001. Unusual floods and droughts in East Africa. *World Climate News* 19: 3–4.

Zadeh, L.A. 1965. Fuzzy sets. *Information and Control* 8: 338–353.

Zadeh, L.A. 1975. The concept of a linguistic variable and its application to approximate reasoning, parts I, II. *Information Science* 8: 199–249, 301–357.

Zadeh, L.A. 1983. A computational approach to fuzzy quantifiers in natural languages. *Computers and Mathematics with Applications* 9: 149–184.

Zhang, X., Friedl, M.A., Schaaf, C.B., Strahler, A.H., Hodges, J.C.F., Gao, F., Reed, B.C. & Huete, A. 2003. Monitoring vegetation phenology using MODIS. *Remote Sensing of Environment* 84: 471–475.

Recent Advances in Remote Sensing and Geoinformation Processing
for Land Degradation Assessment – Röder & Hill (eds)
© 2009 Taylor & Francis Group, London, ISBN 978-0-415-39769-8

The spatial uncertainty of desiccation in the West African Sahel and its implications for land degradation

A. Chappell
Centre for Environmental Systems Research, University of Salford, Manchester, UK

C.T. Agnew
School of Environment and Development, University of Manchester, Manchester, UK

ABSTRACT: For well over three decades controversy has surrounded the characteristics of environmental degradation in the Sahel. One reason for this prolonged debate is the dearth of uncertainty in previous Sahelian rainfall work. Global Historical Climatology Network summer rainfall data between 1930 and 1990 were used in sequential indicator simulations to reproduce global statistics over local accuracy to provide a complete assessment of uncertainty. The traditional area-weighted technique showed the characteristic decline in rainfall but represented the extreme of the simulation mean distribution. Annual estimates of the 5th and 95th percentiles of the simulated rainfall mean distribution quantified uncertainty and showed that there is no longer unequivocal evidence of desiccation in the region between the late 1960s and 1990. The probability of exceeding rainfall was calculated for 200 mm and 500 mm and presented using maps for selected years between 1930 and 1990. The results demonstrated that simple isoline thresholds did not adequately represent the boundaries for vegetation and agriculture, respectively.

1 INTRODUCTION

The summer of 2005 drew world attention to the plight of inhabitants in the West African Sahel (WAS) region of West Africa due to famine, drought and locusts. The FAO and UNICEF reported that in Niger alone some three million people were likely to be affected. Drought related tragedies have been reported several times during the 20th Century and the characteristics of environmental change and environmental degradation in the Sahel have been the focus of scientific interest for well over three decades. Controversy has surrounded the scientific work due to natural environmental variability, the paucity of data, misunderstanding of resilience, and institutional facts (Agnew 1989, 1995). Many of the accounts of environmental change have been challenged and new positions adopted. Examples include possible interactions between climate and land cover changes (Hulme & Kelly 1993, IUCN 1989); the relationships between droughts and human activity (Sivakumar et al. 1991, Glantz 1994) and increasing desiccation as presented by an expanding Sahara (Mainguet 1998, Thomas & Middleton 1994, Tucker et al. 1991, Warren & Agnew 1988). Such studies have led to recognition that the Sahel is environmentally heterogeneous despite significant sources such as UNEP (1992) and IPCC (Houghton et al. 2001) portraying the Sahel as a region. Raynaut (1997) demonstrated that the Sahel is a complex mosaic of environments and human activities. Climatologists have also noted the diversity and variability of the Sahelian environment (Nicholson & Paolo 1993) yet the paradigm of widespread persistent drought and desiccation is so well established (Hulme 2001, LeHouerou 1996) that few question local variations nor the quality of the data used. This is in stark contrast to the work of ecologists, geomorphologists, and others who have challenged the views of Sahelian environmental change established earlier in the 20th Century (Glantz 1994, Garcia 1981, Franke & Chasin 1980).

There is little support for those whose work questions or offer alternative explanations, as forcibly stated by Dai et al. (2004) 'The recent Sahel drought is real'. The paradigm is well established and reinforced by possible explanations of the drying trend. However, drought (a short term, severe reduction in precipitation) is commonly expected in semi-arid regions such as the Sahel whereas desiccation (a long term reduction in precipitation) is an element of climate change which requires an examination of precipitation trends (e.g. Folland et al. 1986, Nicholson 1985, Nicholson & Palao 1993, Zeng et al. 1999, Zeng 2003). Desiccation between the late 1960s and 1990 is explained by comparing empirical evidence or model predictions against 'observations' of Sahelian rainfall (e.g. Giannini et al. 2003, Taylor et al. 2002). However, the outcomes of testing these hypotheses are based upon the assumption that aggregated rainfall observations represent the underlying population despite limitations of the observation network and variability in rainfall. Despite consideration of local interpolation errors (Hulme & New 1997) and spatial and temporal sampling patterns on the estimation of aggregated climatological variables (e.g. Madden et al. 1993, Willmott et al. 1996), there has been no assessment of uncertainty in previous Sahelian precipitation work. Here we provide an assessment of spatial uncertainty between 1930 and 1990. The analyses provide uncertainty about temporal variation of areal rainfall in the region (Chapell et al. 2008) but this is not the main focus. Instead, we consider in detail the spatial variation of the rainfall, its variation and the probability of exceeding thresholds set at 200 mm as an approximation for the edge of the Sahara (Tucker et al. 1991) and at 500 mm as an approximation for the region of rain-fed agriculture (Agnew 1990).

2 DATA AND METHODS

We used the Global Historical Climatology Network (GHCN v.2) rainfall data (Petersen & Vose 1997) and extracted rainfall stations for the years between 1930 to 1990 inclusively, within the west African Sahel (WAS) following the definition of Nicholson (1993; 10–20°N, –20°W to 20°E). Stations locations recorded using latitude and longitude were converted to an equal-area projection using the unit circle. Since we adopted a spatial approach rather than the usual temporal perspective, there was no requirement that the station's data contributing to the analysis should be continuous or nearly continuous (allowing for some missing data). However, we calculated the total summer (June, July, August and September) annual rainfall and therefore any year of data for any station was required to be complete for that period. Therefore, all stations with summer monthly data for a year were extracted from the database. In addition, those extracted data were filtered using the requirements of the calculation of area-weighted anomalies. A reference period of 1961 to 1990 was set to achieve the maximum number of stations with the minimum amount of missing data (29%). Rainfall anomalies were calculated using the method described by Jones & Hulme (1996) and the area-weighting scheme followed that of Dai et al. (1997).

2.1 Area-weighted normalized anomalies of rainfall

The rainfall [P] anomaly ΔP_{ik} is here defined as the deviation of the station rainfall P_{ik} from the station average:

$$\Delta P_{ik} = P_{ik} - \overline{P}_i \tag{1}$$

where P_i is the time-mean of the ith station time series and was calculated over an index time period k (e.g. 30 years) common to all stations (1961–1990). Note that the anomalies were not standardized as suggested by Jones & Hulme (1996), to be consistent with the approach described by Dai et al. (1997). The regional mean of the anomaly time series $\langle \Delta P_R \rangle$:

$$\langle \Delta P_R \rangle = \frac{\sum_{i=1}^{N} w_i \Delta P_{ik}}{\sum_{i=1}^{N} w_i} \tag{2}$$

162

was calculated by the summation of all station rainfall for each k in the region by applying a weight w to each station. The regional station mean $\langle \Delta P_R \rangle$ was calculated by the summation of all station time-means for each k in the region by applying w (equation 2) to each station:

$$\langle \Delta \overline{P}_R \rangle = \frac{\sum_{i=1}^{N} w_i \overline{P}_i}{\sum_{i=1}^{N} w_i} \tag{3}$$

The weights were derived using the distance weighting scheme of Dai et al. (1997) (eq. 3):

$$w_i = 1 - r_i/d \quad \text{if } r_n < r_0$$
$$w_i = 0 \quad \text{if } r_n < r_0. \tag{4}$$

It is important to note that the summation of these weights is only applied to stations within the influence radius. The annual rainfall anomalies ΔP_{ik} and the time mean P_i were interpolated separately onto a regular grid of 2.5° (ca. 270 km) using an influence radius $r_0 = 300$ km and a constant of $d = 880$ km following Dai et al. (1997). The anomaly time series was returned to absolute rainfall units $\langle P \rangle$ by simply adding back the weighted average of the station means thus:

$$\langle P \rangle = \langle \Delta P_R \rangle + \langle \Delta \overline{P}_R \rangle. \tag{5}$$

2.2 Removing systematic spatial variation (drift) of annual rainfall

Matheron (1971) realised that for many spatial variables the local mean values vary predictably or deterministically from one part of a region to another. This may be taken as evidence of drift which violates the assumptions of geostatistics. This systematic variation is common in annual rainfall data because of the strong gradients across longitudes and latitudes. It was decided that drift was present in the rainfall observations. Its form was estimated by fitting low-order (quadratic) polynomials on the spatial co-ordinates of rainfall using least-squares regression (not shown). The goodness of fit of quadratic polynomials was estimated using R^2 and had an average of 0.77 and standard deviation of 0.05. The p-value for the fitted polynomial coefficients was considerably smaller than 0.05 for all years which suggested that the null hypothesis could be rejected and that the coefficients were highly likely to be non-zero. The coefficients of the quadratic polynomials for each year of rainfall observations were used to remove the trend. Subsequent data analysis was performed on the residuals from the trend.

2.3 Modelling spatial variation of rainfall

Despite strong gradients in rainfall across longitudes and latitudes there were insufficient data to reliably quantify the anisotropic spatial variation (Webster and Oliver 1992). Consequently, an omni-directional variogram of rainfall residuals was used to calculate the average variation in all directions. Under the multiGaussian approach the extremes of a distribution are not considered to be spatially correlated (Goovaerts 1997). In such a situation a non-parametric approach may be more applicable and it does not assume any particular shape or expression for the conditional distributions. The function $F(\mathbf{u}; z|(n))$ is modeled through a series of K thresholds values for property z_k that discretise the range of variation of z:

$$F(\mathbf{u}; z_k|(n)) = \text{Prob}\{Z(\mathbf{u}) = z_k|(n)\} \quad k = 1, \ldots, K \tag{6}$$

The indicator data may be coded as a precise measurement of the property as:

$$i(\mathbf{u}; z_k) = 1 \quad \text{if } z(\mathbf{u}_\alpha) = z_k \tag{7}$$
$$i(u; zk) = 0 \quad \text{otherwise } k = 1, \ldots, K$$

However, sample indicator variograms at extremes are not well defined because they depend on the spatial distribution of a few pairs of indicator data. The magnitude and spatial continuity of extremes in rainfall data are particularly important and so the variograms at the other thresholds may be inferred from another threshold using the mosaic model (Journel 1984). This approximation requires that the following two assumptions are jointly met (Goovaerts 1997):

1. The K indicator random functions (RFs) $I(\mathbf{u}; z_k)$ are intrinsically correlated and the variogram models are proportional (f) to a common variogram model $\gamma_{mI}(\mathbf{h})$ for separation distances ($\mathbf{h}$):

$$\gamma_I(\mathbf{h}; z_k) = \phi_k \cdot \gamma_{mI}(\mathbf{h}) \quad \forall k \tag{8}$$

2. All vectors of indicator data are complete (equally sampled case); there are no missing indicator values such as implied by constraint intervals.

The conditional cumulative distribution function (CCDF) was calculated each year between 1930 and 1990 inclusively, for all station's total annual rainfall residuals. The rainfall residual values for $K = 5$ quantiles (10%, 25%, 50%, 75%, 90%) were established each year and were also used in the simulations described below. The thresholds of the quantiles were used to transform the rainfall values into the indicator variable (Eq. 7). The mosaic model avoided calculating and fitting the variogram for all thresholds to reproduce the marginal target statistics and reduced the number of order relation deviations. However, this approximation was known to be at the expense of less flexibility in comparison with the direct indicator coding at several thresholds. The direct indicator variograms for the thresholds were compared with those inferred by the mosaic model to validate the first assumption. No constraint intervals were used and so the second assumption was not applicable in this case. The spatial structure of the median indicator annual rainfall residual data was calculated using omni-directional experimental variograms which varied smoothly with increasing separation distances. These variograms were fitted using non-linear weighted least squares (Genstat 5 Committee 1992) with several authorised models and the models that fitted best, in the least-squares sense, were (spherical and exponential) selected.

2.4 Sequential indicator simulation and post-processing

The estimator of the mosaic model for ordinary kriging:

$$(F(\mathbf{u}; z_k|(n)))^*_{mIK} = \sum_{\alpha=1}^{n(u)} \lambda_\alpha^{OK}(u) I(u_\alpha; z_k) = (F(\mathbf{u}; z_k|(n)))^*_{oICK} \quad k = 1, \ldots, K \tag{9}$$

is called median indicator kriging since the common model $\gamma_{mI}(\mathbf{h})$ is usually inferred from the indicator semivariogram at the median threshold. This is typically because median indicator data are evenly distributed as 0 and 1 values which usually renders the experimental indicator variogram $\hat{\gamma}_I(\mathbf{h}; z_M)$ better defined than at other threshold values. If $\{z^*(\mathbf{u}), \mathbf{u} \in \mathbf{A}\}$ is the set of kriging estimates of attribute z over the study area $\mathbf{A}$, each estimate $z^*(\mathbf{u})$ taken separately is 'best' in the least squares sense because the local error variance $\mathrm{Var}\{Z^*(\mathbf{u}) - Z(\mathbf{u})\}$ is minimised. Kriging was shown to be one of the most reliable two-dimensional spatial estimators (Laslett et al. 1987, Laslett & McBratney 1990, Laslett 1994) and it is expected to produce more reliable estimates than simple methods of interpolation (Webster & Oliver 2001). However, the map of local estimates may not be best as a whole. Interpolation algorithms tend to smooth out local details of the spatial variation of the attribute (e.g. Chappell et al. 2003). Such conditional bias is a serious shortcoming when trying to detect patterns of extremes such as zones of large and small rainfall. The smoothing tends not to be uniform as it depends on the local data configuration (Goovaerts 1997). For example maps of rainfall kriging estimates appear more variable in densely sampled areas than in sparsely sampled areas (e.g. Chappell & Ekström 2005). Thus, kriged maps may display artifact structures.

Instead of a map of local best estimates, stochastic simulation generates a map or a realization of z-values $\{z^{(l)}(\mathbf{u}), \mathbf{u} \in \mathbf{A}\}$ where l denotes the lth realization which reproduces statistics pertinent to the problem e.g. data values are honoured at their locations $z^{(l)}(\mathbf{u}) = z(\mathbf{u}_\alpha) \; \yen \; \mathbf{u} = \mathbf{u}_\alpha, \alpha = 1, \ldots, n$ when the realization is said to be conditional, the histogram of simulated values reproduces the declustered sample histogram and the set of indicator variograms are reproduced (Goovaerts 1997). Unlike sequential Gaussian simulation (Goovaerts 1997), the indicator approach allows one to account for class-specific patterns of spatial continuity using different indicator variograms. The sequential indicator simulation (SIS) approach of a single continuous attribute z at N grid nodes $\mathbf{u}'_j$ conditional on the z-data, proceeds by discretising the range of variation of z into $(K+1)$ classess (8). A random path is defined that visits each node of the grid only once. At each node (Goovaerts 1997):

1. Determine the $\mathbf{u}'$ the K ccdf values $(F(\mathbf{u}'; z_k|(n)))^*$ are determined using median indicator kriging. The conditioning information consists of indicator transforms of neighbouring original z-data and previously simulated z-values.
2. Correct for any order relation deviations and build a complete CCDF model $F(\mathbf{u}'; z_k|(n))$, $\yen \, z$, using an interpolation algorithm (Goovaerts 1997).
3. Draw a simulated value $z^{(l)}(\mathbf{u}')$ from the CCDF.
4. Add the simulated value to the conditioning data set.
5. Proceed to the next node along the random path and repeat steps 1 to 4.

This process is repeated with a different random path for each realisation $\{z^{(l')}(\mathbf{u}'_j, j = 1, \ldots, N)\}, l' \neq l$.

Sequential indicator simulation was used here to generate 300 realisations of rainfall and honour the values of the rainfall stations each year, reproduce the declustered sample histogram and reproduce (approximately) the covariance models for the five thresholds using the median indicator approximation (Deutsch & Journel 1998). The west African Sahel was discretised into a grid with 2208 nodes (92×24) using a grid spacing of 50 km which was approximately $0.4°$ in latitude and longitude. This grid resolution was chosen to be similar to that of gridded climate data currently available. The series of realisations were post-processed and the uncertainty information summarised using the mean per pixel of all realisations, or the E-type estimate, and the variance of the conditional distribution of the realisations. A selection of the E-type maps for several years is provided with the variance between realisations of the same years. In addition, the probability of exceeding rainfall was calculated for two thresholds. The first was set at 200 mm because it was used by Tucker et al. (1991) as an approximation for the edge of the Sahara in their experiment to detect expansion and contraction of the desert. The second was set at 500 mm because Agnew (1990), amongst others, believed it could approximately identify the region of rain-fed agriculture. Maps were selected for the years 1935, 1945, 1955, 1965, 1975 and 1985. The minimum and maximum values of the realisation means for every year between 1930 and 1990 were also plotted over time to provide a conservative estimate of uncertainty.

3   RESULTS

The spatial structure of the median indicator annual rainfall data was calculated using omni-directional experimental variograms which varied smoothly with increasing separation distances. These variograms were fitted using non-linear weighted least squares (Genstat 5 Committee 1992) with several authorised models and the models that fitted best, in the least-squares sense, were (spherical and exponential) selected. Figure 1 provides the temporal variation in the parameters of the models fitted to the omni-directional variograms. The range parameter values decrease over time until approximately 1970 at which point in time the values increase until 1990. The correlated variance decreases over time, whilst the uncorrelated [nugget] variance increases over time.

The inverse-distance weighted [IDW] west African Sahel [WAS] rainfall anomalies for each year of data are shown in figure 2. The long-term [1930–1990] mean annual rainfall of those converted

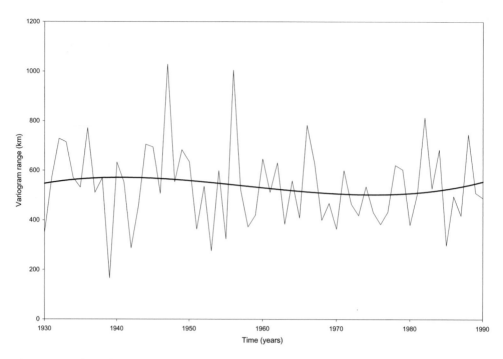

Figure 1a. Values of the range (or effective range) parameter of the models fitted to the median indicator isotropic variograms of west African Sahel annual summer rainfall residuals. The smooth line is used only as a guide to the pattern (Reproduced with permission from Chappell et al. 2008).

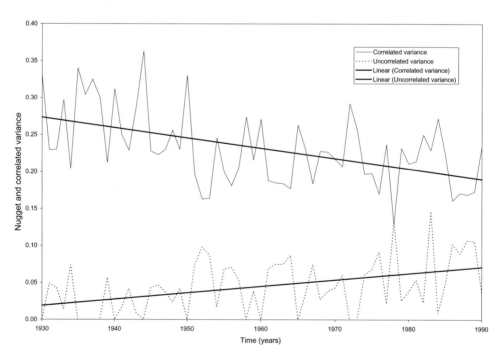

Figure 1b. Values of the correlated and uncorrelated (nugget) variance parameters of the models fitted to the isotropic median indicator variograms of west African Sahel annual summer rainfall residuals. Straight lines are provided as a guide only.

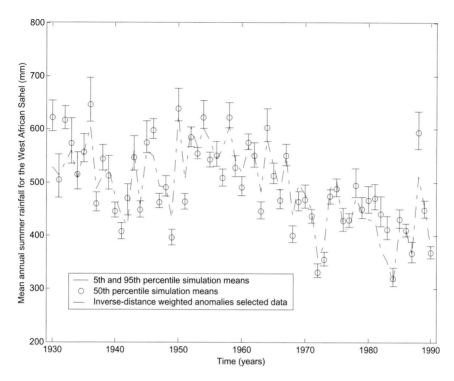

Figure 2.   Annual variation of summer rainfall for ensemble simulation realisations and area-weighted rainfall using inverse-distance weighted anomalies of selected stations. Uncertainty error bars represent the 5th and 95th percentiles of the 300 spatial means (o) of the simulation realizations. (Reproduced with permission from Chappell et al. 2008).

weighted anomalies was 490 mm. This regional rainfall time series has no estimates of uncertainty and shows the characteristic decline in aggregated annual rainfall between the late 1960s and 1990. The rainfall realisation means for each year are also plotted in figure 1. Uncertainty in the mean estimates is represented by the 5th and 95th percentiles of the realisations. Approximately 49% of the IDW mean annual rainfall falls outside the range of the 5th and 95th percentiles for annual simulation realisations means. Nearly half of the IDW rainfall estimates represent an extreme realisation of the time series pattern (Figure 1).

The maps of the per pixel average of the realisations for summer rainfall in the region are shown every ten years between 1935 and 1985 in figure 3 (a–f). The pattern of rainfall for these years is similar. An area of large rainfall exists in the southwest. Notably, that area extends no further east than the border of Guinea. A belt of large (ca. 1000 mm) rainfall extends across the region from are southern Senegal to Nigeria consistently for all years. Similarly a belt of small (<500 mm) rainfall extends from the coast across the region but which terminates at approximately the border with Chad. The low rainfall belt is more or less pronounced in each of the maps and is notably extensive during 1985.

The maps of per pixel conditional variance (Figure 3 (g–l)) reveal that the areas of large rainfall previously identified are highly variable in the simulation realisations. This pattern suggests that there is considerable uncertainty in the rainfall estimates. In general the spatial uncertainty is very patchy. Overall there is more uncertainty in the early part of the sequence (1935 and 1945) than in the latter part. However, in these years the estimation variances are highly spatially variable and notably variances are large across northern Nigeria and into Niger. The speckled nature of the variance maps is caused by the highly spatially variable nature of the variance.

The maps of the per pixel probability of exceeding the 200 mm and 500 mm threshold of summer rainfall for the region are shown every ten years between 1935 and 1985 in figure 4.

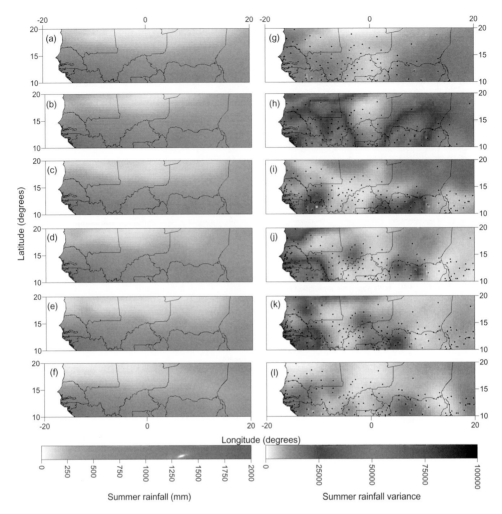

Figure 3. Maps of the per pixel (approximately 0.4°) average of the 300 realisations (a–f) and the variance for the conditional distributions (g–l) of summer rainfall in the West African Sahel for the years of 1935, 1945, 1955, 1965, 1975 and 1985, respectively. The rainfall stations (●) and the political boundaries for countries of the region are also shown using black lines (see colour plate page 389).

Darker shades indicate a tendency to exceed the threshold whilst lighter shades indicate that the threshold is unlikely to be exceeded. The maps of the probability of exceeding the threshold of 200 mm summer rainfall show that the majority of the region has dark shades (Figure 4 (a–f)). However, the pattern does not produce the expected isoline close to the northern boundary of the region. Instead, clusters of lighter shades are identified where parts of several countries have not exceeded the threshold. Notably, in 1985 a large part of the western portion of the region has light shades but the eastern part of the region is expected to exceed the threshold. The speckled nature of the light regions is caused by the rainfall values at the rainfall stations that have been retained in the maps.

In 1935, 1945 and 1955, the areas of the region that exceed the 500 mm threshold of summer rainfall are similar. Large values are found in the low latitudes of the region including southern Senegal, Mali and Burkina Faso and the border between Nigeria and Niger. In years 1965, 1975 and 1985 the area of the region covered by darker shades has visibly decreased and generally only the incursion into Niger of darker shaded pixels persists.

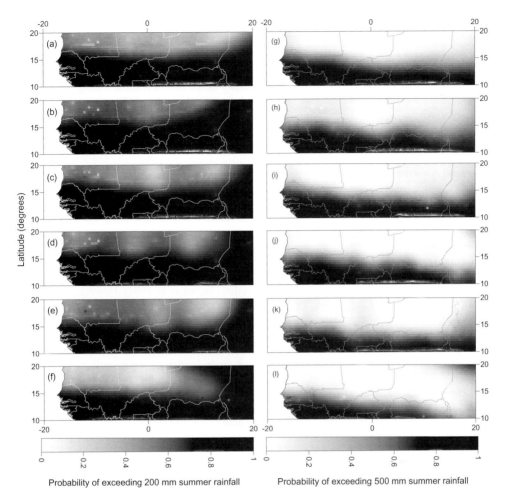

Figure 4. Maps of the probability that the summer rainfall at each pixel (approximately 0.4° in latitude and longitude) in the West African Sahel will exceed the threshold at 200 mm (a–f) and at 500 mm (g–l) for the years of 1935, 1945, 1955, 1965, 1975 and 1985, respectively. The political boundaries for countries of the region are also shown.

4   DISCUSSION

The inverse-distance weighted rainfall anomalies are commonly used to aggregate observations across the west African Sahel (Jones & Hulme 1996, Dai et al. 1997). A decrease in the anomalies between the late 1960s and 1990 is widely accepted as evidence for a drying trend or desiccation in the region (Jones & Hulme 1996). Unfortunately, the inverse-distance weighted approach provides no estimate of uncertainty and is a poor approximation of the rainfall realisation means. The stochastic simulation approach demonstrated that there could be considerable variation in the spatial pattern of rainfall for a given year whilst constrained by global characteristics including, amongst other things, the rainfall measurements. The results showed that 50% of the area-weighted estimates between 1930 and 1990 exceed the 5th and 95th percentiles i.e., the technique provides an extreme average rainfall in 50% of the time period. The large variation each year between the 5th and 95th percentiles of the simulations considerably reduced the certainty of desiccation in the region. The appearance of a drying trend is exaggerated with the inclusion of the wet years of 1936, 1945 and 1948 and the drought years of 1972–1973 and 1983–1984. With the exception of these

169

years the uncertainty estimates between the late 1960s and 1990 overlap considerably with those of the wet phase in the 1950s and early 1960s and those of the dry phase in the 1940s.

The source of the variability in the estimate of annual rainfall for the region is contained in the maps of spatial uncertainty. The mean and the conditional variance per pixel for the 300 realisations provide an estimate of that variation. The heterogeneity of the region is evident from the average of the realisations and it is difficult to appreciate how a single mean, without an expression of accuracy, can reliably estimate that variation. This complexity is further demonstrated by the per pixel variance maps which show that there is considerable variation in the realisations. The spatial uncertainty is different each year but there are consistently occurring patterns. For example, there is consistently large uncertainty in the lowest latitudes of the region i.e., the south-west coastal region and in northern Nigeria. These results identify a source of variation that exceeds the hitherto identified sources of important variation that include local interpolation errors and sampling effects (Hulme & New 1997, Madden et al. 1993, Willmott et al. 1996). That source of variation appears to be highly spatially variable and suggests that the rainfall pattern across the region does have broad bands but also contains clustering.

The per pixel mean maps show consistently occurring clusters of annual rainfall along the edge of the Sahara. This is somewhat surprising since rainfall is expected to decrease across the region in the northerly direction. This is the basis for the use of the 200 mm year^{-1} isoline as an approximation for the boundary between the Sahara zone and the Sahel (Tucker et al. 1991). However, the annual maps for the probability of exceeding that threshold substantiate the pattern detected in the maps of the per pixel mean. The probability maps show distinct clusters of areas that are unlikely to exceed this threshold. Only in 1985 is there a tendency for a strong latitudinal gradient to exist and even this does not extend across the entire region and it occurs around the 15° latitude. There appears to be little evidence that this threshold can be used to identify the boundary between the Sahara and the Sahel. However, there is some support for the use of 500 mm as a threshold that identifies a distinct boundary of rain-fed agriculture (Agnew 1990). The main exception appears to be the consistent occurrence of an area over central and northern Niger that exceeds the threshold. Results suggest that the boundary is at its most northern extent in years 1935 and 1945. It appears to be confined to the most southerly latitudes during the years 1975 and 1985. The probability maps suggest that the use of boundaries based on isolines formed from locally optimal maps are simplistic and hide a considerable amount of variation and uncertainty.

5   CONCLUSION

The well-established desiccation was evident in the decline of area-weighted summer rainfall between the late 1960s and 1990. The technique represented extreme values of the rainfall simulation realisation distribution. The uncertainty in the regional time series was represented by the 5th and 95th percentiles. The results of the stochastic simulations suggest that there is no longer unequivocal evidence of the drying trend or desiccation in the region between the late 1960s and 1990. The maps of spatial uncertainty demonstrate that there is considerable heterogeneity across the region which cannot reasonably be estimated by a single areal mean value that is conventionally used. It is likely that the expectation of being able to estimate rainfall across the region in this manner has reduced the importance of uncertainty in the regional time series statistics and in its reporting in the literature. Reconsideration of its importance for the aggregation of climate observations may make an important contribution to temporal and spatial patterns in this region and others around the globe.

Spatial uncertainty was also expressed as the probability of exceeding thresholds at 200 mm and 500 mm. The results showed that the former threshold was a very poor approximation of the boundary between the Sahara and the Sahel. Evidently, there is a much more complex relationship than was previously recognised between rainfall and vegetation in the detection of the edge of the Sahara desert. At 500 mm there appeared to be a much more pronounced boundary. However, its consistent position near e.g., the border between Nigeria and Niger, suggests that agricultural production in that region was barely viable. It is likely that a simple threshold does not adequately represent the complex spatial heterogeneity of the region or the complexity of rain-fed agriculture. The tendency of the area-weighted technique to over-estimate the amount of rainfall in the region has contributed to the perception of viability and that drought and related land degradation is less common.

Challenging the drying trend and the use of simplistic thresholds in the west African Sahel may appear wholly negative with the intention of undermining for example, the ability to better understand the complexity of agriculture or climate systems. However, the absence of critical checks and perspectives allow assumptions and simplifications to go unchallenged and appear to have become so widely accepted that they are immutable and very difficult to contest. Faced with this situation with respect to west African Sahel rainfall the estimation of spatial uncertainty has robustly questioned the magnitude of rainfall, the clarity of the desiccation pattern and the assumptions for the use of simple thresholds in land degradation assessment.

ACKNOWLEDGEMENTS

We are grateful to NCAR for hosting the provision of the GHCN v. 2 data, to Dr. M. Ekström for numerous useful discussions and the comments of an anonymous referee. Any shortcomings or errors in the manuscript remain the responsibility of the authors.

REFERENCES

Agnew, C.T. 1989. Sahel drought: Meteorological or agricultural? *International Journal of Climatology* 9: 371–382.
Agnew, C.T. 1990. Spatial aspects of drought in the Sahel, *Journal of Arid Environments* 18, 279–293.
Agnew, C.T. 1995. Desertification, drought and development in the Sahel. In: Binns, A.(ed.): People and Environment in Africa, pp. 137–49. John Wiley & Sons, Chichester.
Chappell, A. & Ekström, M. 2005. The importance of de-clustering and uncertainty in climate data: a case study of west African Sahel rainfall. In Leuangthong, O. & Deutsch, C. (Editors) *Quantitative Geology and Geostatistics*, Vol. 14, XXVIII: 1167. Springer.
Chappell, A., McTainsh, G., Strong, C. & Leys, J. 2003. Simulations to optimise sampling of aeolian sediment transport for mapping in space and time. *Earth Surface Processes and Landforms* 28: 1223–1241.
Chappell, A. & Agnew, C. 2008. How certain is desiccation in west African Sahel rainfall (1930–1990)? *Journal of Geophysical Research-Atmospheres* doi:10.1029/2007JD009233 (in press).
Dai, A., Fung, I.Y. & Del Genio, A.D. 1997. Surface observed global land precipitation variations during 1900–88. *Journal of Climate* 10: 2943–2962.
Dai, A., Lamb, P.J., Trenberth, K.E., Hulme, M., Jones, P.D. & Xie, P. 2004. The recent Sahel drought is real. *International Journal of Climatology* 24: 1323–1331.
Deutsch C.V. & Journel, A.G. 1998. *GSLIB Geostatistical Software Library and User's Guide*. Oxford: University Press.
Deutsch, C.V. 1989. DECLUS: A Fortran 77 program for determining optimum spatial declustering weights. *Computers and Geosciences*, 15(3): 325–332.
Folland, C.K., Palmer, T.N. & Parker, D.E. 1986. Sahel rainfall and worldwide sea temperatures, 1901–85. *Nature* 320: 602–606.
Franke, F. & Chasin, B. 1980. Seeds of Famine. New Jersey: Allanheld Osman & Co.
Garcia, R.V. 1981. Drought and Man. Oxford: Pergamon press.
Genstat 5 Committee 1992. *Genstat 5, Release 3, Reference Manual*. Oxford: Oxford University Press.
Giannini, A., Saravanan, R. & Chang, P. 2003. Oceanic forcing of Sahel rainfall on interannual to interdecadal time scales. *Science* 302: 1027–1030.
Glantz, M., (ed.) 1994. *Drought Follows the Plow*. Cambridge: CUP.
Goovaerts P. 1997. *Geostatistics for natural resources evaluation*. Oxford: Oxford University Press.
Houghton, J.T., Ding, Y., Griggs, D.J., Noguer, M., van der Linden, P.J. & Xiaosu, D. 2001. *Climate Change 2001: The Scientific Basis*. Cambridge: Cambridge University Press.
Hulme, M. & Kelly, M. 1993. Desertification and Climate Change. *Environment* 35(6): 39–45.

Hulme, M. & New, M. 1997. Dependence of large-scale precipitation climatologies on temporal and spatial sampling. *Journal of Climate* 10: 1099–1113.

Hulme, M. 2001. Climate perspectives on Sahelian desiccation 1973–1998. *Global Environmental Change* 11:19–29.

International Union for Conservation of Nature 1989. *Sahel Studies*. Nairobi: IUCN.

Isaaks, E.H. & Srivastava, R.M. 1989. *An introduction to applied geostatistics*. Oxford: OUP.

Jones, P.D. & Hulme, M. 1996. Calculating regional climatic time series for temperature and precipitation: methods and illustrations. *International Journal of Climatology* 16: 361–377.

Journel, A.G. 1984. The place of non-parametric geostatistics. In G. Verly, M. David, A.G. Journel and A. Marechal (Eds.) *Geostatistics for Natural Resources Charcaterization* (V1): 307–355. Dordrecht: Reidel.

Laslett, G.M. & McBratney, A.B. 1990. Estimation and implications of instrumental drift, random measurement error and nugget variance of soil attributes-a case study for soil pH. *Journal of Soil Science* 41: 451–471.

Laslett, G.M. 1994. Kriging and splines: an empirical comparison of their predictive performance in some applications. *Journal of the American Statistical Association* 89: 391–409.

Laslett, G.M., McBratney, A.B., Pahl, P.J. & Hutchinson M.F. 1987. Comparison of several spatial prediction methods for soil pH. *Journal of Soil Science* 38: 325–341.

LeHouerou, H.N. 1996. Climate change, drought and desertification. *Arid Environments* 34: 133–85.

Madden, R.A., Shea, D.J., Branstator, G.W., Tribbia, J.J. & Weber, R.O. 1993. The effects of imperfect spatial and temporal sampling on estimates of the global mean temperature: experiments with model data. *Journal of Climate* 6: 1057–1066.

Matheron, M.A. 1971. *The Theory of Regionalized Variables and its Applications*. Cahiers du Centre de Morphologie Mathématique de Fountainebleau no. 5.

Mainguet, M. 1999. *Aridity, Droughts and Human Development*. Berlin: Springer Verlag.

Nicholson, S.E. 1985. Sub-Saharan rainfall 1981–1984. *Journal of Climate Applied Meteorology* 24: 1388–1391.

Nicholson, S.E. 1993. An overview of African rainfall fluctuations of the last decade. *Journal of Climate* 6: 1463–1466.

Nicholson, S.E. & Palao, I.M. 1993. A re-evaluation of rainfall variability in the Sahel. *International Journal of Climatology* 13: 371–389.

Peterson, T.C. & Vose, R.S. 1997. An overview of the Global Historical Climatology Network temperature data base, *Bulletin of the American Meteorological Society* 78: 2837–2849.

Raynaut, C. 1997. *Societies and Nature in the Sahel*. London: Routledge.

Sivakumar, M.V.K., Wallace, J.S., Renard, C. & Giroux, C. 1991. Soil Water Balance in the Sudano-Sahelian Zone. Wallingford: IAHS Publication.

Taylor, C.M., Lambin, E.F., Stephenne, N.R., Harding, J. & Essery, R.L.H. 2002. The influence of land use change on climate in the Sahel. *Journal of Climate* 15: 3615–3629.

Thomas, D.G. & Middleton, N. 1994. *Desertification: Exploding the Myth*. Chichester: J. Wiley.

Tucker, J.C., Dregne, H.E. & Newcomb, W.W. 1991. Expansion and contraction of the Sahara Desert from 1980 to 1990. *Science* 253: 299–301.

United Nations Environment Programme 1992. *Atlas of Desertification*. London: UNEP Edward Arnold.

Warren, A. & Agnew, C.T. 1988. *An assessment of desertification and land degradation in arid and semi-arid areas*. Paper 2, International Institute for Environment and Development, London.

Webster, R. & Oliver, M.A. 1992. Sample adequately to estimate variograms of soil properties. *Journal of Soil Science* 43: 177–192.

Webster, R. & Oliver, M.A. 2001. *Geostatistics for environmental scientists*. Chichester: Wileys.

Willmott, C.J., Robeson, S.M. & Janis, M.J. 1996. Comparison of approaches for estimating time-averaged precipitation using data from the USA. *International Journal of Climatology* 16: 1103–1115.

Zeng, N. 2003. Drought in the Sahel. *Science* 302: 999–1000.

Zeng, N., Neelin, J.D., Lau, K.-M. & Tucker, C.J. 1999. Enhancement of interdecadal climate variability in the Sahel by vegetation interaction. *Science* 286: 1537–1540.

Recent Advances in Remote Sensing and Geoinformation Processing
for Land Degradation Assessment – Röder & Hill (eds)
© 2009 Taylor & Francis Group, London, ISBN 978-0-415-39769-8

Ongoing desertification processes in the sahelian belt of West Africa: An evidence from the rain-use efficiency

Y.C. Hountondji & N. Sokpon
Faculty of Agronomy, University of Parakou, Parakou, Benin

J. Nicolas & P. Ozer
Environmental Sciences Management Department, University of Liege, Arlon, Belgium

ABSTRACT: The implementation of the UNCCD needs the identification of areas that record declining productivity of the vegetation over long-time periods. In this scope, we analyze the state of the vegetation productivity using 1982–1999 time series of NOAA-AVHRR NDVI data and compare it to rainfall data. For this, 354 rain gauges data distributed from yearly average isohyets 100 to 900 mm in five countries of West Africa are analyzed. We use for trends analysis, the ratio between integrated vegetation index (iNDVI) during the growing period (June to October) and the May to October sum of rainfall (RR). This ratio is a proxy of the Rain Use Efficiency is the widely accepted. Overall, 91% and 94% of RR and iNDVI data recorded positive trends over the 1982–1999 periods. Most stations in the Sahel were stable for the iNDVI/RR (49.5%). However, 37.8% showed strong to very strong negative change in the iNDVI/RR ratio, while only 1.3% showed positive trend. These strong negative trends recorded in more than 1/3 of the analyzed stations may reflect ongoing desertification processes in the Sahel and could be a starting point for the identification of hot-spots areas to determine where to take action within the National Action Programs (NAP) or Sub-Regional Action Programs (SRAP) to combat desertification.

1 INTRODUCTION

After about two decades of dramatic rainfall deficits that started in the late 1960s (L'Hôte et al. 2002), the Sahel of West Africa has experienced increasing precipitation since the early 1990s (Ozer et al. 2003, Dai et al. 2004). Although increasing human pressure on the environment over the same period may have enhanced desertification processes, some authors suggested that the Sahel has been greening from the early 1980s to the late 1990s (Eklundh & Olsson 2003, Pearce 2002).

Since the mid-1970s, desertification benefited from a considerable interest from scientists, politics and the public. Land degradation in arid lands is now recognized as one of the major environmental problems for the 21st century (World Bank 2003) and the Sahel of West Africa is often quoted as the most seriously affected region. Yet, desertification processes often evoke an image of advancing desert with moving dunes threatening houses, roads, oasis and fertile lands and leaving behind a barren and sterile environment. The term desertification has been misused for a long time due to the lack of data, objective indicators and rigorous scientific studies, and because of the inexistence of a widely accepted definition (Nicholson et al. 1998). Many authors have experienced a wide range of indicators in order to map the occurrence and severity of desertification (Mouat et al. 1997, Ozer 2000). Although many of these indicators were unsatisfactory because of their prohibitive costs and time-consuming for the process of data collection, low resolution satellite remote sensing data provide a good source of stable, reliable and long-term measurements (Symeonakis & Drake 2004, Prince 2002).

Methods have been developed to assess the vegetation net primary production (NPP) from the Normalized Difference Vegetation Index (NDVI). These techniques were developed in the mid of 1980s (Justice 1986, Prince & Justice 1991) but it is only recently that a medium-term archive has accumulated with a long enough record (20 years) to allow studies at the appropriate time scale (Prince et al. 1998). In addition, previous studies measured a strong, linear relationship between NDVI and primary production or above-ground total dry-matter accumulation in herbaceous vegetation in the Sahel, based on data for one or two successive years (Tucker et al. 1985, Prince 1991). This relationship was established both through a modelling of interactions between radiation and vegetation, and through empirical studies. In this paper, a proxy of the rain use efficiency (as detailed below) is used in order to identify areas suffering from desertification.

2 DATA AND METHODS

2.1 *Meteorological station data*

Monthly rainfall data were made available from an archive assembled by the "Projet Alerte Précoce et Prévision des Productions Agricoles (AP3A)" of the Centre Régional Agrhymet from station observations. Database includes meteorological stations through the nine countries that are grouped in the Inter-States Committee for Drought Control in the Sahel (CILSS). 354 rain gauges with complete monthly rainfall data from 1982 to 1999 were selected in five countries of West Africa (Mauritania, Senegal, Mali, Burkina Faso and Niger). Here we used the total rainfall records during rainy seasons (RR) that extend from May to October. The bioclimatic subdivision as defined by Ozer (2000) is:

- sahelian zone of the north (Z1): yearly rainfall between 100 and 300 mm
- sahelian zone of the central north (Z2): yearly rainfall between 300 and 500 mm
- sahelo-sudanese of the central south (Z3): yearly rainfall between 500 and 700 mm
- sudanese zone in the south (Z4): yearly rainfall between 700 and 900 mm

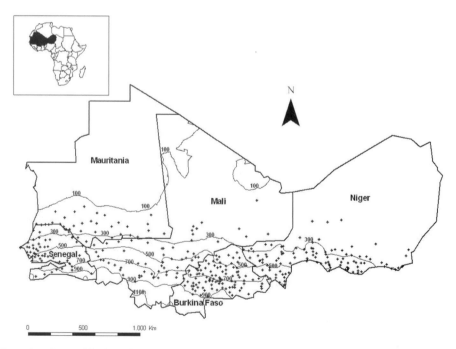

Figure 1.   Geographical position of the 354 studied stations (black crosses) and bioclimatic zones (isohyets defined by kriging on average rainfall value for the 1980–2000 periods).

174

Table 1. Selected stations in West Africa within the 100 to 900 mm isohyets.

Country	Number of selected stations per climatic zone				Total
	Z1: 100–300	Z2: 300–500	Z3: 500–700	Z4: 700–900	
Burkina Faso	0	12	38	49	99
Mali	10	14	16	20	60
Mauritania	27	4	0	0	31
Niger	32	58	16	2	108
Senegal	9	21	18	8	56
Total	78	109	88	79	354

The distribution and repartition of the stations according to four bioclimatic zones (Ozer 2000) are shown in Figure 1 and Table 1.

2.2 Satellite data

For the purpose of analyzing vegetation dynamics at sub-continental scale, satellite data from the NOAA AVHRR sensing system for the period 1982–1999, using the Normalized Difference Vegetation Index (NDVI) were used. The NASA/NOAA Pathfinder AVHRR Land (PAL) dataset (James & Kalluri 1994) has been generated as monthly maximum value composites at a 8×8 km pixel resolution. Noise levels can be very high over many areas in Africa but are comparatively low over arid to sub-humid areas where cloudiness is limited (Chappell et al. 2001). Quality issues of the Pathfinder database have been discussed by Prince and Goward (1996). The NDVI has proved useful in numerous monitoring studies of vegetation and drought. It is calculated as the normalized difference in reflectance between the red band (0.55–0.68 μm) and the near infrared band (0.73–1.1 μm). This quantity is considered to be a "greenness" index. In arid and semiarid regions, it is well correlated with such parameters as leaf area index, greenleaf, biomass, vegetation cover, etc. (Nicholson et al. 1990). Although it is well known that, especially in sparsely vegetated areas, there is considerable background influence, we will not take this criteria into account as the analysis are systematically focused on similar areas where soil composition is not likely changing during the considered period.

In the present study, the time series was restricted to the period 1982–1999 because, after 1999, a systematic shift in the remotely sensed data resulting from a very high shifting solar zenith angles has been detected (Eklundh & Olsson 2003), probably caused by late afternoon overpass of NOAA-14. In addition, year 1994 was not taken into account because of Pinatubo volcanic dust diffusion into the stratosphere (Tanaka et al. 1994). A total of 354 pixels were selected when including a rain gauge station. For all pixels, the seasonal vegetation index integrals (iNDVI) covering the entire growing season from June to October were computed in order to assess the annual net primary production (Diallo et al. 1991). This approach has been preferred to the use of rasterized rainfall data obtained from satellite estimations, because of stormy features and high spatial variation of sahelian rainfall. Previous studies in Sahel (Amani & Lebel 1997) exhibit a significant difference in annual rainfall score within a distance of 10 km. This implies that the spatial rainfall variability on a small scale is more significant than on the survey level scale.

2.3 Method

2.3.1 Indicator of rain use efficiency

The net annual increase of biomass, or net primary production, is a measure of the production of an ecosystem. This quantity bears a direct relationship to photosynthesis and NDVI is strongly correlated with both, particularly in arid lands. Le Houérou (1984) suggests that the ratio of primary production to rainfall, iNDVI/RR (rain use efficiency) is a better parameter to characterize arid and semi arid regions like the Sahel. In the Sahel, the dynamic of the vegetation is strongly linked to the rainfall evolution (Symeonakis & Drake 2004, Hess et al. 1996). For regions laying in the annual rainfall 100 mm to 900 mm, the iNDVI/RR ratio is regarded as a useful proxy for rain-use efficiency (Nicholson et al. 1998, Foody 2003). As mentioned by previous studies in the West African Sahel with spatially comprehensive measurements such as these, the incidence of the individual components of desertification could be detected (Diouf & Lambin 2001, Tottrup & Rasmussen 2004). For this research, we derived this ratio during the growing season. The total rainfall amount taken into account is the sum of the May to October precipitation (hereafter referred as rainfall). This period was used here since it has a strong relationship with the iNDVI calculated during the growing season (Davenport & Nicholson 1989, Hess et al. 1996, Hountondji et al. 2006). Indeed, during that period, the monthly rainfall peak generally occurs between August and September (Hess et al. 1996, Ozer 2000) and previous studies showed that the monthly sum NDVI follows monthly rainfall with a lag of about one month (Justice et al. 1986, Justice et al. 1991, Davenport & Nicholson 1989). As for rainfall, the iNDVI calculation ends October. Thus, the ratio of primary production to rainfall formulation is equivalent to:

$$iNDVI/RR = \frac{\sum_{june}^{october} NDVI}{\sum_{may}^{october} RR} \tag{1}$$

where RR = monthly rainfall, and NDVI = monthly normalized difference vegetation index.

2.3.2 *Trend analysis*

For each station, trends from 1982 through 1999 were estimated by linear regression considering the ratio iNDVI/RR as dependent variable and time (years) as independent variable. Regression slopes were recorded for each station as parameters characterising the global trends either for the rainfall or for NDVI. Moreover, each slope was mapped in seven classes indicating very strong, strong, weak positive or negative and stable trends, adapting a procedure suggested by Eklundh & Olsson (2003). The regression procedure supplies a Student-t test and its resulting significance p-level to analyse the hypothesis that the slope is different from 0. This p-level was used as a criterion to define the class boundaries. The trends, for the iNDVI/RR ratio, were labelled as "very strong" if the p-level exceeded 0.05 for the one-tailed t-test, "strong" if the p-level ranged between 0.05 and 0.1, "weak" if the p-level is between 0.1 and 0.2 and otherwise "stable" if the p-level is up to 0.2. These long-term linear trends for each pixel in the iNDVI/RR ratio may be understood as a combination of a number of interrelated factors including variations of biophysical and human influences. As the iNDVI/RR ratio is thought to remain stable through time, such trends can be interpreted as a measure of possible degradation or improvement of the vegetation growth.

3 RESULTS AND DISCUSSION

3.1 *Rainfall*

A preliminary analysis of rainfall data indicates increasing rainfall in the Sahel of West Africa during the studied period (Fig. 2). Overall, 91.4% of May to October rainfall recorded positive slope over the 1982–1999 period (not shown). In detail, 48.9% of the stations showed a very strong to strong positive trend, and 11% recorded a weak positive change. Stability characterized 39% of the analyzed stations, while less than 1% recorded a weak negative trend. Stations that have recorded the most remarkable positive changes seem to follow a north-south geographic gradient and one station out of two shows a stable tendency in the rainfall fringe between 700–900 mm. Such results were expected especially when considering that the drought period culminated in the early 1980s (L'Hôte et al. 2002, Dai et al. 2004, Nicholson 1985) that is the beginning of the analyzed dataset. In addition, these results are in accordance with recent investigations on rainfall variations in the Sahel suggesting that the drought may have ended in the early 1990s (Ozer et al. 2003).

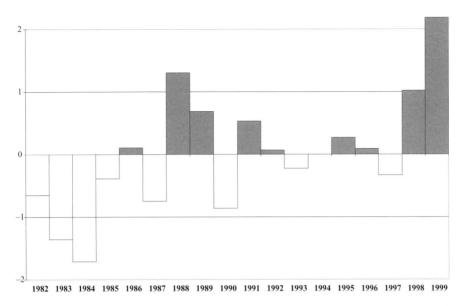

Figure 2. Average Z-scores (expressed as standard deviations) of rainfall for the 354 stations selected throughout the Sahel.

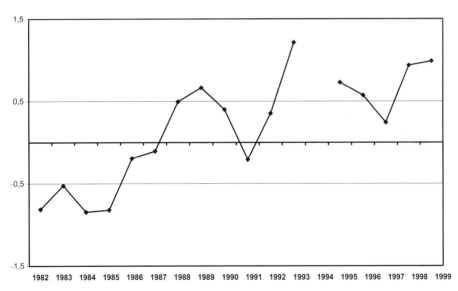

Figure 3. Average Z-scores (expressed as standard deviations) of iNDVI for the 354 stations selected throughout the Sahel. (note that 1994 is not considered because of the Pinatubo effect).

3.2 *The seasonal vegetation index (iNDVI)*

As expected, a substantial increase in the seasonal vegetation index was noticed consecutively to wetter conditions observed during the studied period. Yet, Figure 3 shows a constant below average iNDVI Z-score during the six first years of the analysis (1982–1987) that were among the driest years of the 20th century (L'Hôte et al. 2002, Nicholson 1985) while this value is systematically above average during the eight latest years (1992–1999) logically due to the strong rainfall recovery (Ozer et al. 2003). Overall, throughout the 354 pixels analyzed in the Sahel of West Africa, 62.8%

presented a very strong and strong positive change in iNDVI, while it remained stable in 29.9% of cases (not shown). These results are similar to those of Eklundh & Olsson (2003) who observed a strong increase in seasonal NDVI in the Sahel since 1982 interpreted as vegetation recovery from the drought periods of the 1980s. Other detailed studies in the same area showed that after a long period of strong land degradation, vegetation recovery was observed during the 1990s (Diouf & Lambin 2001, Hiernaux & Turner 2002, Rasmussen et al. 2001).

An unanswered question remains: as the early years of the dataset where exceptionally dry, is this vegetation recovery fully consistent with increasing rainfall?

3.3 The iNDVI/RR ratio

Trend classes for change in the iNDVI/RR ratio are summarized in Table 2 and relative change classes in the Rain Use efficiency considering the bioclimatic zone is shown in Figure 4. Overall, a negative slope is recorded in 68.4% of the 354 analyzed stations, while the remaining stations (31.6%) are characterized by a positive trend.

Detailed analysis shows that, for all considered stations, most areas (61.6%) remained stable during the 1982–1999 periods. However, 11.9% and 13.6% presented a very strong and strong negative change, respectively. 11.9% showed a weak negative trend and only 1.1% showed positive change. About 38% of the analyzed stations have therefore experienced decreasing trends in the iNDVI/RR ratio that may reflect possible degradation of the vegetation growth, and therefore ongoing desertification processes. The geographical distribution of downward weak to strong trends is fuzzy and may indicate localized land degradation processes in some places while few others may experience improved land management. Differences in changes are highlighted when considering phytogeographic areas of West Africa (Fig. 5).

On this graph, the error bars make possible to appreciate for each analyzed year the dispersion of the observations around the average value of the stations: they correspond to a confidence interval of 95% around the average. It is relevant to note that, at high rainfall, plant biomass production remains high as biomass production is a function of both rainfall level and RUE—the decrease in RUE at high rainfall simply indicates that the available rain is used less efficiently.

The linear regressions established with the whole set of the stations for each zone show a highly significant negative slope ($p < 10^{-3}$) for the statistics t of Student in all four bioclimatic zones over the studied period. This situation rises to the fact that the year 1984 was particularly dry in the Sahel, which generates very high values of the iNDVI/RR and thus an increase in the significance of the slope of the regressions. In addition, when one considers only the 1985–1999 periods, the results always show a decrease of the iNDVI/RR ratio in all four bioclimatic zones, with highly significant for Z1 ($p = 0.0005$), for Z2 ($p = 10^{-5}$), for Z3 ($p = 10^{-6}$), and Z4 ($p = 4.10^{-7}$). In the Sahelian belt (<500 mm in annual rainfall), 58.9% of the stations presented a relative stability, 12.7% showed a weak negative change and 28% strongly decreased. The sahelo-sudanese zone (500–700 mm) seems to have been the most affected by land degradation as vegetation resilience to rainfall is the lowest among all phytogeographic areas of the studied zone. Yet, about the half (51.9%) of the 79 stations included in this region show a stable iNDVI/RR ratio, while 20.3% experienced a very strong negative trend and 11.4% a strong negative change. In the Sudanese zone (700–900 mm), 72.2% of the stations appeared to be stable, while 16.5% and 8.9% suffered from a strong and weak negative change, respectively. However, it is worth mentioning that theses regions include the only stations of West Africa that display a weak positive change (2.5%). The obtained results in this southern region should be taken with precaution as previous studies mentioned that the relationship between NDVI and rainfall tends to weaken when annual rainfall is higher than 1000 mm. For sites dominated by woody and perennial herbaceous species, the same overall pattern can be expected. However, the relation may be less obvious because there is less interannual variation in the composition of the vegetation cover, and as woody and perennial herbaceous species can store nutrients and may tap groundwater reserves in deep soil layers. As primary production remains relatively constant, it seems that water is not a major limiting factor in these areas (Nicholson et al. 1990; Davenport & Nicholson 1989).

178

Table 2. Rainfall (RR), integrated NDVI (iNDVI) and rain use efficiency (iNDVI/RR) trends for 354 stations in West Africa (1982–1999).

Parameters	Slopes				Trends significance															Total	%
	POS	%	NEG	%	VSNC	%	SNC	%	WNC	%	S	%	WPC	%	SPC	%	VSPC	%			Total	%
RR																						
100 < P < 300	71	91.0	7	9.0	1	1.3	0	0.0	0	0.0	31	39.7	8	10.3	15	19.2	23	29.5			78	22.0
300 < P < 500	100	91.7	9	8.3	0	0.0	0	0.0	0	0.0	32	29.4	15	13.8	24	22.0	38	34.9			109	30.8
500 < P < 700	82	93.2	6	6.8	0	0.0	1	1.1	0	0.0	33	37.5	8	9.1	16	18.2	30	34.1			88	24.9
700 < P < 900	69	87.3	10	12.7	0	0.0	0	0.0	2	2.5	42	53.2	8	10.1	13	16.5	14	17.7			79	22.3
Total	322	91.0	32	9.0	1	0.3	1	0.3	2	0.6	138	39.0	39	11.0	68	19.2	105	29.7			354	100.0
iNDVI																						
100 < P < 300	76	97.4	2	2.6	0	0	0	0.0	1	1.3	19	24.4	2	2.6	5	6.4	51	65.4			78	22.0
300 < P < 500	100	91.7	9	8.3	0	0	0	0.0	0	0.0	34	31.2	10	9.2	15	13.8	50	45.9			109	30.8
500 < P < 700	82	93.2	6	6.8	0	0	0	0.0	0	0.0	31	35.2	6	6.8	5	5.7	46	52.3			88	24.9
700 < P < 900	77	97.5	2	2.5	0	0	0	0.0	0	0.0	22	27.8	7	8.9	6	7.6	44	55.7			79	22.3
Total	335	94.6	19	5.4	0	0	0	0.0	1	0.3	106	29.9	25	7.1	31	8.8	191	54.0			354	100.0
iNDVI/RR																						
100 < P < 300	14	17.9	64	82.1	6	7.7	13	16.7	10	12.8	49	62.8	0	0.0	0	0.0	0	0.0			78	22.0
300 < P < 500	19	17.4	90	82.6	14	12.8	15	13.8	18	16.5	61	56.0	0	0.0	1	0.9	0	0.0			109	30.8
500 < P < 700	18	20.5	70	79.5	15	17.0	14	15.9	7	8.0	51	58.0	1	1.1	0	0.0	0	0.0			88	24.9
700 < P < 900	61	77.2	18	22.8	7	8.9	6	7.6	7	8.9	57	72.2	2	2.5	0	0.0	0	0.0			79	22.3
Total	112	31.6	242	68.4	42	11.9	48	13.6	42	11.9	218	61.6	3	0.8	1	0.3	0	0.0			354	100.0

Legend: POS: positive; NEG: negative; VSNC: very strong negative change; SNC: strong negative change; WNC: weak negative change; S: stable; WPC: weak positive change; SPC: strong positive change; VSPC: very strong positive change.

Such contrasting evolutions along the bioclimatic gradients of the semi-arid belt West Africa are in accordance with recent findings of Hiernaux & Turner (2002). These authors stated that risks of environmental degradation are moderate and mainly climate-driven in pastoral systems at the drier edge, while they are serious and mainly management-driven in the crop-livestock systems of the sahelo-sudanese zone. Our results indeed show that 58.9% of the stations remained stable in the northern part of the studied area, while about 25% of the analyzed stations suffered from very strong, strong or weak negative changes in the sahelo-sudanese zone in the central north. As a matter of fact, Henry & colleagues (2003) showed that migration flows from the northern ecologically marginal area to the sahelo-sudanese zone were partly explained by unfavourable environmental variables such as high rainfall variability, land degradation, and land availability at the origin, and favourable conditions at the destination for these variables. For this reason, there is currently less pressure in the northern part of the Sahel as people migrated because of drought while migrations are likely to contribute to negative environmental changes at the destination (Lambin et al. 2001).

Recent claims that the Sahel is greening since the 1990s because of improved land management (Mazzucato & Niemeijer 2000, Rasmussen et al. 2001, Niemeijer & Mazzucato 2002, Pearce 2002, Eklundh & Olsson 2003) may be only partly true. As an example, Mazzucato & Niemeijer (2000) closely studied two small areas (Bilanga and Fada-N'Gourma) in Eastern Burkina Faso and suggested that these areas showed no evidence of land degradation as crop yields increased. Our results on these two stations indeed show that if the iNDVI presented positive slopes, the iNDVI/RR ratio experienced a strong negative change in Bilanga and a weak negative change in Fada-N'Gourma, suggesting that the rain-use efficiency of the vegetation has been declining over the last two decades (Figs. 6, 7). In northern Burkina Faso (Gorom-Gorom), Rasmussen et al. (2001) suggested that desertification was in reverse demonstrating that vegetation was reclaiming fossil dunes revitalised during the droughts of the 1970s and 1980s. Our results at this station (Figure 8) suggest that the iNDVI strongly increased and that the iNDVI/RR ratio remained stable during the 1980s and 1990s. In this specific case, it can be accepted that the vegetation is resilient with the rainfall increase observed during the last decade.

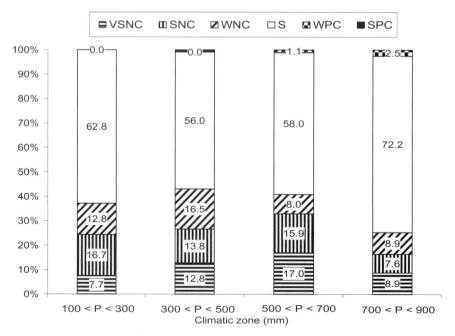

Figure 4.   Relative change classes in the Rain Use efficiency considering the bioclimatic zones (100–900 mm). Legend: VSNC: very strong negative change; SNC: strong negative change; WNC: weak positive change; S: stable; VSPC: very strong positive change; SPC: strong positive change; WPC: weak positive change.

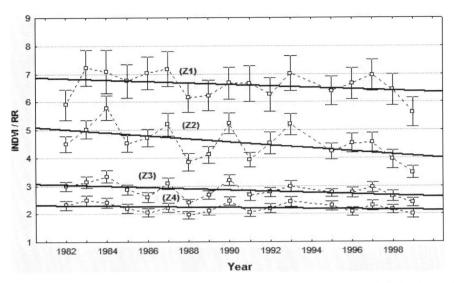

Figure 5.   Rain Use efficiency trends considering the bioclimatic zone (100–900 mm) during 1982–1999 periods in the studied area.

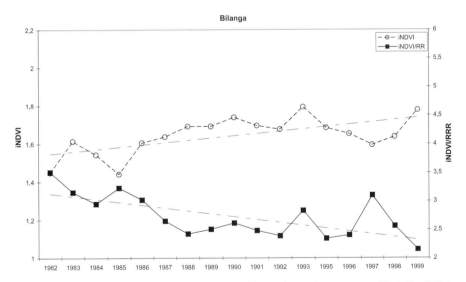

Figure 6.   Interannual variability of integrated NDVI and iNDVI/RR ratio as a proxy of Rain Use Efficiency at Bilanga site (Eastern Burkina Faso) (Hountondji et al. 2006).

Elsewhere, Hein & De Ridder (2006) studying the relationship between RUE and rainfall at Sydenham, South Africa, and in the Ferlo, Senegal, demonstrated how the whole curve shifts towards a lower RUE following degradation of the ecosystem (O'Connor et al. 2001). These authors noticed that the quadratic relation between rainfall and RUE still allows for a decline in RUE following degradation of the vegetation cover. Hence, RUE is a function of both the state of the vegetation cover and annual rainfall, and analysis of degradation with satellite images requires consideration of the rainfall pattern during the time of satellite observations. If further analysis of remote sensing images confirms a process of degradation in the Sahel, this would have important consequences for the ongoing debate on equilibrium vs. non-equilibrium approaches to rangeland dynamics. It would confirm the relevance of the equilibrium approach with respect to the overall

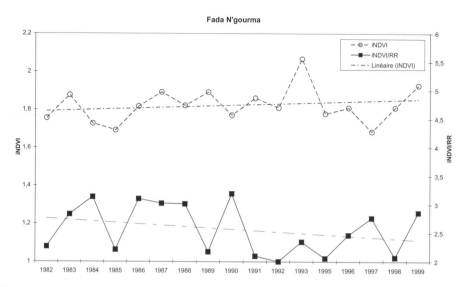

Figure 7.   Interannual variability of integrated NDVI and iNDVI/RR ratio as a proxy of Rain Use Efficiency at Fada N'gourma site (Eastern Burkina Faso) (Hountondji et al. 2006).

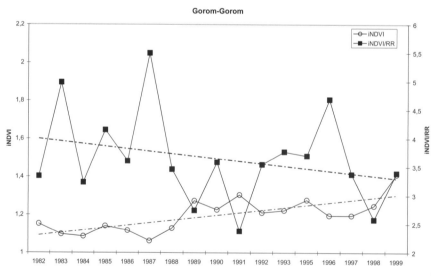

Figure 8.   Interannual variability of integrated NDVI and iNDVI/RR ratio as a proxy of Rain Use Efficiency at Gorom-gorom site (Northern Burkina Faso) (Hountondji et al. 2006).

impact of grazing at the scale of the Sahel under current grazing pressures (cf. Le Houérou 1989; Hérault & Hiernaux 2004). In addition, if man-induced degradation of the Sahel (overgrazing, wood cutting, land mismanagement) is demonstrated, this would have repercussions for the debate on the causes of climate change in the Sahel. Currently, a weakness in the argumentations of Xue & Shukla (1993) and Wang et al. (2004) that anthropogenic-induced land cover changes have contributed to the occurrence of the extreme Sahelian droughts of the last decades of the 20th century is a lack of evidence of degradation from remote sensing data.

Therefore, the conclusion that satellite data do not show long-term degradation in the Sahel is premature (Hountondji et al. 2006, Ozer et al. 2007). This is further illustrated by recent findings on two Sahelian sites for which multiyear data on rangeland productivity were available (Hein & De Ridder 2006). For these sites, the hypothesis that the rainfall pattern would result in a constant RUE in the absence of degradation could be rejected at p levels varying from p = 0.03 to p = 0.08. They concluded that it is likely that the relatively constant RUE found in remote sensing studies indicate a process of human induced degradation of the plant cover in the Sahel. This also implies that there has been no 'greening' of the Sahel beyond the impacts of increasing rainfall, as suggested in Anyamba & Tucker (2005) and Olsson et al. (2005). However, more analysis is required to provide a definite answer as to the existence and rate of degradation in the Sahel. Because of the limited amount of long-term data on phytomass productivity in the Sahel (Le Houérou 1989, Herault & Hiernaux 2003), remote sensing analysis remains the preferred tool.

A limiting factor in this study has been the spatial resolution of the remotely sensed data used for our analysis. In fact, the AVHRR sensor is characterized by shortcomings within the subject of vegetation monitoring because it was not originally designed for this purpose (Teillet et al. 1997; Van Leeuwen et al. 1999, Steven et al. 2003). In addition, the spatial extrapolation from 1 km to 8 km resolution in coarse Pathfinder data could hide some undergoing processes (Hountondji et al. 2006; Niang et al. 2008). It is well known that the relative variability of phytomass production is highly dependent on the spatial scale. And according to Golluscio and colleagues (2005) the variation coefficient of primary production decreases exponentially as the size of plots or pixels increases. Therefore, it appears that the coarse resolution remote sensing estimates of vegetation production may underestimate the temporal variability of production as measured in smaller field sites (Diouf & Lambin 2001). An alternative solution for this weakness could be found through the use of finer spatial resolution imageries such as AVHRR GIMMS as suggested by Fensholt et al. (2006). These authors conclude that the correction for sensor orbital drift in the GIMMS data set has improved the data quality compared to the AVHRR Pathfinder data. Their results suggest that the accuracy of the AVHRR GIMMS NDVI is higher than the AVHRR PAL NDVI and, consequently, GIMMS NDVI should be used for analyses of long-term trends in regional or continental scale NDVI.

As mentioned above, many biophysical and human influences may interfere in the evolution of the iNDVI/RR ratio. Increasing use of fertilizers, better water resources management and land rehabilitation measures have improved over the years. But despite such positive technical evolutions, only four stations out of 354 recorded a weak and strong positive change the iNDVI/RR ratio over the 1982–1999 periods. Another interference in the evolution of the iNDVI/RR ratio may be attributed to climate change due to the ongoing build-up of greenhouse gases. In many regions of the world, extreme precipitation events have significantly increased during the last decades (Houghton 2001). Currently, research on changes in extreme rainfall events specific to Africa, in either models or observations, is limited. However, a general increase in the intensity of high-rainfall events, associated in part with the increase in atmospheric water vapour, is expected in Africa, as in other regions (Christensen et al. 2007). Current works in progress (Ozer and colleagues) show that, despite significant yearly rainfall shortages and large decrease of yearly rainfall days recorded over the 1941–2004 period in Niger and Mauritania, extreme rainfall events frequency remained stable and, sometimes, significantly increasing. This may partly explain a large increase of floods observed lately in several areas of West Africa (Sene & Ozer 2002, Tarhule 2005).

If this trend was confirmed, then the iNDVI/RR ratio may be negatively affected as extreme daily precipitation can not be fully used by vegetation and can further cause erosion and soil crusting.

4 CONCLUSION

Based on observations of increased crop yields and/or NDVI, recent studies have stated that desertification in the African Sahel was in reverse. However, using trends in the iNDVI/RR ratio, our results suggest that about 37.4% of the analyzed stations may have experienced ongoing desertification processes during the 1982–1999 periods. Our findings present an environmental situation that is probably more gloomy than recent papers stated although we are far from the concept of irreversible land degradation that was so fashionable until recently. Nevertheless, these results have to be taken with precaution because of the needs of biophysical significance of the stable trends recorded in the study. It is possible that the coarse spatial resolution of AVHRR data (64 km^2) may hide undergoing trends that are not detectable at this scale. Hence, if new remote sensing analyses confirm anthropogenic degradation (overgrazing, wood cutting, land mismanagement), this would support the hypothesis that degradation of the vegetation layer, in particular through sustained high grazing pressures, has contributed to the occurrence of the 20th century droughts in the Sahel. Furthermore, if degradation of the Sahelian vegetation cover is confirmed, this would indicate that Sahelian pastoralists may be more vulnerable for future droughts than currently assumed. Because degradation of the Sahel in the 1980s and 1990s has been masked by an upward trend in annual rainfall, the consequences of a future drought for the local population could be unexpectedly severe. Otherwise, for more accurate investigations, any remote sensing-based monitoring system of land degradation in such area must be complemented by field data collection, in particular with floristic composition data which is not detectable from space, even at fine spatial resolutions. This could be a starting point for switching from empirical approaches based on vegetation indices, which suggest an improvement of the environmental situation at the regional scale, but that may not reflect real situation.

REFERENCES

Amani, A. & Lebel, T. 1997: Lagrangian kriging for the estimation of Sahelian rainfall at small time steps. *Journal of Hydrology* 192: 125–157.

Anyamba, A. & Tucker, C.J. 2005. Analysis of Sahelian vegetation dynamics using NOAA-AVHRR NDVI data from 1981–2003. *Journal of Arid Environments* 63: 596–614.

Chappell, A., Seaquist, J.W. & Eklundh, L. 2001. Improving the estimation of noise from NOAA AVHRR NDVI for Africa using geostatistics. *International Journal of Remote Sensing* 22: 1067–1080.

Christensen, J.H., Hewitson, B., Busuioc, A., Chen, A., Gao, X., Held, I., Jones, R., Kolli, R.K., Kwon, W.-T., Laprise, R., Magaña Rueda, V., Mearns, L., Menéndez, C.G., Räisänen, J., Rinke, A., Sarr, A. & Whetton, P. 2007. Regional Climate Projections. In S. Solomon, D. Qin, M. Manning, Z. Chen, M. Marquis, K.B. Averyt, M. Tignor & H.L. Miller (eds), *Climate Change 2007: The Physical Science Basis. Contribution of Working Group I to the Fourth Assessment Report of the Intergovernmental Panel on Climate Change.* Cambridge, UK, New York, USA: Cambridge University Press.

Dai, A., Lamb, P.J., Trenberth, K.E., Hulme, M., Jones, P.D. & Xie, P. 2004. The Recent Sahel drought is real. *International Journal of Climatology* 24: 1323–1331.

Davenport, M.L. & Nicholson, S.E. 1989. On the relation between rainfall and Normalized Difference Vegetation Index for diverse vegetation types in East Africa. *International Journal of Remote Sensing* 12: 2369–2389.

Diallo, O., Diouf, A., Hana, N.P., Ndiay, A. & Prevost, Y. 1991. AVHRR monitoring of primary production in Senegal, West Africa. *International Journal of Remote Sensing* 12: 1259–1279.

Diouf, A. & Lambin, E.F. 2001. Monitoring land-cover changes in semi-arid regions: remote sensing data and field observations in the Ferlo, Senegal. *Journal of Arid Environments* 48: 129–148.

Eklundh, L. & Olsson, L. 2003. Vegetation index trends for the African Sahel 1982–1999. *Geophysical Research Letters* 30: 10.1029/2002GL016772.

Fensholt, R., Rasmus N., Thomas T. & Stisen, S. 2006. Evaluation of AVHRR PAL and GIMMS 10-day composite NDVI time series products using SPOT-4 vegetation data for the African continent. *International Journal of Remote Sensing* 27(13): 2719–2733.

Foody, G.M. 2003. Geographical weighting as a further refinement to regression modelling: an example focused on the NDVI-rainfall relationship. *Remote Sensing of Environment* 88: 283–293.

Golluscio, R.A., Perez, J.A., Paruelo, J.M. & Ghersa, C.M. 2005. Spatial heterogeneity at different grain sizes in grazed versus ungrazed sites of the Patagonian steppe. *Ecoscience* 12: 103–109.

Hein, L. & De Ridder, N. 2006. Desertification in the Sahel: a reinterpretation. *Global Change Biology* 12(1–8), doi: 10.1111/j.1365-2486.2006.01135.x

Hérault, B. & Hiernaux, P. 2004. Soil seed bank and vegetation dynamics in Sahelian fallows; the impact of past cropping and current grazing treatments. *Journal of Tropical Ecology* 20: 683–691.

Henry, S., Boyle, P. & Lambin, E. 2003. Modelling inter-provincial migration in Burkina Faso, West Africa: the role of socio-demographic and environment factors. *Applied Geography* 23: 115–136.

Hess, T., Stephens, W. & Thomas, G. 1996. Modelling NDVI from decadal rainfall data in the North East arid zone of Nigeria. *J. Environ. Manag.* 48: 249–261.

Hiernaux, P. & Turner, M.D. 2002. The influence of farmer and pastoralist management practices on desertification processes in the Sahel. In J.F. Reynolds & D.M. Stafford Smith (eds), *Global desertification: Do humans cause deserts?*: 135–148. Dalhem: Dalhem University Press.

Houghton, J.T., Ding, Y., Griggs, D.J., Noguer, M., Van der Linden, P.J., Dai, X., Maskell, K. & Johnson, C.A. 2001. *Climate Change 2001: The Scientific Basis.* New York: Cambridge University Press.

Hountondji, Y-C., Sokpon, N. & Ozer, P. 2006. Analysis of the vegetation trends using low resolution remote sensing data in Burkina Faso (1982–1999) for the monitoring of desertification. *International Journal of Remote Sensing* 27(5): 871–884.

James, M.E. & Kalluri, S.N.V. 1994. The Pathfinder AVHRR land data set: an improved coarse resolution data set for terrestrial monitoring. *International Journal of Remote Sensing* 15: 3347–3364.

Justice, C.O. 1986. Monitoring the grasslands of semi-arid Africa using NOAA-AVHRR data. *International Journal of Remote Sensing* 7: 1383–1622.

Justice, C.O., Digdale, G., Townshend, J.R.G., Narracott, A.S. & Kumar, M. 1991. Synergism between NOAA-AVHRR and Meteosat data for studying vegetation development in semiarid West Africa. *International Journal of Remote Sensing* 12: 1349–1368.

Lambin, E.F., Turner, B.L., Geist, H.J., Agbola, S.B., Angelsen, A., Bruce, J.W., Coomes, O., Dirzo, R., Fischer, G., Folke, C., George, P.S., Homewood, K., Imbernon, J., Leemans, R. Li, X., Moran, E.F., Mortimore, M., Ramakrishnan, P.S., Richards, J.F., Skånes, H., Steffen, W., Stone, G., Svedin, U., Veldkamp, T.A., Vogel, C. & Xu, J. 2001. The Causes of land-use and -cover change: moving beyond the myths. *Global Environ. Change* 11: 261–269.

Le Houérou, H.N. 1984. Rain-Use Efficiency: a unifying concept in arid land ecology. *Journal of Arid Environments* 7: 1–35.

Le Houérou, H.N. 1989. The Grazing Land Ecosystems of the African Sahel. Springer, Berlin.

L'Hôte, T., Mahé, G., Somé, B., & Triboulet, J.P. 2002. Analysis of a Sahelian annual rainfall index from 1896 to 2000; the drought continues. *Hydrol. Sc. J.* 47: 563–572.

Mazzucato, V. & Niemeijer, D. 2000. The cultural economy of soil and water conservation: market principles and social networks in Eastern Burkina Faso. *Development and Change,* 31: 831–855.

Mouat, D., Lancaster, J., Wade, T., Wickham, J., Fox, C., Kepner, W. & Ball, T. 1997. Desertification evaluated using an integrated environmental assessment model. *Environmental Monitoring and Assessment* 48: 139–156.

Niang, A.J., Ozer, A. & Ozer, P. 2008. Fifty years of landscape evolution in Southwestern Mauritania by means of aerial photos. *Journal of Arid Environments* 72: 97–107.

Niemeijer, D. & Mazzucato, V. 2002. Soil degradation in the West African Sahel. How serious is it? *Environment* 44: 20–31.

Nicholson, S.E. 1985. Sub-Saharan rainfall 1981–1984. *Journal of Climate and Applied Meteorology* 24: 1388–1391.

Nicholson, S.E., Davenport, M.L. & Malo, A.R. 1990. A comparison of the vegetation response to rainfall in the Sahel and East Africa, using normalized difference vegetation index from NOAA AVHRR. *Climatic Change* 17: 209–241.

Nicholson, S.E., Tucker, C.J. & Ba, M.B. 1998. Desertification, drought, and surface vegetation: an example from the West African Sahel. *Bulletin of the American Meteorological Society* 79: 815–829.

O'Connor, T.G., Haines, L.M. & Snyman, H.A. 2001. Influence of precipitation and species composition on phytomass of a semi-arid African grassland. *Journal of Ecology* 89: 850–860.

Olsson, L., Eklundh, L. & Ardo, J. 2005. A recent greening of the Sahel—trends, patterns and potential causes. *Journal of Arid Environments* 63: 556–566.

Ozer, P. 2000. Les lithométéores en région sahélienne: un indicateur climatique de la désertification. *GEO-ECO-TROP.* 24: 1–317.

Ozer, P., Erpicum, M., Demarée, G. & Vandiepenbeeck, M. 2003. The Sahelian drought may have ended during the 1990s. *Hydrological Sciences Journal* 48: 489–492.

Pearce, F. 2002. Africans go back to the land as plants reclaim the desert. *New Scientist* 175: 4–5.

Prince, S.D. & Justice, C.O. 1991. Coarse resolution remote sensing in the Sahelian environment. *International Journal of Remote Sensing* 12: 1133–1421.

Prince, S.D. & Goward, S.N. 1996. Evaluation of the NOAA/NASA Pathfinder AVHRR Land Data Set for global primary production modelling. *International Journal of Remote Sensing* 17: 217–221.

Prince, S.D. Brown de Colstoun. E. & Kravitz, L.L. 1998, Increased rain-use efficiencies indicate no extensive Sahelian desertification. *Global Change Biology* 4: 359–374.

Prince, S.D. 2002. Spatial and temporal scales for detection of desertification. In J.F. Reynolds & D.M. Stafford Smith (eds), *Global desertification: Do humans cause deserts?*: 23–40. Dalhem: Dahlem University Press.

Rasmussen, M.S. 1998. Developing simple, operational, consistent NDVI-vegetation models by applying environmental and climatic information: Part I. Assessment of net primary production. *International Journal of Remote Sensing* 19: 97–117.

Rasmussen, K., Fog, B. & Madsen, J.E. 2001. Desertification in reverse? Observations from northern Burkina Faso. *Global Environtal Change* 11: 271–282.

Sene, S. & Ozer, P. 2002. Evolution pluviométrique et relation inondations—événements pluvieux au Sénégal. *Bulletin du Societé Géographique de Liège* 42: 27–33.

Steven, M.D., Malthus, T.J., Baret, F., Xu, H. & Chopping, M.J. 2003. Intercalibration of vegetation indices from different sensor systems. *Remote Sensing of Environment* 88: 412–422.

Symeonakis, E. & Drake, N. 2004. Monitoring desertification and land degradation over sub-Saharan Africa. *International Journal of Remote Sensing* 25: 573–592.

Tanaka, S., Sugimura, T., Harada, T. & Tanaka, M. 1994. 'Pinatubo volcanic dust diffusion into the stratosphere seen from space'. *Advances in Space Research* 14(3): 245–254.

Tarhule, A. 2005. Damaging rainfall and flooding : The other sahel hazards. *Climatic Change* 72(3): 355–377.

Teillet, P.M., Staenz, K. & Williams, D.J. 1997. Effects of spectral, spatial, and radiometric characteristics on remote sensing vegetation indices of forested regions. *Remote Sensing of Environment* 61: 139–149.

Tottrup, C. & Rasmussen, M.S. 2004. Mapping long-term changes in savannah crop productivity in Senegal through trend analysis of time series of remote sensing data. *Agriculture, Ecosystems & Environment* 103: 545–560.

Van Leeuwen, W.J.D., Huete, A.R. & Laing, T.W. 1999. MODIS vegetation index compositing approach: a prototype with AVHRR data. *Remote Sensing of Environment* 69: 264–280.

Wang, G., Eltahir, E.A.B. & Foley, J.A. 2004. Decadal variability of rainfall in the Sahel: results from the coupled GENESISIBIS atmosphere-biosphere model. *Climate Dynamics* 22, 625–637.

World Bank 2003. World development report 2003: *Sustainable development in a dynamic world: transforming institutions, growth, and quality of life*. New Dehli: Oxford University Press.

Xue, Y. & Shukla, J. 1993. The influence of land surface properties on Sahel climate. Part I: desertification. *Journal of Climate* 6: 2232–2244.

Recent Advances in Remote Sensing and Geoinformation Processing
for Land Degradation Assessment – Röder & Hill (eds)
© 2009 Taylor & Francis Group, London, ISBN 978-0-415-39769-8

Vegetation cover and biomass along climatic gradients: The synergy of remote sensing and field studies in two Eastern Mediterranean sites

M. Shoshany

Geo-Information Engineering, Faculty of Civil & Environmental Engineering, Technion,
Israel Institute of Technology, Haifa, Israel

ABSTRACT: Modes of bio-physical transition between the shrub-dominated and desert fringe communities were revealed for North to South and West to East climatic transects in Israel. This was achieved by combining vegetation cover data derived from applying remote sensing techniques on seasonal Landsat TM imagery with field surveys of cover and biomass. Correlation levels between vegetation cover and biomass with precipitation were found to be significant, revealing the non-linearity of these relationships, especially for disturbed ecosystems. Assessment of the resulting patterns when compared to data obtained from existing phytogeographical maps indicated the added value of such synergy from the ecological point of view. The following generalized model for Mediterranean climatic transects is suggested (but needs further assessment): $Biomass = 0.09 * e^{0.045*precipitation}$, where precipitation is measured as cm/year.

1 INTRODUCTION

Desertification resulting from global warming (predicted to reach between 2 and 3 degrees Celsius by the end of this century according to IPCC scenarios), and growing population pressures on use of land resources are most pronounced along sharp climatic gradients. Remote sensing and spatial analysis are instrumental for discovering the intensities and extent of surface property changes and land transformation. However, the task of monitoring spatial changes is most challenging, due to the high spatial heterogeneity in regions of climatic transition in general and Mediterranean regions in particular (e.g., Yaalon 1997). Proper mapping and monitoring of these regions requires synergy of information from field studies, ancillary data, and imagery acquired using different sensors, at different spatial scales and extents and at different periods.

This paper presents the results of the vegetation mapping part within an ongoing interdisciplinary research that was carried out during the last decade along two transects: North to South across the northern Negev (dry land) boundary and West to East across the Judean Desert boundary (Figure 1). Soil conditions, vegetation abundance and composition, and landuse/landcover were surveyed in the field, as well as by aerial photography and satellite imagery (e.g., Shoshany et al. 1996, 1995, Shoshany 2001, 2002, Shoshany & Svoray 2002, Svoray & Shoshany 2004, 2005, Cohen & Shoshany 2004, 2005). The common denominator for all these studies was an attempt to better understand variations in the spatial patterns along these gradients. Of special interest was the identification of threshold zones representing high spatio-temporal rates of change in the environmental conditions.

The first part of this article presents changes in the relative vegetation cover of three main life-forms: Shrubs, dwarf-shrubs, and herbaceous growth. Three main information sources were integrated for this purpose: Remote Sensing, Field Surveys, and existing Phytogeographical Maps. The second part involves estimation of biomass changes along the two transects, based on the characteristic life-form compositions and detailed studies of woody and herbaceous biomass along the Negev Transect.

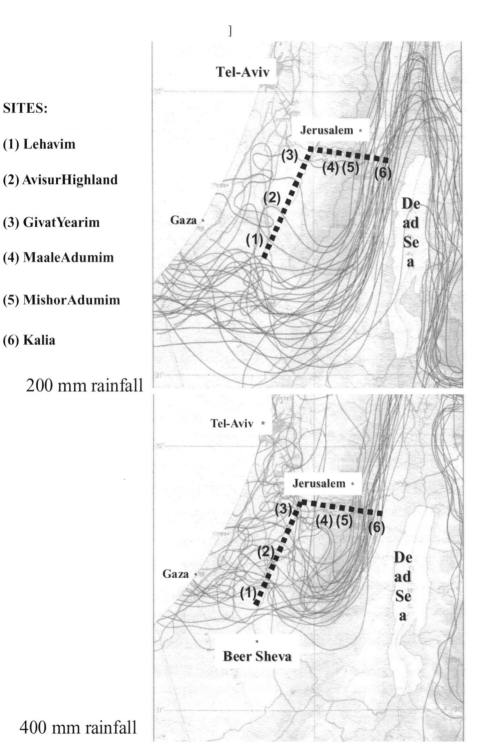

]

SITES:

(1) Lehavim

(2) AvisurHighland

(3) GivatYearim

(4) MaaleAdumim

(5) MishorAdumim

(6) Kalia

200 mm rainfall

400 mm rainfall

Figure 1. Location of the two transect areas with respect to the 200 mm and 400 mm Isohyets' fluctuations between 1930 and 1960 (Atlas of Israel 1964 Edition) (see colour plate page 390).

2   RAINFALL CHANGE ALONG THE TWO CLIMATIC TRANSECTS

Figure 1 presents two maps that appeared in one of the special, early editions of the Atlas of Israel (1964) representing 30 years of rainfall monitoring in Israel. In terms of rainfall, 200 mm/year is regarded as indicative of the arid/semi-arid transition, and 400 mm/year is regarded as indicative of the semi-arid/sub-humid transition. Fluctuations in these two rainfall isolines define the extents of the climatic transition zone. The two lines marked across the humid-to-arid zones represent the central axis of the two study areas reported here.

3   VEGETATION COVER STUDIES

3.1   *The Judean desert transect*

The following three information sources were integrated in order to advance our understanding of vegetation distributions along the Judean Desert gradient:

A. A field survey of vegetation was conducted at four sites (Figure 1): Givaat Yearim, Maaleh Adumim, Mishor Adumim, and Kalia. During the spring of 1991 and 1992, Kutiel et al. (1995) delineated two units of 1000 m^2 at each of these sites, and sampled plant abundances based on Whittaker's method.
B. A remote sensing study was based on a model that links the surface's spectral reflectance with percentage vegetation cover (V-R Model). The model was developed by measuring reflectance using a field spectrometer and by estimating vegetation cover using close range photography (Shoshany et al. 1994, 1995). Implementing this model with calibrated Landsat Images facilitated mapping vegetation cover patterns for a period of two years, representing extreme rainfall conditions: 1991 with 40% below average precipitation, and 1992 with a 60-year record high rainfall. Whereas the winter vegetation cover represents both the herbaceous and the woody vegetation, the spring and summer cover represents the woody vegetation. Thus, subtracting these two quantities yields differentiation between herbaceous growth and wood cover.
C. Phytogeographical mapping by Zohary (1973) and Danin (1975), which provides estimates of woody and herbaceous vegetation cover for the phytogeographic belts according to visual assessment of the habitat conditions (rather than the actual coverage at the survey time).

Figure 2 presents the vegetation cover estimates, together with rainfall for the four sites. Assessment of the data indicates the following:

Woody Cover: all cover estimates highly correlate among themselves (R > 0.9), and all cover estimates highly correlate with average rainfall (R > 0.9). However, the woody cover estimates from the phytogeographical mappings were significantly higher than those given by the two other sources. Furthermore, the along-transect patterns significantly differs: according to the phytogeographical mapping cover estimates, the along-transect pattern decreases gradually eastwards, whereas the field and Remote Sensing estimates suggest the existence of a very sharp threshold east of Mishor Adumim with a decrease by approximately 70% between this area and Givat Yearim to the west.

Herbaceous Cover: although the cover estimates from the phytogeographical mapping correlate well with the Remote Sensing estimates, these two sources do not agree with the field survey. Moreover, all cover sources do not correlate with rainfall intensity along the transect. According to the Phytogeographical and Remote Sensing mappings, the two sites, Mishor Adumim and Maale Adumim, seem to have similar herbaceous cover, decreasing both eastwards and westwards.

3.2   *The Northern Negev transect*

The following three information sources were integrated in order to advance our understanding of vegetation distributions along the Northern Negev Transect:

A. A field survey of vegetation was conducted by botanists from the Department of Surveying and Mapping of the Israeli Ministry of Agriculture (for more details see: Shoshany & Svoray 2002).

Thirty (30) study areas (5 ha size on average) representing a wide variety of vegetation patterns were identified using aerial photographs. The survey methodology at each area was carried out utilizing two sampling techniques: detailed sampling along a central transect, 100-m length (the cover type was read every 50 cm using a slender bar positioned exactly vertical to the ground at each point), and visual estimation along two parallel transects at a distance of 50 m at both sides of the central transect. The following data were gathered: (1) cover proportions of physiognomic categories of herbaceous annuals and perennials (dominant species), dry matter, and shrubs in three height categories; and (2) rock and soil cover.

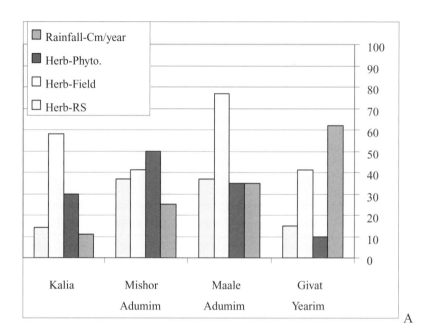

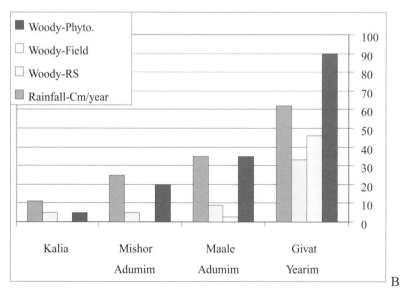

Figure 2. Relative vegetation cover and rainfall for each of the research sites in the Judean Transect: A—Herbaceous Growth; B—Woody Vegetation.

B. A remote sensing study: the new methodology of Adaptive Unmixing was presented for this region by Shoshany & Svoray (2002). This method optimizes endmembers' selection by means of a preceding stage of classifying the region into areas representing characteristic vegetation and soil combinations. For each of these regions, soil, rock, and vegetation image endmembers were defined. Applying the unmixing on three images representing important phyto-phenological stages in the natural vegetation growth yielded three raster information layers describing the spatial distribution of vegetation fractional cover. Based on the Phenological Subtraction Model (PSM), Shoshany & Svoray (2002) defined for each image pixel the relative fractions of three life-forms: Herbaceous growth, Dwarf Shrubs, and Shrubs.

C. Phytogeographical mapping by Sapir (1977) and Danin (1988) was conducted during the 1960s and 1970s and has not been updated since then on a regional scale. Six vegetation associations representing dominant or climax species were mapped north of the Lehavim site. This information source provided species distribution data for each association, but not their relative coverage.

Assessment of the remote sensing estimates using the field survey data yielded the following results (Figure 4):

1. PSM-based estimates for Herbaceous growth cover correlated well with field-estimated values (R2 value of 0.8 and an RMSD of 9%).
2. PSM-based estimates for Shrub cover were highly correlated with field-estimated values (R2 value of 0.95 and an RMSD of 4%).
3. PSM-based estimates for Dwarf Shrub cover were highly correlated with field-estimated values (R2 value of 0.88 and an RMSD of 6%).

Figure 3 presents the results obtained for five sites: Ramat Avisur, Amatzia, Dvira, Lehavim, and Beer Sheva (Figure 1). A generalized northward trend of increased vegetation cover of both herbaceous and woody vegetation was observed, except for the relatively high woody cover in Dvira (which is linked to plantations in this area by the Forest Commission—Jewish National Fund).

According to the phytogeographical map (Sapir 1977, Danin 1988), there are 6 units present in this transect: three of them represent shrub domination and the other three are associated with the dwarf shrub of *Sarcopoterium spinosum*. Both independent information sources are in agreement with regard to the main transition from shrubs to dwarf shrubs south of Amatzia.

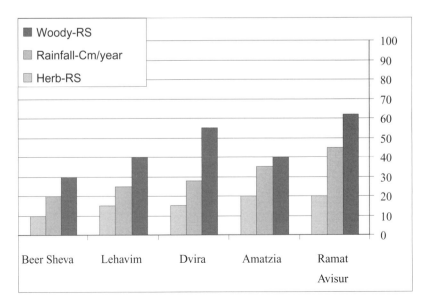

Figure 3.   Relative vegetation cover and rainfall for each of the research sites in the Northern Negev Transect.

4 BIOMASS CHANGE ALONG THE CLIMATIC GRADIENTS

Two field studies were conducted along the Northern Negev Transect: one involved a detailed survey of the biomass characteristics of different woody species in two sites of this transect (Sterenberg & Shoshany 2001, 2002); the second involved a detailed survey of herbaceous vegetation biomass variations in between these two sites (Svoray & Shoshany 2001, 2004).

4.1 *Woody vegetation*

In the first study, 16 quadrats of 10 m × 10 m (eight south-facing and eight north-facing slopes) were established and the vegetation was recorded. Representative branches were cut and weighed, separating the leaves and woody biomass of the dominant 10 shrub species. Using allometric functions based on parameters of area and volume, we estimated the overall biomass characteristics for each shrub. The average woody biomass for Ramat Avisur was estimated to be 1.5 kg/m^2, whereas in Lehavim it was 0.5 kg/m^2.

4.2 *Herbaceous vegetation*

A field survey of dominant homogenous herbaceous vegetation in the study area included, overall, 57 plots (of which 33 were of natural herbaceous vegetation and 24 of wheat and barley), 0.3 ha in size, which were used to estimate aboveground biomass with the 'harvest and assessment' method of Tadmor et al. (1975). Aerial Aboveground Biomass (AAB) was estimated visually at 100 quadrants of 0.625 m^2, located within each of the 57 plots. These estimates were regressed against direct aboveground biomass harvests from 10 random samples of the 100 quadrates. DGPS (differential global positioning system) readings of each plot's center were taken in order to allow assessment with reference to the satellite image data. The results of the field campaigns yielded a wide distribution of AAB levels: natural herbaceous vegetation varied between 0.05 kg/m^{-2} and 0.3 kg/m^{-2} whereas the herbaceous crops had extended the upper range to a value of 0.6 kg/m^{-2}.

Based on these studies, we derived characteristic biomass values for the three life-forms: Shrubs -1.5 kg/m^2, Dwarf Shrubs -0.5 kg/m^2, and Herbaceous vegetation -0.2 kg/m^2. Figure 4 presents the results of applying these characteristic values to vegetation compositions, as estimated using the remote sensing data for the main study locations along the gradients. Whereas in the Northern Negev Transect, the relative fractions of Shrubs and Dwarf Shrubs were estimated explicitly, in the Judean Desert Transect, it was assumed that the woody vegetation is composed of mainly Shrubs in the most western site (Givaat Yearim), and of mainly Dwarf Shrubs in the other sites. Biomass and precipitation are highly correlated ($R^2 > 0.97$) in both transects: in the Negev Transect the relationships were almost linear, whereas they were exponential across the Judean Transect. Significant differences were observed in the productivity of these two transects, with differences over 0.2 kg/m^2 for the same precipitation levels. Applying the characteristic biomass values for the vegetation compositions, as determined from the Phytogeographical mapping, indicated that they follow almost the same relationships as observed along the Negev Transect. These differences between the actually observed values in the Judean Transect and the expected compositions according to the phytogeographical mapping might be attributed to two main reasons: (1) the Judean Desert is a "Rainshadow" desert and (2) possibly the effect of the higher historical anthropogenic pressures in the areas surrounding Jerusalem. The similarity of biomass versus rainfall between the data, according to the phytogeographic mapping and the data obtained for the Northern Negev Transect suggests that it may represent some generalized relationships of the following simple form:

$$\text{Biomass} = 0.09 * e^{0.045 * \text{precipitation}} \tag{1}$$

where precipitation is measured as cm/year.

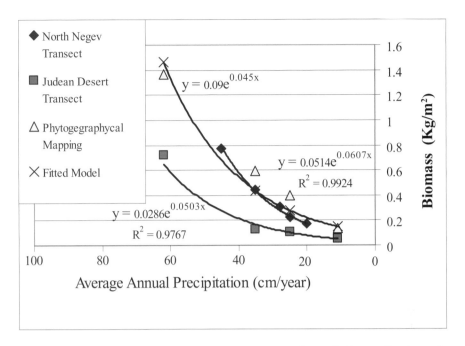

Figure 4. Biomass versus precipitation for the two transects, according to the Remote Sensing studies and according to vegetation cover compositions from the Phytogeographical mapping for the Judean Transect.

5 DISCUSSION AND CONCLUSIONS

Remote sensing, field surveys and Phytogeographical mapping are complementary information sources whose synergy must be based on understanding their similarities and differences (Shoshany et al. 1996). Mapping vegetation cover and biomass over wide regions could be facilitated mainly by using moderate or even low spatial resolution imagery. Deriving general vegetation cover for heterogeneous ecosystems such as Mediterranean regions from such satellite imagery without separating it into their life-form components would yield data of limited ecological significance. In our earlier study, such a separation was obtained by comparing winter and summer interpretations of imagery data. New methods such as adaptive unmixing (Shoshany & Svoray 2002, Kuemmerle et al. 2006) were shown to produce results comparable to those obtained from field surveys. Assessment of the results clearly indicate that there are two transition zones along climatic transects: one from shrubs to dwarf shrub- dominated areas, corresponding to the northern/western extent of the 200 mm/year Isoline (Figure 1), and another from dwarf shrub-dominated bare terrain corresponding to the southern/eastern extent of the 400-mm/year Isoline (Figure 1). Limitations on extracting woody biomass from remote sensing data led us to integrate remote sensing with field assessment of woody and herbaceous biomass. To this end, characteristic biomass values were derived from the field data for the three life-forms and then combined with cover compositions, as derived from remote sensing data and from the phytogeographical mapping. The results of implementing this strategy contributed in two ways: first in better understanding differences in the productivity along these two transects, and second in inferring regarding the non-linearity of the vegetation property changes in relation to precipitation. Non-linearity seems to be weak where there is relatively moderate human disturbance to the natural vegetation: such levels of disturbance characterized the Mediterranean regions for many centuries. However, when the accumulated disturbances increase, such as probably was the case in the Judean transect, the non-linearity increases while mainly reducing the dwarf shrub cover. Comparison of the biomass derived for cover compositions from the phytogeographical maps and those from the remote sensing data indicated that in the Judean transect the disturbance is hypothetically very high also in the most humid site: in Givaat Yearim, where it reaches almost half of the biomass it should produce. The generalized hypothetical model represented in equation (1) may facilitate assessment of disturbance levels; however, this model needs to be further assessed in other sites across the Mediterranean.

ACKNOWLEDGMENTS

An important part of the research reported here was conducted within the framework of the MEDALUS project, supported by the European Research Frameworks 3 & 4.

REFERENCES

Atlas of Israel 1964. *Survey of Israel* (Special Edition).

Cohen, Y. & Shoshany, M. 2002. Integration of remote sensing, GIS and expert knowledge in national knowledge-based crop recognition in Mediterranean environment. *International Journal of Applied Earth Observation and GeoInformation* 4: 75–87.

Cohen, Y. & Shoshany, M. 2005. Analysis of convergent evidence in an evidential reasoning knowledge-based classification. *Remote Sensing of Environment* 96(3–4): 518–528.

Danin, A., Orshan, G. & Zohary, M. 1975. The vegetation of the northern Negev and the Judean desert of Israel. *Israel Journal of Botany* 24: 172.

Danin, A. 1988. Vegetation in the southern Shefelah. In Urman, D. & Stern, E. (eds.), *Man and environment in the southern Shefelah—studies in regional geography and history*: 59–66, Masada, Givataim, Israel (in Hebrew).

Kuemmerle, T., Röder, A. & Hill, J. 2005. Separating grassland and shrub vegetation by multidate pixel-adaptive spectral mixture analysis. *International Journal of Remote Sensing* 27: 3251–3271.

Kutiel, P., Lavee, H. & Shoshany, M. 1995. The influence of a climatic gradient upon vegetation dynamics along a Mediterranean arid transect. *Journal of Biogeography* 22: 1065–1071.

Sapir, G. 1977. *Joudean Lowland vegetation and the affinity between vegetation distribution to habitat conditions*. MSc thesis, Department of Botany, Hebrew University: Jerusalem (in Hebrew, English abstract).

Shoshany, M., Kutiel, P. & Lavee, H. 1994. Remote Sensing of vegetation cover along a climatological gradient. *ISPRS Journal of Photogrammetry and Remote Sensing* 49: 1–8.

Shoshany, M., Kutiel, P. & Lavee, H. 1995. Seasonal vegetation cover changes as indicators of soil types along a climatological gradient: a mutual study of environmental patterns and controls using remote sensing. *International Journal of Remote Sensing* 16: 2137–2151.

Shoshany, M., Kutiel, P. & Lavee, H. 1996. Monitoring temporal vegetation cover changes in Mediterreanean and arid ecosystem using a remote sensing technique: case study of the Judean mountain and the Judean desert. *Journal of Arid Environments* 32: 1–13.

Shoshany, M. 2000. Detection and analysis of soil erodibility patterns using multi-date airphotographs: the case study of Avisur Highland, Israel. In Hassan, M., Slymaker, O. & Berkowicz, S.M. (eds.), *The Hydrology- Geomorphology Interface : Rainfall, Floods, Sedimentation, Landuse*: 127–138. IAHS Publ. No. 261, 127–138.

Shoshany, M. 2002. Landscape fragmentation and Soil erodibility in south and north facing slopes during ecosystems recovery: An analysis from multi-date airphotographs. *Geomorphology* 45: 3–20.

Shoshany, M. & Svoray, T. 2002. Multi-date adaptive spectral unmixing and its application for the analysis of Ecosystems' transition along a climatic gradient. *Remote Sensing of the Environment* 81: 1–16.

Sternberg, M. & Shoshany, M. 2001. Biophysical slope related differences in Mediterranean woody formations: a comparison of a semiarid and arid sites in Israel. *Journal of Ecological Research* 16: 335–345.

Sternberg, M. & Shoshany, M. 2001. Aboveground biomass allocation and water content relationships in Mediterranean trees and shrubs at two climatological regions in Israel. *Plant Ecology* 157: 171–179.

Svoray, T. & Shoshany, M. 2002. SAR based estimation of Aerial Above ground Biomass (AAB) of Herbaceous vegetation in the Semi Arid zone of Israel: A Modification to the Water Cloud Model. *International Journal of Remote sensing* 23: 4089–4100.

Svoray, T. & Shoshany, M. 2003. Herbaceous biomass retrieval in habitats of complex composition: A model merging SAR images with unmixed Landsat TM data. *IEEE Transactions on Geosciences and Remote Sensing* 41(7): 1592–1601.

Svoray, T. & Shoshany, M. 2004. SAR-based mapping of soils by Multi-Scale analysis of drying—rates. *Remote Sensing of Environment* 92(2): 233–246.

Tadmor, N.H., Brieghet, A., Noy-Meir, I., Benjamin, R.W. & Eyal, E. 1975. An evaluation of the calibrated weight-estimate method for measuring production in annual vegetation. *Journal of Range Management* 28: 65–69.

Yaalon, D.H. 1997. Soils in the Mediterranean region: What makes them different? *Catena* 28: 157–169.

Zohary, M. 1973. *Geobotanical Foundation of the Middle East*. Stuttgart: G. Fischer.

*Recent Advances in Remote Sensing and Geoinformation Processing
for Land Degradation Assessment – Röder & Hill (eds)
© 2009 Taylor & Francis Group, London, ISBN 978-0-415-39769-8*

Modelling species distributions with high resolution remote sensing data to delineate patterns of plant diversity in the Sahel zone of Burkina Faso

K. König
J.W. Goethe-University, Frankfurt am Main, Germany

M. Schmidt
Research Institute Senckenberg, Frankfurt am Main, Germany

J.V. Müller
Royal Botanic Gardens, Kew, UK

ABSTRACT: Relationships between patterns of plant species richness and landscape degradation were analysed in the Sahel zone of Burkina Faso, West Africa. To generate predictive models of the distribution of plant species in the study region we used the Genetic Algorithm of Rule-set Production (GARP) modelling system. This system has excellent potential for describing the ecological envelopes and the spatial distribution of species and can be used for many different applications. Furthermore, the spatial pattern of biodiversity can be delineated by combining the distribution maps of many species across different taxonomic and functional groups. In combination with environmental layers we used 12,000 geo-referenced collection localities to model species distributions in the Sahel zone of Burkina Faso. Previous studies have usually been conducted at regional scale using climatic and biomass parameters with a low spatial resolution of environmental variables. However, informed decision-making regarding conservation priorities mainly take place on the local scale. In a new approach we used derivates of high resolution satellite images (LANDSAT ETM+) as environmental input parameters. The potential of these images to differentiate vegetation and soil properties, which determine the distribution of plant species, has been shown in preceding analyses. Modelled distributions were evaluated with independent test data and selected by expert knowledge and statistical analysis. 138 plant species with at least 15 spatially unique occurrence points were modelled with a high accuracy. Consequently we calculated a map of phytodiversity by combining distribution maps of single species. The highest diversities have been predicted for a little disturbed area of dunes and tiger bush vegetation. Our approach has potential for widespread application to assist in conservation priorities and environmental management.

1   INTRODUCTION

Responding to the ongoing drastic decline of biodiversity all over the world and the resulting cuts of ecosystem services, the Johannesburg Summit on Sustainable Development agreed on the goal to "achieve by 2010 a significant reduction of the current rate of biodiversity loss". This tremendous challenge would be an essential step towards the implementation of the United Nations' Convention on Biological Diversity (CBD). Its success will largely depend on timely and sound information on biodiversity patterns and ecosystem functions, their current and future threats, and applicable solutions for the conservation of natural resources and their sustainable use.

   Managing biodiversity in the African Sahel Zone is essentially related to management of natural resources in order to improve livelihood for human beings in a sustainable manner. The sustainable use of natural resources and the conservation of biodiversity require sound scientific information on the geographic distribution of species. In countries with high levels of biodiversity this information is unfortunately often not available in sufficient detail. Therefore, biodiversity management increasingly relies on easily assessable indicators and operational tools to achieve its primary goal of conservation and sustainable utilization of natural plant resources (Balmford et al. 2005). Environmental niche modelling can be one of these operational tools, by predicting potential distribution ranges of species based on documented occurrences (field data, museum specimens), which are spatially extrapolated along the ecological requirements of species and projected onto biotic and abiotic variables. Because of their utility, these models are being increasingly used for priority setting and the evaluation of conservation measures (Ortega-Huerta & Peterson 2004, Elith et al. 2006).

There are various approaches to predict the spatial distribution of species (Guisan & Zimmermann 2000, Seguardo & Araújo 2004). In this study, we used the Genetic Algorithm of Rule-set Production (GARP) modelling system (Stockwell & Nobel 1991, Stockwell & Peters 1999), which has excellent capacities for describing the ecological envelopes and spatial distribution of plants (Arriaga et al. 2004), insects (Levine et al. 2004, Roura-Pascual et al. 2004), reptiles (Raxworthy et al. 2003), birds (Parra et al. 2004, Peterson et al. 2002) and mammals (Anderson et al. 2002, Anderson & Martínez-Meyer 2004, Cicero 2004, Lim et al. 2002). Insights on species distribution were applied for conservation management (Anderson & Martínez-Meyer 2004, Meggs et al. 2004, Ortega-Huerta & Peterson 2004), to model the invasion of neophytes (Arriaga et al. 2004, Underwood et al. 2004) and to predict changes in species distribution according to climate change (Peterson et al. 2002, Roura-Pascual et al. 2004).

The diversity of vascular plants can be seen as a surrogate for the whole biodiversity in a region (being the main structural components and the base of the food chain) and thus be an indicator for the ecological condition of a landscape. In this study, we intend to model the spatial pattern of phytodiversity by combining distribution maps of individual plant species. Austin (1980) suggested three kinds of gradients that influence plant distribution patterns. These were resource, direct and indirect gradients: The resources in resource gradients are directly consumed by plants and include water, light and nutrients. Direct gradients influence plant physiology, but are not consumed, and include pH, temperature and biotic interactions. Indirect gradients have no direct physiological influence on plants, but nevertheless determine the distribution and abundance of plants. Examples are altitude, latitude and longitude. From a mechanistic point of view it is desirable to use direct or resource gradients for predictive modelling (Guisan & Zimmermann 2000). One particularly successful approach of predictive modelling of plant distribution using direct gradients has been using bioclimatic variables to characterize the recorded species observation points (Sindel & Michael 1992, Crumpacker et al. 2001, Schmidt et al. 2005). A disadvantage of this data is that they are usually only available at coarse spatial scales and have no predictive power at finer spatial scales. The effect of biotic and abiotic variables on diversity patterns can only be understood in a broader context if different spatial scales (local, regional) are taken into account. Here, remote sensing data from satellites turn out to be effective tools to quantify environmental parameters (Houssa et al. 1996, Mbow 2000, Couteron et al. 2001).

LANDSAT images have provided a comprehensive view on West African savannas for several decades. They are relatively cheap, widely available and therefore especially valuable in tropical regions where other data sources at fine spatial scales are difficult to find. In the study region LANDSAT images have been applied in characterizing land cover (Thenkabail & Nolte 1996, Mayaux et al. 2004) and its changes (Budde et al. 2004, Nicholson 2001, Mayaux et al. 2005), vegetation (Couteron et al. 2001, Fensholt & Sandholt 2005) and soil properties (Houssa et al. 1996). Furthermore, the images give insights into agricultural land use intensities (Tappan et al. 2000) and can give hints for characterizing pasture and fire regimes (Mbow et al. 2000), Eva & Lambin 2000). Each of these parameters can directly or indirectly reflect distributions of single plant species and phytodiversity (Poschlod et al. 2005). Thus LANDSAT data have the potential "to replace combinations of different resource and direct gradients in a simple way" (Guisan & Zimmermann 2000) and therefore to predict species distributions at high spatial resolutions. Despite its potential, studies, which relate species diversity with LANDSAT reflectance data directly, are relatively rare (but see Seto et al. 2004).

In this study, we evaluate the potential of satellite images for prediction of species distribution patterns in the Sahel zone of West Africa against the background of landscape degradation in this area. We model the distribution of plant species to describe the observed pattern of species distribution in a precise and realistic way to provide a tool to improve the management of natural resources.

2 METHODS

2.1 *Study region*

The study site in Burkina Faso (West Africa) is part of the Sahel Zone. The semi-arid Sahel Zone forms an ecological transition between the Sahara desert to the north and the more humid savannas of the Sudanian and Guinean Zones to the south (White 1983). Characteristic vegetation types are grasslands and shrub savannas. Climatically, the northern limit is defined by the 100–200 mm/year isohyet and the southern limit by the 400–600 mm/year isohyet.

Rainfall can be very variable from year to year. Since the mid 1960s a significant decrease in rainfall and wide spread droughts affecting the whole region has been observed (Tanaka et al. 1975, Bunting et al. 1976, Nicholson 1979, Lamb 1983). This general trend cumulated in a large-scale drought during the 1982–1985 period resulting in famine and consequent migration of the local population (Olsson 1993, Glantz 1994). Since the 1990s rainfall has increased and is comparable to the rainfall amounts of the late 1960s (prior to the drought in the 1970s; Bell et al. 2000, Hulme et al. 2001, Nicholson 2001).

The study region is situated within the "Réserve sylvo-pastorale et partielle de la faune du Sahel". It is the largest legally protected area in the country, encompassing 16,000 km². The main landscape units are dunes and pediplain, interspersed by inselbergs and usually temporary watercourses and seasonal lakes. Anthropogenic impact in the form of pastoralism and agriculture has been shaping the vegetation for more than 5000 years. Due to the current population growth (annual net increase ca. 2.5%), the pressure on natural resources is steadily increasing. Numerous

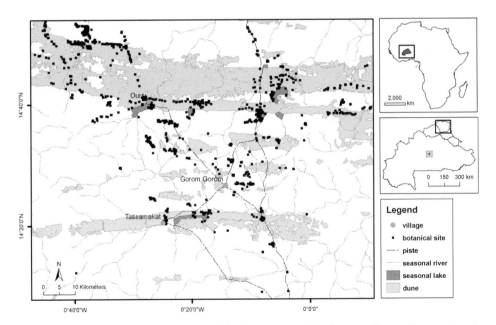

Figure 1. Map of the study area in north Burkina Faso near Oursi and Gorom-Gorom. Grey quadrangles indicate the location of phytosociological sample sites. Pleistocene sand dunes (light grey) were determined with a supervised image classification of LANDSAT images from the rainy season.

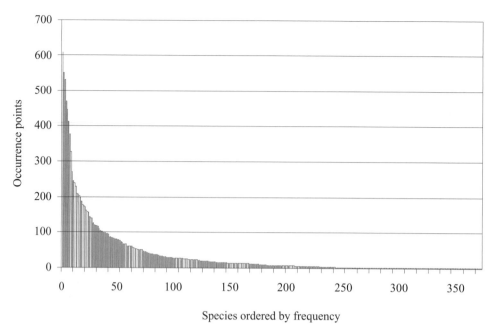

Figure 2. Occurrences per species. Only species with more than 15 occurrence points (i.e. occurring at 15 or more sites) were considered in the modelling process. The two most frequent species are the legumes *Zornia glochidiata* and *Alysicarpus ovalifolius*.

studies on different aspects of land degradation and desertification (Lindqvist & Tengberg 1993, Lykke et al. 1999, Rasmussen et al. 2001, Albert et al. 2004) have focused on this area.

For the whole country, a recent compilation of records from literature and herbarium collections (Schmidt 2006) counts 1629 species, but the low collection intensity and comparisons with similar floras allow for estimations of about 2000 species (Madsen et al. 2004), with the highest species richness in the Sudanian zone (Schmidt et al. 2005). White (1983) estimates 1200 species for the entire Sahel zone. There are no endemics known for Burkina Faso and only a few for the Sahel as a whole (White 1983). Due to the long dry season, more than half of the region's flora is composed of therophytes (Schmidt et al. 2005).

2.2 *Species data*

Our data set comprises 1461 phytosociological sample sites from the Sahel zone of Burkina Faso distributed over all landscape and vegetation units. Furthermore, 800 specimens from the West Africa Collection of the Herbarium Senckenbergianum (FR) completed the data, resulting in 13,380 spatially unique species occurrence points for 354 species. All botanical date were georeferenced with GPS coordinates with a spatial accuracy between 5 and 10 m. For the determination of plant species, we used Hutchinson & Dalziel (1954–1972).

2.3 *Environmental data*

Our objective is to predict the spatial distribution of plant species and delineating pattern of plant diversity at large scales and high spatial resolutions. Coverage of spatially continuous variables, such as climatic parameters or soil properties, which influence the distribution of plant species and therefore could be used for modelling, are only available at much coarser scales for the study region. Climatic data are measured at several field stations and are used to characterize the climatic conditions in the study region. However, they cannot be used to interpolate spatially continuous layers of climatic data in sufficient detail. Similarly, geological and soil maps are only available at regional scales (1:500,000). Initially, a maximum number of 25 environmental data layers were considered for inclusion in models. Land cover data were derived from two LANDSAT satellite images (spatial resolution 30 meters) from the year 2000 in the middle of the rainy and the beginning of the dry season. The digital signatures of bands three to six were converted into reflectance values according to Chavez (1996). To account for vegetation properties three vegetation indices, the normalized vegetation index (NDVI), the transformed vegetation index (TNDVI), and the mean soil adjusted vegetation index (MSAVI) were calculated. Additionally, the first three components of the tasselled cap transformation (TC) and four first components of a principal component transformation (PC) were calculated. Information on soil was derived from the ratio of band 5/band 7 and band 5/band 4, which reflect properties of clay and ferrous minerals. Spatial heterogeneity of the landscape was assessed by using the spectral variance of the reflectance values, which were calculated in a moving window (size 3×3 pixels) and written in the centre pixel. The inclusion of topographic data was obtained from the digital elevation model of the Shuttle Radar Topographic Mission (SRTM). All layers were resampled to a spatial resolution of 60 meters. Combinations of layer were tested with multiple linear regression analysis to determine which layer contributes significantly to the predictions of the spatial distribution of species. In a jack-knifing process with a reduced species number (20 species) combinations of layers were tested to check which layers had a significant impact on the accuracy of the models. In a multiple linear regression extrinsic omission and commission errors were used as dependent variables, and inclusion or exclusion respectively of coverage were used as independent variables. In a stepwise procedure only those layers, which improve the ability of the models to predict species distribution, remain.

2.4   *Modelling approach*

To generate predictive models of species distribution, we used the Genetic Algorithm of Rule-set Production (GARP) modelling system (Stockwell & Nobel 1991, Stockwell & Peters 1999, Stockwell 1999). GARP is a well established modelling method, which was utilized successfully in predicting species' potential distributions under a wide variety of applications (Anderson et al. 2003, Ortega-Huerta & Peterson 2004, Stockwell & Peters 1999). In the modelling process species occurrences points and environmental data coverages are used to develop models that describe the species distribution in ecological space. A set of rules, which is created by artificial intelligence algorithms, relates the species occurrence and environmental variables via several methods such as environmental range detection and logistic regression. The rules are determined in an iterative process of generation, testing and selecting rules, which are successful in maximizing accuracy measures based on independent data. Ecological envelopes define the environmental ranges where a species is likely to be found. Consequently the multidimensional envelopes are then projected onto the whole landscape to predict geographical species distributions. Because of the inherent nature of genetic algorithms GARP does not produce singular results. For each species multiple solutions of the spatial distribution will be produced. Solutions, which do not reflect the species distribution, pattern, very well have to be identified with appropriate error measures and excluded in a subsequent evaluation process. Two types of errors can be distinguished: underprediction error (omission error) and overprediction error (commission error) (Fielding & Bell 1997). GARP calculates both error types in an iterative resampling process by comparing model results with intrinsically or extrinsically defined test points (for a detailed description see (Anderson et al. 2003, Stockwell & Peters 1999). We preferred extrinsic testing with statistically independent test points. Therefore we equally divided the species occurrence points in two statistically independent data sets: one for modelling and one for testing. We only modelled plant species with at least 15 spatially unique occurrence points (138 plant species with 12,438 spatially unique occurrence points), to achieve reliable model results following the recommendations of Stockwell & Peterson (2002). Due to the stochastic nature of genetic algorithms, GARP produces several solutions with the same optimisation criteria. To account for the variability of possible solutions we created

20 incidence maps representing absence (0) and presence (1) for each species. Consequently, a careful evaluation of the resulting distribution maps was conducted following the proposed methodology of Anderson et al. (2003) to select optimal models. Therefore we first defined the upper limit of the extrinsic omission error. After visual interpretation of model results an upper limit of 20% seemed to be acceptable. All models with higher omission error values were rejected. In a second step the true proportion of the species' potential distribution was approximated as the mean commission value for those models, which fall below the defined omission limit. An optimal region of commission values was defined (mean commission value $+/-0.1$). Consequently we ranked the remaining models according to their omission and commission values and superimposed the best five models to create a composite map showing the number of optimal models in each pixel. For 15 species additional visual interpretation and evaluation of modelled maps were conducted. To reduce commission error we applied various thresholds of concordance among the best five models. A threshold of at least three models predicting presence was determined as appropriate for gaining a suitable balance between omission and commission errors. Consequently we added together single incidence maps to obtain the distribution pattern of plant diversity. Processing of maps written in ASCII format was conducted with visual basic scripts implemented in Access 2000 software.

3 RESULTS

3.1 Floral composition

The flora of the study region comprises 354 species according to our synthesis of collection and observation data. More than half of these are therophytes, frequent especially among the four largest families Poaceae, Fabaceae, Cyperaceae and Convolvulaceae. Phanerophytes are mainly from Mimosaceae (especially *Acacia*), Capparidaceae, Combretaceae and Euphorbiaceae, chamaephytes (including some stem succulents) are a small portion of several families. Most geophytes and hemicryptophytes are from Poaceae and Cyperaceae.

3.2 Model performance

In total, 2760 distribution maps of 138 plant species were obtained. The evaluation of environmental data layers led to the selection of 12 layers. Finally, principal components one to four of the rainy and dry season image, TNDVI from the rainy season image and channel ratios 5/7 and 5/4 of the dry

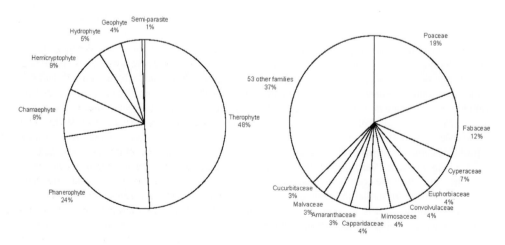

Figure 3.   Life form and family composition of the flora of the Sahel of Burkina Faso (354 spp.).

Figure 4. Distribution of plant diversity in the study region.

season image and altitude were included. Models regarding measures of landscape heterogeneity had no better predictive ability in terms of commission and omission values. Consequently these parameters were rejected. The average value of producer accuracy of all models reached 76% (corresponding to 24% omission error). Average user accuracy reached 58% (corresponding to 42% commission error). The visually evaluated distribution maps of fifteen species showed a high variability in their proportion of area where they were predicted to be present. After the evaluation and selection procedure variability of model results was reduced considerably following the recommendations of Anderson et al. (2003). Average omission error was reduced to 10%.

3.3 *Delineating patterns of phytodiversity*

The combination of the 138 single species models led to the map of plant diversity of the study region. The highest values of plant species richness are found north of the village Oursi, an area with dunes and tiger bush vegetation that is only sparsely grazed because water holes and rivers are lacking. Tiger bush provides many different microhabitats with its alternating stripes of dense thickets and bare soil. Lowest values occur on heavily grazed and often incrusted pediplains in the southern and eastern part of our study area, covered with *Acacia* and grass savanna. Watercourses and depressions usually display higher diversity values than their surroundings; the same is true for the edges of the dunes.

4 DISCUSSION

Predictive models belong to the class of empirical models in the sense of the classification of Levins (1966). In contrast to analytical and mechanistic models, empirical models are designed to maximize two intrinsic properties of the model performance: precision and reality. Empirical models are not expected to describe "cause and effect" relationships between explanatory parameters and predicted response (Guisan & Zimmermann 2000) but to predict observed patterns of species distribution in a precise and realistic way. In this study we use LANDSAT satellite images and field data to predict pattern of plant diversity in the Sahel region of Burkina Faso. We demonstrate the ability of widely available satellite data to model species distributions with high accuracy and at fine spatial resolutions. All species could be modelled with low omission error. Evidence from commission error has to be interpreted with caution. Commission error can reflect either real over-prediction of species distribution, but also apparent overprediction in regions where a particular species has not been sampled but is in fact there. Therefore we decided not to use the model with the lowest commission error, but to find an appropriate balance between omission and commission error. The resulting pattern of plant diversity shows highest values in an area covered to a large extent by tiger bush. This is partly in accordance with Hiernaux & Gérard (1999), whose findings reveal a higher herbaceous plant richness of tiger bush, as compared to the Sahel average, but lower woody species richness. The area is also less influenced by grazing, because of the lack of watering points, and this might also contribute to a higher diversity, since overgrazing and degradation are serious threats to dryland biodiversity (Darkoh 2003). The higher diversity along watercourses and depressions may be due to the fact that water is the limiting factor for many species and the neighbouring, more humid Sudanian Zone has a much richer flora (White 1983, Schmidt et al. 2005). It is also affected by the size of our grid cells, which encompass a whole range of habitats along the watercourses, and these will all contribute with various species sets. Using the proposed approach several drawbacks have to be considered. The first concerns the sampling quality of species occurrence, the second the capability of satellite images as explanatory variables and the third the conducted modelling approach.

Two properties of species occurrence points can greatly affect the predictive quality and accuracy of distribution models: 1) number of occurrence points and 2) data completeness. Below a certain threshold the number of species occurrence points does not allow meaningful model training and testing. Even the necessary number is also affected by inherent properties of the species e.g. habitat specificity, niche width and frequency (Kadmon et al. 2003). As a rule of thumb at least 10–20 points are needed for reliable predictions (Stockwell & Peterson 2002). In most studies on species diversity species frequency distribution looks similar, as in Figure 2. A relatively small proportion of species is sampled with high frequency. However, the great majority of species are sampled with very low individual numbers. Thus a majority of species are not modelled with high predictive accuracy. If the modelling of rare species is the main objective, e.g. for conservation purposes, this approach will prove to be rather unsatisfactory. The previous point stresses the need for adequate field sampling especially designed for modelling plant distribution patterns. To make predictive modelling operational, adequate sampling design may help to reduce sampling efforts by maintaining predictive quality of models. Inadequate sampling of species can seriously bias modelling results if the range of possible environmental values is not completely covered. In this study sampling was conducted with high effort and complete coverage of environmental gradients is likely to have occurred. Whether or not this is actually the case will be tested in future studies.

Scale is of considerable importance for environmental modelling. Coarse scale environmental data layer like interpolated temperature or precipitation maps cannot account for local variations in species distribution. High resolution satellite images can assess the spatial and temporal variability of environmental gradients directly. Therefore the extrapolation of local species' point data to landscape scale can be achieved with high spatial accuracy and resolution. LANDSAT satellite images can reflect a considerable proportion of the environmental envelopes of species. However, relationships between causal factors determining species distributions and the signal of satellite images remain unclear, and use of indirect gradient data could only lead to robust results if indirect gradients are directly correlated with causal gradients (Guisan & Zimmermann 2000). Direct relationships between signals of satellite images and direct environmental gradients are usually assumed but rarely tested. Another problem that arises with the utilization of indirect gradient data is that findings can only be applied within a discretely defined area. Extrapolation of findings from one area to another must be performed with caution. This requires species occurrence data across the whole range of environmental gradient values. In a recent review Elith et al. (2006) compared the predictive ability of well-established modelling methods such as GARP with methods which have been developed recently to modelling species' distributions. They concluded that the novel methods consistently outperformed more established methods and therefore should be regarded in future studies. To cope with the huge quantity of modelled distribution maps, we applied an unsupervised evaluation approach to select the best performing models (Anderson et al. 2003). The underlying assumption of an inverse association between omission and commission error was not tested for all modelled species in this study. This assumption is, for example, not fulfilled for species with a very small potential distribution compared to the whole study area (e.g. aquatic plants). Most modelled species are widely distributed and are thus likely to be predicted with high accuracy (Kadmon et al. 2003). This has to be examined in more detail in future studies. The results of our analysis show that species distribution modelling with satellite data has potential for widespread application to assist in conservation priorities and environmental management.

ACKNOWLEDGMENTS

This work has been supported by the BIOTA-West (Biodiversity Monitoring Transect Analysis, West Africa) initiative of the Federal Ministry of Research and Education (BMBF). We thank Henning Sommer, University of Bonn for assistance and useful advice in implementing GARP. We thank Julia Krohmer and the Herbarium Senckenbergianum, Frankfurt for the contributions of species' distribution data.

REFERENCES

Albert, K.-D., Müller, J., Ries, J.B. & Marzolff, I. 2004. Aktuelle Landdegradation in der Sahelzone Burkina Fasos. In: Albert, K.-D., Löhr, D. & Neumann, K. (eds.): *Mensch und Natur in Westafrika. - Ergebnisse aus dem Sonderforschungsbereich "Kulturentwicklung und Sprachgeschichte im Naturraum Westafrikanische Savanne"*: 289–330. Weinheim: Wiley Verlag.

Anderson, R.P, Gómez-Laverde, M. & Peterson, A.T. 2002. Geographical distributions of spiny pocket mice in South America. insights from predictive models. *Global Ecology & Biogeography* 11: 131–141.

Anderson, R.P. & Martínez-Meyer E. 2004. Modeling species' geographic distributions for preliminary conservation assessments. an implementation with the spiny pocket mice (Heteromys) of Ecuador. *Biological Conservation* 116: 167–179.

Anderson, R.P., Lew, D. & Peterson, A.T. 2003. Evaluating predictive models of species' distributions. criteria for selecting optimal models. *Ecological Modelling* 162(3): 211–232.

Arriaga, L., Castellanos, A.E., Moreno, E. & Alarcon, J. 2004. Potential ecological distribution of alien invasive species and risk assessment. a case study of buffel grass in arid regions of Mexico. *Conservation Biology* 18(6): 1504–1514.

Austin, M.P. 1980. Searching for a model for use in vegetation analysis. *Vegetatio* 42: 11–21.

Balmford, A., Crane, P., Dobson, A., Gree, R.E. & Mace, G.M. 2005. The 2010 challenge. data availability, information, needs and extraterrestrial insights. *Phil. Trans. R. Soc. B* 360: 221–228.

Bell, G.D., Halpert, M.S., Schnell, R.C., Higgins, R.W., Lawrimore, J., Kousky, V.E., Tinker, R., Thiaw, W., Chelliah, M., Artusa, A. 2000. Climate Assessment for 1999. *Bulletin of the American Meteorological Society* 81: 1328–1328.

Brown, D.G., 1994. Predicting vegetation types at treeline using topography and biophysical disturbance variables. *Journal of Vegetation Science* 5: 641–656.

Budde M.E., Tappan G. & Rowland J. 2004. Assessing land cover performance in Senegal, West Africa using 1-km integrated NDVI and local variance analysis. *Journal of Arid Environments* 59(3): 481–498.

Bunting, A.H., Dennett, M.D., Elston, J., Milford, J.R. 1976. Rainfall trends in the West African Sahel. *Quarterly Journal of the Royal Meteorological Society* 102: 59–64.

Chavez, P.S. 1996. Image-based atmospheric corrections—revisited and revised. *Photogrammetric Engineering and Remote Sensing* 62(9): 1025–1036.

Cicero, C. 2004. Barriers to sympatry between avian sibling species (Paridae: Baeolophus) in local secondary contact. *Evolution* 58(7): 1573–1587.

Couteron P., Deshayes M. & Roches C.A. 2001. Flexible approach for woody cover assessment from SPOT HRV XS data in semi-arid West Africa. Application in northern Burkina Faso. *International Journal of Remote Sensing* 22(6): 1029–1051.

Crumpacker, D.W., Box E.O. & Hardin, E.D. 2001. Implications of climatic warming for conservation of native trees and shrubs in Florida. *Conservation Biology* 15(4): 1008–1020.

Darkoh, M.B.K. 2003. Regional perspectives on agriculture and biodiversity in the drylands of Africa. *Journal of Arid Environments* 54: 261–279.

Elith, J., Graham, C.H., Anderson, P., Dudik, M., Ferrier, S., Guisan, A., Hijmans, J., Huettmann, F., Leathwick, R., Lehmann, A., Li, J., Lohmann, G., Loiselle, A., Manion, G., Moritz, C., Nakamura, M., Nakazawa, Y., Overton, C.M., Townsend Peterson, A., Phillips, J., Richardson, K., Scachetti-Pereira, R., Schapire, E., Soberon, J., Williams, S., Wisz, S. & Zimmermann, E. 2006. Novel methods improve prediction of species' distributions from occurrence data. *Ecography* 29: 129–151.

Eva, H.D. & Lambin, E.F. 2000. Fires and land-cover change in the tropics. a remote sensing analysis at the landscape scale. *Journal of Biogeography* 27(3): 765–776.

Fensholt, R. & Sandholt, I. 2005. Evaluation of MODIS and NOAA AVHRR vegetation indices with in situ measurements in a semi-arid environment. *International Journal of Remote Sensing* 26(12): 2561–2594.

Fielding, A.H. & Bell, J.F. 1997. A review of methods for the assessment of prediction errors in conservation presence/absence models. *Environment Conservation Journal* 24: 38–49.

Glantz, M.H. 1994. Drought Follows the Plow: Cultivating Marginal Areas. Cambridge University Press.

Guinko, S. 1984. Végétation de la Haute Volta. Dissertation, Université de Bordeaux III.

Guisan, A. & Zimmermann, N.E. 2000. Predictive habitat distribution models in ecology. *Ecological Modelling* 135: 147–186.

Guisan, A., Theurillat, J.-P. & Kienast, F. 1998. Predicting the potential distribution of plant species in an alpine environment. *Journal of Vegetation Science* 9: 65–74.

Guisan, A., Weiss, S.B. & Weiss, A.D. 1999. GLM versus CCA spatial modeling of plant species distribution. *Plant Ecology* 143: 107–122.

Hiernaux, P. & Gerard, D.B. 1999. The influence of vegetation pattern on the productivity, diversity and stability of vegetation. The case of 'brousse tigree' in the Sahel. *Acta Oecologica* 20: 147–158.

Houssa, R.; Pion, J.C. & Yesou, H. 1996. Effects of granulometric and mineralogical composition on spectral reflectance of soils in a Sahelian area. *Photogrammetric Engineering and Remote Sensing* 51(6): 284–298.

Hulme, M. 2001. Climatic perspectives on Sahelian desiccation: 1973–1998. *Global Environmental Change-Human and Policy Dimensions* 11: 19–29.

Hutchinson & Dalziel 1954–1972. Flora of West Tropical Africa. London: The Whitefriars Press.

Kadmon, R., Farber, O. & Danin, A. 2003. A systematic analysis of factors affecting the performance of climatic envelope models. *Ecol. Appl.* 13(3): 853–867.

Lamb, P.J. 1983. Sub-Saharan rainfall update for 1982: continued drought. *Journal of Climatology* 3: 419–422.

Lebrun, J.P., Toutain, B., Gaston, A. & Boudet, G. 1991. Catalogue des plantes vasculaires du Burkina Faso. Maisons-Alfort, France: èd. I.E.M.V.T.

Levine, R.S., Peterson, A.T. & Benedict, M.Q. 2004. Geographic and ecographic and ecologic distributions of the *Anopholes gambiae* complex predicted using a genetic algorithm. *Am. J. Trop. Med. Hyg.* 70: 105–109.

Levins, R. 1966. The strategy of model building in population ecology. *Am. Sci.* 54: 421–431.

Lim, B.K., Peterson, A.T. & Engstrom, M.D. 2002: Robustness of ecological niche modelling algorithms for mammals in Guyana. *Biodiversity and Conservation* 11: 1237–1246.

Lindqvist, S. & Tengberg, A. 1993. New Evidence of Desertification from Case Studies in northern Burkina Faso. *Geografisker Annaler* 75: 127–135.

Lu, D. 2005. Aboveground biomass estimation using Landsat TM data in the Brazilian Amazon. *International Journal of Remote Sensing* 26(12): 2509–2525.

Lykke, A.M., Fog, B. & Madsen, J.E. 1999. Woody vegetation changes in the Sahel of Burkina Faso assessed by means of local knowledge, aerial photos, and botanical investigations. *Geografisk Tidsskrift, Danish Journal of Geography* Special Issue: 57–68.

Madsen, J.E., Lykke, A.M., Boussim, J. & Guinko S. 2004. Floristic composition of two 100 km(2) reference sites in West African cultural landscapes. *Nord. J. Bot.* 23: 99–114.

Mayaux, P., Bartholomé, E., Fritz, S. & Belward, a. 2004. A new land-cover map of Africa for the year 2000: *Journal of Biogeography* 31: 861–877.

Mayaux, P., Holmgren, P., Achard, F., Eva, H., Stibig, H.-J. & Branthomme, A. 2005. Tropical forest cover change in the 1990's and options for future monitoring. *Philosophical Transactions of the Royal Society Series B* 360(1454): 373–384.

Mbow, C., Nielsen, T.T. & Rasmussen, K. 2000. Savannah fires in east-central Senegal. Distribution patterns, resource management and perceptions. *Human Ecology* 28(4): 561–583.

Meggs, J.M., Munks, S.A., Corkrey & R., Richards K. 2004. Development and evaluation of predictive habitat models to assist the conservation planning of a threatened lucanid beetle, Hoplogonus simsoni, in north-east Tasmania. *Biological Conservation* 118: 501–511.

Nicholson, S.E. 1979. The methodology of historical climate reconstruction and its application to Africa. *Journal of African History* 20: 31–49.

Nicholson S.E. 2001. Climatic and environmental change in Africa during the last two centuries. *Climate Research*. 17(2): 123–144.

Olsson, L. 1993. On the causes of famine—Drought, desertification and market failure in the Sudan. *Ambio* 22: 395–403.

Ortega-Huerta, M.A. & Peterson, A.T. 2004. Modelling spatial patterns of biodiversity for conservation prioritization in North-eastern Mexico. *Diversity and. Distributions* 10(1): 39–54.

Parra, J.L., Graham, C.C. & Freile, J.F. 2004. Evaluating alternative data sets for ecological niche models of birds in the Andes. *Ecography* 27: 350–360.

Peterson, A.T., Ortega-Heuerta, M.A., Bartley, J., Sánchez-Cordero, V., Soberón, J., Buddemeier, R.H. & Stockwell, D.R.B. 2002. Future projections for Mexican faunas under global climate change scenarios. *Nature* 416: 626–629.

Poschlod, P., Bakker, J.P. & Kahmen, S. 2005. Changing land use and its impact on biodiversity. *Basic and Applied Ecology* 6(2): 93–98.

Rahman, M.M., Csaplovics, E & Koch, B. 2005. An efficient regression strategy for extracting forest biomass information from satellite sensor data. *International Journal of Remote Sensing* 26(7): 1511–1519.

Rasmussen, K., Fog, B. & Madsen, J.E. 2001. Desertification in reverse? Observations from northern Burkina Faso. *Global Environmental Change-Human and Policy Dimensions* 11: 271–282.

Raxworthy, C.J., Martinez-Meyer, E., Horning, N., Nussbaum, R.A., Schneider, G.E., Ortega-Huerta, M.A. & Peterson, A.T. 2003. Predicting distributions of known and unknown reptile species in Madagascar. *Nature* 426: 837–841.

Roura-Pascual, N., Suarez, A.V., Gomez, C., Pons, P., Touyama, Y., Wild, A.L. & Peterson, A.T. 2004. Geographical potential of Argentine ants (Linepithema humile Mayr) in the face of global climate change. *Proceedings of the Royal Society of London, Series B* 271: 2527–2534.

Schmidt, M., Kreft, H., Thiombiano, A. & Zizka, G. 2005. Herbarium collections and field data-based plant diversity maps for Burkina Faso. *Divers. Distrib.* 11: 509–516.

Schmidt, M. 2006. *Pflanzenvielfalt in Burkina Faso—Analyse, Modellierung und Dokumentation*. Dissertation, J.W. Goethe-Universität Frankfurt am Main.

Seguardo, P. & Araújo, M.B. 2004. An evaluation of methods for modelling species distributions. *Journal of Biogeography* 31: 1555–1568.

Seto, K.C., Fleishman, E. & Fay, J.P. 2004. Linking spatial patterns of bird and butterfly species richness with Landsat TM derived NDVI. *International Journal of Remote Sensing* 25(20): 4309–4324.

Sindel, B.M. & Michael, P.W. 1992. Spread and potential distribution of Senecio-Madagastariensis Poir (Fireweed) in Australia. *Aust. J. Ecol.* 17(1): 21–26.

Stockwell, D.R. & Nobel, I.R. 1991. Induction of sets of rules from animal distribution data. A robust and informative method of data analysis. *Mathematics and Computers in Simulation* 32: 249–254.

Stockwell, D.R. & Peters, D. 1999. The GARP modelling system. problems and solutions to automated spatial prediction. *International Journal of Geographical Information Science* 32: 143–158.

Stockwell, D.R.B. & Peterson, A.T. 2002. Effects of sample size on accuracy of species distribution models. *Ecological Modelling* 148: 1–13.

Stockwell, D.R.B. 1999. Genetic algorithms II. In. Fielding, A.H. (Ed.), *Machine Learning Methods for Ecological Applications*: 123–144. Boston: Kluwer Academic Publishers.

Tanaka, M., Weare, B.C., Navato, A.R. & Newell, R.E. 1975. Recent African rainfall patterns. *Nature* 255, 201–203.

Tappan, G.G., Hadj, A. & Wood, E.C. 2000. Use of Argon, Corona, and Landsat imagery to assess 30 years of land resource changes in west-central Senegal. *Photogrammetric Engineering and Remote Sensing* 66(6): 727–735.

Thenkabail, P. S. & Nolte, C. 1996. Capabilities of Landsat-5 Thematic Mapper (TM) data in regional mapping and characterization of inland valley agroecosystems in West Africa. *International Journal of Remote Sensing*. 17(8): 1505–1538.

Underwood, E.C., Klinger, R. & Moore, P.E. 2004. Predicting patterns of non-native plant invasions in Yosemite National Park, California, USA. *Diversity and Distribution* 10: 447–459.

White, F., 1983. *The Vegetation of Africa*. Paris: UNESCO.

White, J.D., Ryan, K.C., Key, C.C. & Running, S.W. 1996. Remote sensing of forest fire severity and vegetation recovery. *International Journal of Wildland Fire* 6(3): 125–136.

Recent Advances in Remote Sensing and Geoinformation Processing
for Land Degradation Assessment – Röder & Hill (eds)
© *2009 Taylor & Francis Group, London, ISBN 978-0-415-39769-8*

Retrieving rangeland vegetation characteristics through constrained inverse reflectance modelling of earth observation satellite imagery

J. Hill & A. Röder
I Remote Sensing Department, FB VI Geography/Geosciences, University of Trier, Trier, Germany

W. Mehl
European Commission, Joint Research Centre, Institute for Environment and Sustainability (IES), Ispra (Va), Italy

G.M. Tsiourlis
Research Institute, Laboratory of Ecology, National Agricultural Research Foundation (NAGREF), Vasilika-Thessaloniki, Greece

ABSTRACT: Rangeland ecosystems are important elements of Mediterranean landscapes and provide a variety of goods and services, ranging from resources for livestock production systems to conservation of flora and fauna. The sustainable use of areas dominated by livestock grazing requires finding the sometimes delicate balance between productivity and corresponding stocking rates. However, given the land use change processes of the past decades, it is not sufficient to define average regional stocking rates. The access to rangelands often occurs with such a spatial diversification that environmental degradation might occur although average stocking rates are not exceeded. This not only affects the local carrying capacity, but might impose feedbacks on the hydrological functioning of headwater areas and downstream valleys. This is but one example illustrating the need for spatially differentiated information on ecological state variables for supporting the required spatial management of resources.

This paper explores options for deriving leaf area index as a major biophysical variable from earth observation satellite data. Inverse reflectance modelling of Landsat-TM data is constrained with fractional cover estimates obtained from very high spatial resolution Quickbird imagery, thereby increasing the possibilities to obtain woody vegetation LAI estimates from open canopies. Results underline the complexity of canopy reflectance signals in heterogeneous Mediterranean landscapes. The spatial diversity and variation was well reproduced by the model inversion, and woody LAI integrated to site level was found consistent with field-based measurements. On the other hand, the estimation shrub LAI was found to be strongly affected from the background signal introduced in the model, which affected results in particular in regions with low woody vegetation cover.

Keywords: Rangeland ecology, biophysical parameters, LAI, inverse reflectance modelling

1 INTRODUCTION

As with many other fields of vegetation science also rangeland ecology has a need for quantitative vegetation parameters for thoroughly estimating available resources and their change over time. This is an important prerequisite for a wide range of tasks, ranging from understanding carbon dynamics within these ecosystems to assessing adequate stocking rates for sustainable grazing systems. Given the spatial extension of rangelands and the limited resources for ground surveys, monitoring of important rangeland parameters is only feasible by extrapolating limited field data with support of additional means such as remote sensing. In particular against the background of ongoing degradation and desertification processes in many parts of the world (Reynolds & Stafford-Smith

2002), retrospective monitoring approaches which use available remote sensing archives have the potential to reveal important information on whether past management decisions have potentially triggered degradation processes. However, such monitoring strategies require standardised remote sensing data interpretation schemes to ensure their consistency across time and space.

A number of studies on land degradation processes in Europe, were carried out in the past, which assessed the development of rangeland resources over long time periods in comparison to changing socio-economic boundary conditions. For instance, the mountainous ecosystem of Central Crete (Greece), a traditional grazing area, experienced increasing pressure on natural resources through quite excessively expanding grazing activities during the past decades. These were mainly triggered by an increasing demand for animal products (partially by tourism), agricultural subsidies, and the construction of access roads after Greece joined the European Union in 1981 (Hill et al. 1998; Hostert et al. 2003a, b). In comparison, the area of Lagadas County is located in Northern Greece, in the immediate neighbourhood of the rapidly expanding urban agglomeration of Thessaloniki. Corresponding to increasing animal numbers in the recent past, regional authorities claimed that range resources had notably deteriorated and action should be taken. Yet geospatial analyses did not reveal a general decrease of vegetation cover as an indicator of range condition. Rather, a differentiated pattern of stability, recovery and degradation became evident, which was shown to coincide with the prevalent grazing scheme (Röder et al. 2007, 2008b).

Horizontal and vertical canopy densities, as measured by fractional cover and leaf area, represent important indicators which might be directly fed into sustainable range management concepts. In the studies introduced before (Röder et al. 2007, 2008b), monitoring approaches were built on linear spectral mixture analysis (SMA) of atmospherically corrected Landsat imagery (Smith et al. 1990). The vegetation abundance estimates provided improvements in comparison to empirical relations between proportional cover and various vegetation indices, and the results in fact proved very useful to unveil the spatial diversity of degradation processes through this consistent indicator. However, many SMA-based studies remain confined to linear mixture, although non-linear mixing was shown to be an important factor, even when low plants in arid environments are considered (Ray & Murray 1996). In general, SMA-derived estimates (Elmore et al. 2000) only appear to be in good correspondence to ground-measured cover proportions as long as canopy height (i.e., volume) is not too variable. If canopy structure becomes highly variable, SMA-derived abundance estimates rather tend to represent a combination of proportional cover and canopy volume (Ray & Murray 1996).

It is thus advisable to explore additional options for deriving information on the biophysical characteristics of rangelands which can be more directly interfaced with resource assessment and management concepts. Fang et al. (2005) have advocated the use of vegetation indices, surface broadband albedo, leaf area index and the fraction of photosynthetically active radiation (FPAR) absorbed by the vegetation. In particular leaf area index is an important state variable with direct relation to biomass as well as exchange of mass and energy. Leaf area is often derived from empirical relationships to vegetation indices (Lacaze 2005). The major disadvantages of these approaches emerge from the difficulty to apply predictive equations in areas where they were not established, or where the available data are not sufficiently standardised and calibrated. There is thus a preference to estimate LAI through the inversion of radiative transfer (RT) models, but the complicated structure of rangelands with their mixture of green and senescent canopy elements, woody components, litter and soil (White et al. 2000, Asner & Heidebrecht 2002) impose quite substantial problems.

In this study we were particularly interested in mapping the proportional cover and the leaf area index of woody vegetation communities in Mediterranean rangelands of Northern Greece based on an inversion of the GeoSail reflectance model (Huemmrich 2001) and a combination of late summer Landsat-TM corresponding Quickbird images.

In contrast to deriving canopy properties using vegetation indices and empirical relations, reflectance models provide more robust assessments as they are physical-based and explicitly take into account the interaction of radiation with diverse canopy components at foliage and canopy levels. Thus, such models provide a more efficient characterization of different biophysical canopy-related parameters. Furthermore, they are less affected by problems associated with many vegetation indices, such as background influence or saturation effects at low and high cover, and better suited to be transferred to other regions (e.g. Kimes et al. 2000, Jacquemoud et al. 2000), the inversion of such models is frequently affected by the so-called "ill-posed problem", i.e. various combinations of biophysical parameters (e.g. "proportional cover" as the surface area covered by vegetation when considering a planar projection, and "leaf area index" as the one-sided green leaf area per unit ground surface) may generate similar or even identical reflectance signatures. To reduce this effect introduced from the combination of horizontal and vertical canopy densities, the GeoSail model (inverted on Landsat TM data) was constrained with vegetation cover proportions derived from very high spatial resolution Quickbird imagery.

2 TEST AREA AND OBJECTIVES

The test area is characterized by a heterogeneous mosaic of a variety of ecosystems and land uses, which is typical for Mediterranean rangelands (Di Castri 1981). Surrounding the central Mygdo-nia valley with its two large freshwater lakes, agricultural areas are interwoven with rangelands, which are mainly used for livestock husbandry. Shrublands of varying density are dominating these rangelands, which can be subdivided into three distinctive units: dense (shrub/tree-fraction of 70–100%), medium dense (shrub/tree-fraction of 40–70%) and sparse (shrub/tree-fraction of 10–40%), the latter forming a transition to grasslands. With its resistance to grazing due to its spiny small leaves, its resilience to fires and its competitiveness due to fast resprouting and coppicing abilities, the kermes oak (*Quercus coccifera*) is by far the dominant species. Also *Quercus ilex* is present in high percentages. They are accompanied by characteristic Mediter-ranean evergreen species such as *Pistacia lentiscus, Phillyrea media, Olea europaea var. sylvestris, Erica arborea,* constituting the association Oleo-Lentiscetum. In the higher altitudes of this zone (300–500 m asl) deciduous Mediterranean shrubs like *Spartium junceum* and *Crataegus monogyna* are also encountered. Herbaceous species are numerous in this area, and the grasses consist of a high number of annual species together with some perennial grasses (Konstantinidis & Tsiourlis 2003).

Climate of the area is typical Mediterranean, with precipitation ranging from about 400 mm at lower altitudes to 650 mm at the station of the locality Arnaia at 564 m elevation. Maximum rainfall is received in late spring and autumn, but total amount and distribution are modified by relief and elevation. In the rangeland areas, bedrock is largely dominated by metamorphic rocks, and gneiss as well as quartzite are abundant. Upon these, shallow soils in a rather low development stage can be found. As an exception, accumulation zones are characterized by fertile alluvial soils, which are frequently used for agricultural purposes.

Like in many Mediterranean countries, Greece becoming a member of the European Union in 1981 significantly affected rural activities, and resulted in widespread land abandonment in less accessible rural areas, especially in the higher zones. While in some regions numbers of grazing animals dramatically increased as a result of the European Common Agricultural Policy, the communities of the Lagadas area show a more stable behaviour. Nonetheless, grazing regimes have been reported to have strongly changed towards sedentary livestock keeping and intensification in concentrated areas.

Given their physical properties, the rangeland areas form an excellent test case to assess potentials and limitations of geometric-optical reflectance modeling in heterogeneous environments. In the frame of this study, this was addressed by pursuing three major objectives:

- Upscale field-based optical measurements of LAI to provide woody site LAI estimates for a larger area using a geometric-optical canopy reflectance model
- Assess options to constrain the modeling approach to circumvent the "ill-posed problem"
- Characterize the influence of the different parameters and elements considered by the coupled model

3 DATA PROCESSING AND ANALYSIS

3.1 *Satellite imagery*

A Landsat-5 TM scene acquired on the 24th of July, 2003 formed the basis for the model inversion. After orthorectification, the image was radiometrically corrected using a modified 5S radiative transfer code (Tanré et al. 1990) with published sensor calibration factors (Thome et al. 1997, Teillet et al. 2001). This code was extended to correct terrain-induced illumination variations, where direct and diffuse radiance terms were separately considered (Hill et al. 1995). The required parameters characterizing the sun-surface-sensor constellation at the time of image acquisition were calculated using a digital elevation model supplied by Geoapikonisi Ltd. (Athens, Greece) with 30 m spatial resolution. As a result, surface reflectance data were available free of topography-induced variations.

To complement the higher spectral resolution by a higher level of spatial detail and supporting the scale transition from ground to landscape level, a Quickbird bundle image was available, comprising panchromatic and multispectral information at 0.6×0.6 m^2 and 2.4×2.4 m^2, respectively, where the latter was resampled to a resolution of 3×3 m^2. The image was supplied as four tiles acquired on 21st of July and 13th August, 2003, matching the late summer conditions of the Landsat image where grass and herbaceous vegetation could be expected to be largely senescent, such that the signal is mostly determined by the woody vegetation component. All tiles were orthocorrected using a combination of rationale polynomial coefficients (RPC) and field-measured positions and incorporating digital elevation data and finally mosaicked. The test area is described in detail in Röder et al. (this volume).

The Quickbird data set served to prepare a spatially explicit representation of woody vegetation cover of the test area, which was to be used in the further analysis for stratification. It draws from the fact that the panchromatic and multispectral layers represent woody vegetation elements in two complementary ways. In the panchromatic image, shrubs and trees appear as dark objects in the open grassland/shrubland areas, in particular where they occur in isolated or clumped patches. In these regions, they can be mapped using thresholding. In particular where canopies become denser in the transition zone towards forests, directional reflectance effects cause bright signals such that these regions are not adequately mapped in this manner (Figure 1, middle). On the other hand, the NDVI, despite its known limitations (Elmore et al. 2000), captures the part of the signal driven by vegetation also in these conditions. Hence, NDVI was calculated from the multi-spectral bands and woody cover was mapped applying a threshold to the resulting index values. Notwithstanding, in locations where grasses and herbaceous vegetation are not completely senesced, they may still contribute to the overall signal. Hence, a combination of both approaches was chosen, making use of thresholds applied to the panchromatic and NDVI-images and combining both maps. Figure 1 shows the individual components of this process flow (compare figure 1).

In order to provide an overall woody cover mask, the panchromatic binary image was resampled to match the resolution of the NDVI-threshold image, and the binary values were averaged. Then, a Boolean comparison was executed to yield the final cover mask: where the NDVI based mask indicated 'no cover', the value of the panchromatic mask was selected; else, the pixel was assigned the NDVI-based value (i.e. 1), indicating a fully covered pixel. Since the NDVI threshold was conservatively set to mark only areas with a high degree of photosynthetic activity (figure 1), this approach yielded an accurate map of woody vegetation cover at a resolution of 3×3 m^2. In a final step, this dataset was averaged to 30×30 m^2 resolution, thus representing a map of woody vegetation cover matching the Landsat data for subsequent integration in the reflectance model.

3.2 *Ground-based measurements*

In May 2002 and September 2006, two field campaigns were carried out in the test area. During the first campaign, major emphasis was put on the collection of spectral reflectance measurements of major surface materials using an ASD Pro Fieldspec spectroradiometer, which was used to record soil and bedrock samples, as well as vegetation both through integrated shrub measurements and

| Quickbird MS | Quickbird PAN | PAN Threshold | MS-NDVI Threshold | Combined Threshold |

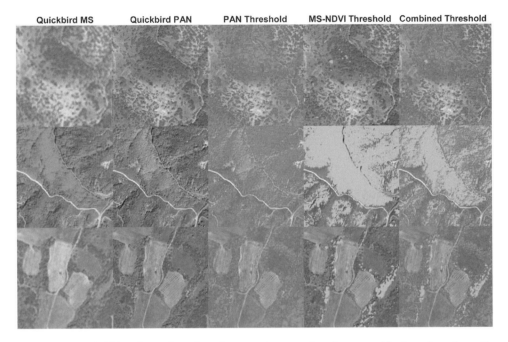

Figure 1. Threshold-based mapping of woody vegetation cover based on a combination of panchromatic and multispectral (NDVI) information for three locations: open shrubland (top), partially forested (middle), pastures with shrub edges (bottom) (see colour plate page 391).

leaf samples. The measurements were carried out in field as well as under controlled conditions in the laboratory. In addition, a large number of *Quercus coccifera* leafs were clipped, cooled and brought to the laboratory for a chemical analysis of constituents, such as chlorophyll, leaf water content, dry matter content etc., as well as the respective leaf area. These measurements served to initialize the PROSPECT model component (Jacquemoud et al. 1990, section 4).

The second campaign aimed at characterizing the range of woody leaf area index in the test site. Canopy LAI is often difficult to assess through direct methods, such as stratified clipping techniques, litter collections, and allometric relationships. These methods are resource-intensive, often site specific and have a limited spatial and temporal validity (Soudani et al. 2002). To overcome such difficulties, indirect optical methods have been developed. Ground-based LAI measurements using optical instruments have become a standard as they allow for intensive spatial and temporal sampling over large areas (Soudani et al. 2002). The LAI-2000 plant canopy analyser (Li-Cor Inc., Lincoln, USA) (Welles & Norman 1991) has an optical sensor that consists of five detectors, arranged in concentric rings, which measure radiation (below 490 nm) from different sections of the sky. Canopy transmittance for these different sections is calculated as the ratio between below-canopy and above-canopy readings for each detector ring (Stenberg 1996). From these transmittance values (gap fraction) LAI is estimated using an approximation of the Lambert-Beer law. This estimation is based on the assumption of randomly distributed leaves in the crown and does not consider other shading elements such as stems. Therefore, LAI estimates from optical methods are effective plant area indices, which account neither for clumping nor for separating leaves from non-green material, nor for distinguishing species (Soudani et al. 2002, Jonckheere et al. 2003). From a remote sensing perspective, the effective or plant area index may even be advantageous compared to the true LAI, as it is often better correlated to vegetation indices and was found to be very stable over the course of a year (Chen & Cihlar 1996). However, it is possible to convert effective into true LAI and vice versa using specific correction factors (Fernandes et al. 2002).

Using the Quickbird images, 15 plots were established, covering areas with a homogeneous distribution of shrubs and trees at areas between 0.3 and 9 hectares. In each of these plots, single

Table 1.   Average, standard deviation, minimum and maximum values for LAI measurements of woody species for 15 field plots.

Site-ID	Average	Stdv
LAI_1	2.89	0.62
LAI_2	2.83	0.58
LAI_3	3.20	0.47
LAI_4	4.09	1.18
LAI_5	3.59	0.76
LAI_6	2.99	1.12
LAI_7	4.14	0.46
LAI_8	2.98	0.37
LAI_9	3.95	0.23
LAI_10	4.33	0.42
LAI_11	4.05	0.64
LAI_12	3.56	0.46
LAI_13	3.07	1.24
LAI_14	4.42	0.39
LAI_15	3.84	1.07

woody species or groups of specimens representing the physiognomic range were selected and LAI was measured for each of them using a LAI-2000 device. In 3 sites of the transition zones to forests, closed canopies were measured. A total of 121 individual shrubs and shrub/tree groups were measured, where the LAI finally recorded for each consisted of 2 to 3 measurements from different positions, which in turn comprised 5 to 10 separate readings. Overall average leaf area index was 3.6, with values ranging from 1.1 to 5.9 and a standard deviation of 0.95. Table 1 shows the statistical parameters for each of the plots.

In parallel, overall site woody species LAI was estimated using field mapping protocols. For the 15 sites, the relative distribution of major plant types (mostly *Quercus coccifera*) height and ground cover was mapped visually and by projection to the ground, respectively. Results were then integrated with an extensive data base of species-specific allometric relations with biomass and LAI to yield the latter two parameters for each of the 15 plots.

The data base includes a total of 122 samples of different biomass related parameters as well as height. Above ground biomass was measured on 1-m^2 quadrants into different classes of height (0–1 m, 0–2 m, 0–3 m, 0–4 m). Shrubs in every height class were cut at the ground level. Fresh weight was determined in the field using a heavy weight balance. Samples of approximately 500gr (fresh weight) were brought to the laboratory and oven-dried at 105°C for 48 hours. They were separated to their leaves, woody parts, reproductive parts and dead parts (necromass). The woody parts were further separate in field into classes according to their diameter (0–1 cm, 1–3 cm, 3–5 cm, etc). Due to high biomass of samples in each diameter class, representative sub—samples were taken to the laboratory for further treatment. Furthermore the woody parts were separated into bark and wood according to class diameter. Using the field-based measurements and the laboratory results, allometric relationships were established, linking height to total, leaf, total stand and litter biomass (Tsiourlis & Sklavou, 2004). Also, species-specific relations to LAI were established as shown in table 2.

Based on these results, a preliminary analysis was carried out, linking site LAI derived from mapping/allometry with site LAI derived from optical measurements, where the average shrub/tree LAI (from LAI-2000 measurements) was multiplied by woody cover estimates for the plots derived from the Quickbird analysis to attain comparable measures, i.e. average woody cover LAI for the given plot (figure 2).

Despite a certain amount of scatter, value pairs arrange close to the 1:1 line, the gain of the regression function is close to 1 and the offset is relatively small. This shows that, despite a potential bias introduced by the large amount of opaque branches and twigs present in *Quercus* shrubs, the combination of field-based optical measurements of LAI and satellite image-based estimations

Table 2.   Average LAI for major woody vegetation species.

Species	LAI/m²
QC 1 (*Quercus coccifera* 1 m H)	2,13
QC 2 (*Quercus coccifera* 2 m H)	3,35
QC 3 (*Quercus coccifera* 3 m H)	4,38
PA2 (Pyrus amygdaliformis 2 m H)	1,12
PA3 (*Pyrus amygdaliformis* 3 m H)	1,46
QP1/QF (*Quercus pubescens/frainetto* 1 m H)	1,00
QP3(2−5) (*Quercus pubescens/frainetto* 2−5 m H)	2,00
QP5−6 (*Quercus pubescens/frainetto* 5−6 m H)	3,00
QP7−8 (*Quercus pubescens/frainetto* 7−8 m H)	3,50
QP10−12 (*Quercus pubescens/frainetto* 10 m H)	4,50
QP/QF10−14 (*Quercus pubescens/frainetto* 10−14 m H)	4,00
FS1−2 (*Fagus sylvatica* 1−2 m H)	1,00
FS2−4 (*Fagus sylvatica* 2−4 m H)	2,00
FS3−5 (*Fagus sylvatica* 3−5 m H)	2,50
FS12−14 (*Fagus sylvatica* 12−14 m H)	4,50
FS20−22 (*Fagus sylvatica* 20−22 m H)	5,00
CO4 (Cornus mas 4 m H)	2,00
Shrubs (other shrubs)	1,68

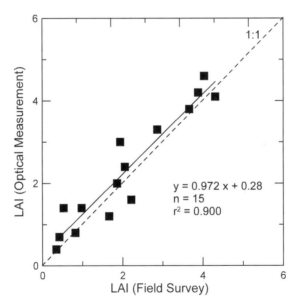

Figure 2.   Site LAI derived from a combination of optical (LAI-2000) and remote sensing (Quickbird) measurements vs. field-based estimate.

of proportional woody cover yields a characterization of spatial diversity that is congruent with traditional, field-based estimations.

3.3   *The relationship between satellite and field data*

An initial understanding of the sensitivity of the atmospherically corrected Landsat reflectance towards LAI and proportional cover data can be gained by comparing them with the Landsat-derived NDVI for the different field sites. Although one would not necessarily expect that the evergreen

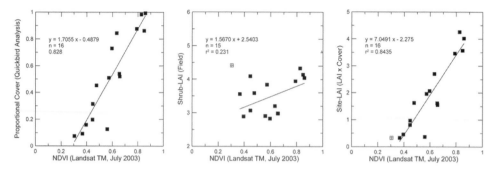

Figure 3. Relationship between Landsat-TM-derived NDVI and proportional cover (left), shrub-LAI (centre) and site-LAI (right).

Quercus shrubs with their dense wooden support structure will generate a strong greenness-related spectral signal, it becomes evident that the Quickbird-derived proportional cover estimates have a strong correlation with NDVI (figure 3); the relationship between NDVI and the average shrub LAI for the sites is comparatively weak. The fact that the relationship between integrated site-LAI (i.e. shrub-LAI multiplied with cover) is only moderately better than with proportional cover alone, however, allows only one conclusion: proportional vegetation cover already explains a large part of the spectral information content of Landsat observations over heterogeneous canopies, while leaf area index appears to add only a minor contribution to the satellite-measured signal. This is in good agreement with previous research, showing that cover has the closest relation to reflectance, while the relationship between cover and LAI is asymptotic (e.g. Lefsky & Cohen 2003). It implies that analysing shrub-LAI by means of reflectance modelling needs to explore very small information quantities and can thus be expected to be very sensitive to radiometric noise and depend on an exact model parameterisation. In open canopies the background reflectance characteristics will have a particularly important impact on the model inversion; even in areas where the background signal is as homogeneous as in our study site (during summer primarily a mixture of soil and dry grass) one could expect substantial disturbances.

4 RADIATIVE TRANSFER MODELLING APPROACH

The consistency of site LAI estimations for woody vegetation cover attained using different method-ologies suggests the procured data base is appropriate for parameterization of the canopy reflectance model. The different model components and processing steps are described in the following section along with results, which will be discussed and commented in section 5.

4.1 *Model components and parameterization*

In case of heterogeneous canopies, such as they prevail in the Lagadas rangelands, it is mandatory to use geometrical models (Jasinski 1996) or hybrid approaches which couple the simulation of three-dimensional canopy elements with leaf and turbid media reflectance models. In this study we have used a modified version of the GeoSail model (Huemmrich 2001) which is based on the combination of a geometric model (Jasinski 1994, Jasinski & Eagleson 1989, 1990) to calculate the amount of shadowed and illuminated components in a scene, with an early version of the SAIL model (Verhoef 1984) from which the reflectance and transmittance of canopy elements is obtained.

Leaf optical properties are injected via the PROSPECT model of Jacquemoud and Baret (1990). The geometric part of GeoSail uses the model developed by Jasinski for the limiting case where the shadows cast by clumps of vegetation are very small relative to the size of the area observed

Table 3. Statistics of Prospect Retrievals (n = 31, 2 outliers removed due to ambiguous measurements).

N		Cab		Cw		Cm	
Average	Stdv	Average	Stdv	Average	Stdv	Average	Stdv
1.38	0.49	45.24	17.402	0.025	0.0076	0.012	0.009

(Jasinski 1994, Jasinski & Eagleson 1989, 1990). Scene reflectance is determined in GeoSail by calculating an area-weighted average of three landscape components: illuminated canopy, illuminated background, and shadowed background. The model considers a scene made up of geometric solids scattered over a plane with a Poisson distribution; the solids are identical in size and shape and cast shadows on the background plane, but do not overshadow each other. Its simplicity is a result of the assumption that single values for the reflectance and transmittance of light by canopy clumps are enough to provide a reasonable overall description of scene reflectance and absorption.

In order to characterise biochemical properties of leaves, spectroradiometric measurements of leave samples (single leaves) were subjected to a numerical inversion of the PROSPECT model. Given the many successful applications of the model (e.g. Jacquemoud et al. 1996, Gastellu-Etchegorry & Bruniquel-Pinel 2001, Schlerf & Atzberger 2006, Colombo et al. 2008), it can be assumed that the retrievals provide a good indication of leaf properties for the late spring situation, during which the samples had been collected and measured. Table 3 provides an overview of results attained for the PROSPECT structure parameter N, Chlorophyll a & b (cab), leaf water (cw) and dry matter content (cm) as derived from the numerical inversion of PROSPECT for the leaf spectra from the Lagadas ground data collection. Apart from single outliers the retrieved values are well within the expected range for this type of woody vegetation.

However, careful adaptations of the leaf parameters were applied to reflect the summer situation in which the Landsat data were acquired (see section 4.3).

4.2 Forward modelling

The reflectance models were coupled in the following way: since the Landsat image was recorded in late July 2003 the parameterisation of PROSPECT was slightly modified in comparison to the results from the ground data collection on leaf optical properties (section 3); with regard to the much drier summer situation, chlorophyll concentration was reduced to 25 μgcm^{-2}, EWT lowered to 0.01 while the dry matter concentration was kept unchanged. The modeled leaf reflectance was then assimilated into SAIL for simulating reflectance properties of turbid canopy components (trees and shrubs of *Quercus coccifera*). The GeoSail part was parameterised for such canopy elements with cylindrical shape and a base-to-height-ratio of 1.5 which is adequate for a wide range of woody species with predominantly planophile leaf angle distribution.

Taking into account the observation geometry of the Landsat scene from 24 July 2003 scene reflectance for the fifteen field control sites was modelled by combining the corresponding LAI measurements of the individual shrubs (average for each site) and the Quickbird-derived proportional cover estimates for the same locations. Comparison with the true Landsat-derived site reflectance confirms that the coupled leaf-canopy reflectance model is suited to generate adequate reflectance signature when driven with adequate canopy descriptors (figure 4).

The slope of the regression model ($r^2 = 0.941$) is close to one and has only a negligible offset. There is only a single value which qualifies itself as an exceptional outlier; it comes from one of the dense mountain forests which exhibit an exceptionally high near-infrared reflectance. These results suggest that retrieving LAI-estimates through inverting the GeoSail reflectance model on Landsat-TM reflectance might be feasible. Nonetheless, the associated root mean squared error of ±0.032 indicates a substantial sensitivity of the model, requesting careful analysis of the contribution of the different driving parameters to retrieval of LAI estimates.

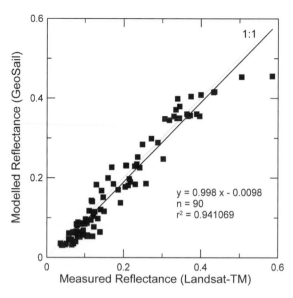

$y = 0.998 x - 0.0098$
$n = 90$
$r^2 = 0.941069$

Figure 4. Comparison between modelled and TM-measured scene reflectance (in the six TM channels) for 15 control sites with known LAI and proportional cover.

4.3 *Model inversion*

The GeoSail inversion was performed by means of matching compliant model signatures to the Landsat reflectance of each pixel. The GeoSail signature database was produced by feeding modelled leaf reflectance (N = 1.4, cab = 25, cw = 0.01, cm = 0.01) into SAIL for simulating the reflectance properties of turbid canopy components (trees and shrubs of *Quercus coccifera*) with LAI values ranging from 0.5 to 5.5 (in steps of 0.25); leaf inclination was parameterized with an average leaf inclination angle of 20 degrees (using an elliptic leaf angle distribution function). Finally, the geometric part of GeoSail was used to produce for each of these canopy spectra the reflectance associated with a range of cover proportions (0.05 to 1, in steps of 0.05); the shape of the canopy elements was considered to be best approximated as a cylinder with a base/height ratio of 1.5 (i.e. slightly broader than high). To calculate the scene reflectance the GeoSail model weighs the reflectance of each component by its fractional area. The illuminated background reflectance is the same background reflectance used in the SAIL model, the shadow reflectance is the product of the transmittance through the canopy, calculated by the SAIL model, and the background reflectance. This formulation of scene reflectance allows for multiple scattering within tree canopies but assumes no multiple scattering between trees (Huemmrich 2001). With regard to the dominant abundance of dry grass at the time of image acquisition, the background and inter-space reflectance was not represented by a soil spectrum but a typical mixture between soil and dry vegetation spectra available from previous field measurement campaigns with an ASD FieldSpec II instrument. The simulated spectral data base thus includes 420 simulated reflectance spectra at TM resolution.

A substantial complication in inverting coupled leaf-canopy reflectance models results from so-called 'ill-posed' problems, were specific reflectance signatures can be equally well approximated by multiple parameter combinations (Colombo et al. 2003, Lewis & Disney 2006). Various regularisation strategies have been proposed for efficiently handling this problem (Combal et al. 2003, Koetz et al. 2005). Here, in order to reduce the ambiguity between the spectral response of canopies with high LAI and low coverage on one hand, and low to medium LAI in high cover situations on the other side, the retrieval of LAI values was constrained by integrating the Quickbird-derived proportional cover estimates into the model inversion. This implies that the model inversion for each Landsat pixel only had to consider the simulated LAI-range for the known canopy cover proportion at this position.

4.4 Results

Following the methodology described in the previous sections, a map of site LAI was derived by retrieving shrub LAI for each pixel using a lookup table approach, and multiplying this with the Quickbird-derived cover values (figure 5).

The resulting map shows a distinct spatial pattern and reflects the distribution of woody vegetation elements in the test area. Throughout the rangeland areas, low site LAI values are visible, while higher estimates are found in the narrow valleys. These correspond to regions with favourable water supply and a reduced pressure by grazing animals, thus supporting denser canopies of woody vegetation (Röder et al. 2007, 2008b). The highest LAI values are being found in spatially compact arrangements. They pertain to the forest zones in the northeastern and southeastern regions of the test area, while in the West an area with aforestation schemes emerges as a prominent patch of very high LAI values. Thus, the visual assessment of the model results underlines its capacity to adequately characterize the spatial differentiation of woody vegetation elements in heterogeneous landscape predominantly covered by grass- and shrubland.

Comparison between the retrieved site-LAI values and the corresponding field assessments (already presented in figure 2) suggests that the model inversion is well capable to reproduce these gradients (figure 6). However, it is also seen that quite substantial scatter is affecting the medium range of LAI values (LAI = 2−3), and that the model appears to underestimate the low LAI range.

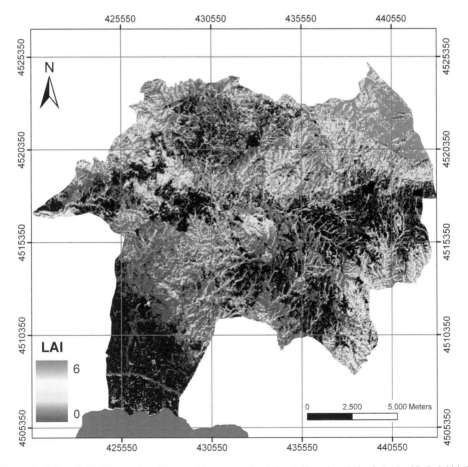

Figure 5.   Map of site-LAI retrieved by combing constrained GeoSail-retrieved shrub-LAI with Quickbird-derived cover estimates (agricultural areas and settlements in black, parts of Lake Koronia visible in blue in the South) (see colour plate page 392).

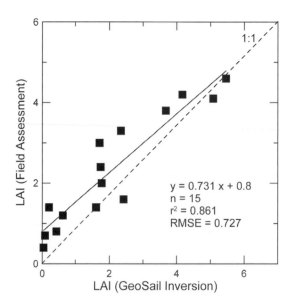

Figure 6.   Site LAI derived from GeoSail inversion in comparison to the field-based estimate.

We also tested the model inversion in an unconstrained mode, allowing the algorithm to map both, LAI and cover. This yielded a regression function of $y = 0.513x + 1.212$ (r^2 0.721) and a larger deviation from the 1:1 line. In particular, cover values mapped from the lookup tables showed a strongly deviating pattern from the reference derived using the Quickbird-based approach.

5   DISCUSSION AND PERSPECTIVES

Summarizing, different elements request special attention when implementing modelling approaches in heterogeneous Mediterranean rangelands, as they determine the sensitivity of the model and potentially limit the accuracy of results that may be attained. First, all steps involved in satellite image processing are likely sources of error. These include potential geometric mismatch between the Landsat and Quickbird data, as well as sensor calibration and radiative transfer modelling. However, the involved steps have been shown to perform in an accurate and robust manner in a wealth of studies (e.g. Hostert et al. 2003b, Röder et al. 2008a, b). Limitations in the positional accuracy of GPS positions utilized to locate field measurement positions are mitigated by assigning the measurements to larger patches comprising various pixels.

A second important component is the comparability of phenological situations as represented by image acquisition dates as well as dates when field campaigns were carried out. Generally, care has been taken to reflect only late summer conditions when senescence of grasses and herbaceous vegetation had already been completed, although minor variations can not entirely be ruled out. For instance, an overestimation of shrub LAI would result if in the Landsat image grasses would still be phenologically active and contribute to the vegetation portion of the signal, while they were not mapped in the Quickbird data set.

The third set of critical issues relates to the methodologies employed during the field surveys. All reference plots have been characterized through extensive optical LAI measurements. These are subject to a certain degree of error in cases where stems and branches strongly contribute to the shrub/tree structure and exert a dominant influence on the gap fraction compared to leave elements. This would cause an overestimation of field-based LAI. In addition, representing the individual plots through averaging of individual measurements might introduce bias where a large variance of measured values is encountered (compare Table 1).

In addition, the coupled model and its sub-elements itself request a critical appraisal. Numerous studies have proven the potential of different coupled canopy reflectance models (e.g. Jacquemoud et al. 2000); yet, these are often focused on homogeneous environments, such as agricultural areas or forests (e.g. Gemmel et al. 2002, Zarco-Tejada et al. 2005, Schlerf & Atzberger 2006, González-Sanpedro et al. 2008). In contrast, the test site addressed here comprises an irregular spatial arrangement of woody vegetation elements. In addition, the representation of shrub/tree species by parameters calibrated to *Quercus coccifera* shrubs is a simplified assumption, which does not hold in slightly modified situations characterized by an intricate mixture of different shrub and tree species, as well as the higher elevations covered by thermophilous forests. This is likely to introduce errors in LAI retrieval in such areas; on the other hand, there is no feasible means of circumventing this problem given the heterogeneous nature of the test area, while including parts of the forest zone helps to provide a larger dynamic range for model testing.

Finally, the model is strongly driven by the selected background signal in cases of sparse woody vegetation cover. In heterogeneous rangeland areas, this is a major potential source of error, as "background" might comprise a substantial range of spectrally different surfaces that occur in manifold arrangements and mixtures, such as soils in different stages of development, different bedrock types, as well as grasses and herbaceous vegetation. The impact of background reflectance on the retrieval accuracy of the constrained model becomes evident when comparing the GeoSail estimates of shrub-LAI with regard to the averaged, field-based measurements for the 15 plots (figure 7). While the comparison for all sites exhibits tremendous scatter, substantial correlation becomes evident when all sites with proportional cover below 40% are eliminated. Improvements would require that, besides constraining the model by introducing estimates of woody vegetation cover, a similar spatially explicit map of background signatures would be available to further constrain the inversion process.

Based on the results of this study, and considering the number of crucial factors involved in collecting and processing required input data as well as parameterizing the model, two critical conclusions emerge. First of all, inverting a canopy reflectance model over environments such as the one covered in our study, appears only feasible if the model can be adequately constrained. To this end, utilizing very high resolution data, such as Quickbird, Ikonos or WorldView-1 images, was shown to be an efficient approach. Wherever possible, it is important to stratify the input data and consider different vegetation types in the model parameterization as well as background situations as much as possible. However, in cases of high spatial heterogeneity this is often not possible.

Second, the analysis has shown that in rangeland ecosystems such as the one covered here, an exceptionally high degree of the signal recorded by remote sensing systems is actually determined by proportional cover of vegetation, while the leaf area of woody vegetation elements only contributes

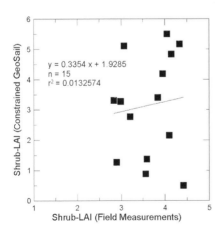
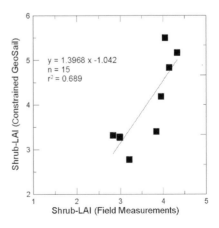

Figure 7. Comparison between field-measured average shrub-LAI and the constrained GeoSail inversion results for all sites (left) and those with proportional cover above 40% (right).

a minor portion. Given the sensitivities and limitations discussed before, this casts some doubts on the routine and operational derivation of LAI for a wide range of similar ecosystems, sometimes even using coarse-scale satellite images (e.g. Knyazikhin et al. 1998, Zhang et al. 2005, Hill et al. 2006). With regard to the incorporation of information on woody vegetation canopies for local range management, it can be concluded that information on proportional cover appears to reflect most of the variety of the vegetation-related signal. Thus, focusing on a precise derivation of proportional woody vegetation cover appears the preferable solution for many ecological applications.

ACKNOWLEDGEMENTS

This study was partially carried out in the frame of the projects 'GeoRange' (Geomatics in the Assessment and Sustainable Management of Mediterranean Rangelands) and DeSurvey-IP (A Surveillance System for Monitoring and Assessing of Desertification), which have been funded by the European Union, DG Research. The authors wish to thank Samuel Bärisch for support in initial data processing and the student staff at Aristotle University of Thessaloniki for support in field data collection.

REFERENCES

Asner, G.P. & Heidebrecht, K.B. 2002. Spectral unmixing of vegetation, soil and dry carbon cover in arid regions: comparing multispectral and hyperspectral observations. *International Journal of Remote Sensing*, 23: 3939–3958.

Boer, M. & Puigdefabregas, J. 2005. Assessment of dryland condition using spatial anomalies of vegetation index values. *International Journal of Remote Sensing* 26(18): 4045–4065.

Boer, M. 1999. Assessment of dryland degradation: linking theory and practice through site water balance modelling. Nederlandse Geografische Studies 251, Utrecht.

Chen, J.M. & Cihlar, J. 1996. Retrieving leaf area index of boreal conifer forests using Landsat TM images. *Remote Sensing of Environment* 55: 153–162.

Colombo, R., Bellingeri, D., Fasolini, D. & Marino, C.M. 2003. Retrieval of leaf area index in different vegetation types using high resolution satellite data. *Remote Sensing of Environment* 86(1): 120–131.

Colombo, R., Meroni, M., Marchesi, A., Busetto, L., Rossigni, M., Giardino, C. & Panigada, C. 2008. Estimation of leaf and canopy water content in poplar plantations by means of hyperspectral indices and inverse modeling. *Remote Sensing of Environment* 112: 1820–1834.

Combal, B., Baret, F., Weiss, M., Trubuil, A., Mace, D., Pragnere, A., Myneni, R., Knyazikhin, Y. & Wang, L. 2003. Retrieval of canopy biophysical variables from bidirectional reflectance—using prior information to solve the ill-posed inverse problem. *Remote Sensing of Environment* 84(1): 1–15.

Di Castri, F. 1981. Mediterranean type shrublands of the world. In Di Castri, F., Goodall, D.W. & Specht, R.L. (eds), Mediterranean-type shrublands, ecosystems of the world 11: 1–42. Amsterdam, Oxford, New York: Elsevier.

Elmore A.J., Mustard, J.F., Manning, S.J. & Lobell, D.B. 2000. Quantifying vegetation change in semiarid environments: Precision and accuracy of spectral mixture analysis and the normalized difference Vegetation index. *Remote Sensing of Environment*, 73: 87–102.

Fang, H., Liang, S., McClaran, M.P., van Leeuwen, W.J.D., Drake, S., Marsh, S.E., Thomson, A.M., Izaurralde, R.C. & Rosenberg, N.J. 2005. Biophysical characterization and management effects on semiarid rangeland observed from Landsat ETM+ data. *IEEE Transactions on Geoscience and Remote Sensing* 43(1): 125–134.

Fernandes, R., Miller, J.R. & Hu, B. 200.: A multi-scale approach to mapping effective leaf area index in boreal Picea marina stands using high spatial resolution CASI imagery. *International Journal of Remote Sensing* 23(18): 3547–3568.

Gastellu-Etchegorry, J.P. & Bruniquel-Pinel, V. 2001. A modeling approach to assess the robustness of spectrometric predictive equations for canopy chemistry. *Remote Sensing of Environment* 76: 1–15.

Gemmell, F., Varjo, J., Strandstrom, M. & Kuusk, A. 2002. Comparison of measured boreal forest characteristics with estimates from TM data and limited ancillary information using reflectance model inversion. *Remote Sensing of Environment* 81: 365–377.

González-Sanpedro, M.C., Le Toan, T., Moreno, J., Kergoat, L. & Rubio, E. 2008. Seasonal variations of leaf area index of agricultural fields retrieved from Landsat data. *Remote Sensing of Environment* 112(3): 810–824.

Hill, J., Hostert, P., Tsiourlis, G., Kasapidis, P., Udelhoven, T. & Diemer, C. 1998. Monitoring 20 Years of increased grazing impact on the Greek island of Crete with earth observation satellites. *Journal of Arid Environment* 39: 165–178.

Hill, J., Mehl, W. & Radeloff, V. 1995. Improved forest mapping by combining corrections of atmospheric and topographic effects. In J. Askne (Ed.), *Sensors and Environmental Applications of Remote Sensing* (pp. 143–151). Proceedings of the 14th EARSeL Symposium, Göteborg, Sweden, 6–8 June 1994. Rotterdam, Brookfield: A.A. Balkema.

Hill, M.J., Senarath, U., Lee, A., Zeppel, M., Nightingale, J.M., Williams, R.J. & McVicar, T.R. 2006. Assessment of the MODIS LAI product for Australian ecosystems. *Remote Sensing of Environment* 101 (4): 495–518.

Hostert, P., Röder, A. & Hill, J. 2003b. Coupling spectral unmixing and trend analysis for monitoring of long-term vegetation dynamics in Mediterranean rangelands. *Remote Sensing of Environment* 87: 183–197.

Hostert, P., Röder, A., Hill, J., Udelhoven, T. & Tsiourlis, G. 2003a. Retrospective studies of grazing-induced land degradation: a case study in central Crete, Greece. *International Journal of Remote Sensing* 24: 4019–4034.

Huemmrich, K.F. 2001. The GeoSAIL model: a simple addition to the SAIL model to describe discontinuous canopy reflectance. *Remote Sensing of Environment* 75(3): 423–431.

Jacquemoud, S. & Baret, F. 1990. PROSPECT: a model of leaf optical properties spectra. *Remote Sensing of Environment* 34: 75–91.

Jacquemoud, S., Ustin, S.L., Verdebout, J., Schmuck, G., Andreoli, G. & Hosgood, B. 1996. Estimating leaf biochemistry using the PROSPECT leaf optical properties model. *Remote Sensing of Environment* 56:194–202.

Jacquemoud, S., Bacour, C., Poilve, H. & Frangi, J.P. 2000. Comparison of four radiative transfer models to simulate plant canopies reflectance: Direct and inverse mode. *Remote Sensing of Environment* 74(3): 471–481.

Jasinski, M.F. 1996. Estimation of subpixel vegetation density of natural regions using satellite multispectral imagery, *IEEE Trans. Geoscience and Remote Sensing* 34(3): 804–813.

Jasinski, M.F. & Eagleson, P.S. 1989. The structure of red-infrared scattergrams of semivegetated landscapes. *IEEE Transactions on Geoscience and Remote Sensing* 27(4): 441–451.

Jasinski, M.F. & Eagleson, P.S. 1990. Estimation of subpixel vegetation cover using red-infrared scattergrams. *IEEE Transactions on Geoscience and Remote Sensing* 28(2): 253–267.

Jasinski, M.F. 1994. Sensitivity of the normalised difference vegetation index to subpixel canopy cover, soil albedo, and pixel scale. *Remote Sensing of Environment* 32: 169–187.

Jonckheere, I., Fleck, S., Nackaerts, K., Muys, B., Coppin, P., Weiss, M. & Baret, F. 2003. Review of methods for in situ leaf area index determination: Part I. Theories, sensors and hemispherical photography. *Agricultural and Forest Meteorology* 121(1–2): 19–35.

Kimes, D., Knyazikhin, Y., Privette, J.L., Abuelgasim, A.A. & Gao, F. 2000. Inversion methods for physically-based models. *Remote Sensing Reviews* 18: 381–439.

Knyazikhin, Y., Martonchik, J.V., Myneni, R.B., Diner, D.J. & Running, S.W. 1998. Synergistic algorithm for estimating vegetation canopy leaf area index and fraction of absorbed photosynthically active radiation from MODIS and MISR data. *Journal of Geophysical Research* 103 (24): 32257–32276.

Koetz, B., Baret, F., Poilvé, H. & Hill, J. 2005. Use of coupled canopy structure dynamic and radiative transfer models to estimate biophysical canopy characteristics. *Remote Sensing of Environment* 95: 115–124.

Konstantinidis, P. & Tsiourlis, G.M. 2003. *Description—Analysis and Mapping of Vegetation Units (habitats) of Lagadas County (Thessaloniki, Greece)*. NAGREF—Forest Research Institute, GeoRange project report (unpublished). Brussels: European Commission

Lacaze, B. 2005. Remotely-sensed optical and thermal indicators of land degradation, In Oluic, M. (Ed.): *Proc. 24th Symp. EARSeL, Dubrivnik, Croatia, 25–27 May 2004:* 211–217, Millpress: Rotterdam.

Lefsky, M.A. & Cohen, W.B. 2003. Selection of remotely sensed data, In Wulder, M. & Franklin, S.E. (eds.): Methods and Applications for Remote Sensing: Concepts and Case Studies: 13–46, Kluwer Academic Publishers: Dordnecht.

Lewis, P. & Disney, M. 2006. Spectral invariants and scattering across multiple scales from within-leaf to canopy. *Remote Sensing of Environment* 109(2): 196–206.

Ray, T.W. & Murray, B.C. 1996. Nonlinear spectral mixing in desert vegetation. *Remote Sensing of Environment* 55: 59–64.

Reynolds, J.F. & Stafford-Smith, M.D. 2002. Global desertification. Do humans cause deserts? Dahlem workshop report 88. Berlin: Dahlem University Press.

Röder, A. 2005. A remote sensing based framework for monitoring and assessing Mediterranean rangelands. PhD Thesis, University of Trier, Germany.

Röder, A., Hostert, P., Tsiourlis, G., Kasapidis, P. & Hill, J. 2001. Resource Assessment to Support the Sustainable Management of Mediterranean Ecosystems. An Approach Integrating Remote Sensing and Ecology. In Belward, A., Binaghi, E., Brivio, PA, Lanzarone, GA and Tosi, G. (eds.), *Proc. Int. Workshop Geo-Spatial Knowledge Processing for Natural Resource Management, 28th-29th June 2001*: 303–309, Varese, Italy.

Röder, A., Kuemmerle, T., Hill, J., Papanastasis, V.P. & Tsiourlis, G.M. 2007. Adaptation of a grazing gradient concept to heterogeneous Mediterranean rangelands using cost surface modelling. *Ecological Modelling* 204: 387–398.

Röder, A., Udelhoven, T., Hill, J., Del Barrio, G. & Tsiourlis, G.M. 2008. Trend analysis of Landsat-TM and – ETM+ imagery to monitor grazing impact in a rangeland ecosystem in Northern Greece. *Remote Sensing of Environment* 112(6): 2863–2875.

Rosema, A., Verhoef, W., Noorbergen, H. & Borgesius, J.J. 1992. A new forest light interaction model in support of forest monitoring. *Remote Sensing of Environment* 42: 32–41.

Schlerf, M. & Atzberger, C. 2006. Inversion of a forest reflectance model to estimate structural canopy variables from hyperspectral remote sensing data. *Remote Sensing of Environment* 100: 281–294.

Smith, M.O., Ustin, S.L., Adams, J.B. & Gillespie, A.R. 1990. Vegetation in deserts: I. A regional measure of abundance from multispectral images. *Remote Sensing of Environment*, 31: 1–26.

Soudani, K., Trautmann, J. & Walter, J.M. 2002. Leaf area index and canopy stratification in Scots pine stands. *International Journal of Remote Sensing* 23(18): 3605–3618.

Teillet, P.M., Barker, J.L., Markham, B.L., Irish, R.R., Fedosejevs, G. & Storey, J.C. 2001. Radiometric cross-calibration of the Landsat-7 ETM+ and Landsat-5 TM sensors based on tandem data sets. *Remote Sensing of Environment* 78: 39–54.

Thome, K.J., Markham, B., Barker, J., Slater, P. & Biggar, S. 1997. Radiometric calibration of Landsat. *Photogrammetric Engineering & Remote Sensing* 63(7): 853–858.

Tsiourlis, G.M. & Sklavou, P.S. 2004. Allmetric relationships of biomass of kermes oak (*Quercus coccifera* L.). In Sfouggaris, A. (ed.), *Book of abstracts of the 4tg Panhellenic Rangeland Congress, Volos, 10–12 November 2004:* 30, Volos, Greece.

Verhoef, W. 1984. Light scattering by leaf layers with application to canopy reflectance modelling: the SAIL model. *Remote Sensing of Environment* 16: 125–141.

Welles, J.M. & Norman, J.M. 1991. Instrument for indirect measurement of canopy architecture. Agronomy Journal, 83: 818–825.

White, M.A., Asner, G.P., Nemani, R.R., Privette, J.L. & Running, S.W. 2000. Measuring fractional cover and leaf area index in arid ecosystems. Digital camera, radiation transmittance, and laser altimetry methods. *Remote Sensing of Environment* 74: 45–57.

Zarco-Tejada, P.J., Berjón, A., López-Lozano, R., Miller, J.R., Martín, P., Cachorro, V., González, M.R. & de Frutos, A. 2005. Assessing vineyard condition with hyperspectral indices: leaf and canopy reflectance in a row-structured discontinuous canopy. *Remote Sensing of Environment* 99: 271–287.

Zhang, Q., Xiao, X., Braswell, B., Linder, E., Baret, F. & Moore III, B. 2005. Estimating light absorption by chlorophyll, leaf and canopy in a deciduous broadleaf forest using MODIS data and a radiative transfer model. *Remote Sensing of Environment* 99(3): 357–371.

Recent Advances in Remote Sensing and Geoinformation Processing
for Land Degradation Assessment – Röder & Hill (eds)
© 2009 Taylor & Francis Group, London, ISBN 978-0-415-39769-8

Using reflectance spectroscopy and Landsat data to assess soil inorganic carbon in the Judean Desert (Israel)

T. Jarmer
Technion—Faculty of Civil and Environmental Engineering, Transportation and Geo-Information Engineering Unit, Haifa, Israel

H. Lavée & P. Sarah
Department of Geography, Bar-Ilan University, Ramat Gan, Israel

J. Hill
Remote Sensing Department, FB VI Geography/Geosciences, University of Trier, Trier, Germany

ABSTRACT: The spatial distribution of soil inorganic carbon was investigated with remote sensing methods. The relationship between laboratory spectral measurements and inorganic carbon content was accomplished with respect to the characteristic absorption features of carbonate and C.I.E. color coordinates. An empirical model for the spectral detection of carbonate content was generated which allows the prediction of soil inorganic carbon content with a cross-validated r^2 of 0.957. In a second step the established model was modified to allow its application to Landsat images. Since C.I.E. color coordinates were found to be well suitable parameters for predicting the inorganic carbon content of soils under laboratory conditions, the reflectance values of the Landsat-TM bands were transformed into C.I.E. color coordinates. Subsequently, the C.I.E. based model approach was adapted to a Landsat image with low vegetation cover from July 1998 to predict spatial distribution of the soils inorganic carbon content. Transferring the regression model to the satellite image allowed the prediction of the inorganic carbon content in the spatial dimension. The concentrations predicted from satellite data correspond well with the concentration range of the chemical analysis. They reflect the geographic conditions and show a dependence on the annual rainfall amount. A general trend to increasing concentrations of inorganic carbon can be stated with increasing aridity. Furthermore, local conditions are well reflected by the predicted concentrations.

1 INTRODUCTION

Almost one third of the earth's surface is covered by drylands, and especially the semi-arid ecosystems provide important land resources for adapted land use and livestock farming. But the pressure on arid and semi-arid areas resulting from climatic variability, climate change, demands of increasing stocking rates and population development was probably never as high as nowadays (White 2000). Consequently, a broad monitoring of the status of these arid and semi-arid areas is of enormous importance, since these regions are often considered as risk areas in the context of global climate change and desertification dynamics. Against this background the soils of these ecosystems are of outstanding relevance.

Soils are a substantial part of the terrestrial ecosystems and their physical and chemical properties are fundamental indicators for the land condition status. Soil loss resulting from a combination of physiographic and human factors is an important sign of degradation processes. In semi-arid and arid environments the inorganic carbon content in soils developed on carbonatic parent material is one major indicator of soil development. While low contents of inorganic carbon point to well developed soils, high inorganic carbon contents indicate weakly developed or degraded soils.

The detection of the spatial heterogeneity of soil inorganic carbon demands a large number of soil samples to be analyzed in the laboratory which is time-consuming and expensive. With respect to the great importance of the spatial detection of soil inorganic carbon in the context of monitoring land degradation and sub-recent morphodynamics, the development of a remote sensing based approach for spectral determination of inorganic carbon content is necessary.

Consequently, the aim of this study is the spatial assessment of the inorganic carbon content of soils. An analysis of the relationship between laboratory spectral measurements and inorganic carbon content is accomplished with respect to the characteristic absorption features of carbonate. Additionally, the investigation of the influence of inorganic carbon on the soil color is expedient, since soil color is related to important soil properties. Further on, a method for spectral detection of inorganic carbon content for laboratory bi-directional reflectance measurements is generated based on empirical-statistical analysis.

But even with laboratory spectroscopy a spatial detection of soil properties for broader and inaccessible areas exclusively based on terrestrial inquiry is rarely feasible. Operational EOS are in particular suitable for the environmental monitoring of these sensitive ecosystems (Hill 2000). But these systems are usually restricted by a relatively limited number of spectral channels with a comparatively small spectral resolution which necessitate the use of these few spectral bands or parameters derived from them for the assessment of soil properties and soil development (e.g. Pickup & Nelson 1984, Madeira et al. 1997, Hill & Schütt 2001).

The established empirical model for prediction of inorganic carbon content based on laboratory reflectance measurements will be modified to allow spatial analysis of the inorganic carbon content of soils based on Landsat images. Subsequently, an up-scaling of the model approach will be adapted to a Landsat image in order to predict spatial distribution of the soils inorganic carbon content for the investigated region.

2 MATERIAL AND METHODS

2.1 The Study site

Research was conducted along a transect following a hypsometric rainfall gradient from east of Jerusalem to the Dead Sea covering Mediterranean, semi-arid and mildly-arid climate zones and extending to the east towards the arid area near the Dead Sea (figure 1). The transect shows a strong decrease in elevation, ranging from 650 m a.s.l. in the Judea-Samaria mountains, in the west, to 60 m below sea level in the Judean Desert.

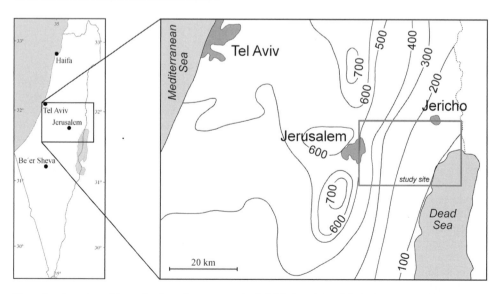

Figure 1.   Location of the study site and mean annual rainfall in Israel (modified after Survey of Israel 1985).

Mean annual rainfall decreases on a distance of only 33 km from 620 mm in the west to about 120 mm in the east (Lavée et al. 1998, Sarah 2001). The rainfall is almost exclusively limited to winter months but annual precipitation varies considerable between the years. In addition to the rainfall amount and its distribution within the year especially the rainfall intensity is very meaningful. In general the percentage of rainfall events with intensities above 30 mm/h are considerably higher in the eastern part of the study site. Such extreme events often cause flooding and give rise to intense soil erosion (Gat & Karni 1995, Morin et al. 1998).

All sample sites are located on limestone parent material. The soils vary according to the amount of rainfall. In the semi-arid part of the higher altitudes of the Judean Desert shallow brown rendzinas dominate, but wide areas are captured by calcareous bedrock or crusts (Nari, Calcrete). Shallow to medium pale rendzinas are typical for the slopes where the calcareous crusts have been eroded. The more developed soils in the valleys (partly Grumosols) in places are irrigated for olive and orchard plantations. The central part of the study site mostly compasses steep, stony and eroded slopes which are characterized by brown lithosols and desert lithosols. Further to the east the brown lithosols become shallower and desert lithosols and outcrops of bedrock material dominate (Dan et al. 1976, Dan & Smith 1981). However, general conditions are often modified by local conditions. Micro-scale catchment areas on knolls and gentle slopes which result and are protected from surrounding outcrops of bedrock material, thus allow in semi-arid and arid areas the development of soils of more humid regions.

2.2 Field sampling

The general conditions within the study area are provided by the rainfall gradient but at meso-microscale level physiogeographic factors considerable modify climatic conditions. In this context the relief is of special importance because it influences local conditions in different ways. Irradiation is changing both in total and in daily and seasonal distribution strongly depending on aspect and slope. South-oriented slopes show higher irradiation amounts with stronger variations (see also

Figure 2. Varying physiogeographic conditions along a sampling transect in the semi-arid part of the study site (upper left: top of hill; upper right: upper slope; lower left: lower slope; lower right: foot of slope).

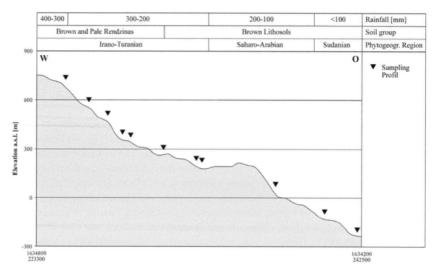

400-300	300-200	200-100	<100	Rainfall [mm]
Brown and Pale Rendzinas		Brown Lithosols		Soil group
Irano-Turanian		Saharo-Arabian	Sudanian	Phytogeogr. Region

Figure 3.   Location of sampling transects and relationships between climate, soils and vegetation in the study site.

Boer et al. 1996). Water supply is varying due to surface flow between middle slope and top or valley bottom. Run off from bedrock outcrops at protected areas at hill tops increases rainfall amount substantial and allows the development of soils typical for wetter conditions than under given climatic conditions at these mini-catchments (see figure 2, upper left). Hence, the different physiogeographic conditions at micro- to meso-scale show substantial effects on soil development.

Within the study site sampling of the soil surface (upper 2 cm) material was conducted along eleven topographical transects covering north- and south- facing slopes and valley bottoms to consider the different physiogeographic conditions (figure 3). In total 53 samples were collected at the different climatic regions.

2.3   Chemical analysis

The soil samples were air-dried and gently crushed in order to pass a 2 mm-sieve. Subsamples were carefully homogenized for an enhancement of spectral features and to allow better reaction with the chemical reagents. For all samples organic and inorganic carbon content were analyzed by an infrared cell in a high-frequency induction oven (LECO). The content of inorganic carbon varies between 1.57 and 9.24% with a mean of 5.22% and a standard deviation of 1.96.

2.4   Laboratory spectral reflectance measurements

Inherent spectral characteristics of varying combinations of mineral components, organic matter and the soil moisture affected the spectral reflectance of soils (Baumgardner et al. 1985). Consequently, in many studies the relations of spectral reflectance and soil properties, like grain size distribution, soil moisture, iron oxides, carbonate content and organic matter, have been outlined (e.g. Krishnan et al. 1980, Kosmas et al. 1984, Jarmer & Schütt 1998, Hill & Schütt 2000, Udelhoven et al. 2003, Jarmer 2005). The relationships between soil reflectance and carbonate content have been established using statistical methods such as multiple regression analysis (Ben-Dor & Banin 1990, 1994).

The bi-directional reflectance measurements of the soil samples were acquired in the laboratory with an ASD FieldSpec-II spectroradiometer. Spectral readings were taken in 1 nm steps between 350 nm and 2500 nm using a reflectance standard of known reflectivity (Spectralon). The optical head of the spectroradiometer was mounted on a tripod in nadir position with a distance of 10 cm to

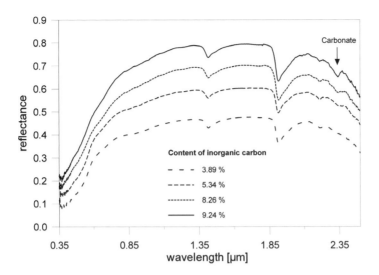

Figure 4. Reflectance spectra of soil samples with different content of inorganic carbon (Jarmer et al. 2001).

the sample. For illumination, a 1000 W quartz-halogen lamp set in a distance of approximately 30 cm and with a zenith angle of 30° was used. Absolute bi-directional reflectance spectra were obtained by multiplying the raw reflectance spectra by the certified reflectivity of the Spectralon panel.

The spectral reflectance of soils which have developed on carbonatic bedrock under relatively dry climatic conditions is substantially influenced by their content of inorganic carbon. In these soils often an increase of reflectance with rising contents of inorganic carbon can be seen (figure 4).

Since this is also given for the visible domain, it was decided to include color information into the further spectral analysis. Instead of using soil scientists Munsell color chart the Color system of the "Commission Internationale de 'Eclairage [C.I.E.]" from 1931 was adapted. For the calculation of color parameters, the reflectance measurements were converted into trichromatic specifications, and then expressed in terms of the C.I.E. color notation according to the following equations:

$$X = \sum_{380\ nm}^{770\ nm} \rho(\lambda) * S(\lambda) * \bar{x}(\lambda) \tag{1}$$

$$Y = \sum_{380\ nm}^{770\ nm} \rho(\lambda) * S(\lambda) * \bar{y}(\lambda) \tag{2}$$

$$Z = \sum_{380\ nm}^{770\ nm} \rho(\lambda) * S(\lambda) * \bar{z}(\lambda) \tag{3}$$

and

$$x = \frac{X}{X+Y+Z} \qquad y = \frac{Y}{X+Y+Z} \qquad z = \frac{Z}{X+Y+Z} \tag{4}$$

with

$$x + y + z = 1 \tag{5}$$

$\rho(\lambda)$ is the object reflectance weighted by the relative spectral energy distribution $S(\lambda)$ while $\bar{x}, \bar{y}$ and $\bar{z}$ are the color-matching functions (Wyzecki & Stiles 1982).

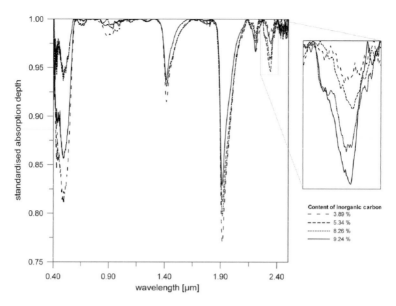

Figure 5.   Continuum removal normalized reflectance spectra (right) of soil samples with different inorganic carbon content (Jarmer et al. 2001).

In this color system the color intensity is characterized by the tristimulus value "Y" (luminance) which represents the brightness of color while "x" and "y" are the chromaticity coordinates (Wyzecki & Stiles 1982). The C.I.E. color coordinates were already found to contain substantial information for spectroradiometric detection of soil properties (e.g. Jarmer et al. 2001, Jarmer 2005).

Carbonates show diagnostic vibrational absorption bands in the Near Infrared at 2.30–2.35 μm and at 2.50–2.55 μm caused by the planar CO_3^{-2} ion (figure 4). Three weaker absorption bands occur at 1.85–1.87 μm, 2.12–2.16 μm and at 1.97–2.00 μm. The band positions in carbonates vary with the composition of different carbonates (Hunt & Salisbury 1971, Gaffey 1986). With increasing content of Mg-carbonate the wavelength position of maximum absorption is shifted to the shorter wavelength (Clark 1999).

Individual absorption features for the absorption band at 2.30–2.35 μm were calculated by continuum removal (figure 5). Derived parameters for this absorption feature are band position, maximum absorption depth, absorption width and area integrals.

Considering derived parameters of the carbonate absorption band and C.I.E. color coordinates a multiple regression model was performed to predicted inorganic carbon content. Validation of the model results were carried out by cross-validation. Cross-validated statistics were calculated based on estimates derived according to the 'leave-one-out-method' which means that each sample is estimated by means of an empirical-statistical model which was calibrated using the remaining (n-1) samples. To assess the cross-validated prediction accuracy of the approach the coefficient of determination (r_{cv}^2) and the root mean square error ($RMSE_{cv}$) were calculated. For the adequacy of the implemented regression model it was proved by t-test if the mean of the residuals is zero. To prove the normal distribution of the residuals Kolmogorov-Smirnov-test was applied and auto-correlation of model residuals was excluded by a Durbin-Watson-test.

2.5   Satellite image pre-processing

Reflectance spectrometry is accepted as a non-destructive and fast screening tool to assess soil properties (e.g. Couillard et al. 1997) but even with this technique the spatial detection of soil properties for broader and inaccessible areas exclusively based on terrestrial inquiry is rarely feasible. Hence, one major objective of this study is the spatial analysis of inorganic carbon content in the Judean Desert based on EOS data. The used Landsat-TM 5 image from 27th July 1998 was selected considering available meteorological data. A pronounced dry period prior to image acquisition was mandatory to guarantee optimized conditions for the detection of inorganic carbon since long dry periods insure minimum vegetation cover in semi-arid and arid regions. Furthermore, the influence of varying soil moisture contents on the reflectance signal can be rated as very low.

The satellite image has been geometrically corrected to the New Israeli Grid coordinate system based on the topographic map of the region at scale 1:50,000. A digital elevation model has been included to correct for pixel-dislocation due to elevation. The image was corrected for atmospheric effects by applying a radiative transfer model based on a modified 5S code which also accounts for terrain-induced illumination effects (Hill et al. 1995).

The focus of this study is the spatial detection of inorganic carbon. The spectral reflectance signature of soils is seriously influences by vegetation cover which modifies the signal of the litho- and pedological components. Since these components show a different spectral behavior than vegetation, an estimation and elimination of the spectral component of vegetation at the reflectance signal is well feasible by spectral mixture analysis (SMA). Therefore, based on SMA areas with vegetation cover of more than 20 percentages were excluded from further analysis.

3   RESULTS AND DISCUSSION

3.1   *Estimation of inorganic carbon from laboratory reflectance spectra*

Modeling inorganic carbon content of soil samples requires comparable lithological conditions for soil development, to guarantee that spectral characteristics resulting for carbonate occur due to varying concentrations instead of different chemical composition of the bedrock material. The study site is characterized by relatively homogeneous cretaceous bedrock and the developed soils show a similar soil texture (mostly silt and loam) (Jarmer 2005).

The intensity and the characteristics of an absorption band are mainly depending on the concentration of the mineral responsible for the absorption. To insure that no influence of different carbonates is given the absorption position was determined (table 1). Gaffey (1986) documented the absorption positions for Calcite (2.333–2.340 μm) and Dolomite (2.312–2.322 μm). The observed absorption position of the investigated soil samples stated that the carbonate exists as calcium carbonate (Calcite) (Jarmer 2005). Since the carbonate absorption is clearly detectable at the characteristic wavelength position, a relationship between carbonate content and the absorption feature is assumed.

Increasing inorganic carbon contents lead to an increase of the intensity of the carbonate absorption. A strong relationship is observed between the inorganic carbon content and the normalized absorption depth. While this relationship seems to be linear for lower concentrations, it is nonlinear when considering all soil samples (figure 6, left). Assuming a single reflection, the absorption intensity should with every involved mineral grain increase by the same amount and hence, a linear relationship should be the result. But this assumption does not correspond to reality since multiple reflections at the sample surface occur which do not have an additive component of given extent.

Table 1.   Wavelength position in the range of the inorganic carbon absorption band at 2.33 μm.

	n	position [μm]	mean [μm]	std.-dev.
all samples	53	2.333–2.355	2.342	$4.529 * 10^{-3}$
$C_{inorg} > 2\%$	50	2.335–2.353	2.341	$3.586 * 10^{-3}$
$C_{inorg} > 3\%$	44	2.335–2.346	2.341	$2.537 * 10^{-3}$
$C_{inorg} > 4\%$	37	2.335–2.345	2.340	$2.484 * 10^{-3}$
$C_{inorg} > 5\%$	27	2.335–2.345	2.340	$2.449 * 10^{-3}$

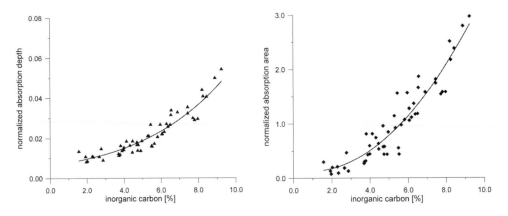

Figure 6. Inorganic carbon vs. normalized absorption depth (left) and vs. absorption area (right).

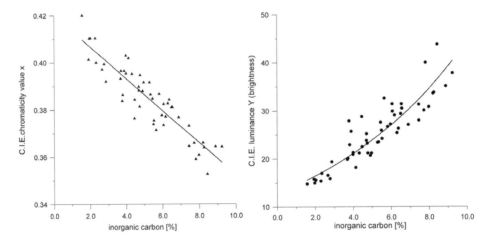

Figure 7. Inorganic carbon vs. C.I.E. chromaticity value x (left) and vs. C.I.E. tristimulus value Y (luminance) (right).

In this case the relationship between the concentration of a given mineral and its absorption intensity is codetermined by the spectral characteristics of its surroundings (Geerken 1991).

Analyzing the relationship between the inorganic carbon content and the calculated normalized absorption area leads to similar results examined for the normalized absorption depth. A similar non-linear relationship is documented for the absorption area of the carbonate absorption feature. Obviously, the absorption area increases disproportionately with higher C_{inorg}-concentrations (figure 6, right) due to the same reason described already.

In a first attempt the inorganic carbon contents in the soil samples were predicted based on derived absorption features. The best result was obtained using the "normalized absorption depth". This approach allowed to estimate inorganic carbon with a cross-validated accuracy of $r^2_{cv} = 0.876$ ($RMSE_{cv} = 0.674$). This accuracy already shows the potential of spectral determination of inorganic carbon.

In addition to the described effects of inorganic carbon concentrations on the intensity and characteristics the influence of inorganic carbon on the soil color was investigated. The relative amount of carbonate in soils influences soil brightness substantially (Ben-Dor & Banin 1994). Increasing inorganic carbon concentrations result in exponentially higher C.I.E. luminance values Y. Additionally, strong linear relationships were found for the C.I.E. chromaticity values x and y which substantially decrease with higher inorganic carbon concentrations (figure 7). The higher the content of inorganic

234

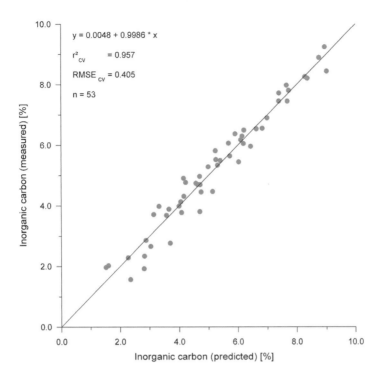

Figure 8. Scatterplot for cross validation of the generated model for predicting the inorganic carbon content.

Table 2. Comparison of predicting inorganic carbon results from laboratory reflectance measurements for the models based on absorption features only, on C.I.E. color coordinates and the combined approach.

C_{inorg} [%]	Absorption feature only	C.I.E. only	Absorption feature + C.I.E.
r^2_{cv}	0.876	0.851	0.957
$RMSE_{cv}$	0.674	0.749	0.405

carbon the closer the C.I.E. chromaticity values x and y are to the 'equal-energy stimulus' of the C.I.E. color system which has, of course the coordinates x = y = 1/3 (Judd & Wyszecki 1975).

These findings lead to the integration of C.I.E. color coordinates into the modeling. It was expected that the integration of the luminance "Y" representing the soil brightness would improve the modeling result substantially. However, it was found that the integration of the luminance "Y" did not have a positive effect on the modeling results while the integration of the C.I.E. chromaticity value "x" yielded a significantly higher fit. The final model is based on the variables "normalized absorption depth" and "C.I.E. chromaticity value x" and results in a r^2_{cv} of 0.957 with a $RMSE_{cv}$ of 0.405 for the cross validation:

$$C_{inorg}[\%] = 0.591 * \log(\text{absorption depth}) - 0.446 * \text{C.I.E.-x} \qquad (6)$$

The result of the cross-validation is shown in figure 8. The regression for the cross-validation is very close to the 1:1-line with a gain of almost one and an offset that can be neglected. This supplies evidence for the high accuracy of the model approach.

In table 2 the different predicting results for inorganic carbon from laboratory reflectance measurements are listed. As mentioned above best modeling results were obtained by the combination of absorption features and C.I.E. color information ($r^2_{cv} = 0.957$; $RMSE_{cv} = 0.405$). It is remarkable that based only on the C.I.E. color coordinates a high predicting accuracy with a r_{cv}^2 of 0.851 ($RMSE_{cv} = 0.749$) is achieved. This is especially important in the context of modeling the spatial distribution of inorganic carbon concentrations using a satellite image, because the C.I.E. color coordinates can be calculated from Landsat-TM channels as described in the following section.

3.2 *Mapping the spatial distribution of inorganic carbon from satellite data*

The limited spectral resolution of the Landsat image does not allow the exploitation of the carbonate absorption feature at 2.30–2.35 µm. Since an estimation of inorganic carbon based on C.I.E. color coordinates showed sound results (see section 3.1), these color coordinates form a suitable alternative for the spatial modeling of inorganic carbon. But to transform the satellite data into the C.I.E. color system some intermediate steps are necessary.

The reflectance values in the Landsat-TM channels TM1, TM2 and TM3 do not correspond to the C.I.E. tristimulus values B, G and R and need to be transformed by empirical relationships. These relationships are depending on the surveyed soils and application to other data sets is limited. While the calculation of C.I.E. color coordinates is done based on the color matching functions, the color information of Landsat-TM is determined by the sensitivity of the detectors (figure 9).

The needed empirical equations for transforming the satellite data into C.I.E. color coordinates were determined based on the laboratory spectra. For this the reflectance values of the three Landsat channels were calculated from the continuous reflectance measurements.

Based on this approach the C.I.E. tristimulus values X, Y and Z can be calculated from the satellite data using the following equations (Escadafal et al. 1988):

$$X = 2,7659R + 1,7519G + 1,1302B \tag{7}$$

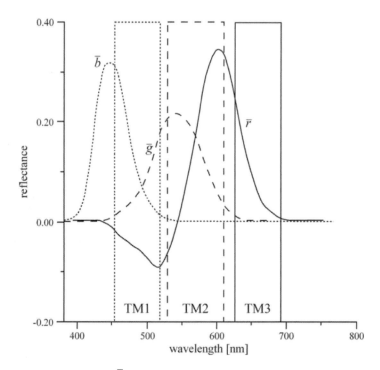

Figure 9.   Color matching functions $\bar{b}$, $\bar{g}$ and $\bar{r}$ of the C.I.E. standard observer and the wavelength ranges of the Landsat channels TM1, TM2 and TM3.

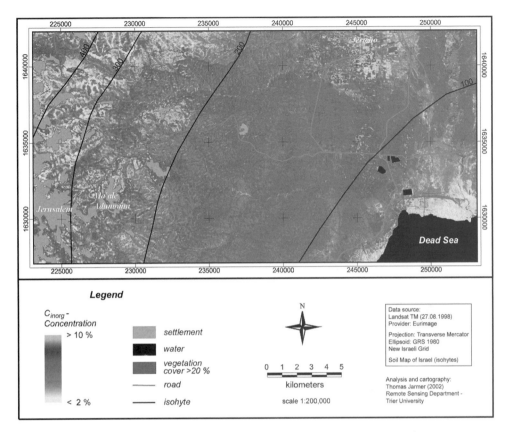

Figure 10.   Inorganic carbon contents at the study site derived from Landsat image applying the C.I.E. based model (see colour plate page 392).

$$Y = R + 4,5909G + 0,06012B \qquad (8)$$

$$Z = 0,0565G + 5,5944B, \qquad (9)$$

which allows the calculation of the C.I.E. chromaticity values x and y (see eq. 4).

It is remarkable that based only on the C.I.E. color coordinates a high predicting accuracy with a r_{cv}^2 of 0.851 ($RMSE_{cv} = 0.749$) is achieved. This is especially important in the context of modeling the spatial distribution of inorganic carbon concentrations using a satellite image, because the C.I.E. color coordinates can be calculated from Landsat-TM channels as described in the following section.

Since the carbonate absorption depth could not be calculated from Landsat-TM channels due to the limited spectral resolution, a model for predicting inorganic carbon concentration using only C.I.E. color coordinates was established:

$$C_{inorg}[\%] = -0.509 * C.I.E.-x + 0.444 * \log(C.I.E.-Y) \qquad (10)$$

The integration of the logarithm of the C.I.E. luminance Y instead of the original value takes into consideration the non-linearity of its relationship with the inorganic carbon concentration. Based on equation 10 the inorganic carbon concentration for the study site was predicted (figure 10). The explained variance of this model is distinctly lower than the model fit achievable when considering also the absorption feature parameters to predict inorganic carbon contents but the cross-validated

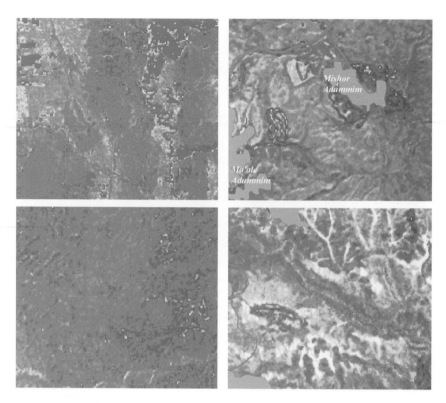

Figure 11. Inorganic carbon contents for a subset showing the Jordan valley (upper left), the settlement of Ma'ale Adummim and the industrial area of Mishor Adummim (upper right), the central part of the study area (lower left) and Wadis in the north-western part of the study site (lower right) (scale approx. 1:75.000; legend see figure 10) (see colour plate page 393).

determination coefficient of 0.851 is still very high. However, the $RMSE_{cv}$ is almost double compared to the original model. This can mainly be attributed to the wide concentration range with a maximum exceeding nine percent.

The model was directly applied to the satellite data and allows good estimates of inorganic carbon concentrations from the satellite image and represents the spatial dynamics of the study site. The concentration ranges estimated from the image data are in accordance with the concentration range of the chemical analysis (see section 2.3).

The weathering of limestone and the corresponding removal of carbonate is mainly influenced by the available amount of water. This process is accelerated by carbonic acid and organic acids as well as high temperatures. This explains the general trend of increasing inorganic carbon content with decreasing rainfall amount in the study site (see also Makhamreh 2006). This is shown in figure 10, but it also has to be considered that in the eastern part of the study site are quartary sediments (alluvials, Lisan marls) are the bedrock material. In this region (light blue) concentrations of above ten percentages are due to the different bedrock material. The inorganic carbon concentrations of two reference samples were determined with 5.5 percent. Since the introduced model was calibrated for soil samples developed on limestone parent material it has only limited significance for the real inorganic carbon concentrations of the surface soil material. Consequently, the local geogenic conditions cause an overestimation of inorganic carbon contents for the soils developed on quartary sediments.

A totally different situation is given for the area under agricultural use (south of Jericho and the Jordan valley) where predicted inorganic carbon contents with three to six percentages are relatively low (figure 11, upper left). Due to irrigation and water from the Jordan river more carbonate can be dissolved and discharged. This process leads to a reduction of inorganic carbon concentrations. Outside the area benefiting from the Jordan river water concentrations of inorganic carbon escalate.

Highest concentrations occur apart from the west-east-gradient in new build-up areas where the soil has been removed and the bare bedrock appears at the surface. Such high concentrations are given for the industrial area of Mishor Adummim and the extending settlement of Ma'ale Adummim. In the case of Ma'ale Adummim the newly set up street system is clearly visible (figure 11, upper right). In these areas in the maximum concentrations between ten and eleven are predicted which are comparable to the inorganic carbon concentrations of pure bedrock samples of around eleven percentages.

Beside such anthropogenic influenced locations, mainly the mountain ridge and hill tops in the center of the study site are characterized by high to very high estimates (figure 11, lower left). This seems to be contradictory to field survey results because in these positions on top of the hills often the best developed soils were found (compare figure 2). At the same time in these locations a high amount of stones and bedrock outcrops have been observed in these locations, which reaches shares between 45 and 80%. The bedrock material has the function of a sealed surface and results in a lateral surplus of rainfall which allows higher weathering rates and better soil development. Apart from this, well developed soils areas are very limited as most of the surface is covered by stones and bedrock material. As a consequence the reflectance signal of these areas is dominated by the limestone which results in high estimates of inorganic carbon concentrations of >8% maximum.

In the western part of the study site mainly lower to medium concentrations prevail with lowest estimates occurring in the Wadis, sometimes even below 3% which correspond to concentrations determined by chemical analysis of well developed soils in the study site. The reason is the better water supply and the resulting higher soil moisture compared to the slopes, which allows a substantially better soil development in the valleys (figure 11, lower right).

4 CONCLUSIONS

Inorganic carbon concentrations were predicted with high accuracy from laboratory reflectance measurements ($r_{cv}^2 > 0.95$) by combining the normalized absorption depth of the carbonate absorption feature at 2.33 μm and the C.I.E. chromaticity value x. It was found that the adaptation of the regression model to C.I.E. color values only also allows sound estimates of inorganic carbon contents ($r_{cv}^2 > 0.85$). The adaptation of the model made an up-scaling to satellite data possible and reflects the spatial pattern in the study site very well. Furthermore, the regression model allows sound estimates of the spatial distribution of inorganic carbon concentrations in the study site. Finally it is assumed that the introduced approach can also be used for other soil parameters and is applicable to other semi-arid and arid regions with similar physiogeographic conditions.

ACKNOWLEDGMENTS

This work was financially supported through the ERMES-II project (ENV4-CT95-0181), funded by the European Commission and the Forschungsfonds of the University of Trier. The computer code for calculating the C.I.E. colour values was kindly provided by R. Escadafal (Cesbio, France). This support is gratefully acknowledged.

REFERENCES

Baumgardner, M.F., Silva, L.F., Biehl, L.L. & Stoner, E.R. 1985. Reflectance properties of soils. *Advances in Agronomy* 38: 1–44.
Ben-Dor, E. & Banin, A. 1990. Near-infrared reflectance analysis of carbonate concentration in soils. *Applied Spectroscopy* 44(6): 1064–1069.

Ben-Dor, E. & Banin, A. 1994. Visible and near-infrared (0.4–1.1 μm) analysis of arid and semiarid soils. *Remote Sensing of Environment* 48: 261–274.

Boer, M., del Barrio, G. & Puigdefábregas, J. 1996. Mapping soil depth classes in dry Mediterranean areas using terrain attributes derived from digital elevation model. *Geoderma* 72: 99–118.

Clark, N. 1999. Spectroscopy of rocks and minerals and principles of spectroscopy. In A.N. Rencz (ed.), Remote Sensing for the Earth Sciences; Manual of Remote Sensing, Volume 3 (3rd Edition): 3–58. New York, Chichester, Weinheim, Brisbane, Singapore: Wiley.

Couillard, A., Turgeon A.J., Westerhaus, M.O. & Shenk, J.S. 1997. Determination of soil separates with near infrared reflectance spectroscopy. *Journal of Near Infrared Spectroscopy* 4: 201–212.

Dan, J., Yaalon, D.H., Koyumdjisky, H. & Raz, Z. 1976. The soils of Israel (with map 1:500.000). Bet Dagan, Israel.

Dan, J. & Smith, H. 1981. Soils of the Judean Desert, with special reference to those along the Teqoa-Mizpe Shalem road. In J. Dan, R. Gerson, H. Koyumdjisky, H. & D.H. Yaalon (eds.), Arid soils of Israel—Properties, genesis and management (=Agricultural Research Organization. Institut of Soils and water. Special Publication No. 190): 51–106. Bet Dagan, Israel.

Escadafal, R., Girard, M.-C. & Courault, D. 1988. Modelling the relationship between Munsell soil color and soil spectral properties. *International Agrophysics* 4(3): 249–261.

Gaffey, S.J. 1986. Spectral reflectance of carbonate minerals in the visible and near infrared (0.35–2.55 microns): calcite, aragonite, and dolomite. *American Mineralogist* 71: 151–162.

Gat, Z. & Karni, O. 1995. Climate and agrometeorology of the Jordan valley, adjacent Samaria slopes and Dead sea regions as a basis for agricultural planning and operation (ed.: Israel Meteorological Survice). Bet Dagan, Israel (in Hebrew).

Geerken, R. 1991. Informationspotential von hochauflösenden Fernerkundungsdaten für doe Identifizierung von Mineralen und Gesteinen. Laborversuche und Anwendungsbeispiele in der Geologie. Stuttgart.

Hill, J., Mehl, W. & Radeloff, V. 1995. Improved forest mapping by combining corrections of atmospheric and topographic effects in Landsat TM imagery. In J. Askne (ed.), Sensors and environmental applications in remote sensing; Proceedings of the 14th EARSeL Symposium, Göteborg, Sweden, 6–8 June 1994: 143–151.

Hill, J. & Schütt, B. 2000. Mapping complex patterns of erosion and stability in dry mediterranean ecosystems. *Remote Sensing of Environment* 74: 557–569.

Hill, J. 2000. Semiarid land assessment: monitoring dry ecosystems with remote sensing. In R.A. Meyers (ed.), Encyclopedia of Analytical Chemistry: 8769–8794. Chichester: Wiley.

Hunt, G.R. & Salisbury J.W. 1971. Visible and near-infrared spectra of minerals and rocks: II. Carbonates. *Modern Geology* 2: 23–30.

Jarmer, T. & Schütt, B. 1998. Analysis of iron contents in carbonate bedrock by spectroradiometric detection based on experimentally designed substrates. In M. Schaepmann, D. Schläpfer and K. Itten (eds.), 1st EARSeL Workshop on Imaging Spectroscopy: 375–382. Zurich.

Jarmer, T., Lavée, H., Hill, J. & Sarah, P. 2001. Spectral detection of inorganic carbon content along a semi-arid to hyper-arid climatic gradient in the Judean desert (Israel). In Proc. 2nd EARSeL Workshop on Imaging Spectroscopy, Enschede, 11–13. July 2000 (on CD).

Jarmer, T. 2005. Der Einsatz von Reflexionsspektrometrie und Satellitenbilddaten zur Erfassung pedochemischer Eigenschaften in semi-ariden und ariden Gebieten Israels. (=Trierer Geographische Studien, 29). Trier.

Judd, D.B.; Wyszecki, G. 1975. Color in business, science and industry. New York.

Kosmas, C.S., Curi, N., Bryant, R.B. & Franzmeier, D.P. 1984. Characterisation of iron oxide minerals by second-derivative visible spectroscopy. *Soil Sci. Soc. Am. J.* 48: 401–405.

Krishnan, P., Alexander, D.J., Butler, B. & Hummel, J.W. 1980. Reflectance technique for predicting soil organic matter. *Soil Sci. Soc. Am. J.* 44: 1282–1285.

Lavée, H., Imason, A.C. & Sarah, P. 1998. The impact of climate change on geomorphology and desertification along a mediterranean-arid transect. *Land degradation & development* 9: 407–422.

Morin, J., Sharon, D. & Rubin, S. 1988. Rainfall intensity in Israel, selected stations (=revised version of Research Report I/94). Bet Dagan, Israel.

Madeira, J., Bédidi, A., Cervelle, B., Pouget, M. & Flay, N. 1997. Visible spectrometric indices of hematite (Hm) and hoethite (Gt) content in lateritic soils: the application of a Thematic Mapper (TM) image for soil-mapping in Brasilia, Brazil. *IJRS* 18(13): 2835–2852.

Makhamreh, Z. 2006. Evaluation of soil quality and development stage using spectral reflectance of soils: case study in eastern Mediterranean region. In Proceedings of the International Conference "Soil and Desertification—Integrated Research for the Sustainable Management of Soils in Drylands", 5–6 May

2006, Hamburg, Germany (http://desertnet.de/proceedings/content/Makhamreh.pdf [access 15th August 2007])

Pickup, G. & Nelson, D.J. 1984. Use of Landsat radiance parameters to distinguish soil erosion, stability, and deposition in arid central Australia. *Remote Sensing of Environment* 16: 195–209.

Sarah, P. 2001: Soluble salts dynamics in the soil under different climatic regions. *Catena* 43(4): 307–321.

Survey of Israel (eds.) 1985: The Atlas of Israel (3rd ed.) (English-Hebrew). Tel Aviv, New York, London.

Udelhoven, T., Emmerling, C. & Jarmer, T. 2003. Quantitative analysis of soil chemical properties with diffuse reflectance spectrometry and partial-least-square regression: A feasibility study. *Plant and Soil*, 251(2): 319–329.

White, D.H. 2000. Drought policy, monitoring and management in arid lands. *Annals of Arid Zone* 39(2): 105–129.

Wyzecki, G. & Stiles, W.S. 1982. Color science. Concepts and methods, quantitative data and formulas. New York, London, Sidney: John Wiley & Sons.

Recent Advances in Remote Sensing and Geoinformation Processing
for Land Degradation Assessment – Röder & Hill (eds)
© 2009 Taylor & Francis Group, London, ISBN 978-0-415-39769-8

Simulating Multi-angle Imaging Spectro-Radiometer (MISR) sampling and retrieval of soil surface roughness and composition changes using a bi-directional soil spectral reflectance model

A. Chappell
School of Environment & Life Sciences, University of Salford, Manchester, UK

J.F. Leys
Department of Natural Resources, Gunnedah, NSW, Australia

G.H. McTainsh & C. Strong
Faculty of Environmental Sciences, Griffith University, Brisbane, Queensland, Australia

T.M. Zobeck
USDA, Cropping Systems Research Laboratory, Lubbock, Texas, USA

ABSTRACT: Soil surface changes due to soil erosion can be detected by ground-based, hyper-spectral measurements of angular reflectance and a bi-directional soil spectral reflectance model. The next generation of wind and water erosion models should incorporate directional remote sensing data to improve large area assessment more frequently in time. The utility of this approach was investigated by simulating the angular sampling of pre-defined soil surface spectral reflectance models using the configuration of the Multi-angle Imaging Spectro-Radiometer (MISR) sensor. At least two solar zenith angles, regardless of the number of solar azimuth angles, were required to simulate MISR overpasses and match the 'true' values. The simulated MISR parameter values were used to detect soil surface change after rainfall and aeolian abrasion. The coarse spectral resolution and range of the simulated MISR wavebands limited the inferences that were made about the soil surface changes compared to earlier work.

1   INTRODUCTION

The work on remote sensing of soil and in particular land degradation and soil erosion appears to have developed slowly despite a demonstrated need for large scale assessment of the soil resource for environmental protection and amelioration and which is perhaps best embodied by the considerable amount of literature on field measurements (Nickling et al. 1999, Zobeck et al. 2003) and on the development of wind erosion models (e.g., Wind Erosion Prediction System; Hagen et al. 1995). Shao et al. (1996) suggested that the main limitation of wind erosion models is their inability to incorporate the evolution of surface soil conditions. They provided one of the first physically-based wind erosion models to operate over scales from the field to the continent (Australia) because of, amongst other things, the inclusion of Normalised Difference Vegetation Index from remote sensing. There remains considerable potential for the inclusion of remote sensing data in future generations of soil erosion models. Many authors have shown (e.g., Baumgardner et al. 1985, Huete & Escadafal 1991, Ben-Dor et al. 1999, Galvão et al. 2001, Leone & Sommer 2000) that properties such as organic matter content, moisture content and mineral composition can be determined by examining the spectral response of soils. This can be achieved by examining the intrinsic optical properties of the soil whereby such measurements produce wavelength-specific absorption of electromagnetic radiation, yielding diagnostic reflectance spectra for the properties under investigation. Furthermore, there have been several key papers which have attempted to find spectral characteristics that were related to known soil erosion surfaces (Seubert et al. 1979, Latz et al. 1984). Later work has attempted to establish the relationship between spectral reflectance and soil properties by obtaining samples of both properties from the field for calibration. Many studies that derive soil physical and chemical properties from reflectance are based on empirical relationships between various waveband combinations and the properties of interest. There have been few studies that have considered the importance of the structural information about the surface that is available from directional sensors (Pinty et al. 1989). It appears that in most soil remote sensing studies that the soil properties are obtained from an unspecified support and to some depth. This approach is despite knowledge that the optical spectral reflectance is limited to the characteristics of the soil surface. However, it is well-known within soil science and soil geomorphology that the soil surface responds to a range of soil processes (e.g., eluviation, illuviation, evaporation etc.) and may cause the characteristics of the surface to be very different from those underlying and is evident from recent work on soil crusts and infiltration (Karnieli et al. 1999, Goldshleger et al. 2001, Ben-Dor et al. 2003). Perhaps more fundamentally, water and wind erosion are highly selective processes that remove preferentially a size of aggregate related to the fluid shear stress. Not only does this create a highly variable mosaic of soil surfaces, the processes upstream contribute to the formation of soil surfaces downstream.

Thus, there is a need for the next generation of wind and water erosion models to more easily incorporate remote sensing data and improve large area assessment more frequently in time. This demand requires information about the composition and structure of the soil surface and an understanding of how wind and water erosion and soil forming processes change the soil surface. In recent years several workers have made use of directional reflectance measurements and existing bi-directional soil spectral reflectance models to characterise the soil surface change during controlled wind tunnel and rainfall simulation experiments in the laboratory (Chappell et al. 2005, 2006) and in situ (Chappell et al. 2007). That work demonstrated the additional variability in soil spectral reflectance of a given soil type due to changes in the soil surface. It also demonstrated the ability to detect and associate changes at the soil surface with variation in model parameters. Their work was ground-based and used a fine spectral resolution radiometer and consequently there remain at least two fundamental questions surrounding the generalisation of their work for sensors on airborne or satellite platforms: (1) can the angular reflectance of spectral bands available from current directional sensors (e.g., Multi-angle Imaging SpectroRadiometer; MISR) provide the same explanations of soil surface processes as those established using ground-based angular hyperspectral information? (2) do the established ground-based relationships between angular reflectance and soil surface processes remain valid for larger pixel sizes typical of those available from airborne and satellite platforms. The latter question will not be addressed here. The former question is the focus of this paper and will be addressed by using the bi-directional soil surface spectral reflectance model parameter values for a range of soil surfaces from that previous work (Chappell et al. 2007) to simulate the MISR sensor viewing angles and wavebands and several combinations of solar azimuth and zenith angles. By regarding those predictions as reflectance samples by the MISR sensor, they will be inverted against a bi-directional soil reflectance model to retrieve the parameter values. The optimised parameter values will be compared against the 'true' optimised values and a relationship with soil surface treatments will be established and compared to that of our previous work using hyperspectral measurements (Chappell et al. 2007). An outline of experimental methods, measurements and models used in that previous in situ work is provided here to assist the reader in assessing the performance of the simulated MISR information.

2 DATA AND METHODS

The approach used here was developed to include simulations of natural environmental processes that were known to alter the soil surface (erodibility) condition. This provided an alternative to the traditional approach in which a relationship is developed between soil properties and spectral wavebands. The approach used here focuses attention on the soil surface and processes operating above and beneath it. Straightforward experiments involving rainfall simulators and wind tunnels were used to illustrate differences in soil material and soil surfaces during a treatment.

2.1 *In situ soil treatments*

The study area for these experiments was the Lake Constance claypan (or playa) in the Diamantina National Park (DNP), western Queensland, Australia (Figure 1). The playa lies within a currently active aeolian transport region (Nickling et al. 1999) and was the focus of a number of wind erosion experiments by the authors (e.g., McTainsh et al. 1999, Chappell et al. 2003a, b).

During this study (August and September 2002) the playa was largely bare with few areas of plant cover. The playa soil was classified as a brown vertosol with occasional alluvial and aeolian deposits of sand on top of the vertosol. Unconsolidated material on the surface formed a fine sandy veneer derived from alluvial deposition and near the playa margins dunes provided coarser red sands. Despite its propensity for slaking, the playa surface had a complex continuum of crusts that varied in thickness, strength and porosity from extremely loose, aerated and fragile at one extreme to thick, massive and hard packed at another. Two sites of bare soil on the claypan were chosen to characterise the extremes in the continuum of crusts and therefore in the erodibility. Details of the study area and the soil properties of the chosen sites can be found elsewhere (Chappell et al. 2007).

A field portable wind tunnel and rainfall simulator were transported to the sites and were used to reproduce aeolian abrasion and rainfall under controlled conditions. These treatments were known

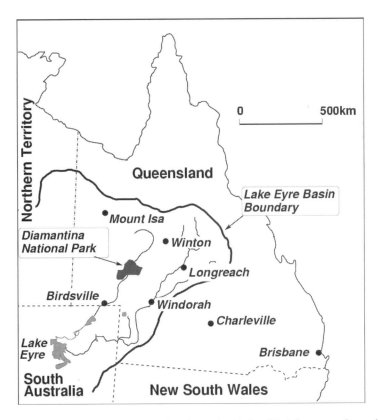

Figure 1.   The location of the study area within the Diamantina National Park in western Queensland, eastern Australia (modified from Butler et al. 2005).

245

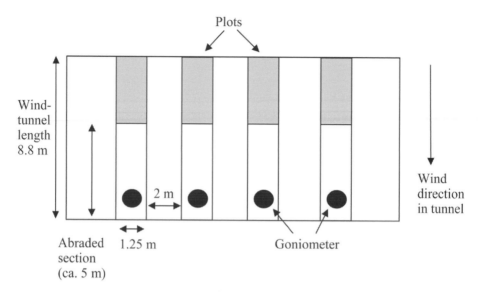

Figure 2. Schematic representation of the experimental plots at each of the two locations in the study area (modified from Chappell et al. 2007).

to alter the soil surface conditions and therefore to control the susceptibility of the surface to wind erosion. At each site, four plots were defined for use with the treatments (Figure 2).

The plots were used to consider the response to different combinations and intensities of the treatment. The wind tunnel was operated for 30 minutes at 14 m s^{-1} (measured at 0.3 m height) with abrader added at a constant rate (19 g m^{-1} s^{-1}). A rotating disc rainfall simulator was used to deliver rain drops to the plot surfaces over a period of five minutes. Collected rain water was applied continuously at 110 mm h^{-1} or 45 mm h^{-1} and created ponding at the surface, henceforth referred to as high and low intensity ponding rainfall (R110P and R45P, respectively). This type of treatment was used to represent the range of rainfall common in the study area. A second type of treatment was designed to avoid ponding of water at the surface using discontinuous application of rainfall and an increasing delay between applications to enable sufficient time for infiltration of water at the surface to occur. In this case the rainfall simulator was operated for a total of 5 minutes at 110 mm h^{-1} or 45 mm h^{-1}, respectively, but with a delay after each minute of rainfall. The delay between bursts of rainfall was for 2, 4, 8 and 16 minutes and these delays were kept constant regardless of the rainfall intensity. These treatments are henceforth referred to as high and low intensity non-ponding rainfall (R110NP and R45NP, respectively). The untreated soil surface was dry prior to conducting the experiments. After the rainfall treatments the soil was allowed to dry naturally for at least 48 hours after rainfall simulation. The clear skies and slightly windy conditions ensured highly evaporative conditions that dried rapidly the soil surface and ensured that the soil surface moisture was consistently small during the angular reflectance measurements.

2.2 *Directional reflectance measurements*

A spectroradiometer was used in the in situ experiments. It had a spectral range of 350–2500 nm and spectral sampling of 1.4 nm between 350–1050 nm and 2 nm between 1000–2500 nm. An eight degree field of view was used and measurements were conducted under clear sky conditions between approximately 1000 and 1600 hrs local time, each day. The exact time and location of measurement was recorded so that the solar zenith and azimuth angle could be determined. The solar zenith angle varied between 20° and 65° depending on the time of measurement.

A goniometer was used to allow repeatable and consistent measurements of multi-angular reflectance of the soil surface from a constant distance (ca. 35 cm) and enabled several view zenith and azimuth angles. Some view zenith angles were omitted to avoid the inclusion of shadow cast by the goniometer. Details of the measurement geometries for directional spectral data can be found elsewhere (Chappell et al. 2007). A calibrated Spectralon panel was used to reduce the sensitivity of the processed measurements to the characteristics of the source of illumination. Two Spectralon reflectance reference measurements were made immediately before and after the target measurements under the same conditions as the measurement. Conversion to spectral reflectance was conducted by dividing the reflectance spectra of the soil samples by the spectra of the highly reflecting white Spectralon reference panel.

Since the measurements were performed in the field, reflectance at wavebands between 1350–1470 nm and between 1790–1980 nm were removed from the analysis because they were influenced by atmospheric water vapour and heat fluctuations. Diffuse irradiance should also be removed from the measurements otherwise the retrieved values of the parameters might incorporate information on the soil bi-directional reflectance and the illumination conditions and hence would not be atmospherically invariant (Privette et al. 1995). However, the inclusion of an isotropic diffuse irradiance formulation added to the soil model of Jacquemoud et al. (1992) was found by Privette et al. (1995) to be of limited use. Consequently diffuse irradiance was not included in the model used here and the measurement of reflectance on clear sky days was assumed to contribute little to the retrieved values of the bi-directional reflectance parameters.

2.3 *Directional reflectance model and parameters*

Using the fundamental principles of radiative transfer theory, Hapke (1963, 1981, 1986) derived an analytical equation for the bi-directional reflectance function of a medium composed of dimensionless particles. Pinty et al. (1989) extended that work to describe soil surfaces on Earth where individual particles have non-uniform angular distributions. Jacquemoud et al. (1992) provided a simplified formulation which required six parameters and attempted an explanation of both backward and forward scattering (the specular effect) of light by smooth soils. Thus, the model and its parameters are likely to be useful for characterising the soil surface and changes to it. It is used here and described below.

Hapke's model assumed that a plane surface at $z = 0$ contained irregular and randomly orientated particles that are large compared with the wavelength. The bi-directional reflectance r of a surface illuminated with a zenith angle i, viewed from a zenith angle e and normalised with respect to the reflectance of a perfectly reflecting Lambertian surface under the same conditions of illumination and observation is given by:

$$r(i, e, \phi) = \frac{\omega}{4} \frac{1}{4 \cos i + \cos e} \left\{ [1 + B(g)]P(g, g') + H(\cos i)H(\cos e) - 1 \right\} \tag{1}$$

where,

$$\cos g = \cos i \cos e + \sin i \sin e \cos \phi,$$

$$\cos g' = \cos i \cos e - \sin i \sin e \cos \phi,$$

$$B(g) = \frac{1}{1 + (1/h)\tan(g/2)},$$

$$P(g, g') = 1 + b \cos g + c \frac{3\cos^2 g - 1}{2} + b' \cos g' + c' \frac{3\cos^2 g' - 1}{2},$$

and

$$H(x) = \frac{1 + 2x}{1 + 2\sqrt{(1 - \omega)x}}.$$

247

In equation 1, ϕ is the viewing azimuth relative to the Sun's azimuth, ω is the single scattering albedo (the ratio of the scattered energy to the total energy either scattered or absorbed by the particle), g is the phase angle between the incoming and outgoing light directions, g' is the angle between the specular and the outgoing light directions. The function $B(g)$ explains backscattering of light as a function of g and a roughness parameter h which may be related to the grain size distribution, the porosity of the medium and gradient of compaction with depth (Hapke 1963). In other words, as the surfaces become smooth, h increases. The type of scattering is related to the surface roughness and the nature of the particles. The phase function $P(g,g')$ describes the angular distribution of the light scattered by a terrestrial surface. The term $H(\cos i)H(\cos e)$-1 approximates the contribution from multiple scattering following Pinty et al. (1989) and x is used to indicate the substitution of the cos i and cos e terms into the equation, respectively.

2.4 Application of the model

If the radiometric properties of a bare soil surface can be described by equation 1 the interpretation problem consists of finding the values of the six parameters such that the computed value of r best approximates the actual observations. Following previous workers (Pinty et al. 1990, Jacquemoud et al. 1992), a non-linear least squares fitting procedure was used to solve the inverse problem:

$$\delta^2 = \sum_{k=1}^{n} [r_k - r(i_k, e_k, g_k)]^2 \tag{2}$$

where r_k is the measured bi-directional reflectance of the surface for the relative geometry of illumination and observation defined by i_k, e_k, g_k and r is the predicted bi-directional reflectance. The problem then reduces to finding the optimal values of the parameters which minimises δ^2 for a set of observations. The performance of the optimisation is judged using the square root of mean squared difference (RMSE) $\sqrt{\delta^2/N_f}$, where N_f is the number of degrees of freedom, which is the number of independent data points minus the number of parameters estimated by the procedure. This inverse modelling problem was solved using either a Gauss-Newton or Levenberg-Marquardt method, depending on performance subject to fixed upper and lower bounds on the independent variables using function values alone.

The sensitivity of the model parameters for bi-directional reflectance using this model inversion approach was investigated by Pinty et al. (1989) and followed by Jacquemoud et al. (1992). The final sensitivity experiment performed by Pinty et al. (1989) was replicated by Chappell et al. (2006) and showed that although some uncertainty remained, the highly resolved spectral information and a high overall accuracy of the estimation procedure (0.012 average of the RMSE) was believed to be sufficient for the accurate retrieval and interpretation of the model parameter values.

2.5 Simulating Multi-angle Imaging Spectro-Radiometer (MISR) angular reflectance

Angular sampling of previously modeled reflectance was conducted in a similar manner to that of Pinty et al. (1989). They used arbitrary models and sampled the reflectance repeatedly to consider the average behaviour of the sampling configuration. Here we used the bi-directional soil spectral reflectance models previously obtained for each soil surface using ground-based hyperspectral data (Chappell et al. 2007). The soil surface models were sampled using the configuration of the MISR sensor which has nine view angles (0, 26.1, 45.6, 60.0 and 70.5° in both directions). In addition, five solar zenith angles (SZA; 30, 40, 50, 60 and 70°) and azimuth (relative to the view azimuth) angles (SAA; 0, 45, 90, 135 and 180°) were sampled. The SZA and SAA angles were varied systematically but all view angles were included in each sampling configurations. The use of additional SZA and SAA angles are a simulation of the inclusion of multiple images from repeated overpass with this satellite sensor. Reflectance was sampled using the MISR wavebands (425–467, 544–573, 661–683, 846–886 nm) centred at 446, 558, 672 and 867 nm.

2.6 *Canonical ordination (Redundancy analysis) of soil spectral reflectance*

Redundancy analysis (RDA) was used here to establish the relations between model parameter values, soil type, soil treatment and the single scattering albedo for simulated MISR wavebands. The technique was used by Chappell et al. (2005, 2006) with considerable success to simplify the relationship between spectral reflectance changes with treatment. As with that previous analysis, the program CANOCO (version 4.02, ter Braak 1988) was used for RDA. The model parameter values were not transformed. Samples were standardised and wavebands were centred and standardised. A focus on scaling of inter-sample distances was used to interpret the relationships among model parameter values from the ordination diagram. The model parameter values were divided (after extraction of their axes) by their standard deviations so that the ordination diagram displayed standardised reflectance data and correlations instead of covariances. A vector drawn from the origin of the ordination diagram to each model parameter represents the fit (correlation) with the ordination axis, i.e., vectors close to an axis are highly correlated with the information extracted by that axis. The treatment indicators were represented on the ordination diagram by plotting the centroids of their samples. Similarity between the model parameter vectors and those of the treatment centroids quantify the correlation between the two types of data. Vectors oriented in approximately the same direction indicate strong positive correlation, whilst those oriented in opposite directions imply strong negative correlation. Vectors orthogonal to one another were uncorrelated.

The most significant predictive variables were selected after the removal of redundant variables with large multi-collinearity. A forward selection procedure which used a Monte Carlo permutation test at each selection-step was used to test the statistical significance of the variance explained by each variable added to the model (ter Braak 1988).

3 RESULTS

3.1 *Simulating Multi-angle Imaging Spectro-Radiometer (MISR) angular reflectance*

An example of the sampling of solar zenith angle only is shown in figure 3. The hyperspectral model values for all surfaces are plotted against the simulated MISR model values for each parameter (b, c, h, b', c'). A linear regression model was fitted using least squares to each sampling configuration of the model parameters. The proximity of that model to a 1:1 line and the intercept on the y-axis provides an impression of bias in the configuration and the R^2 statistic also provides an assessment of scatter in the reflectance model values about the regression model. Table 1 shows some of the sampling configurations for SZA and SAA and goodness of fit statistics that exceed an $R^2 = 0.5$. Also included in that table are F statistics and p-values for the regression model parameters. These last statistics show that almost all of the regression model parameters are highly statistically significant at the 95% level of significance. Those most significant parameters are found for three SZAs (30, 40 and 50°) and one SAAs (0°, principal plane) and coincide with large R^2 values and have slope gradients close to 1 and small intercepts on the y-axis. In other words this sampling configuration appears to be the best at matching the original bi-directional soil surface spectral reflectance model parameter values.

3.2 *Optimised bi-directional soil spectral reflectance values*

The 'best' configuration of soil surface view and solar zenith and azimuth angles identified in the previous sampling experiment was used in the model inversion procedure. The samples were stratified using site (soil type), plot and treatment. Figure 4 shows the results of the measured spectra (MISR bands) at site A and site B against the calculated spectra for all treatments. The results showed a very good agreement (RMSE = 0.01) and plot along the 1:1 line. The optimised values of each model parameter for each site, plot and treatment and an assessment of accuracy are shown in Table 2. The values of the bi-directional reflectance parameters represent the response of the soil surfaces to the controlled processes of aeolian abrasion, rainsplash and evaporation etc.

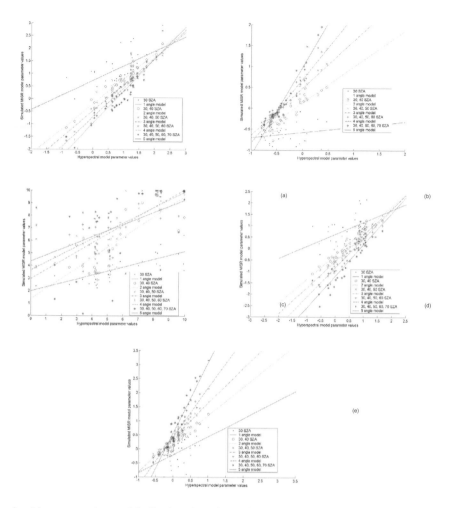

Figure 3. Linear regression models fitted to the values of bi-directional soil spectral reflectance model parameters b (a), c, (b), h (c), b′ (d), c′ (e) from hyperspectral data and simulated MISR view angles and solar zenith angles (SZA) (see colour plate pages 394 and 395).

This assumes that the soil reflectance model is appropriate and that the effect of reflectance sampling and illumination conditions is removed.

Values of the *h* parameter appear to be sensitive to the treatments. In general, values of *h* are larger (indicating smoother surfaces) for untreated soil surfaces, or those treated with rainfall, than those affected by abrasion. The values of *h* appear to identify the untreated surfaces of plots at site A as much smoother (larger values of *h*) than those at site B. In general, rainfall appeared to make the soil surfaces at both sites much smoother than the previous treatment. However, there were some plots in which rainfall caused the surface to become rougher than the previous treatment presumably as a consequence of the type of rainfall and the combination of antecedent conditions. It is these subtle changes in the soil surfaces which the canonical ordination was used to elucidate.

3.3 Single scattering albedo (SSA) spectra

The single scattering albedo (SSA; ω) spectra for simulated MISR wavebands calculated by the model for both sites prior to any treatment are shown in figure 3a and 3b. Also shown in figure 5 are the SSA spectra for the calculated hyperspectral curves. As expected the spectra for the MISR

Table 1. Indicators of locations, plots and treatments for samples used as 'environmental' variables for the canonical ordination (redundancy analysis).

Sample	Site A	B	Plots 1	2	3	4	Treatments A1	A1	R1	R2	R3	R3	R4	R4	N
110	1	0	1	0	0	0	0	0	0	0	0	0	0	0	0
111	1	0	1	0	0	0	1	0	0	0	0	0	0	0	0
112	1	0	1	0	0	0	1	0	1	0	0	0	0	0	0
113	1	0	1	0	0	0	1	1	1	0	0	0	0	0	0
114	1	0	1	0	0	0	1	1	1	0	1	0	0	0	0
115	1	0	1	0	0	0	1	1	1	0	1	0	0	0	1
120	1	0	0	1	0	0	0	0	0	0	0	0	0	0	0
121	1	0	0	1	0	0	1	0	0	0	0	0	0	0	0
122	1	0	0	1	0	0	1	0	0	1	0	0	0	0	0
123	1	0	0	1	0	0	1	1	0	1	0	0	0	0	0
124	1	0	0	1	0	0	1	1	0	1	0	0	1	0	0
125	1	0	0	1	0	0	1	1	0	1	0	0	1	0	1
130	1	0	0	0	1	0	0	0	0	0	0	0	0	0	0
131	1	0	0	0	1	0	0	0	0	0	1	0	0	0	0
132	1	0	0	0	1	0	1	0	0	0	1	0	0	0	0
133	1	0	0	0	1	0	1	0	0	0	1	1	0	0	0
135	1	0	0	0	1	0	1	0	0	0	1	1	0	0	1
140	1	0	0	0	0	1	0	0	0	0	0	0	0	0	0
141	1	0	0	0	0	1	0	0	0	0	0	0	1	0	0
142	1	0	0	0	0	1	1	0	0	0	0	0	1	0	0
143	1	0	0	0	0	1	1	0	0	0	0	0	1	1	0
144	1	0	0	0	0	1	1	1	0	0	0	0	1	1	0
145	1	0	0	0	0	1	1	1	0	0	0	0	1	1	1
210	0	1	1	0	0	0	0	0	0	0	0	0	0	0	0
211	0	1	1	0	0	0	1	0	0	0	0	0	0	0	0
212	0	1	1	0	0	0	1	0	1	0	0	0	0	0	1
220	0	1	0	1	0	0	0	0	0	0	0	0	0	0	0
221	0	1	0	1	0	0	1	0	0	0	0	0	0	0	0
222	0	1	0	1	0	0	1	0	0	1	0	0	0	0	1
230	0	1	0	0	1	0	0	0	0	0	0	0	0	0	0
231	0	1	0	0	1	0	0	0	0	0	1	0	0	0	0
232	0	1	0	0	1	0	1	0	0	0	1	0	0	0	1
240	0	1	0	0	0	1	0	0	0	0	0	0	0	0	0
241	0	1	0	0	0	1	0	0	0	0	0	0	1	0	0
242	0	1	0	0	0	1	1	0	0	0	0	0	1	0	1

A1 = Wind tunnel operated for 30 minutes at 14 m s-1 (at 30 cm height) with 19 g m^{-1} s^{-1}. R1 = Rainfall simulator set at 110 mm h^{-1} and operated discontinuously for 5 minutes at 1 minute intervals with increasing delays between simulations (i.e. 2, 4, 8 and 16 minute delays). R2 = Rainfall simulator set at 45 mm h^{-1} and operated discontinuously for 5 minutes at 1 minute intervals but with increasing delays between simulations (i.e. 2, 4, 8 and 16 minute delays). R3 = Rainfall simulator set at 110 mm h^{-1} and operated continuously for 5 minutes. R4 = Rainfall simulator set at 45 mm h^{-1} and operated continuously for 5 minutes. N = Breakdown in the radiometric equipment curtailed the experiments and allowed undefined natural environmental processes to act on the surface before the reflectance was repeated.

bands are similar to those of the hyperspectral data. At site A (Figure 5a) the MISR spectra of the untreated plots resemble those of the hyperspectral data; plot 1 and plot 3 are very different from plots 2 and 4. This spectral difference has occurred despite the plots being adjacent to each other and demonstrates the highly spatially variable nature of the soil surface, the sensitivity of the ground-based spectroradiometer to differences at the surface and the potential of the MISR sensor to detect such differences. However, the simulated MISR spectra do not display the same magnitude of variation evident in the hyperspectral data. This underestimate is particularly evident at site B (Figure 5b) where the calculated SSA hyperspectral data for plot 1 are considerably larger than that of the simulated MISR bands.

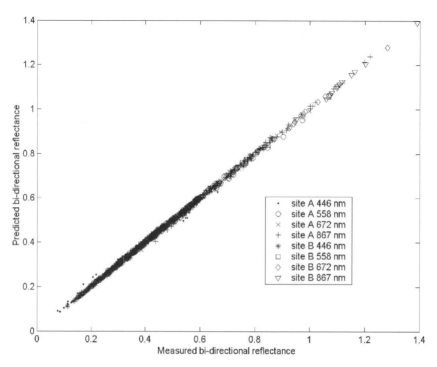

Figure 4. Comparison between measured and calculated bi-directional spectra (450–2450 nm) for soil at site A (a) and B (b) for all treatments.

3.4 Canonical ordination (Redundancy analysis)

Ordination analyses with hyperspectral waveband SSA and model parameter values showed that information from the untreated stage of the plots and from the final and much later reflectance measurements diminished the strength of the relationships (Chappell et al. 2007) and they were removed from the final analyses described here. The results of the redundancy analysis (RDA) between simulated MISR waveband SSA, model parameter values and the remaining soil surface treatments calculated separately for each combination and for both sites (A and B) are shown in Table 4.

The eigenvalues measure the importance of each of the canonical axes. In the case of the simulated MISR waveband SSA, and the soil treatments, the first axis explains 49% of the variation whilst the second axis explains only 7%. Thus, 53% of the variation in the data is explained by the first two axes. In addition, the relationship between the wavebands and the treatments for each axis is very strong. The amount of variation between the wavebands and the treatments is explained by each axis and is given as a cumulative percentage on the bottom line of Table 3. Approximately 88% of the variation is explained by the first axis and the second axis provides only an additional 12%.

In the case of the model parameter values and the soil surface treatments, the first axis explains 37% of the variation whilst the second axis explains 22%. Thus, 59% of the variation in the data is explained by the first two axes. Approximately 59% of the variation is explained by the first axis and the second axis provides an additional ca. 33%. When the simulated MISR waveband SSAs and model parameter values are combined and compared against the soil surface treatments the same amount of variation in the data is explained as the previous RDA. However, a larger amount of variation in the treatment is explained by the first axis (71%) but a smaller amount is explained by the second axis (ca. 17%) relative to the previous RDA. It appears from these results that when combined, the simulated MISR wavebands and the model parameters contained redundant

Table 2. Goodness of fit results of sampling previously calculated bi-directional soil surface spectral reflectance models using three solar zenith angle (SZA) and solar azimuth angle (SAA) combinations. Bold indicates the best fit using a combination of regression parameter values and significant test results.

Number of angles		Model	Regression parameters			Significance test	
SZA	SAA	parameters	intercept	gradient	R^2	F-statistic	p-value
30, 40	0	b	−0.20	0.99	0.93	458.9	0.00
30, 40	0	c	0.13	0.86	0.86	195.7	0.00
30, 40	0	h	1.34	0.82	0.67	67.6	0.00
30, 40	0	b′	−0.27	1.06	0.93	434.7	0.00
30, 40	0	c′	0.19	1.01	0.88	232.9	0.00
30, 40	0, 45	b	−0.33	0.88	0.92	370.9	0.00
30, 40	0, 45	c	0.26	0.75	0.88	250	0.00
30, 40	0, 45	h	2.86	0.69	0.73	87.9	0.00
30, 40	0, 45	b′	−0.28	0.79	0.92	373.2	0.00
30, 40	0, 45	c′	0.29	0.82	0.93	437.4	0.00
30, 40	0, 45, 90	b	−0.29	0.69	0.79	123.1	0.00
30, 40	0, 45, 90	c	0.31	0.62	0.92	362.9	0.00
30, 40	0, 45, 90	h	3.76	0.60	0.73	88.6	0.00
30, 40	0, 45, 90	b′	−0.16	0.52	0.68	70.3	0.00
30, 40	0, 45, 90	c′	0.28	0.63	0.94	482.1	0.00
30, 40, 50	0	b	**−0.47**	**1.04**	**0.98**	1511.5	0.00
30, 40, 50	0	c	**0.48**	**1.13**	**0.98**	1873.2	0.00
30, 40, 50	0	h	**2.96**	**0.71**	**0.64**	57.7	0.00
30, 40, 50	0	b′	**−0.55**	**1.15**	**0.98**	1381.3	0.00
30, 40, 50	0	c′	**0.40**	**1.24**	**0.97**	1240.5	0.00
30, 40, 50	0, 45	b	−0.53	0.94	0.94	514.7	0.00
30, 40, 50	0, 45	c	0.47	0.91	0.98	1525.8	0.00
30, 40, 50	0, 45	h	4.16	0.60	0.70	78.2	0.00
30, 40, 50	0, 45	b′	−0.51	0.91	0.94	541.4	0.00
30, 40, 50	0, 45	c′	0.41	1.04	0.97	966	0.00
30, 40, 50	0, 45, 90	b	−0.42	0.66	0.81	136.2	0.00
30, 40, 50	0, 45, 90	c	0.46	0.62	0.97	965.5	0.00
30, 40, 50	0, 45, 90	h	5.15	0.52	0.70	77	0.00
30, 40, 50	0, 45, 90	b′	−0.33	0.57	0.75	101.4	0.00
30, 40, 50	0, 45, 90	c′	0.41	0.71	0.94	546.1	0.00

information. The RDA also identified only four treatments as the important variables explaining the variation in the model parameter values (Table 5).

Figure 6 provides a visual explanation for the model parameter values against soil treatments. The model parameter b', c' and h are aligned with axis 1. This axis is itself largely explained by the continuous application of low intensity continuous rainfall and high intensity discontinuous rainfall. The second axis is very strongly correlated with the b and (to a lesser extent) c model parameters. The interpretation of the second axis is complicated by the absence of any strong alignments of the treatments. However, abrasion and high intensity continuous rainfall appear to be important.

4 DISCUSSION

4.1 Simulating Multi-angle Imaging Spectro-Radiometer (MISR) soil surface samples

The combination of nine view angles, three solar zenith angles and one solar azimuth angle (relative to the view azimuth angle) of the MISR sensor appears to provide the best match of the bi-directional soil surface spectral reflectance model parameter values. There is only a small reduction in goodness of fit between the 'true' model parameter values and those optimised with the simulated MISR characteristics when two solar azimuth angles are used.

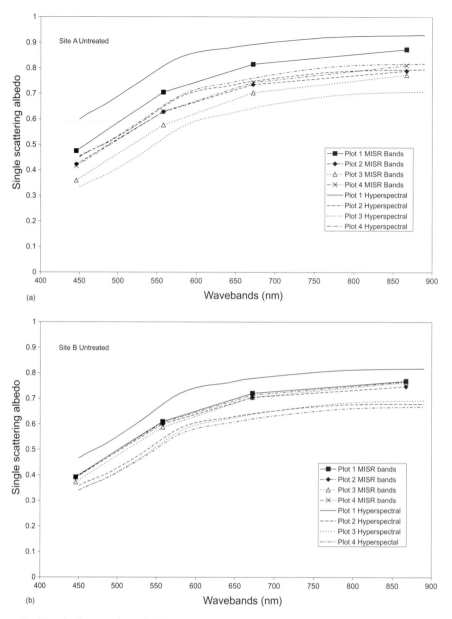

Figure 5. The single scattering albedo (SSA) estimated using the bi-directional model of soil spectral reflectance for the untreated surface of soil at site A (a) and site B (b).

The results suggest that a single overpass of the MISR sensor, using nine view angles and only one solar zenith and azimuth angle provides a poor retrieval of the 'true' reflectance model parameter values and that at least two overpasses (with different solar zenith and azimuth angles) are required. It is notable that more solar azimuth angles do not improve the retrieval of the roughness (*h*) parameter as expected. The reason for this is uncertain and may not be a generally applicable result but a consequence of the soil surface reflectance models used here.

The reflectance model parameter values optimised for the best MISR sampling configuration were used to reconstruct the single scattering albedo (SSA) and were compared to those of the hyperspectral ground-based soil surface reflectance models (Figure 5). As suggested by the reflectance

Table 3. Optimised values of the parameters from Jacquemoud et al. (1992) bi-directional soil spectral reflectance model for the treatments applied to each plot at two sites using simulated MISR spectra.

Sample	Site	Treatment	Bi-directional reflectance model parameters					
			b	c	h	b$'$	c$'$	RMSE
110	A	Untreated	0.85	2.22	3.42	2.65	1.36	0.01
111	A	Abrasion	1.62	0.09	1.39	1.74	0.82	0.00
112	A	R110NP	2.82	− 0.35	3.29	0.29	1.18	0.00
113	A	Abrasion	1.13	− 0.27	0.98	1.57	1.11	0.00
114	A	R110P	− 0.02	2.63	4.37	0.31	2.62	0.02
115	A	Natural	1.91	− 0.08	1.72	2.81	0.01	0.00
120	A	Untreated	2.62	0.10	2.79	0.66	1.32	0.00
121	A	Abrasion	1.30	0.27	1.34	2.04	0.69	0.00
122	A	R45NP	1.15	1.63	2.71	1.17	1.94	0.01
123	A	Abrasion	− 0.56	1.55	1.96	2.37	0.97	0.01
124	A	R45P	2.47	0.27	3.52	− 0.31	1.92	0.01
125	A	Natural	1.76	1.29	2.98	0.60	2.38	0.01
130	A	Untreated	1.58	0.04	1.40	1.95	0.83	0.00
131	A	R110P	2.08	1.46	2.82	2.46	1.40	0.01
132	A	Abrasion	1.42	− 0.43	1.16	1.92	0.66	0.00
133	A	R110P	1.15	0.68	1.48	1.26	0.88	0.00
135	A	Natural	2.19	− 0.31	2.33	0.39	0.86	0.00
140	A	Untreated	3.00	− 0.17	2.84	1.03	1.13	0.01
141	A	R45P	3.13	− 0.60	3.61	0.72	0.67	0.00
142	A	Abrasion	2.91	1.26	3.81	0.93	2.97	0.01
143	A	R45P	3.30	0.23	4.01	0.11	2.48	0.01
144	A	Abrasion	1.35	0.01	1.25	1.64	0.93	0.00
145	A	Natural	0.93	0.92	1.42	1.82	0.74	0.01
210	B	Untreated	1.83	0.77	2.10	1.34	1.53	0.01
211	B	Abrasion	2.23	− 0.12	1.65	1.80	0.91	0.00
212	B	R110NP+Natural	1.10	1.62	2.87	0.95	2.17	0.01
220	B	Untreated	1.45	0.04	1.30	2.03	0.65	0.00
221	B	Abrasion	1.78	− 0.26	1.32	2.34	0.25	0.00
222	B	R45NP+Natural	2.11	− 0.24	1.88	0.83	0.95	0.00
230	B	Untreated	1.83	− 0.21	1.40	2.00	0.58	0.00
231	B	R110P	3.19	1.79	3.90	3.00	2.23	0.01
232	B	Abrasion+Natural	2.19	− 0.71	1.69	1.05	0.65	0.00
240	B	Untreated	1.96	− 0.34	1.51	2.34	0.12	0.00
241	B	R45P	4.02	− 0.11	4.69	0.31	2.37	0.01
242	B	Abrasion+Natural	2.10	− 0.59	1.55	1.00	1.02	0.00

Abrasion = saltation of particles in a wind tunnel operated for 30 minutes at 14 m s-1 (at 30 cm height) with 19 g m^{-1} s^{-1}. R110NP = Rainfall simulator set at 110 mm h^{-1} and operated discontinuously for 5 minutes at 1 minute intervals with increasing delays between simulations (i.e. 2, 4, 8 and 16 minute delays). R45NP = Rainfall simulator set at 45 mm h^{-1} and operated discontinuously for 5 minutes at 1 minute intervals but with increasing delays between simulations (i.e. 2, 4, 8 and 16 minute delays). R110P = Rainfall simulator set at 110 mm h^{-1} and operated continuously for 5 minutes. R45P = Rainfall simulator set at 45 mm h^{-1} and operated continuously for 5 minutes. Natural = Breakdown in the radiometric equipment curtailed the experiments and allowed undefined natural environmental processes to act on the surface before the reflectance was repeated.

comparisons shown in figure 4, there is good similarity in the SSA. However, there is a tendency for the MISR SSA to underestimate the variability inherent, particularly between plots, in the hyperspectral data. The implication of this result is that the retrieval process from the MISR sensor is likely to provide conservative estimates of the variation in soil surface spectral reflectance. This finding is also supported by the redundancy analysis which showed little benefit of including the SSA values for the simulated MISR wavebands (Table 5). Furthermore, the RDA of wavebands against treatment provided a limited and one-dimensional (single axis) explanation of the variation. In contrast, the reflectance model parameters spread the explanation across both axes and therefore provided greater dimensionality than using only the wavebands. The model parameters

Table 4.   Ordinary Redundancy Analysis to explain separately the variation in selected wavebands and soil spectral model parameters using sites (soil type) and treatments.

	Wavebands and treatments		Model parameters and treatments		Wavebands, parameters and treatments	
	1	2	1	2	1	2
Axes						
Eigenvalues	0.49	0.07	0.37	0.22	0.37	0.22
Treatment correlations	0.73	0.88	0.91	0.76	0.92	0.76
Cumulative percentage variance						
of waveband/parameter data	48.9	55.7	37.3	58.8	51.5	64.1
of treatment relations	87.5	99.5	58.5	92.2	71.3	88.7

Table 5.   Explained variance, P-values and F statistics in RDA with forward selection removal of multi-collinearity for predicting model parameter values (ranked according to the variance explained during forward selection). Shading indicates those variables that are considered not statistically significant by using a p-value threshold around 10%.

Treatment variables	Explained variance (%)	P-value	F statistic
1st R45P	0.16	0.02	3.19
1st R110P	0.12	0.02	3.93
2nd Abrasion	0.12	0.05	3.10
R110NP	0.10	0.10	2.44
R45NP	0.11	0.12	2.50
1st Abrasion	0.01	0.68	0.53
Site A	0.01	0.87	0.22
2nd R45P	0.00	0.93	0.13
2nd R110P	0.01	0.96	0.08

and the treatments explained a large proportion (59%) of the variation in the soil surface and this is consistent with the findings of earlier work (Chappell et al. 2006, Chappell et al. 2007).

The detailed interpretation of the redundancy analysis supports the earlier observations of the variation in the model parameters. High intensity discontinuous rainfall and low intensity continuous rainfall appears to produce a smooth surface. This is indicated by a proportional relationship between the h parameter and microscopic roughness (large h values are smooth). It is reasonable to expect that aeolian abrasion is inversely proportional to the h parameter. However, that is not the case here. The abrasion treatments and those rainfall treatments (continuous high intensity) that also produced a rough surface are closely aligned to axis-2 (orthogonal and unrelated to axis-1). Chappell et al. (2007) found a similarly poor relationship between the h parameter and abrasion which they believed was due to the scale of erosion features (macroscopic) beyond that of the h parameter. Abrasion, particularly the second occurrence after a rainfall treatment, appears to be positively correlated with the c parameter and hence inversely correlated with the b parameter. In the study of Chappell et al. (2007) the b parameter was very strongly positively correlated with axis-1 and the h parameter was negatively correlated with the same axis. These are the only major departures of the ordination analysis performed with the simulated MISR parameter values from that of the previous work. In general, the pattern of the results is similar to that previous study. This suggests that the simulated MISR reflectance parameter values can be used to detect differences in the soil surface. However, the discrepancies between the ordination analyses (RDA) performed by Chappell et al. (2007) using the 'true' parameter values and those available from the simulated MISR parameter values suggests that interpretations about the nature of the soil surface change may

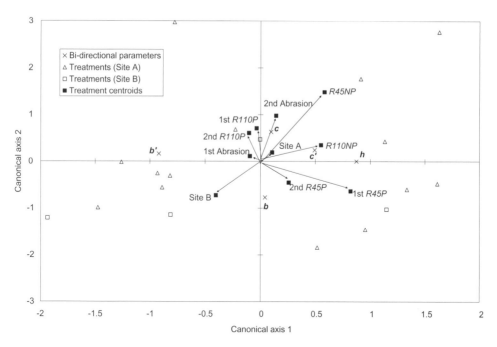

Figure 6. Ordination diagram of redundancy analysis showing the relationships between simulated MISR model parameter values and the ordination axes and the correlations between soil surface treatments and ordination axes.

vary. Furthermore, the simulated MISR parameters, particularly single scattering albedo (SSA), are limited in the waveband range. Chappell et al. (2007) were able to develop a detailed interpretation of the soil surface change by making inferences about mineralogical differences at the soil surface using hyperspectral information between 350–2500 nm. The waveband information provided by the MISR sensor is limited in this respect and similar types of interpretations are difficult to make.

5 CONCLUSION

Soil surface changes due to wind erosion, rainsplash, evaporation etc. can be detected by ground-based, hyperspectral measurements of angular reflectance and a bi-directional soil spectral reflectance model. This approach recognizes that variation in soil surface reflectance can vary considerably within a soil type or a bulked soil sample as a consequence of soil forming processes (eluviation, crusting etc.). The utility of this approach for large area assessment using directional remote sensing data was investigated by simulating the angular sampling of predefined soil surface spectral reflectance models using the configuration of the Multi-angle Imaging Spectro-Radiometer (MISR) sensor. The issue of scale associated with the satellite platform of the sensor was not considered here. The nine view angles of the sensor were used consistently, whilst solar zenith (SZA) and solar azimuth angles (SAA) were allowed to vary in the sampling configurations. The angular sampling configurations for all soil surfaces were inverted against a bi-directional soil spectral reflectance model to optimize the values of the model parameters. The optimized values were compared with the 'true' values of the parameters from the soil surface reflectance models. A single simulated overpass of the MISR sensor with one SZA was not adequate to retrieve the 'true' values of the model parameters. At least two SZAs, regardless of the number of SAAs, were required to simulate MISR overpasses and match the 'true' values. The angular sampling configuration that produced the best match, defined using statistically significant regression model parameters, between the simulated MISR parameter values and the 'true' values was used in canonical ordination (RDA) to elucidate the relationship between the model parameters and the soil surface treatments. The RDA demonstrated that the simulated MISR parameter values could be used to detect soil surface change after rainfall and aeolian abrasion. The results gave slightly different process-based interpretations compared to earlier work. Furthermore, the coarse spectral resolution and range of the simulated MISR wavebands limited the inferences that could be made about the soil surface changes. There is a demonstrated need for large area assessment of soil surface characteristics using remote sensing and therefore further investigation is required to determine the extent to which angular reflectance from airborne and satellite platforms can be used. The inevitable upscaling issues of any arising developments need further consideration.

ACKNOWLEDGMENTS

This work was funded by an award to AC from the UK Natural Environmental Research Council (NER/M/S/2001/00124). We are grateful to Griffith University for constructing the goniometer and to S. Heidenrich for technical assistance with the operation of the wind tunnel. Assistance from S. Jacquemoud with the bi-directional soil spectral reflectance model and support from P. Lewis is also gratefully acknowledged. The comments of an anonymous reviewer improved the presentation of the paper for which we are grateful. Any omissions or inaccuracies that remain in the paper are the sole responsibility of the authors.

REFERENCES

Baumgardner, M.F., Silva, L.F., Biehl, L.L. & Stoner, E.R. 1985. Reflectance properties of soils. *Advances in Agronomy* 38: 1–44.

Ben-Dor E, Irons J.R. & Epema, G. 1999. Soil reflectance. In A.N. Renzc (ed.), *Remote Sensing for the Earth Sciences* (3): 111–188. New York: Wiley.

Ben-Dor, E., Goldshleger, N., Benyamini, Y., Agassi, M. & Blumberg, D.G. 2003. The spectral reflectance properties of soil structural crusts in the 1.2 to 2.5 μm spectral region. *Soil Science Society American Journal* 67: 289–299.

Chappell, A., McTainsh, G., Leys, J. & Strong, C. 2003a. Simulations to optimise sampling of aeolian sediment transport for mapping in space and time. *Earth Surface Processes and Landforms* 28: 1223–1241.

Chappell, A., McTainsh, G., Leys, J. & Strong, C. 2003b. Using geostatistics to elucidate temporal change in the spatial variation of aeolian sediment transport. *Earth Surface Processes and Landforms* 28: 567–585.

Chappell, A., Zobeck, T. & Brunner, G. 2005. Induced soil surface change detected using on-nadir spectral reflectance to characterise soil erodibility. *Earth Surface Processes and Landforms* 30(4): 489–511.

Chappell, A., Zobeck, T.M. & Brunner, G. 2006. Using bi-directional soil spectral reflectance to model soil surface changes induced by rainfall and wind-tunnel abrasion. *Remote Sensing of Environment* 102(3–4): 328–343.

Chappell, A., Strong, C., McTainsh, G. & Leys, J. 2007. Detecting induced *in situ* erodibility of a dust-producing playa in Australia using a bi-directional soil spectral reflectance model. *Remote Sensing of Environment* 106: 508–524.

Galvão, L.S., Pizarro, M.A. & Epiphanio, J.C.N. 2001. Variations in reflectance of tropical soils: Spectral-chemical composition relationships from AVIRIS data. *Remote Sensing of Environment* 75: 245–255.

Goldshleger, N., Ben-Dor, E., Benyamini, Y., Agassi, M., & Blumberg, D.G. 2001. Characterization of soil's structural crust by spectral reflectance in the SWIR region (1.2-2.5 μm). *Terra Nova* 13:12–17.

Hagen, L.J., Zobeck, T.M., Skidmore, E.L. & Elminyawi, I. 1995. *WEPS Technical Documentation: Soil Submodel.* SWCS WEPP/WEPS Symposium. Ankeny, IA.

Hapke, B.W. 1963. A theoretical photometric function for the lunar surface. *Journal of Geophysical Research* 68: 4571–4586.

Hapke, B.W. 1981. Bidirectional reflectance spectroscopy 1. Theory. *Journal of Geophysical Research* 86: 3039–3054.

Hapke, B.W. 1986. Bidirectional reflectance spectroscopy 4. The extinction coefficient and the opposition effect. *Icarus* 67: 264–280.

Huete, A.R. & Escadafal, R. 1991. Assessment of biophysical soil properties through spectral decomposition techniques. *Remote Sensing of Environment* 35: 149–159.

Jacquemoud, S., Bater, F. & Hanocq, J.F. 1992. Modelling spectral and bidirectional soil reflectance. *Remote Sensing of Environment* 41: 123–132.

Karnieli, A., Kidron, G.J., Glaesser, C. & Ben-Dor, E. 1999. Spectral characteristics of cyanobacteria soil crust in semi-arid environments. *Remote Sensing of Environment* 69: 67–75.

Latz, K.R.A., Weismiller, G.E., Van Scoyoc, G.E. & Baumgardner, M.F. 1984. Characteristic variations in spectral reflectance of selected eroded Alfisols. *Soil Science Society American Journal* 48: 1130–1134.

Leone, A.P. & Sommer, S. 2000. Multivariate analysis of laboratory spectra for the assessment of soil development and soil degradation in the southern Apennines (Italy), *Remote Sensing of Environment* 72: 346–359.

McTainsh, G.H., Leys J.F. & Nickling, W.G. 1999. Wind erodibility of arid lands in the Channel Country of western Queensland, Australia. *Zeitschrift für Geomorphologie Supplementband* 116: 113–130.

Nickling, W.G., McTainsh, G.H. & Leys, J.F. 1999. Dust emissions from the Channel Country of western Queensland, Australia. *Zeitschrift für Geomorphologie Supplementband* 116: 1–17.

Pinty, B., Verstraete, M.M. & Dickinson, R.E. 1989. A physical model for predicting bidirectional reflectances over bare soil. *Remote Sensing of Environment* 27: 273–288.

Privette, J.L., Myneni, R.B., Emery, W.J. & Pinty, B. 1995. Inversion of a soil bidirectional reflectance model for use with vegetation reflectance models. *Journal of Geophysical Research* 100 (D12): 25497–25508.

Seubert, C.E., Baumgardner, M.F. & Weismiller, R.A. 1979. Mapping and estimating aerial extent of severely eroded soils of selected sites in Northern Indiana. *Institute of Electrical and Electronic Engineers*: 234–239.

Shao, Y., Raupach, M.R. & Leys, J.F. 1996. A model for prediction of aeolian sand drift and dust entrainment on scales from paddock to region. *Australian Journal of Soil Research* 34: 309–342.

ter Braak, C.J.F. 1988. CANOCO: *A FORTRAN Program for Canonical Community Ordination by [partial] [detrended] [canonical] Correspondence Analysis and Redundancy Analysis* (version 2.1). Agricultural Mathematics Group, Wageningen, The Netherlands.

Zobeck, T.M., Sterk, G., Funk, R., Rajot, J.L., Stout, J.E. & Van Pelt, S. 2003. Measurement and data analysis methods for field scale wind erosion studies and model validation. *Earth Surface Processes and Landforms* 28: 1163–1188.

*Recent Advances in Remote Sensing and Geoinformation Processing
for Land Degradation Assessment – Röder & Hill (eds)*
© *2009 Taylor & Francis Group, London, ISBN 978-0-415-39769-8*

Mapping land degradation risk: Potential of the non-evaporative fraction using Aster and MODIS data

M. García, S. Contreras, F. Domingo & J. Puigdefábregas
Estación Experimental de Zonas Áridas (CSIC), Almería, Spain

ABSTRACT: Land degradation is associated with decreases in resources retention by ecosystems. In water-limited environments this loss of functionality has an impact in the water use efficiency which should be reflected in the partition of surface energy fluxes through the actual evapotranspiration or latent heat. Remote-sensing indicators of land degradation based on energy ratios, such as the non-evaporative fraction, could help to monitor land condition. In this study, we test a simple operational model for calculating energy fluxes in a semiarid mountainous region at two different spatial resolutions (90 m and 1 km), using Aster and MODIS data on 18-07-2004. Results show that Aster and MODIS results are comparable within reported instrumental errors. However, the lost of detail is remarkable. If the processes of land degradation related with changes in the surface energy balance are explicit at 1 km, which needs further elucidation, MODIS is an adequate tool to perform regional assessments by means of its high temporal and spatial coverage. Comparisons with field data show low net radiation errors and large errors for sensible heat but within the ranges obtained by other authors. The spatial patterns for the ratio non-evaporative fraction (NEF) proposed as an indicator of land condition are coherent with the surface type. Using NEF and NDVI, a reliable identification of disturbed sites with high risk of degradation and non-degraded sites is accomplished. Non-degraded sites are better identified using NDVI rescaled only, while degraded sites are better classified using only NEF. Therefore, despite the fact that vegetation cover is a clear symptom of land degradation, indicators not directly related with vegetation cover, based on the surface energy balance, such as the NEF (non-evaporative fraction), can reveal important information about ecosystem functioning. Because of the limited verification data and dates, these results are preliminary and need further testing.

1 INTRODUCTION

Currently, there is a lack of standard and operational procedures for monitoring land degradation over large regions (Puigdefábregas & Mendizábal 2003). Land degradation processes cause disturbances in structure and functioning of landscapes leading to decreases in its resources retention capacity (Ludwig & Tongway 1997). A theoretical continuum of functionality can be established from landscapes that effectively trap, store, concentrate and utilize resources to those characterized by a severe degradation stage in which all the resources are lost (Ludwig & Tongway 2000). Desertification has been associated with higher spatio-temporal heterogeneity of water and other resources (Schlesinger et al. 1990). To monitor landscape condition and measure the magnitude of land degradation processes, simple indicators based on structural properties of the ecosystem or more complex indicators associated with water (Sharma 1998, Boer & Puigdefábregas 2003, 2005) and energy fluxes can be used (Wang & Takahashi 1998). The indicators based on the water or energy balance are linked through the latent heat flux (λE) or evapotranspiration. Calculation of surface energy ratios requires spatially disaggregated estimates of surface energy balance components at temporal scales compatible with the temporal scales of land degradation processes.

The law of conservation of energy states that the available energy reaching a surface is dissipated mainly as latent heat (λE) and sensible heat (H):

$$Rn - G = \lambda E + H \qquad (1)$$

being Rn = net radiation; G = soil heat flux; and Rn−G = available energy. To select an appropriate energy balance indicator, it is important to consider the low value of latent heat fluxes and the evaporative fraction (EF) during several months in semi-arid areas. For this reason, we explore the non-evaporative fraction, NEF, as an indicator for land degradation as in (2).

$$NEF = 1 - EF = 1 - \frac{\lambda E}{\lambda E + H} = 1 - \frac{\lambda E}{Rn - G} = \frac{H}{Rn - G} \qquad (2)$$

The NEF should present a wider range of variability than EF and a higher signal-to-noise ratio. For instance, in the study region, latent heat is within error level of models during several days (Domingo et al. 2001).

The algorithms to calculate these energy components use information in the solar and thermal range, being remote sensing the only data source providing radiometric temperature and vegetation cover observations over large extents. This is crucial as these variables explain most of the partition of the available energy into sensible and latent heat (Kustas & Norman, 1996).

The most vulnerable areas to land degradation are located in arid regions (Safriel et al. 2003 where the development of an operative system would require data such as MODIS (Moderate Resolution Imaging Spectrometer), available at high temporal resolution but with 1 km pixel size. This questions the validity of models originally designed for agricultural areas or almost ideal conditions when these models are applied over arid regions and heterogeneous sites with sparse vegetation covers (Chehbouni et al. 1997, Wassenaar et al. 2002).

In this study, we test two simple operational models (Jackson et al. 1977, Seguin & Itier 1983, Carlson et al. 1995, Roerink et al. 2000) for calculating energy fluxes in a semi-arid region at two different spatial resolutions (90 m and 1 km). We expect that the increase in bare soil due to decreases in vegetation cover taking place as a consequence of land degradation should increase surface temperature and albedo causing increases in H, and decreases in Rn similarly to results presented by Dolman et al. (1977) in the Sahel. Feedback effects, such as those occuring between albedo and surface temperature might counterbalance some of these impacts (Phillips 1993). This work aims to elucidate some of these aspects. The specific objectives are:

1. Evaluate two simple daily energy balance algorithms in a mountainous semi-arid region in SE Spain characterized by its high land cover heterogeneity and fragmentation.
2. Compare the land surface energy fluxes estimates result from the application of two spatial resolution data: MODIS with 1 km. and Aster (Advanced Spaceborne Thermal Emission and Reflection Radiometer) with 90 m.
3. Map land degradation risk in a semiarid mountainous region, Sierra Gador, using the non-evaporative fraction (NEF) as an indicator for land degradation.

Aster is currently the only sensor collecting multispectral thermal infrared data at high spatial resolution being very appropriate for testing of models and direct ground comparisons (French et al. 2005). On the other hand, both sensors, Aster and MODIS, are on board the Terra platform, allowing analyses of scale. Recent work has been done to compare both sensors showing good agreement (<1 K) (Jacob et al. 2004). It is desirable to extend this type of comparisons to other variables and regions.

2 STUDY SITE AND DATA

The study region (Fig. 1) located in South East Iberian Peninsula (Almería, Spain) comprises 3600 km² (36.95° N, 2.58°W). This region is characterized by its heterogeneity and the abrupt

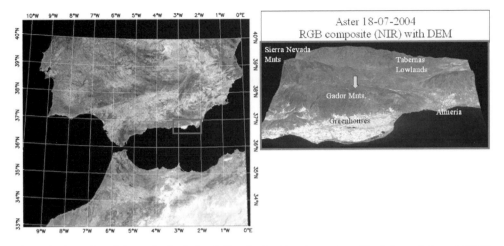

Figure 1. Location of the study site in South East Spain, Almeria province. The left panel shows MODIS True Color Composite corresponding to the study site and 18-07-2004. The right panel shows the study site with a false colour composite for ASTER 18-04-07 image with image relief. Location of the eddy covariance system is shown by the arrow (see colour plate page 393).

relief changes with altitudinal gradients ranging from sea level up to 2800 m (a.s.l.) in Sierra Nevada mountain. Precipitation and temperature regimes present wide contrasts driven by the orography (Lopez-Bermúdez et al. 2005). Annual precipitation is the lowest in the Tabernas desert, with less than 200 mm, while in the mountains can be enough to sustain forest growth, ranging between 400 mm up to 700 mm.

In the center of the study site, Sierra Gador mountain-range covers 552 km^2. During the 18th and 19th centuries, this range area was subjected to an intense and widespread deforestation for ship construction and mining purposes. The original vegetation composed of oaks (*Quercus ilex* L. and *Quercus faginea* Lam.), olive trees (*Olea europaea* L.) poplars (*Populus* L. spp.) and strawberry-trees (*Arbutus unedo* Lam.) was extensively cut down and nowadays is dominated by a sparse shrubland dominated by *Genista cinerea* Vill. and mixed with rock outcrops, bare soil or grassland vegetation mainly dominated by *Festuca scariosa* Lag. Around 73% of Sierra de Gádor presents this pattern with vegetation cover lower than the 50%. The shrubland with a small cover of dense pine (*Pinus* L. spp.) woodland represents the second natural land cover type (12% of the area). Only a 1.5% of the land is covered by dense woodlands of reforestation pines. Agriculture covers the 9% of Sierra de Gádor and is dominated by a mixture of unirrigated and irrigated lands (almonds and olive trees) (Contreras 2006).

The rest of the image includes part of the Natural Park of Sierra Nevada including pine forests, oak relicts, and shrublands. In the northeast is the natural desert of Tabernas with a complex topography comprising an area of badlands. Along the Andarax ephemeral river which flows by Almeria city, there is a mosaic of citrus orchards and vinegrapes. One of the most outstanding features of the scene is the large plastic greenhouse area covering more than 330 km^2. This unique combination of land covers and uses allows using this area as a pilot site where to test a simple model for calculating surface energy fluxes that could be extended to regional and global scales.

For this study, we have used Aster and MODIS data acquired on July-18th-2004 at 11.00 UTC. The Aster products used were 2AST07 which is surface reflectance at 15 m (VNIR) and 30 m (SWIR), and 2AST08 kinetic temperature at 90 m. Two MODIS products were used. To decrease errors due to cloud coverage and bidirectional reflectance, we took the 8 day surface reflectance at 500 m (MOD09). During this 8 day period, albedo and vegetation indices can be assumed to be constant. Finally, daily land surface temperature product (MOD11) at 1 km was used. In the case of MODIS, surface temperature errors range from 1–3 K with no incidences for ASTER, where the reported absolute precision is 1–4 K.

A digital elevation model (DEM) from USGS (United States Geological Survey) available at 30 m resolution and a digital orthophoto (from the Andalusian Regional Government) at 0.5 m were used at different stages of the study.

Instrumental field data are acquired continuously at Llano de los Juanes experimental field site since 2003. They were used to compare with model results. The eddy covariance technique was used to measure latent and sensible heat flux using a three dimensional sonic anemometer CSTAT3 and a krypton hygrometer KH_2O, both from Campbell Scientific Inc., USA. Llano de los Juanes is a representative flat area of $\sim$2 km^2 located at 1600 m in the well developed karstic high plain of Sierra de Gador. The vegetation cover is 50%–60% and consists mainly of patchy dwarf perennial shrubs (30%–35%) dominated by *Genista pumilla*, *Thymus serpylloides* Bory and *Hormathopylla spinosa* L. and grasses (20%–25%) dominated by *Festuca scariosa* Lag. and *Brachypodium retusum* Pers. (Li et al. in press). Mean NDVI measured in Llano de los Juanes with a Dycam camera in June of 2004 was 0.296. Net radiation (NR-LITE; Kipp & Zonen, Delft, Netherlands), relative humidity (thermohygrometer HMP 35 C, Campbell Scientific, Logan, UT, USA) sonic and soil temperature (SBIB sensors) are also continuously measured at the site. Besides air temperature measurements at Llano de los Juanes field site, air temperature was measured from other 11 meteorological stations at the time of the satellite overpass (11.00 UTC). At Sierra Gádor mean ± standard deviation values derived from Aster data for July-18th-2004 for albedo and NDVI were respectively: 0.19 ± 0.030 and 0.29 ± 0.08. At Llano de los Juanes experimental field site, mean values were: 0.186 (albedo), 0.269 (NDVI), very similar to the Dycam measurement from one month ago, for what it is expected that vegetation cover at the time of satellite overpass will be also very similar to measured values in June (around 50%–60%).

3 ESTIMATION OF THE NON-EVAPORATIVE FRACTION USING THE H/RN RATIO

The NEF (non-evaporative fraction) that could be linked to land degradation processes, was estimated with Aster data using the ratio of sensible heat versus net radiation, H/Rn (Seguin & Itier 1983, Carlson et al. 1995). At a daily scale, soil heat flux (G) is negligible in relation to the other components of the balance (Kustas & Norman 1995) in this case:

$$\text{NEF} = 1 - \text{EF} = 1 - \frac{\lambda E}{Rn - G} = \frac{H}{Rn - G} = \frac{H}{Rn} \qquad (3)$$

3.1 Sensible heat flux (H)

H can be estimated by a model of turbulent transport from the surface to the lower atmosphere based on surface layer similarity of mean profiles of temperature and wind speed using a resistance form:

$$H = \rho \cdot C_p \frac{T_s - T_{air}}{r_h} \qquad (4)$$

where: T_s is land surface temperature; T_{air} is air temperature, both at the time of image acquisition; r_h is the atmospheric resistance to the transfer of H; ρ and C_p are air density and specific heat at constant pressure respectively. B can be defined as an exchange coefficient to sensible heat transfer (Jackson et al. 1997, Seguin & Itier 1983) and is calculated as:

$$B = \frac{\rho \cdot C_p}{r_h} \qquad (5)$$

Being r_h a turbulent exchange coefficient dependent on wind velocity, aerodynamic roughness length, roughness length for heat transfer and Monin-Obukov length (Brutsaert 1982).

As having estimates of these variables at large scales is difficult, more operational parameterizations have been proposed. Seguin & Itier (1983) found a global mean value for B = 0.25 for unstable conditions ($T_s - T_{air} > 0$) consistent with analytical calculations. In addition, we test the approach from Carlson et al. (1995). They shown that the main factor affecting conductance to heat transfer is vegetation cover and established a linear relation between the exchange coefficient to sensible heat transfer B and fractional cover. At this date, NDVI from bare soil at the study site was: 0.16 ± 0.012 (mean $\pm$ standard deviation) and from complete vegetation cover: 0.68 ± 0.20. Mean values from bare soil and complete vegetation were taken to calculate B.

3.1.1 *Air temperature (T_{air})*

Air temperature is used in the estimation of sensible heat flux, H, and also of net radiation, Rn. To avoid relying on meteorological information, air temperature was estimated from the images using the triangle NDVI-T_s proposed by Carlson et al. (1995). The apex of the NDVI-T_s space (high NDVI and low temperature) should correspond to pixels with high NDVI located at the wet edge of the triangle, and can be assumed to be at the air temperature (Idso & Jackson 1969). The apex is selected in a supervised manner. To get T_s at the apex, minimum surface temperature areas are located in the scene, then those with highest NDVIs corresponding to forest patches are selected, and the average T_s for that selected region is calculated. Due to the high altitudinal gradients at the study area, it is necessary to apply a correction to air temperature considering as a reference altitude that of the pixels at the region forming the apex. Afterwards, positive corrections for altitude are made for pixels below the base altitude and vice-versa for pixels above considering a lapse rate of 6.5 °C each 1000 m. This is better than approaches considering a unique air temperature for the whole area assuming constant meteorological conditions at the blending height (Carlson et al. 1995; Czajkowski et al. 2000) and also worked better than the Bastiaanssen et al. (1998) dry and wet pixel approach, probably more suited for flat areas, evaluated in preliminary tests (results not shown).

3.2 *Net radiation (Rn)*

Rn is calculated as the balance between incoming ($\downarrow$) and outgoing fluxes ($\uparrow$) of shortwave (Rs) and longwave (Lw) radiation. By agreement, incoming fluxes are positive and outgoing negative. This can be expressed as the sum of shortwave (Rns) and longwave net radiation (Lnw)

$$Rn = Rs \uparrow + Rs \downarrow + Lw \downarrow + Lw \uparrow = Rns + Lnw \tag{6}$$

3.2.1 *Shortwave Net radiation*

The shortwave net radiation using remote sensing is calculated as:

$$Rns = Rs \downarrow (1 - \alpha) \tag{7}$$

where α is the broadband surface albedo estimated according to Liang (2000) for 6 Aster and MODIS bands. Rs$\downarrow$ (incoming solar radiation) was estimated using a solar radiation model from Fu & Rich (2002).

3.2.2 *Longwave net radiation*

The longwave energy components are related to surface and atmospheric temperatures through the Stephan- Boltzmann law. The longwave net radiation is calculated as in (8):

$$Lnw = -\varepsilon_s \sigma Ts^\wedge 4 + Lw \downarrow \tag{8}$$

Where broadband emissivity for the surface, ε_s, was estimated based on a logarithmic relationship with NDVI (van de Griend & Owe, 1993) and radiometric surface temperature, Ts, was directly obtained from Aster and MODIS LST (Land Surface Temperature) products calculated with the

TES (Temperature Emissivity Separation) algorithm for Aster and the day/night LST algorithm for MODIS. An empirical function is used for the incoming longwave radiation Lw↓ (Idso & Jackson, 1969).

4 ESTIMATION OF THE NON-EVAPORATIVE FRACTION FROM S-SEBI

The non-evaporative fraction, NEF, was derived from S-SEBI (Simplified Surface Energy Balance Index) model (Roerink et al. 2000). S-SEBI estimates directly the evaporative fraction, EF, on a pixel basis considering the relationship between surface temperature and albedo. It assumes that the atmospheric conditions remain relatively constant across the study region and that enough wet and dry pixels are present in the scene. Taking into account these assumptions, we have:

$$EF = \frac{T_{obs} - T_{LE}}{T_H - T_{LE}} \tag{9}$$

Where Tobs is the observed temperature, T_{LE} is the temperature at the lower boundary function or evaporation controlled domain and T_H is the temperature at the upper boundary function or radiation controlled domain. Both boundary functions, the lower and the upper, are calculated by quantile regression (Koenker & Hallock, 2001) from the albedo vs. surface temperature scatter-plot for the study region and using the 5% and the 95% quantiles respectively. When calculating S-SEBI, Roerink et al. (2000) used Valor and Caselles (1996) emissivity which provides a theoretical explanation for van de Griend & Owe (1993) NDVI-emissivity relationship. It requires an a-priori knowledge of vegetation and soil emissivities. However, when this information is not available, errors in emissivity estimates are very similar between both methods (Valor & Caselles 1996). For this reason and also to get comparable results in in terms of model performance with previous NEF estimations using H/Rn ratio, the more operational van de Griend & Owe (1993) emissivity has been used.

5 MAPPING THE RISK OF LAND DEGRADATION IN SIERRA GADOR

A prior step when evaluating land condition is the choice of a reference state corresponding to optimum or non-degraded status. In this case, we assumed that in Sierra Gádor, there is enough variability as to find degraded and non-degraded sites acting as reference levels.

Two variables were employed as inputs for classification: NEF and NDVI. They were previously rescaled according to the aridity index to make comparisons across different climatic regions.

The aridity index was calculated in the study area as the ratio between the long-term annual average values of potential evapotranspiration and precipitation (Contreras 2006).

The Hargreaves-Samani equation (Hargreaves & Samani 1982) (see equation 9 below), was used to estimate spatially-distributed potential evapotranspiration (PET). This method is appropriate for semiarid environments (Vanderlinden et al. 2004, Gavilán et al. 2006) and when meteorological information is scarce as in Sierra de Gádor:

$$PET = aR_a(T_{avg} + b)(T_{max} - T_{min})^{0.5} \tag{10}$$

R_a is the solar radiation in equivalent evaporated water-depth (mm month^{-1}), T_{avg} the monthly average temperature (°C), T_{max} and T_{min} are monthly maximum and minimum average temperatures (°C) respectively, and *a y b* are regionally-calibrated parameters.

Monthly precipitation and temperature maps were obtained by interpolating data from 35 and 16 meteorological stations respectively by using multiple regression with altitude, longitude and distance to the sea. R^2 ranged between 0.6 to 0.87 for precipitation and between 0.6–0.97 for temperature.

Spatially-distributed monthly radiation values were estimated using POTRAD5 irradiance model (van Dam 2000) developed on PCRaster (van Deursen & Wesseling 1992). Monthly reference evapotranspiration from Penman-Montheith of 6 meteorological stations at the study site were used for calibration. The values obtained for the empirical coefficients a and b were 0.00317 and 36.45 respectively ($R^2 = 0.97$) (Contreras 2006).

For the two variables selected for classification: NDVI, and NEF, there is a range of variation from optimum or reference state (low NEF or high NDVI) to degraded state (low NDVI and high NEF) for each aridity class.

To find the boundary functions corresponding to reference levels of NEF and NDVI, quantile regression for the 5% and 95% was applied in a similar way as with S-SEBI.

The rescaled NEF is:

$$NEF_{resc} = \frac{NEF_{obs} - NEF_{5\%}}{NEF_{95\%} - NEF_{5\%}} \qquad (11)$$

where:

NEF_{obs} = observed value of NEF in the pixel.
$NEF_{5\%}$ = value of the NEF lower boundary function for the aridity level corresponding to that pixel.
$NEF_{95\%}$ = value of NEF upper boundary function for the aridity level corresponding to that pixel.

Classification thresholds for NEF_{resc} and $NDVI_{resc}$ were established based on the histograms (mode ± standard deviation).

Classification performance was evaluated using ground truth sites from field visits and photointerpretation with areal photographs. Non-degraded validation sites include oak relicts with three density levels and a dense reforested pine area. Ground truth areas with risk of land degradation included disturbed sites: a strong burnt scar from 2002, an active limestone quarry, an intensively mining area, and almond orchards plowed for weeds. Therefore, disturbed sites can be sites with risk of degradation that are related, in this case study, with decreases in vegetation cover and increase in albedo.

6 RESULTS

6.1 Comparison of air temperature with field data

Eleven meteorological stations were used to evaluate air temperature (T_{air}) estimations at the time of the Terra satellite overpass (11.00 UTC). First, the temperature of the apex according to (8) was selected (Fig. 2). Afterwards, corrections for altitude improved the overall error to less than 2°C (Table 1).

The overall adjustment is good, but T_{air} estimates are subjected to local errors. One concern is that altitude is not the only factor affecting T_{air}. Nevertheless, using this approach presents the advantage of relief from using ancillary data. Also any systematic error in T_s estimation will propagate in the T_{air}; therefore, these errors should cancel out when calculating differences $T_s - T_{air}$ for estimating sensible heat flux. Difference in reference altitude between Aster and MODIS is due to the fact that MODIS covers an additional area of Sierra Nevada of high altitude but not included in the Aster scene. In any case, T_{air} results between both sensors are similar in terms of MAE. In both cases, the region of interest contributing to calculate apex temperature were dense pine forests located in the Sierra Nevada mountains.

6.2 Model performance at the Llano de los Juanes field site

At Llano de los Juanes field site, Rn and surface energy fluxes were measured and compared with Aster estimates (Table 2). The B values modeled at Llano de los Juanes are reasonable and similar to the fixed value of 0.25 proposed by Seguin & Itier (1983) for dryland and irrigated crops in

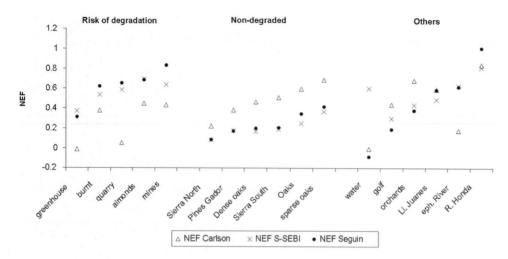

Figure 2. NEF calculated by three methods for different surface types. Surface types have been grouped in three categories: sites with risk of degradation or disturbed sites. Non-degraded sites with dense vegetation and others, including miscellaneous covers. NEF Carlson corresponds to the ratio H/Rn calculated according to Carlson et al. (1995) approach. NEF S-SEBI corresponds to 1-EF calculated by Roerink et al. (2000) and NEF Seguin corresponds to the ratio H/Rn using a fixed value of 0.25 for the exchange coefficient for sensible heat flux, provided by Seguin & Itier (1983) for unstable conditions.

Table 1. Air temperature estimates at the study site. "MAE" or Mean Absolute Error is the absolute average difference of the residuals between estimated and measured air temperature at the 11 stations. MAE after adjustment corresponds to the temperature after considering a correction lapse rate of 6.5 °C every 1000 m. Reference altitude is the altitude of the pixels for which the apex temperature has been extracted.

	ASTER	MODIS
R^2 (Tair$_{stations}$ and altitude)	0.61	
MAE before adjustment (°C)	4.31	2.45
MAE after adjustment (°C)	1.96	1.91
T apex (°C)	24.0	28.5
Reference altitude (m)	1800	800

Table 2. Comparisons between measured and estimated values at the Gádor Mountain range field site. B is the exchange coefficient for heat transfer, Rn is net radiation.

	Measured	Aster	Error diff	Error %
Rn (W m^{-2})	179.72	182.44	2.72	1.5
B (W m$^{-2\circ}$C^{-1})	0.37*	0.250	−0.06	−24.0

*Estimated empirically from field data at Llano de los Juanes field site.

France and much lower than the fixed value of 0.64 of Jackson et al. (1977) in irrigated wheat fields in Arizona. Errors are very low for Rn (around 1.5% compared to field data) and therefore within reported net radiometer errors (±10% and directional error of <25 W m^{-2}). Measured B is higher than estimated B for this day which can be due to the influence of other factors not considered in the model such as wind speed.

In Llano de los Juanes, Aster results underestimate H when compared with eddy covariance measurements. Although reported errors for T_s are within acceptable quality levels, it would be necessary to evaluate its influence in H and Rn estimates in future studies. Error in H propagates into the NEF = H/Rn. NEF values calculated from Seguin and Carlson methods provide similar results than the measured value while NEF value from S-SEBI provides a higher error (Table 3). We also have to be aware that the eddy covariance technique is subjected to uncertainty levels of 20%–30% (Baldocchi et al. 2001). Moreover, in semi-arid areas with sparse vegetation cover, error in energy fluxes tend to be on the higher side of this range around 30% (Were 2005).

In general, the range of errors reported by other authors in H flux is very variable. Seguin et al. (1999) consider around 50 W m^{-2} an acceptable error for H. In the literature, errors in the best cases are around 22 W m^{-2} (Kustas & Norman 1996) and can reach up to 50% even with sophisticated models when the parameterizations are not good. Using a fixed value for kB^{-1} in agricultural areas produces errors as high as 150 W m^{-2} or 43–64 W m^{-2} (Seguin et al. 1999). Humes et al. (2000) obtained a RMSE value of 43.35 W m^{-2} for sensible heat while Laymon & Quattrochi (2000) in the Great Basin desert got errors in H around 40% similarly to the worse results in this study.

6.3 Analysis of the NEF (non-evaporative fraction) spatial patterns

The assessment of model performance for mapping land degradation risk cannot be made solely on the basis of an eddy covariance value. To evaluate the spatial coherence of results, relative NEF values among different surface types were compared for each method (Fig. 2). Several sites inside and in the surroundings of Sierra Gador were selected by photo-interpretation and field knowledge.

In general, NEF values from Seguin and S-SEBI methods give similar results in terms of trends and also absolute values. Carlson method overestimates NEF for vegetation corresponding to non-degraded sites and underestimates NEF for degraded sites (e.g. oaks and pines vs quarry or mining zone). NEF from Seguin method provides the most consistent results according to surface types. The three methods yield similar results at Llano de los Juanes field site. Sites where we cannot be sure about results are greenhouses and the Andarax ephemeral river. S-SEBI does not model well free water.

In order to spot sites with high risk of degradation, either NEF$_{Seguin}$ or NEF$_{S-SEBI}$ should work, with the exception of greenhouse areas. Rescaling of NEF values according to climate variations should be performed, so that more arid areas are not classified as degraded sites due to higher NEF or lower NDVI values.

The relationship between NDVI, as a proxy for vegetation cover, and NEF (Figure 3) shows that for the selected surface types (disturbed and non-degraded sites in Sierra Gádor), high NDVI values are associated with low NEFs. In the case of NEF$_{Seguin}$ and NEF$_{S-SEBI}$ the NDVI-NEF trend is linear and low NDVI values are associated with high NEF, which is logic, considering also that we are in summer.

It seems that there is an albedo feedback effect which decreases net radiation at low vegetation covers (Bastiaanssen et al. 1998, Roerink et al. 2000) apparent by a negative relationship between

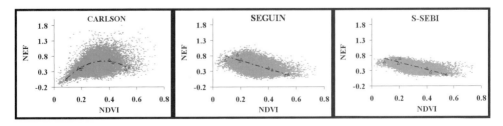

Figure 3.   Non-evaporative fraction (NEF) versus NDVI calculated according to Carlson, Seguin, and S-SEBI methods in Sierra Gador. Grey dots include all pixels within Sierra Gador. Black markers represent validation sites only within Sierra Gador with risk of degradation (burnt, limestone quarry, almond, mines from Fig. 2) and non-degraded sites (pines gador, dense oaks, oaks, sparse oaks). Seguin and S-SEBI best fit with those selected sites within Gador corresponds to linear regressions while Carlson fits a second order polynomial degree function.

Table 3. Comparisons between measured (with eddy covariance technique) and estimated (with three different methods) values of sensible heat, H, and non-evaporative fraction, NEF, in Llano de los Juanes. Seguin method assumes a fixed value for the exchange coefficient for heat transfer, B, while Carlson method uses a B value related with NDVI. S-SEBI is the simplified SEBI method developed by Roerink et al. (2000). N/A (not applicable to this method).

Variable		Eddy	Seguin	Carlson	S-SEBI
H (Wm^{-2})	Mean	149.59	106.70	107.58	N/A
	Error diff	0	−42.89	−42.01	N/A
	Error %	0	−28.67	−28.08	N/A
NEF	Mean	0.83	0.59	0.59	0.49
	Error diff	0	−0.24	−0.24	−0.34
	Error %	0	−28.92	−28.92	−40.96

NEF and NDVI values lower than 0.25, thus corresponding to bare soil, for the upper NEF values of the cloud in Figure 3.

It has been described in a semi-arid region in Spain (Castilla la Mancha) that pixels at the wet edge (evapotranspiring at potential rates) present a positive relationship between T_s or T_s–T_{air} with albedo and negative with NDVI with no inflection points. However, pixels at the dry edge (with no available water for evapotranspiration) present a positive relationship between T_s or T_s–T_{air} with albedo up to an inflection point. Similarly occurs with the relationship between NDVI and T_s or T_s–T_{air}. The albedo for which a feedback effect on temperature occurs corresponds to the NDVI inflection point as well (Garcia et al. in press). For this reason, we can assume that in this case albedo is decreasing surface temperature and NEF values at low NDVI levels.

In any case, this effect on surface temperature should not be as strong as to decrease NEF to the level of oaks or orchards as it is the case in NEF$_{Carlson}$. In this case, for bare soil (NDVI < 0.2), there is a strong decrease in NEF that does not seem realistic. It was shown previously in Figure 2 that this model underestimates surface roughness and sensible heat for bare soil. If this model is used as input for a land degradation classification, sites where a vegetation cover loss has occurred will be confused with high vegetated areas.

When considering all the pixels at the study site, a significant scattering between NDVI and NEF appears (Fig. 3). This reveals that although NDVI and NEF are correlated to some extent, they provide different information. There are many factors involved in surface resistance to sensible heat transfer (Brutsaert 1982), and Carlson et al. (1995) simplification in this semi-arid conditions does not work. The error commited by estimating surface resistance as a function of NDVI is worse than not estimating it at all. There is uncertainty in selection of NDVI maximum and minimum values. The errors are amplified for bare soil conditions. Not considering surface resistance variations across the landscape, will be similar to consider a "potential" or "reference" sensible heat flux, approach that might still be valid for mapping land degradation sites. Future efforts will be devoted to a precise estimation of surface resistance by considering vegetation height, wind speed, and other variables, which will allow to evaluate the impact of not considering it on the land degradation indicator.

Considering this and previous results, the model finally selected for use at a regional scale is based on calculating NEF values as H/Rn and according to Seguin et al. (1989).

6.4 Comparison between MODIS and ASTER performance

Comparisons between Aster and MODIS were performed by resampling the Aster pixel size of 90 m using a cubic convolution filter into MODIS 1 km pixel size. Figure 4 shows results for output variables using Aster and MODIS. The main structure of the surface fluxes explained by the NEF is maintained from MODIS to Aster scene, but level of detail is not comparable. For this reason, features with a spatial resolution lower than 1 km are not resolved (e.g. the mosaic of irrigated citrus and bare soil by the Andarax ephemeral river). Therefore, if the processes of land degradation related with changes in the surface energy balance are explicit at 1 km grains, which needs further

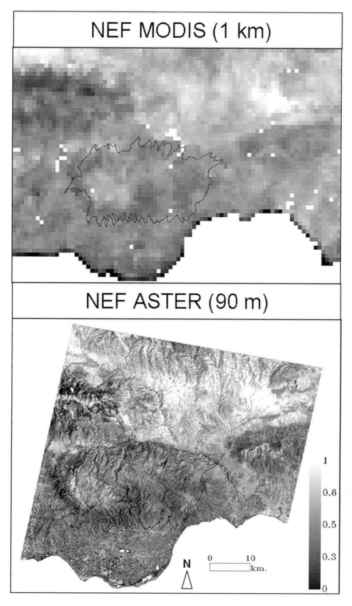

Figure 4.   Comparison of spatial patterns nd NEF (H/Rn Seguin) for ASTER (90 m) and MODIS (1 km) over the study site on 18-07-2004. Sierra Gador is outlined in black.

elucidation, MODIS can be consider an adequate tool to perform regional assessments by means of its high temporal and spatial coverage.

Results in Table 4 show that for the set of input variables, MODIS values tend to be slightly lower than Aster with the exception of albedo. The RMSE value in T_s is 2.4°C. These differences are translated to the output variables causing underestimations of MODIS with respect to Aster. All the variables present R^2 greater than 0.76, the lowest corresponds to NDVI, which means that the variables aggregate linearly but there is a remaining ~20% of the variance responding to non-linear aggregation effects combined with sensor differences in sensor performance and correction algorithms employed.

Table 4. Input and output variables for MODIS and Aster: T_s (Land Surface Temperature), H Seguin (sensible heat flux calculated according to Seguin), Rn (net radiation), NEF Seguin (non-evaporative fraction calculated according to Seguin). RMSE (Root Mean Square Error) between both variables, R^2 (Pearson coefficient after aggregation of Aster to 1 km pixel), Slope (slope of regression between Aster -independent and Modis-dependent).

Variables		RMSE	R^2	Slope
INPUT	T_s (°C)	2.4	0.80	0.81
	Broadband albedo	0.04	0.86	1.13
	NDVI	0.053	0.76	0.93
	Emissivity	0.011	0.86	0.77
OUTPUT	Rn (W m^{-2})	9.08	0.80	0.86
	H Seguin (W m^{-2})	18.24	0.78	0.71
	NEF Seguin	0.096	0.83	0.74

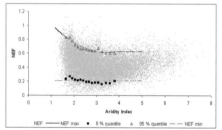 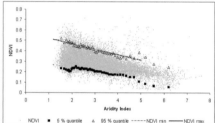

Figure 5. Quantile regression for the upper (95%) and lower (5%) quantiles between (a) aridity index and NEF and (b) aridity index and NDVI in Sierra Gador.

In any case, both RMSE values and mean differences are within reported net radiometer precision for Rn. RMSE results for H are comparable to some closure errors in energy balance using eddy covariance technique (Baldocchi et al. 2001). This means that despite some differences between Aster and MODIS sensor performance, combined with the impact of change in spatial resolution, the two sensors provide similar results, for spatial features of 1 km or larger, comparable to repeatability of instrumental data.

6.5 Case study: mapping the risk of land degradation in Sierra Gador

The selected NEF model (H/Rn Seguin) and the NDVI were rescaled by aridity levels, using a polynomic function for the lower boundary (5% quantile) and a linear function for the upper boundary (95% quantile) (Fig. 5).

The classification method was based on histogram thresholding corresponding to the mode ± standard deviation (Fig. 6). In this way, values at the extremes corresponding to near to optimum status and very far from optimum are selected.

Three classification approaches were compared (a) using NEF and NDVI rescaled as input variables, (b) using only NEF rescaled and (c) only NDVI rescaled. Output maps were compared with ground truth sites (Table 5). They show that classes associated with degraded sites correspond to (a) ↓NDVI-↑NEF, (b) ↑NEF alone, and (c) ↓NDVI alone. The mapping was better, in terms of correct identification of pixels using only (b) ↑NEF for degraded sites and (c) ↑NDVI for non-degraded sites (Table 5). Missclassification was negligible in all cases. The three aproaches identified all the hot spots selected as validation sites but some of the methods did not cover the complete ground truth sites resulting in non-classification. It was prefered a conservative classification than an over-classification, to decrease the number of false alarms.

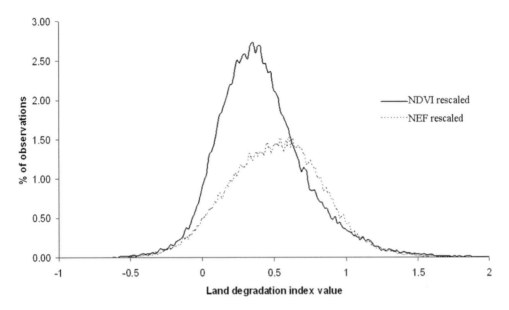

HISTOGRAMS NDVI rescaled and NEF rescaled

Figure 6. Histograms for NEF and NDVI after being rescaled according to the aridity index in Sierra Gador. The thresholds used for classification were: a) for NDVI rescaled, 0.35 ± 0.31; and b) for NEF rescaled, 0.60 ± 0.25 corresponding to the mode ± standard deviation.

Table 5. Confusion matrix for three classificaction methods: using NEF rescaled and NDVI rescaled by aridity levels, using only NEF rescaled and only NDVI rescaled. Non-degraded sites, n = 172; degraded site, n = 317.

Ground Truth (%)

Method	Class	Non degraded	High risk of degradation	Total
NEF$_{resc}$ & NDVI$_{resc}$	↑NDVI ↑NEF	0	0	42.75
	↓NDVI ↓NEF	0	0.32	28.12
	↓NDVI ↑NEF	0	53	10.03
	↑ NDVI ↓NEF	62.79	0	6.45
	Unclassified	37.21	46.69	12.66
	Total	100	100	100
NEF$_{resc}$	↓ NEF	69.77	0.63	24.95
	↑NEF	0	82.65	53.58
	Unclassified	30.23	16.72	21.47
	Total	100	100	100
NDVI $_{resc}$	↑NDVI	82.56	0	29.04
	↓ NDVI	1.16	60.57	39.67
	Unclassified	16.28	39.43	31.29
	Total	100	100	100

Figure 7 shows the final map of classification of areas with high risk of degradation in Sierra Gador mapped using ↑NEF, and non-degraded areas mapped using ↑NDVI.

Currently, only heavily disturbed sites are identified. Ongoing research is focused on the assessment intermediate classes of land degradation (e.g. shrublands replacing oak forests) and also to include more tests sites expanding the region of analysis. This issue is particularly complicated as

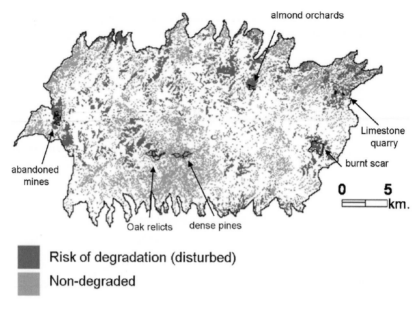

almond orchards

Limestone
quarry

burnt scar

abandoned
mines

0   5

km.

Oak relicts   dense pines

■ **Risk of degradation (disturbed)**

□ **Non-degraded**

Figure 7.   Classification map of areas with high risk of degradation (red) and low risk of degradation (green) in Sierra Gador (Almeria, Spain). Validation sites for non-degraded sites are composed of oak relicts with three density levels and pine forests (blue polygons), and disturbed sites considered at high risk of degradation are mapped as black polygons (burnt scar from 2002, limestone quarry, almond orchards, and abandonded mines) (see colour plate page 396).

intermediate levels of degradation can present a reversible trend that a single or few images can not capture and are even hard to identify and define in field validations.

7   CONCLUSIONS

A reliable identification of disturbed sites presenting high risk of degradation and non-degraded sites is accomplished using the non-evaporative fraction (NEF) and the NDVI. Non-degraded sites are better identified using NDVI rescaled only, while degraded sites are better classified using only NEF. Both indicators, related with loss of vegetation cover and decrease of evapotranspiration levels, are correlated to some extent. However, they provide different information as shown by the significant scattering between them and by the fact that classification using both variables produces poorer results than classification using each variable separately. Therefore, despite the fact that vegetation cover is a clear symptom of land degradation, other indicators not directly related with vegetation cover, based on the surface energy balance, such as the NEF, can reveal important information about ecosystem functioning.

The spatial patterns of the non-evaporative fraction are coherent with the type of surface using the S-SEBI model and especially using the H/Rn ratio when a constant value of the exchange coefficient to sensible heat based on Seguin & Itier (1989) is considered. Comparison with field data using eddy covariance technique at Llano de los Juanes experimental field site shows very low net radiation errors (within net radiometer precision). Sensible heat provides larger errors but within the ranges obtained by other authors, within the threshold of 50 Wm^{-2} proposed by Seguin et al. (1999) for the three methods. Air temperature can be extracted from the images using the NDVI-T_s space relationship (Carlson et al. 1995) corrected by altitude with an acceptable overall error ($<2°C$).

Comparison between MODIS and Aster spatial patterns of surface energy balance components in summer reveals that the main spatial structure is maintained from Aster to MODIS. However, the loss of detail is remarkable and features smaller than 1 km are not resolved. Regarding RMSE values between Aster and MODIS, net radiation values are within reported net radiometer precision and sensible heat flux are comparable to closure errors in energy balance using eddy covariance technique. Therefore, if the processes of land degradation related with changes in the surface energy balance are explicit at grains of 1 km, which needs further elucidation, MODIS is an adequate tool to perform regional assessments by means of its high temporal and spatial coverage.

Establishing a robust and simple land degradation index needs additional analyses including several dates to fully understand the interaction between water and energy at different surfaces. Efforts should be devoted to scale daily results to temporal scales compatible to the time scale of land degradation processes. Along this line, a remaining question is the spatial and temporal scales in which alterations of energy balance fluxes related with land degradation are explicit.

ACKNOWLEDGMENTS

This research has been developed within the EU project DeSurvey: "A Surveillance System for Assessing and Monitoring of Desertification" (FP6-00.950). The authors wish to thank Dr. del Barrio, M. San Juan and Dr. Lázaro for help during this work. Computer support from R. Ordiales and S. Vidal is greatly appreciated. An anonymous reviewer is acknowledged for very useful comments that contributed to improve this paper.

REFERENCES

Brutsaert, W. 1982. *Evaporation into the atmosphere. Theory, history, and applications*. Dordrecht: Holland, D.Reidel Publishing Company.

Baldocchi, D.D., Falge, E., Gu, L., Olson, R., Hollinger, D., Running, S., Anthoni, P., Bernhofer, Ch., Davis, K., Fuentes, J., Goldstein, A., Katul, G., Law, B., Lee, X., Malhi, Y., Meyers, T., Munger, J.W., Oechel, W., Pilegaard, K., Schmid, H.P., Valentini, R., Verma, S., Vesala, T., Wilson K. & Wofsy, S. 2001. FLUXNET: A new tool to study the temporal and spatial variability of ecosystem-scale carbon dioxide, water vapor and energy flux densities. *Bulletin of the American Meteorology Society* 82: 2415–2434.

Bastiaanssen, W.G.M., Menenti, M., Feddes, R.A. & Holtslag, A.A.M. 1998. A remote sensing surface energy balance algorithm for land (SEBAL) 1. Formulation. *Journal of Hydrology* 212–213(1–4): 198–212.

Boer, M.M. & Puigdefábregas, J. 2003. Predicting potential vegetation index values as a reference for the assessment and monitoring of dryland condition. *International Journal of Remote Sensing* 24: 1135–1141.

Boer, M.M. & Puigdefábregas, J. 2005. Assessment of dryland condition using spatial anomalies of vegetation index values. *International Journal of Remote Sensing* 26: 4045–4065.

Carlson, T.N., Capehart, W.J. & Gillies, R.R. 1995. A new look at the simplified method for remote sensing of daily evapotranspiration. *Remote Sensing of Environment* 54: 161–167.

Chehbouni, A., Loseen, D., Njoku, E.G., Lhomme, J.P., Monteny, B. & Kerr,Y.H. 1997. Estimation of sensible heat flux over sparsely vegetated surfaces. *Journal of Hydrology* 89(1–4): 855–868.

Contreras, S. 2006. Spatial distribuition of the annual water balance in semi-arid mountainous regions: application to Sierra Gádor (Almería, SE Spain) Ph.D. Thesis (in Spanish). Almería: Servicio Publicaciones Universidad de Almería.

Czajkowski, K.P., Goward, S.N., Mulhern, T., Goetz, S.J., Walz, A., Shirey, D., Stadler, S., Prince, S.D. & Dubayah, R.O. 2000. Estimating environmental variables using thermal remote sensing. In Quattrochi, D.A. & Luvall, J.C. (eds), Thermal remote sensing in land surface processes: 11–32. Boca Raton, Florida: CRC Press.

Dolman, A.J., Gash, J.H.C., Goutorbe, J.P., Kerr, Y., Lebel, T., Prince, S.D. & Stricker, J.N.M. 1997. The role of the land surface in Sahelian climate: HAPEX-Sahel results and future research needs. *Journal of Hydrology* 89(1–4): 1067–1079.

Domingo, F., Villagarcía, L., Boer, M., Alados-Arboledas, L. & Puigdefábregas, J. 2001. Evaluating the long-term water balance of arid zone stream bed vegetation using evapotranspiration modelling and hillslope runoff measurements. *Journal of Hydrology* 243: 17–30.

French, A.N., Jacob, F., Anderson, M.C., Kustas, W.P., Timmermans, W., Gieske, A., Su, B., Su, H., Mccabe, M.F., Li, F., Prueger, J.H. & Brusnell, N. 2005. Surface energy fluxes with the advanced spaceborne thermal emission and reflection radiometer (ASTER) at the Iowa 2002 SMACEX site (USA). *Remote Sensing of Environment* 99(1–2): 55–65.

Fu, P. & Rich, M. 2002. A geometric solar radiation model with applications in agriculture and forestry. *Computers and Electronics in Agriculture* 37: 25–35.

García, M., Palacios-Orueta, A., Puigdefábregas, J., Contreras, S., Del Barrio, G., Fernandez, F.J. & Moreno, M. in press. Estimating evapotranspiration from TVDI: Towars a land degradation index for regiona analysis. In *Proceeding of the 2nd Symposium on Quantitative Advances in Remote Sensing:* 664–669, Valencia (Spain) Sept. 24–29.

Gavilán, P., Lorite, I.J., Tornero, S. & Berengena, J. 2006. Regional calibration of Hargreaves equation for estimating reference ET in a semiarid environment. *Agricultural Water Management* 81(3): 257–281

Hargreaves, G.H. & Samani, Z.A. 1982. Estimating potential evapotranspiration. *ASCE Journal of Irrigation and Drainage Engineering* 108: 225–230.

Humes, K., Hardy, R., Kustas, W., Prueger, J. & Starks, P. 2000. High spatial resolution mapping of surface energy balance components with remotely sensed data. In Quattrochi, D.A. & Luvall, J.C. (eds), *Thermal Remote Sensing in Land Surface Processes*: 110–132. Boca Raton, Florida: CRC Press.

Idso, S.B. & Jackson, R.D. 1969. Thermal radiation from the atmosphere. *Journal of Geophysical Research* 74: 5397–5403.

Jackson, R.D., Reginato, R.J. & Idso, S.B. 1977. Wheat canopy temperature: a practical tool for evaluating water requirements. *Water Resources Research* 13(3): 651–656.

Jacob, F., Petitcolin, F., Schmugge, T., Vermote, E., French, A. & Ogawa, K. 2004. Comparison of land surface emissivity and radiometric temperature derived from MODIS and ASTER sensors. *Remote Sensing of Environment* 90(2): 137–152.

Koenker, R. & Hallock, F.F. 2001. Quantile regression. *Journal of Economic Perspectives* 15(4): 143–156.

Kustas, W.P. & Norman, J.M. 1996. Use of remote sensing for evapotranspiration monitoring over land surfaces. *Hydrological Sciences Journal* 41(4): 495–516.

Laymon, C.A. & Quattrochi, D.A. 2000. Estimating spatially-distributed surface fluxes in a semi-arid great basin desert using Landsat TM thermal data. In Quattrochi, D.A. & Luvall, J.C. (eds), *Thermal Remote Sensing in Land Surface Processes*: 110–132. Boca Raton, Florida: CRC Press.

Li, X.-Y., Contreras, S. & Solé-Benet, A. Spatial distribution of rock fragments in dolines: a case study in a semiarid Mediterranean mountain-range (Sierra Gador, SE Spain). *CATENA*, in press.

Liang, S. 2000. Narrowband to broadband conversions of land surface albedo in algorithms. *Remote Sensing of Environment* 76: 213–238.

López-Bermúdez, F., Boix-Fayos, C., Solé-Benet, A., Albaladejo, J., Barberá, G.C., del Barrio, G., Castillo, V., García, J., Lázaro, R., Martínez-Mena, M.D., Mosch, W., Navarro-Cano, J.A., Puigde-fábregas, J. & Sanjuán, M. 2005. Landscapes and desertification in South-East Spain: overview and field sites. Field Trip Guide A-5. 6th International conference on geomorphology. Sociedad Española de Geomorfología. 40p.

Ludwig, J.A. & Tongway, D.J. 1997. A landscape approach to rangeland ecology. In Ludwig, J.A., Tongway, D.J., Freudenberger, D.O., Noble, J.C. & Hodgkinson, K.C. (eds), Landscape ecology, function and management: principles from Australia's rangelands: 1–12. Melbourne: CSIRO Publishing.

Ludwig, J.A. & Tongway, D.J. 2000. Viewing rangelands as landscape systems. In Arnalds, O. & Archer, S. (eds), Rangeland desertification: 39-52. Dordrecht, Netherlands: Kluwer Academic Publishers.

Phillips, J.D. 1993. Biophysical feedbacks and the risks of desertification. *Annals of the Association of American Geographers* 83(4): 630–640.

Puigdefábregas, J. & Mendizábal, T. 2004. Prospects for desertification impacts in Southern Europe. In A. Marquina (ed), Environmental challenges in the Mediterranean 2000–2050: 155–172. Boston, Kluwer Academic Publisher.

Roerink, G.J., Su, Z. & Menenti, M. 2000. S-SEBI: a simple remote sensing algorithm to estimate the surface energy balance. *Physics and Chemistry of the Earth (Part B)* 25: 147–157.

Safriel, U., Adeel, Z., Niemeijer, D., Puigdefábregas, J., White, R., Lal, R., Winslow, M., Ziedler, J., Pronce, S., Archer, E. & King, C. 2003. Dryland Systems. In "Ecosystems & human well-being. Current state and trends", Millenium Ecosystem Assessment Series, Vol 1, Cap 22, pp. 623–662, World Resources Institute, Washington, DC.

Seguin, B. & Itier, B. 1983. Using midday surface temperature to estimate daily evaporation from satellite thermal IR data. *International Journal of Remote Sensing* 4: 37–383.

Seguin, B., Becker, F., Phulpin, T., Gu, X.F., Guyot, G., Kerr, Y., King, C., Lagouarde, J.P., Ottle, C., Stoll, M.P., Tabbagh, A. & Vidal, A. 1999. IRSUTE: a minisatellite project for land surface heat flux estimation form field to regional scale. *Remote Sensing of Environment* 68: 357–369.

Schlesinger, W.H., Reynolds, J.F., Cunningham, G.L., Huenneke, L.F., Jarrell, W.M., Virginia, R.A. & Whitford, W.G. 1990. Biological feedbacks in global desertification. *Science* 247: 1043–1048.

Sharma, K.D. 1998. The hydrological indicators of desertification. *Journal of Arid Environments* 39: 121–132.

Valor, E. & Caselles, V. 1996. Mapping Land Surface Emissivity from NDVI: Application to European, African, and South American Areas. *Remote Sensing of Environment* 57(3): 167–184.

Van Dam, O. 2000. Modelling incoming potential radiation on a land surface with PCRaster: POTRAD5.MOD manual. Utrech Centre for Environment and Landscape Dynamics, Utrecht University.

Van Deursen, W.P.A. & Wesseling, C.G. 1992. The PC-Raster Package. Dept. of Physical Geography, Utrecht University. PCRaster webpage: http://pcraster.geo.uu.nl/index.html.

Van de Griend, A.A. & Owe, M. 1993. On the relationship between thermal emissivity and the normalized difference vegetation index for natural surfaces. *International Journal of Remote Sensing* 14(6): 1119–1131.

Vanderlinden, K., Giráldez, J.V. & van Meirvenne, M. 2004 Assesing reference evapotranspiration by the Hargreaves method in southern Spain. *Journal of Irrigation and Drainage Engineering* 130(3): 184–191.

Wassenaar, T., Olioso, A., Hasager, C., Jacob, F. & Chehbouni, A. 2002. Estimation of evapotranspiration on heterogeneous pixels. In Sobrino, J.A. (ed), Recent advances in quantitative remote sensing, Proc. 1st intern. symp.: 319–328. Valencia, September 2002.

Were, A. 2005. Spatial aggregation of evapotranspiration in semi-arid climate. Ph.D. Thesis (in Spanish). Universidad de Almeria, Spain.

Wang, Q.X. & Takahashi, H. 1998. Regional hydrological effects of grassland degradation in the Loess Plateau of China. *Hydrological Processes* 12: 2279–2288.

Part 4
Stories behind pixels: process-based assessment of geospatial data

*Recent Advances in Remote Sensing and Geoinformation Processing
for Land Degradation Assessment – Röder & Hill (eds)*
© 2009 Taylor & Francis Group, London, ISBN 978-0-415-39769-8

Geomatics-based characterization of spatial and temporal trends in heterogeneous Mediterranean rangelands of Northern Greece

A. Röder & J. Hill
Remote Sensing Department, FB VI Geography/Geosciences, University of Trier, Trier, Germany

T. Kuemmerle
Department of Geomatics, Humboldt University Berlin, Germany

G. del Barrio
Estación Experimental de Zonas Aridas, Consejo Superior de Investigaciones Cientificas, Almeria, Spain

V.P. Papanastasis
Faculty of Forestry and Natural Environment, Laboratory of Range Ecology, Aristotle University Thessaloniki, Greece

G.M. Tsiourlis
Forest Research Institute, Laboratory of Ecology, National Agricultural Research Foundation (NAGREF), Vasilika-Thessaloniki, Greece

ABSTRACT: The county of Lagadas in northern Greece is typical for heterogeneous Mediterranean rangelands embedded within a patchwork of land use types. In front of the spatial configuration of the landscape, major changes in rural economies were experienced in past years which have caused competing demands of different stakeholders on the utilization of natural resources. Local-scale remote sensing data archives provide opportunities to assess the present state of resources and retrospectively characterize their development. In the present study, temporal and spatial trends were analyzed using linear trend analysis of Landsat-TM and—ETM+ imagery, and grazing pressure and its effects were investigated using a cost surface modeling approach. It was found that stability, degradation and regeneration of vegetation are present in close proximity and their spatial pattern was found to be largely determined by socio-economic factors. Most importantly, the different interpretation approaches were found to contribute to a general understanding of the various feedback loops operating in the watershed of Mygdonia valley.

1 INTRODUCTION

1.1 *Land degradation in Mediterranean rangelands*

In the Mediterranean Basin, rangelands relate to non-arable, marginal lands that serve numerous uses, including forage production, and encompass a variety of highly heterogeneous ecosystems. Compared to rangelands found in North and South America or in Australia, they are often small structured and interwoven with cultivated areas (Di Castri 1981). As a result of their history of utilisation spanning back over millennia, they have often reached an equilibrium state with the demands and pressures exerted on them. This equilibrium was often attained within a semi-natural environment, as man had already widely replaced Mediterranean forests with evergreen and sclerophyllous vegetation types following the Neolithic revolution (Di Pasquale et al. 2004).

In general, degradation may be perceived as the loss of the natural potential of an ecosystem to provide goods and services of various kinds. This may result from a combination of climatic background, ecological conditions and socio-economic determinants (e.g. Stafford-Smith & Reynolds 2002, Mulligan et al. 2004). While natural conditions in the Mediterranean explain its general disposition to degradation, it is most frequently human interventions that initiate a degradation of resources (e.g. Hobbs et al. 1995, Blondel & Aronson 1995, Gomez-Sal 1998). In particular in recent times, significant land-use transformations have occurred that often manifest in different directions of utilisation intensity (Hobbs et al. 1995). Conflicts often arise from competing demands on rangelands, for example between past and present uses following land use transitions or between ecological and economic priorities (Gomez-Sal 1998, Noy-Meir 1998).

In this context, especially the soil and vegetation domains require major attention (e.g. Thornes 1985, Francis & Thornes 1990). Bare soil is most prone to linear and sheet erosion, or soil sealing through splash on bare soil, reduced infiltration and increased overland flow (e.g. Thornes 1985). On the other hand, besides being an important resource in itself, vegetation has shown to reduce these effects by stabilising the soil, slowing down overland flow, reducing splash impact, and preventing capillary rise effects in the soil (Francis & Thornes 1990). In addition, increased vegetation cover is associated with increases in evapotranspiration rates, which is an important factor for hydrological properties and local water balances (Boer 1999).

1.2 The role of grazing

Grazing is both an important agent of land degradation and a vital element of rural economies. It has been among the main factors contributing to the degradation of many forest features and shaping today's appearance of Mediterranean semi-natural landscapes (Le Houérou 1981).

In the developed countries of the Northern Mediterranean, the early 20th century was characterised by shrinking animal numbers and land abandonment in many poorer regions (Le Houérou 1981). The political and socio-economic framework has largely changed in the last decades as a result of the accession of Mediterranean countries to the European Community and later European Union (Greece: 1981). Most of these countries have been net receivers of funds ever since and have benefited from different European funding schemes supporting rural infrastructure, such as the European Regional Development Fund, the European Social Fund, and the Agricultural Guidance and Guarantee Fund. Especially the Common Agricultural and Rural Policy (CAP) has made livestock grazing profitable again through per-capita subsidies, leading to re-increasing animal numbers and semi-intensive to intensive production systems (Dubost 1998).

Grazing affects plants and plant communities in various ways. While plant function is immediately disrupted by grazing through the reduction in photosynthetic leaf area, there are environmental modifications for surviving plants, which include reduced competition, increased photosynthetic irradiance, or the opening of gaps (Briske & Noy-Meir 1998). Plant communities frequently show an increase in the relative abundance of grazing-resistant or –tolerable species, such as for instance *Quercus coccifera* shrubs which dominate shrublands in many Mediterranean rangelands (Le Houérou 1981, Alados et al. 2004). Given the availability of both woody and herbaceous vegetation, sheep act as feeders and prefer the latter while goats act as browsers and are particularly suited to forage on woody plants, where they prefer young shoots and seedlings (Le Houérou 1981, Carmel & Kadmon 1999). On a broader scale, overgrazing triggers the reduction of vegetation cover, thus increasing the risk of soil erosion. This is often aggravated by the creation of animal trails, which are potential starting features for linear erosion. The compaction of soil and the loss of stratum may lead to reduced water availability, reduced fertility and a change in soil texture (Thornes 1990).

1.3 Remote sensing and geoinformation processing

Land degradation or the impact of grazing are not directly detectable from remote sensing data (Hill et al. 2004). However, proportional vegetation cover is a suitable indicator because it is closely related to above-ground biomass (Tsiourlis 1998, Chiarucci et al. 1999) and may be related to processes of accelerated erosion, degradation, increase of flammable biomass etc. (e.g. Thornes 1990).

Spectral Mixture Analysis (SMA, Smith et al. 1990) allows to decompose the signal into proportions of reference materials based on a linear mixture of spectral reflectance measurements of these materials or image-derived spectra ('endmembers'). This approach is superior to traditional vegetation indices in sparsely vegetated areas (Elmore et al. 2000) and was successfully applied in semi-arid environments (Roberts et al. 1998, Shoshany & Svoray 2002, Garcìa-Haro et al. 2005). While degradation effects related to grazing may be characterised as temporal trends in vegetation cover (Hostert et al. 2003a, Röder et al. 2008), they may also manifest in spatial trends. For instance, systematic changes in photosynthetic vegetation or soil cover with increasing distance from watering holes could be related to animal distribution patterns (Pickup and Chewings 1988, Sparrow et al. 1997, Röder et al. 2007). These trends are often not discernible visually, or can be masked by strong vegetation response to rainfall variations. Consequently, statistical techniques have to be employed to describe grazing gradients and prove decreasing pressure with increasing distance from watering holes (Pickup and Chewings 1994, Pickup et al. 1998, Karnieli et al. 2006).

2 TEST SITE AND OBJECTIVES

2.1 *Physical setting*

The county of Lagadas belongs to the Region of Central Macedonia and is situated north of the Chalkidiki peninsula and east of the city of Thessaloniki. The landscape is clearly structured into different elevation zones and transitions between these. While natural vegetation partially corresponds to these zones, there is a mosaic of land uses of various types and intensities according to which different pressure systems operate. There are some major zones of intensive agriculture, but it is especially the mosaic of rangelands used for grazing, with embedded smaller agricultural plots, which requires major attention (Fig. 1).

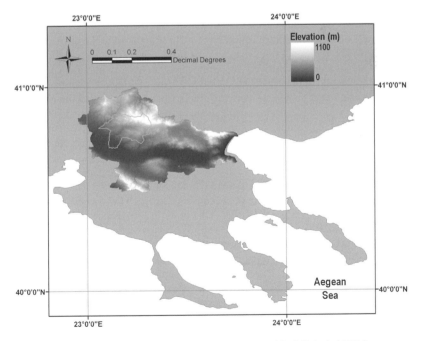

Figure 1. Topography of the Lagadas test area (full area represented by hillshaded DEM. *Source*: Geoapikonisi Ltd.; perimeter indicates core study area referred to in section 6).

While the highest areas are mostly covered by thermophilous forests and the two major plains are used for intensive agriculture, it is especially the undulating areas in between that can be characterised as rangelands and that are at the focus of this study.

Climate of the area is typical Mediterranean, with maximum precipitation being received in spring and autumn, although foramination of the trade-wind inversion may cause major precipitation events at other times. Total amounts range from 410 mm to 685 mm, depending on elevation and orientation relative to atmospheric currents. Despite its pronounced topographic structure, bedrock of the area is relatively homogenous, with metamorphic rocks being abundant, accompanied by pyroclastic rocks and limestones in limited areas. Upon these, shallow soils are most frequent, which can be characterised as leptosols and regosols, complemented by cambisols where conditions sustain a thorough pedologic development. In larger depressions and intermediate plain areas, quaternary deposits and alluvial soils dominate which are mainly used for agriculture.

The highest elevations of the County are dominated by sub-mountainous beech forests and thermophilous oak forests, although the latter are frequently observed in a degraded state due to grazing and forestry activities. Beside agricultural areas, the remaining surface is characterised by typical Mediterranean rangelands, found interwoven with agricultural areas in a patchwork mosaic. These rangelands are constituted by grasslands and shrublands which are dominated by *Quercus coccifera* shrubs and trees. These are highly adapted due to their spiny small leaves, their resilience to fires and their fast resprouting and coppicing abilities (Di Pasquale et al. 2004). Herbaceous species, especially *Labiatae* and *Fabaceae,* are numerous in the shrubland areas. In the grasslands, annual species together with some perennial grasses occur (Konstantinidis & Tsiourlis 2003). As these grasslands are a result of intense grazing activities, the presence of isolated specimen of thorny shrubs is characteristic, such as *Quercus coccifera, Crataegus monogyna, Juniperus oxycedrus* and *Prunus spinosa*.

2.2 *Socio-economic setting*

In the county of Lagadas, rural economy is strongly dependant on agricultural production, fishery at the coast and in Lake Volvi, and livestock grazing in the rangeland areas. Besides, the proximity of the industrial and economical centre of the city of Thessaloniki and the tourist centres of the Chalkidiki peninsula are additional sources of income in the region.

Livestock breeding of almost exclusively sheep and goats has been an important economic factor for centuries, mainly carried out for meat production as well as milking and manufacturing of dairy products (Hadjigeorgiou et al. 1998). Historically, the husbandry system can be characterised as a shepherded, extensive grazing system, where shepherds guide their animal flocks, while in the winter period animals are occasionally held in paddocks (Hadjigerogiou et al. 1998, Yiakoulaki et al. 2002). The historic development of grazing and animal numbers follows largely the general fluctuations described before (fig. 2).

The grazing system of the Lagadas area has undergone severe changes in the past decades compared to the traditional transhumance system (Yiakoulaki et al. 2002). Especially the recent trend towards milk production led to a more sedentary grazing system, where flocks are usually mixed with sheep and goats (Legg et al. 1998, Oba et al. 2000). These changes limit the area that is actually being grazed, leading to over-grazing for regions where livestock is concentrated and often animals are held in sheds for considerable time. In addition, there is an increasing use of feedstuffs, which contributed to the intensification of irrigated agriculture in the basin of Mygdonia (Yiakoulaki et al. 2002). Due to rural abandonment and urbanisation, shepherding is mainly carried out by men of the older generation, which has significant implications for the grazing scheme (Hadjigeorgiou et al. 1998). Grazing is confined to the morning and late afternoon hours during the summer months, and animals are returning to their sheds during the hottest hours, while in winter most time of the day is spent grazing if meteorological conditions permit (Papanastasis 1998). This seasonality is superimposed by modifications according to the respective elevation.

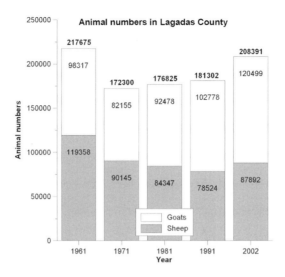

Figure 2.   Development of animal numbers in the county of Lagadas.
Source: National Statistical Survey of Greece.

As a contrast to the intensification of grazing in concentrated areas, undergrazing of less accessible locations that were formerly grazed occurs, which results in a thickening of shrublands with
a implication on species composition hydrological cycles, fire risk, etc. (Noy-Meir 1998).

2.3   *Objectives*

This study undertakes to assess dynamic processes in the rangelands and investigate if the reported
changes in the socio-economic framework have triggered effects that are tangible using remote
sensing and ancillary data. Land degradation is a 'slow' process (Mulligan et al. 2004) that may
effectuate temporal and spatial trends; hence, different methodological and conceptual approaches
are being employed to analyze the presence of degradation effects at various levels. The overall
goal is to link quantitative and qualitative spatial and non-spatial information to provide a coherent
interpretation of the state and development of the Lagadas rangelands.

3   DATA SETS AND PROCESSING

3.1   *Satellite data*

A long time-series of 15 Landsat-5 TM and Landsat-7 ETM+ data was acquired for the test area,
covering the years 1984 to 2000 with one image per year. These were selected to represent the
period of maximum photosynthetic activity in the rangeland area. In addition, a late summer scene
was acquired for 2000 to represent only woody vegetation. All scenes were supplied from Eurim-
age © as system-corrected products. A digital elevation model (DEM) at 30 m grid resolution was
made available from Geoapikonisis Ltd., which had been derived from photogrammetric analysis
of digital aerial photographs. Finally, a standard Quickbird image product from late summer 2003
for a smaller subset of the test area was available, consisting of four individual tiles from two dates.

The Landsat images were geometrically corrected to match the standard Greek reference system.
First, a master image was referenced based on a map scale 1:50,000. Subsequently, large sets of
ground control points were retrieved using a procedure based on correlation windows (Hill & Mehl
2003). All images were corrected to sub-pixel accuracy and non-systematic distortions introduced
by local relief conditions were accounted for using the DEM.

The Quickbird tiles were geometrically corrected using the sensor model and the rationale polynomial coefficients supplied with each tile. The DEM was included in the correction process and sub-pixel accuracy was attained by incorporating field-measured GPS positions. The resulting data set provided a geometric resolution of 0.6 m for the panchromatic band and 2.4 m for the multi-spectral bands.

All Landsat scenes were fully radiometrically corrected to ensure that differences emerging from the multi-temporal analyses do result from changes in surface properties (Röder et al., 2008). This comprised sensor calibration and full modeling of the radiative transfer including a correction of topography-induced illumination variations. Landsat-5 TM calibration was calculated similar to the approach proposed by Teillet & Fedosejevs (1995) to account for decaying detector sensitivity. Based on vicarious calibration experiments, a time-dependant function for gain factors was derived (Thome et al. 1997a, b). For more recent acquisition dates, estimates provided by Teillet et al. (2001) were employed. Since notation of these factors does not identify offset parameters, these were set to post-launch values given by Markham & Barker (1986). Calibration of the ETM+ images was based on parameters published by the US Geological Survey (USGS, http://landsat7.usgs.gov/cpf/). Subsequently, a radiative transfer model based on the 5S Code by Tanré et al. (1990) was parameterized for each scene. Atmospheric transmission factors were calculated making use of the Modtran-4 code (Berk et al. 1999), which allowed accounting for different water vapor concentrations. The scattering behavior of aerosols at different viewing angles was characterized by a set of phase functions according to Aranuvachapun (1985). Given the ruggedness of the local terrain, consideration of its influence on illumination conditions was of utmost importance, which was directly incorporated in the radiative transfer calculation. Considering the sun-sensor-surface constellation for each pixel, the visible sky portion could be calculated, while the anisotropy index proposed by Hay & McKay (1985) and Hay et al. (1986) was used to independently characterize diffuse and direct radiance fluxes (Hill et al. 1995). In a first step, a master image was processed and validated using ground reflectance measurements. Subsequently, a set of pseudo-invariant targets was used to iteratively parameterize the radiative transfer models and ensure the quantitative consistency of the data set.

Spectral Mixture Analysis (SMA) was used to infer proportional cover of photosynthetically active green vegetation for each pixel and all scenes (Smith et al. 1990). Assuming that the spectral variance in an image may be represented by a limited set of reference surface types (spectral endmembers), a library of spectral reflectance measurements carried out in field and under laboratory conditions was used. With reference to the spectral contrast in the scenes and the spectral dimensionality of Landsat-TM/ETM+ (Small 2004), a variable four endmember model was set up, consisting of a spectrum for green vegetation, developed soil and the abundant gneiss bedrock. In addition, an artificial shade spectrum was introduced to account for albedo variations. In the adopted pixel-adaptive SMA approach, the vegetation and shade components were forced to be included in the model for each pixel, while the soil and bedrock spectrum were used alternatively to solve equation (1):

$$R_i = \sum_{j=1}^{n} F_j \cdot RE_{ij} + \varepsilon_i \quad \text{and} \quad \sum_{j=1}^{n} F_j = 1 \tag{1}$$

where R_i = reflectance of the mixed spectrum in band I; RE_{ij} = reflectance of the endmember spectrum j in band I; F_j = fraction of endmember j; n = number of spectral endmembers; and ε_i = residual error in band i.

In the next step, a shade normalization was carried out, according to

$$f = \frac{1}{(1 - F_{Shade})} \quad \text{and} \quad \sum_{j=1}^{(n-1)} F_j \cdot f = 1 \tag{2}$$

where f = shade normalisation factor; F_{Shade} = fraction estimate for the shade endmember; and F_j = fraction estimate for endmember j.

The results were validated by assessing individual histograms, band residuals and RMSE estimates and relating image-based proportional vegetation cover to ground surveys (Röder 2005, Röder et al. 2008).

Using time series analysis techniques, spatial patterns of temporal trends may be derived from multi-temporal data sets. Such trends involve transient and cyclic components as deterministic elements, and memory effects as well as stochastic elements (Schlittgen & Streitberg 1999). Given the available long-term time series, the transient component was calculated for each pixel by means of a linear regression, such that the temporal development of SMA-derived green vegetation cover for each pixel could be characterized using the gain and offset of the resulting function according to

$$y_t = g \cdot t + o \tag{3}$$

where y_t = vegetation cover at date t; t = date of image acquisition (e.g. in days since launch of sensor); g = regression coefficient (gain); and o = regressions constant (offset).

To assess the statistical robustness of the trend, the correlation coefficient (r), coefficient of determination (r^2), standard deviation (sd) and a two-sided t-test were calculated. As locations with stable vegetation cover may exhibit low variations and hence not show a significant trend, the root mean squared error (RMSE) was introduced to account for such cases. Furthermore, a threshold analysis was included in the time series analysis to identify major changes in temporal development, which may for instance result from short-term disturbances, such as fires, clearings etc.

3.2 Ancillary data

From the DEM, major landscape mesoforms were inferred using a methodology proposed by del Barrio et al. (1996). This methodology was refined and adapted to conditions of the Lagadas test site. Different topographic variables were computed from the 30 m DEM: slope, profile and plan curvature, catchment size, wetness index, sediment transport index, distance to the nearest stream, and insolation factor. An unsupervised classification of the resulting matrix yielded 23 preliminary classes, which were aggregated to 10 superclasses using sequential agglomerative hierarchical non-overlapping (SAHN) cluster analysis (Sneath & Sokal 1973). They were then related to landscape mesoforms or relief categories (Röder et al. 2008).

A detailed habitat map was derived from the visual interpretation of digital aerial photographs acquired at a scale of 1:20,000 in 1980, which was complemented and updated by numerous field surveys. This vector data set provided information on major vegetation formations following the Natura 2000 habitat code, which was extended by structural information on the average cover of the different habitats (Konstantinidis & Tsiourlis 2003).

In addition to the spatial data layers, animal census information for sheep, goats and cattle was available for the communities of Lagadas County at decadal intervals from 1961 to 1991 and for 2002. This was complemented by numerous field surveys and documented interviews with local people and representatives of local administrations.

4 COMMUNITY AGGREGATION LEVEL

4.1 Adopted approach

In rangelands dominated by livestock production systems, proportional vegetation cover is strongly related to grazing activities; hence, it was assumed that a negative relationship exists between the development of vegetation cover and the development of animal numbers. To evaluate this hypothesis, animal figures were related to communities and areas not suitable for grazing were masked out. This enabled the calculation of effective stocking rates (animals/ha) for 1981, 1991 and 2002. In order to suppress the influence of phenological variations (e.g. triggered by exceptional climatic conditions preceding single satellite image acquisitions), the linear trend function was employed to calculate vegetation cover estimates for these dates. Excluding areas not used for grazing, these estimates were averaged for the communities (Röder et al. 2008). The resulting temporal trajectories illustrate the relationship between the temporal development of vegetation cover and stocking rates of grazing animals for 45 communities.

4.2 Results and discussion

According to Tsiouvaras et al. (1998) and Chouvardas & Papanastasis (2004), stocking rates can be classified into lightly grazed (0–0.5 an/ha), properly grazed (>0.5–1.5 an/ha), heavily grazed (>1.5–2.5 an/ha) and very heavily grazed (>2.5 an/ha). In general, there is a spatially diverse development ranging from constant, low stocking rates in communities close to the Mediterranean Sea to high and very high stocking rates exceeding local carrying capacities, which are found in many communities. Figure 3 shows representative trajectories for the major categories.

Assessing the different plots, four major groups of trajectories can be differentiated. Group A represents communities with a continuous negative relation between stocking rate and vegetation cover. Here, increasing stocking rates imply decreasing average cover and vice versa. This behaviour can be found for the villages of Sohos and Adam. In the community of Sohos more than 12,600 sheep and goats were recorded. Although the community area is large, high stocking rates are noted here as a considerable part of the community area is not suitable for grazing. The increase in stocking rates during the two decades assessed is accompanied by a moderate decrease in vegetation cover. The village of Adam shows the same relation, but here a decrease in stocking rates and a corresponding increase in vegetation cover are present. In this case, stocking rates developed from the 'heavy grazing' to the 'proper grazing' classes, explaining the stronger gradient compared to Sohos, where stocking rates indicate heavy grazing for the full period.

Group B also relates to a continuous development, but with an adverse relation. In this case, increasing stocking rates coincide with increasing vegetation cover, while the complementary effect is not observed. The community of Evagelsimos shows a strong increase in stocking rates which fall in the 'heavy grazing' category throughout the observation time. Notwithstanding, an increase in vegetation cover can be noted. It must be assumed that concentration processes and/or a diet strongly based on feedstuffs are responsible for this pattern. Especially the good road infrastructure in the valley of Mygdonia makes the provision of nutrition very convenient. In addition, arable

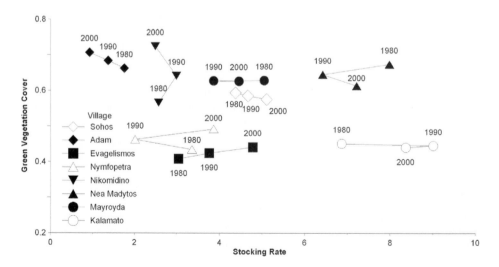

Figure 3.   Temporal trajectories relating vegetation cover and stocking rate at three decades (adapted from Röder et al. 2008).

land occupies a significant portion of the area, such that grazing of animals on harvested fields may also play a role.

A third major group of trajectories (group C) is characterized by a significant change in the direction of the relation between stocking rate and vegetation cover, indicating a transition from type A to type B or vice versa. Different factors may contribute to this behaviour, which mostly relate to the degree of 'artificiality' of the local grazing system or the general level of stocking rates. In the first case, increasing stocking rates are (over-) compensated for by the provision of additional feedstuffs and/or the local concentration of animals (Nymfopetra). This is especially important where villages are located in the valley of Mygdonia or at major transport routes and the good accessibility and infrastructure supports the concentration and intensification of grazing systems. The second major factor is the general level of stocking rates. Where these are very high, even a change in the direction does not necessarily translate in a corresponding regeneration of vegetation. This is further emphasized where agriculture is the dominant land use and only little generic grazing areas exist. In these cases, rangelands can not be set aside for regeneration purposes and overgrazing of the remaining areas results (Nea Madytos). On the other hand, vegetation cover may increase independently of minor changes in stocking rates if these are generally at a level that is sustained by the local environment (Nikomidino).

In the last group (D), vegetation cover remains stable regardless of changes in stocking rates (Mayroyda). Where this effect occurs at high stocking rate levels, it must be assumed that the modern feeding and animal keeping practices complementing the traditional pastoral system have led to the creation of an equilibrium that sustains high stocking rates within a given range (Kalamato).

As a result of the present grazing scheme, expected relations between the development in stocking rates and vegetation cover are can not fully be unveiled at the community aggregation level. The results from the temporal trajectory analysis underline this and suggest that degradation effects induced by livestock grazing manifest on finer scales and are also driven by a combination of socio-economic factors (e.g. Oba et al. 2000).

5   TREND ANALYSIS

5.1   *Adopted approach*

Assuming that temporal trends in rangelands show a distinct, spatially differentiated pattern, the results of the time series analysis were addressed at the per-pixel level. Given the dominance of gradual processes apparent from the time series analysis, agricultural and settlement areas as well as the lakes were eliminated from this analysis. The gain factor derived from the linear regression allows to identify positive and negative developments in vegetation cover as well as stable conditions. Concerning the general climatic conditions, peculiar meteorological conditions in single years may induce fluctuations in the time series. On the other hand, no overall changes in precipitation patterns were observed in past years (Röder 2005). Consequently, calculated trends are expected to be free of impact from changing climate patterns, while singular fluctuations are balanced by the linear trend analysis. As the same trend may be rated differently depending on the overall level on which it manifests, both factors were combined in a degradation index (Hostert et al. 2003b, Röder et al. 2008).

5.2   *Results and discussion*

For the whole county area, the resulting distribution of degradation index values shows that less than half of the area shows a neutral behavior given an uncertainty estimate of ±5%. For positive and negative trends, the majority of index values belong to the moderate trend classes, while trends manifesting at mean cover levels dominate. Although this may partially be triggered by averaging over a long period, it is also an indication that developments may still be reversible (table 1). The spatial representation shows a highly differentiated picture, with stability, increase and decrease of vegetation cover operating in close proximity. The spatial distribution of degradation

Table 1. Areal statistics for different degradation classes.

Trend	Vegetation cover level	Degradation index	Area (ha)	Percent[1]	Percent[2]
Strong negative < −15%	Low	11	609.030	0.56	5.86
	Medium	12	5263.74	4.85	
	High	13	487.35	0.45	
Negative < −5–15%	Low	21	1450.08	1.34	17.948
	Medium	22	16444.71	15.16	
	High	23	1560.51	1.44	
Neutral −5–5%	Low	31	2211.21	2.04	42.42
	Medium	32	34739.19	32.03	
	High	33	9048.15	8.34	
Positive >5–15%	Low	41	958.32	0.88	27.42
	Medium	42	19597.41	18.07	
	High	43	9182.70	8.47	
Strong positive >15%	Low	51	212.22	0.19	6.36
	Medium	52	4795.29	4.42	
	High	53	1888.29	1.74	
Total			108442.20	100	100

[1]percentage for each class; [2]percentage aggregated for trend classes.

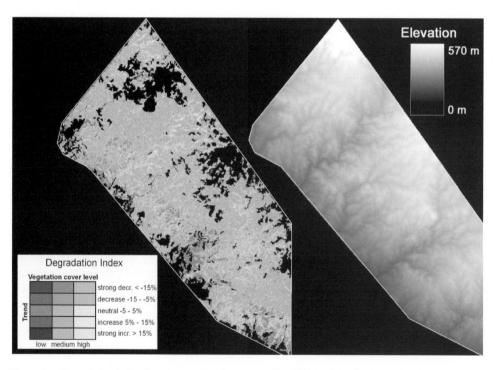

Figure 4. Degradation index for subset area and corresponding DEM-subset (interpretation in the text) (see colour plate page 397).

index values was investigated using ancillary information on elevation zones and landscape meso-forms and by identifying typical temporal profiles in vegetation cover development responsible (Röder 2005).

These specific properties and possible driving factors are illustrated for a representative subset within the major rangeland zone (Fig. 4).

The spatial representation shows a matrix of neutral index values, within which positive index values arrange in a reticulate pattern, while negative values appear rather clumped. Comparing this spatial arrangement with the structure depicted by the digital elevation model, degradation index values seem to coincide with the relief structure. In particular the valleys and channels cut deeply in the plain areas are prominent. The patch indicating strong positive development in the North corresponds to a recently established reforestation area.

In order to further exploit this pattern, the results from the trend analysis and the degradation index values were assessed in relation to major landscape mesoform classes.

Figure 5 depicts the average trend gain per landscape mesoform class. A gradient between major classes results, with negative average gains occurring on the lower divides; neutral gains in gentle areas, upper slopes and high elevation ridges; and positive gains on steep slopes and in channels. An ANOVA ($F_{6, 673072, \alpha = 0.05}$) and subsequent post-hoc Scheffé tests confirmed that the impact of mesoforms on the distribution and the differences between classes are statistically significant (Röder et al. 2008).

These results confirm the expected impact today's sedentary, semi-intensive grazing scheme on the landscape structure. Flocks are being led by shepherds in concentrated areas for limited time. In addition, urbanization and changing job opportunities have led to a higher average age of shepherds. Bearing in mind these peculiar properties, elevation and relief categories are here assumed to act as a proxy for the 'accessibility' of locations for shepherds and their flocks. From this point of view, the charts suggest a direct relation between temporal trends and accessibility of specific locations. At an average level, areas that are potentially easy to reach show a negative or no trend at all, such as the 'Low elevation divides', 'Gentle areas', 'Convex upperslopes' and 'High elevation ridges and divides' classes. In contrast, areas which are presumed to require higher efforts to be reached, such as the 'Steep mid and low slopes' and 'Channels' classes, show positive average trends.

These analyses illustrate the effects of 'selective grazing' driven by the shepherd's behaviour as determined by physical factors. Beside pastoral quality, convenience of access is a major factor for shepherds who do not roam with their flock over days but follow this as a daytime profession. It can hence be assumed that the open plains and ridges between the deep valleys are most attractive, while steep slopes or the bottoms of narrow valleys are not favored. As a consequence, these categories are grazed at increased or decreased intensities, while others experience 'intermediate' pressure, explaining their neutral behavior or total gains close to zero.

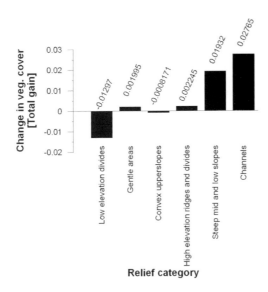

Figure 5.   Average total gain from linear trend analysis for the relief categories in the focus area.

6 SPATIAL TRENDS

6.1 *Adopted approach*

The previous analyses have yielded evidence that the dynamics of rangelands can be shown to result to a large degree from human interventions. In particular, recent changes in the grazing system were shown to affect the distribution of livestock and hence their impact on the environment. The analysis of spatial trends was pursued to validate this hypothesis, assuming that a determining role of grazing should effectuate detectable impacts on the landscape structure. The grazing gradient approach introduced by Pickup & Chewings (1994) and Pickup et al. (1998) makes use of the concept of piospheres (e.g. Forman and Gordon 1981) and assumes large, homogeneous areas, where livestock grazing can be approximated by circular buffers around watering holes, indicating a stochastic distribution of animals. As these presumptions can not be maintained for the specific properties of the Lagadas rangelands, the approach was modified and adapted for five communities that cover the major grazing areas within the County (Röder 2005).

Firstly, animal sheds were considered the relevant points of livestock concentration (PLC), which could be conveniently mapped from Quickbird very high resolution imagery. As an indicator of grazing pressure, woody vegetation cover was used (Le Houérou 1981, Carmel & Kadmon 1999), which was derived from a late summer Landsat-TM image using spectral mixture analysis following the approach described before. Most importantly, a spatial parameterisation of the distribution of livestock was required that models the combination of environmental conditions, animal behaviour and influence of the socio-economic framework. This was implemented using the concept of cost surfaces (Eastman 1989, Tomlin 1990). The approach is based on regular grids, with one layer for each specific factor. Each cell value in the individual 'friction surface' grids represents the effort required to cross this cell according to the respective factor, i.e. the 'cost'. Based on these, the 'cost distance' is calculated to represent the accumulated costs away from starting features, e.g. points of livestock concentration. This model was used to stratify the target information, i.e. to derive the spatial gradient based on woody vegetation cover.

6.2 *Results and discussion*

In a first step, major factors determining grazing patterns as a result of the preferences of shepherds and their animals were identified with the help of local experts (Röder 2005). In accordance with conclusions from the previous sections, four major factors were identified: distance, topography, attractiveness and accessibility. Using a set of functional hypotheses, these needed to be 'translated' into regular grids of friction or cost values scaled to the same range (Röder et al. 2007). These represent the effort required for animals and shepherds to travel through these cells under consideration of a given factor (Eastman 1989).

The factor 'distance' incorporates knowledge about common walking distances of shepherds and their flocks. This implies that the probability of moving further depends on the distance already accumulated between their current position and the shed. An exponential function was applied to a distance grid calculated around the PLC. The function was scaled to reach a cost of 100 at a distance of 2000 m, but in this case higher costs were allowed as shepherds may still walk further. The factor 'topography' directly incorporates knowledge of shepherds' customs as discussed in the previous sections. This factor was modeled in a straightforward manner based on slope calculated from the DEM and steep areas were assigned highest costs.

While the first two factors are more related to the influence of shepherd decisions, the 'attractiveness' and 'accessibility' factors directly reflect animal's grazing behavior. Based on knowledge of the preferred diet of goats and sheep, information available through the GIS data could be translated into friction values, where lowest values indicate the highest nutritional value. Given the different demands of goats (browsers) and sheep (feeders), mixed rangelands with an open structure composed of shrubs and herbaceous vegetation are most attractive to mixed flocks. Hence, sparse shrublands were rated most attractive ahead of grasslands and recently abandoned fields, while beech and thermophilous oak forests were assigned highest friction values. Also, villages, barren fields or agricultural areas were assigned a maximum friction value of 100. Accessibility might be related to different parameters, such the accessibility of patches, the network of roads and tracks etc. In this study, 'accessibility' was used to model the openness of the landscape. It is exclusively related to the structural land surface type, accounting for the fact that animals are deterred by dense vegetation. Again, the necessary information was derived from the habitat GIS, leading to low costs for grasslands, which increase with increasing shrub cover and attain maximum values for agricultural fields and villages. Although these are masked from the gradient analysis, it is important to ensure they act as barriers in the model.

Once the four factors were spatially represented, an integrated friction surface was calculated by averaging. Using the PLC as starting features, a pushbroom algorithm (Eastman 1989) was applied to calculate for each pixel the accumulated cost to reach the nearest PLC (Fig. 6). The resulting data set can be considered a spatial representation of the "cost" of movement of shepherds and their animals. Hence, high accumulated costs correspond to a low probability of a location being grazed.

In a next step, a categorization using a threshold of 100 cost units allowed to convert this accumulated cost distance surface into discrete zones approximating the expected spatial distribution of livestock animals. These could be intersected with spatial information on the target indicator of grazing pressure to assess the presence of spatial trends. In this study, a late summer Landsat-5 TM image acquired in 2000 was used to infer information on green vegetation cover. Due to the acquisition date, it was presumed that only shrubs as an indicator of grazing are still photosynthetically active. After masking agricultural and urban areas, average cover, standard deviation and the number of pixels contained within a zone were extracted from this image for each buffer zone (Fig. 7).

The result shows a distinct gradient of increasing cover estimates with increasing cost distance away from the PLC, from which different zones can be inferred. The first represents the 'sacrifice zone' (Thrash & Derry 1999) in the direct vicinity of the sheds where the effects of grazing are most prominent due to an almost permanent grazing pressure (zone 10). This pressure impacts the environment through trampling effects as well as the fact that grazing animals prefer young shoots and smaller plants. This effectively prevents regeneration of vegetation in the direct surroundings of sheds. Adjacent, the almost indifferent section corresponds to the major grazing

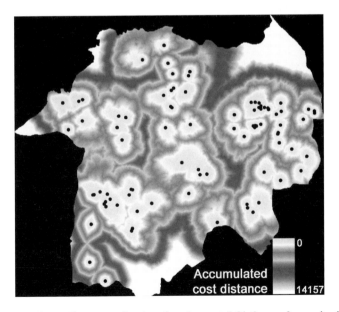

Figure 6.   Accumulated cost distance surface based on integrated friction surface and calculated using a pushbroom algorithm starting from points of livestock concentration (PLC depicted as black dots; adapted from Röder et al. 2007) (see colour plate page 397).

293

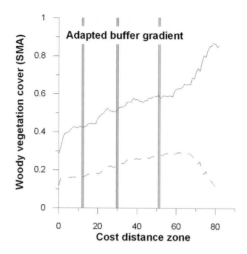

Figure 7. Gradient of woody vegetation cover (solid line) and standard deviation (dashed line) derived from piospheric analysis. The bars indicate the minimum, average and maximum cost distance that may be reached at a Euclidian distance of 2000 m.

area most frequently visited (zone 10–22). Here, a continuous increase in cover is noted. Grazing animals only manage to feed on shrubs until they reach a certain height and often graze compact shrub canopies from the outside. Hence, shrubs in this region frequently display umbrella-like structures and the landscape often exhibits a high degree of fragmentation. It is also the zone from which shepherds still manage to return their flocks to the shed during the hot midday hours. The next section of the gradient corresponds to the areas that are more remote from sheds and/or difficult or unattractive to reach (zone 22–60). Consequently, the gradient shows again a strong increase in cover estimates directly related to decreasing grazing pressure. Especially at high altitudes, this absence of grazing may result in the development of dense *Quercus coccifera* shrublands or even the evolvement of open forests, such as the open forest zone in the Northeast of the pilot area (beyond zone 60).

7   DISCUSSION AND CONCLUSIONS

The previous analyses demonstrated the use of different remote sensing and geomatics techniques to assess dynamic processes and identify degradation patterns in the rangelands of Lagadas County. It was shown that different spatial and temporal scales and different aggregation levels provide complementary information. At the community aggregation level, despite increasing animal numbers no overall degradation of resources could be detected. Rather, the results imply that the significant modification of traditional grazing schemes has introduced a high degree of artificiality, causing a departure from simple correlations. Notwithstanding, the analysis of temporal trends proofs the existence of local areas where a degradation of vegetation takes place. At the same time, distinct patterns of increasing vegetation cover are present. Again, these patterns may be related to socio-economic factors, particularly grazing as the major driver. The related hypotheses concerning the human impact on rangeland structure through the established grazing scheme were further assessed by using cost surface modeling to approximate grazing livestock distribution and assessing spatial trends of woody vegetation cover. Again, the impact of grazing on the rangelands was confirmed although it has to be noted that this is a mutual dependency. While the location of sheds as points of livestock concentration strongly affects the apparent landscape structure, the landscape structure in turn always determined the establishment of sheds. Thus, the Lagadas rangelands are an excellent example for cultural landscapes that have been used for long periods of time and which are in a constant feedback loop with their (changing) boundary conditions.

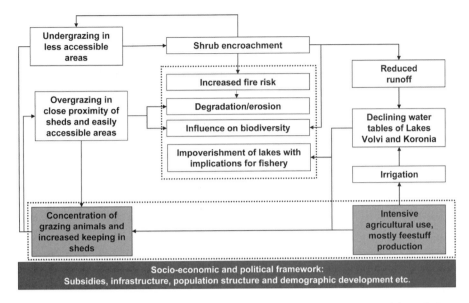

Figure 8.   Ecosystem processes and dynamic interactions in Lagadas County (adapted from Röder et al. 2008).

The actual rangeland areas are interspersed with other land uses within the clearly structured landscape units in Lagadas County. When interpreting the findings from this study with local knowledge of the area and other studies, numerous processes and mutual interactions between and within these units may be identified. Figure 8 aims at placing the results from this study in a wider physical and socio-economic framework. The latter encompasses the system of subsidies granted by different European funding schemes, demographic tendencies, the availability of different income sources etc.

Grazing as a major factor of the rural economy is nowadays being managed in a semi-intensive to intensive manner. This includes a concentration of grazing grounds, long periods of keeping animals in sheds and extensive provision of feedstuffs. This has led to widespread overgrazing in easily accessible areas in the proximity of sheds, while at the same time causing undergrazing in less accessible areas. This causes two major feedback loops—severe degradation in the first case, necessitating even further concentration. In the second, it effectuates the encroachment of shrubs, rendering such areas even more unsuitable for grazing and thus inducing a second feedback loop. Although an increase in vegetation cover may be perceived positively at first, it may result in an increase in fire risk, reduced water availability in the aquifers and reduced runoff to the lakes. All the same, the destruction of vegetation cover through fire frequently accelerates erosion of bare soil, such that undergrazing may ultimately contribute to the same process loop as do overgrazing and trampling. Further to these evident consequences, all of these processes clearly impact on biodiversity. Although the sign of this impact may vary depending on the observation scale and target, studies of avifauna in the area have shown that the maximum variety and number of species are found in sparse to medium rangelands, which are most threatened by unsustainable use of rangelands (Papoulia et al. 2003).

On the other hand, frequent land use conversion towards arable land and irrigated agriculture was noted in past years (Stellmes et al. 2007), which is partially due to the increased production of feedstuffs. Added to by reduced runoff in channels covered by woody vegetation, the uptake of water from the lakes is affecting water tables and fishery stocks in both lakes (Stellmes et al. 2007), and Lake Koronia has already experienced a reduction in water table of almost 90% for the period from 1972 to 1998 (Karavokyris et al. 1998). In addition, fertilizers, pesticides and other chemicals deployed on the fields are additionally causing damage to the lake ecosystems.

As a result of these findings and the evidence of specific aspects provided by geospatial data analysis, different recommendations for an enhanced management of the resources emerge. These encompass the implementation a more extensive system of rotating grazing schemes with different intensities, which combines conservation of floral diversity with pastoral quality (Noy-Meir 1998, Rowntree et al. 2004). Recent modifications of the subsidy schemes of the European Union provide the necessary instruments, as payments are de-coupled from production numbers, while incentives may be provided for good agricultural practices that meet the demands of ecological soundness and sustainability. Furthermore, mechanically clearing dense shrubs in the narrow valleys and conversion of irrigated agriculture in the valley of Mygdonia to less intensive uses might improve the quality of the lake ecosystem. This could be added by reinforcing its status as protected areas and promote additional sources of income other than intensive agricultural use or grazing, such as recreation or sustainable tourism.

The pattern of processes and mutual impacts illustrated here depicts the present state consensus, thus reflecting one particular situation within a dynamic system. Ongoing studies may require modifications, while changes in socio-economic and physical frameworks may affect the whole system by causing single elements to change. In particular the predicted impacts of global climate change on the Mediterranean Basin (Jeftic et al. 1994, 1996) are expected to severely affect rangeland-dependant economies in the future, necessitating appropriate mitigation strategies.

Concluding, this study demonstrated different methodological pathways to characterizing environmental dynamics based on specific processing and analysis of remote sensing and geospatial data. Given a sound conceptual knowledge of processes operating in a target area, these could be used to provide spatially explicit evidence for specific phenomena, thus representing an important basis for the formulation of sustainable management schemes adapted to the local needs.

ACKNOWLEDGEMENTS

This study was partially carried out in the frame of the projects 'GeoRange' (Geomatics in the Assessment and Sustainable Management of Mediterranean Rangelands) and DeSurvey-IP (A Surveillance System for Monitoring and Assessing of Desertification), which have been funded by the European Union, DG Research.

REFERENCES

Alados, C.L., El Aich, A., Papanastasis, V.P., Ozbek, H., Navarro, T., Freitas. H., Vrahnakis, M., Larossi, D. & Cabezudo, B. 2004. Change in plant spatial patterns and diversity along the successional gradient of Mediterranean grazing ecosystems. *Ecological Modelling* 180(2004): 523–535.

Aranuvachapun, S. 1985. Satellite remote sensing of atmospherical optical depth spectrum. In E.M. Rollin, M.D. Steven & P.M. Mather (eds.), Atmospheric Corrections for Remote Sensing. *Proceedings of a Remote Sensing Workshop on Atmospheric Corrections*, 22nd May 1985. University of Nottingham.

Berk, A., Anderson, G.P., Acharya, P.K., Chetwynd, J.H., Bernstein, L.S., Shettle, E.P., Matthew, M.W. & Adler-Golden, S.M. 1999. *MODTRAN4 user's manual*. Hanscom AFB, MA 01731-3010: Air Force Research Laboratory, Space Vehicles Directorate, Air Force Material Command.

Blondel, J. & Aronson, J. 1995. Biodiversity and ecosystem function in the Mediterranean basin: human and non-human determinants. In G.W. Davis, & D.M. Richardson (eds.), *Mediterranean-type ecosystems*. (Ecological Studies 109) (pp. 43–104). New York, Berlin, Heidelberg: Springer.

Boer, M. 1999. *Assessment of dryland degradation: linking theory and practice through site water balance modelling*. Nederlandse Geografische Studies, 251, Utrecht.

Briske, D.D. & Noy-Meir, I. 1998. Plant responses to grazing: a comparative evaluation of annual and perennial grasses. In V.P. Papanastasis & Peter, D. (eds.), Ecological basis of livestock grazing in Mediterranean ecosystems (pp. 13–26). Proceedings of the International Workshop held in Thessaloniki (Greece), October 23–25, 1997. Luxembourg: Office for Official Publications of the European Communities.

Carmel, Y. & Kadmon, R. 1999. Effects of grazing and topography on long-term vegetation changes in a Mediterranean ecosystem in Israel. *Plant Ecology* 145: 243–254.

Chiarucci, A., Wilson J.B., Anderson, B.J. & De Dominicis, V. 1999. Cover versus biomass as an estimate of species abundance: does it make a difference to the conclusions? *Journal of Vegetation Science* 10, 35–42.

Chouvardas, D. & Papanastasis, V.P. 2004. Stocking rate evolution in Lagadas County. Unpublished Presentation at the GeoRange Science Meeting, March 11th–12th, 2004, Ispra, Italy.

Del Barrio, G., Boer, M. & Puigdefábregas, J. 1996. Selecting representative drainage basins in a large research area using numerical taxonomy on topographic and climatic raster overlays. In M. Rumor, M.C. Millan & H.F.L Ottens (eds.), *Geographical information, from research to application through cooperation* 1, pp. 398–407, Amsterdam, IOS Press.

Di Castri, F. 1981. Mediterranean type shrublands of the world. In F. Di Castri, D.W. Goodall & R.L. Specht (eds.), *Mediterranean-type shrublands, ecosystems of the world*, 11 (pp. 1–42). Amsterdam, Oxford, New York: Elsevier.

Di Pasquale, G., Di Martino, P. & Mazzoleni, S. 2004. Forest history in the Mediterranean region. In S. Mazzoleni, G. Di Pasquale, M. Mulligan, P. Di Martino & F. Rego (eds.), *Recent dynamics of Mediterranean vegetation and landscape* (pp. 13–20). New York: Wiley & Sons.

Dubost, M. 1998. European policies and livestock grazing in Mediterranean ecosystems. In Papanastasis, V.P. & Peter, D. (eds.), *Ecological basis of livestock grazing in Mediterranean ecosystems* (pp. 298–311). Proceedings of the International Workshop held in Thessaloniki (Greece) on October 23–25, 1997. Luxembourg: Office for Official Publications of the European Communities.

Elmore A.J., Mustard, J.F., Manning, S.J. & Lobell, D.B. 2000. Quantifying vegetation change in semiarid environments: Precision and accuracy of spectral mixture analysis and the normalized difference Vegetation index. *Remote Sensing of Environment* 73: 87–102.

Forman, R.T. & Godron, M. 1981. Patches and structural components for landscape ecology. *Bioscience* 31: 733–740.

Francis, C.F. & Thornes, J.B. 1990. Runoff hydrographs from three Mediterranean vegetation cover sites. In J.B. Thornes (ed.), Vegetation and geomorphology (pp. 363–385). London: Wiley & Sons.

Garcìa-Haro, F.J., Sommer, S. & Kemper, T. 2005. A new tool for variable multiple endmember spectral mixture analysis (VMESMA). *International Journal of Remote Sensing* 26 (10): 2135–2162.

Gomez-Sal, A. 1998. Relationships between ecological and socio-economic evaluations of grazing in Mediterranean ecosystems. In Papanastasis, V.P. & Peter, D. (eds.), *Ecological basis of livestock grazing in Mediterranean ecosystems* (pp. 275–286). Proceedings of the International Workshop held in Thessaloniki (Greece) on October 23–25, 1997, Luxembourg: Office for Official Publications of the European Communities.

Haddjigeorgiou, I., Vallerand, F., Tsimpoukas, K. & Zervas, G. 1998. The socio-economics of sheep and goat farming in Greece, and the implications for rural development. *Proceedings of the LSIRD Conference*, Bray, Dublin, Dec. 3rd to 5th, 1998.

Hay, J.E. & McKay, D.C. 1985. Estimating solar irradiance on inclined surfaces: A review and assessment of methodologies. *International Journal of Solar Energy* 3: 203–240.

Hay, J.E., Perez, R. & McKay, D.C. 1986. Addendum and errata to the paper estimating solar irradiance on inclined surfaces: A review and assessment of methodologies. *International Journal of Solar Energy* 4: 321–324.

Hill, J. & Mehl, W. 2003. Geo- und radiometrische Aufbereitung multi- und hyperspektraler Daten zur Erzeugung langjähriger kalibrierter Zeitreihen. *Photogrammetrie, Fernerkundung, Geoinformation* 1/2003: 7–14.

Hill, J., Hostert, P. & Röder, A. 2004. Long-term observation of mediterranean ecosystems with Satellite Remote Sensing. In S. Mazzoleni, G. Di Pasquale, M. Mulligan, P. Di Martino, & F. Rego (eds.), *Recent dynamics of Mediterranean vegetation and landscape* (pp. 29–39). New York: Wiley & Sons.

Hill, J., Mehl, W. & Radeloff, V. 1995. Improved forest mapping by combining corrections of atmospheric and topographic effects. In J. Askne (Ed.), *Sensors and environmental applications of remote sensing* (pp. 143–151). Proceedings of the 14th EARSeL Symposium, Göteborg, Sweden, 6–8 June 1994. Rotterdam, Brookfield: A.A. Balkema.

Hobbs, R.J., Richardson, D.M. & Davis, G.W. 1995, Mediterranean-type ecosystems: opportunities and constraints for studying the function of biodiversity. In G.W. Davis, & D.M. Richardson (eds.), *Mediterranean-type ecosystems. The function of biodiversity* (pp. 1–42). Berlin, Heidelberg, New York: Springer.

Hostert, P., Röder, A. & Hill, J. 2003a. Coupling spectral unmixing and trend analysis for monitoring of long-term vegetation dynamics in Mediterranean rangelands. *Remote Sensing of Environment* 87: 183–197.

Hostert, P., Röder, A., Hill, J., Udelhoven, T. & Tsiourlis, G. 2003b. Retrospective studies of grazing-induced land degradation: a case study in central Crete, Greece. *International Journal of Remote Sensing* 24(20): 4019–4034.

Jeftic, L., Milliman, J.D. & Sestini, G., 1994. *Climatic change and the Mediterranean: 1.* Chichester: Wiley & Sons.

Jeftic, L., Keckes, S. & Pernetta, J.C., 1996. *Climatic Change and the Mediterranean: Environmental and societal impacts of climatic change and sea level rise in the Mediterranean region: 2.* Chichester: Wiley & Sons.

Karavokyris, G. 1998. *Environmental Rehabilitation of Lake Koronia, Greece. A Master Plan.* Final Report (with annexes). European Commission, Directorate General XVI, Regional Policy and Cohesion.

Karnieli, A., Gilad, U., Ponzet, M. & Svoray, T. 2006. Satellite image processing and geo-statistical methods for assessing land degradation around watering points in the central Asian deserts. In: Röder, A. & Hill, J. (eds.), *Proceedings of the 1st International Conference on Remote Sensing and Geoinformation Processing in the Assessment and Monitoring of Land Degradation and Desertification (RGLDD)*, Trier, 7.-9. Sept, 2005; 301–307 (http://ubt.opus.hbz-nrw.de/volltexte/2006/362/).

Konstantinidis, P. & Tsiourlis, G., 2003. *Description—analysis and mapping of vegetation units (habitats) of Lagadas County (Thessaloniki, Greece).* NAGREF—Forest Research Institute, GeoRange project report. Thessaloniki, Greece.

Le Houérou, N.H. 1981. Impact of man and his animals on Mediterranean vegetation. In F. Di Castri, W. Godall, & R.I. Specht (eds.), *Ecosystems of the world, 11: Mediterranean-type shrublands (* pp. 479–521). New York: Elsevier.

Legg, C., Papanastasis, V.P., Heathfield, D., Arianoutsou, M., Kelly, A., Muetzelfeldt, R. & Mazzoleni, S. 1998. Modelling the impact of grazing on vegetation in the Mediterranean: the approach of the ModMED project. In Papanastasis, V.P. & Peter, D. (eds.), *Ecological basis of livestock grazing in Mediterranean ecosystems* (pp. 189–199). Proceedings of the International Workshop held in Thessaloniki (Greece) on October 23–25, 1997. Luxembourg: Office for Official Publications of the European Communities.

Mulligan, M., Burke, S.M. & Ramos, C. 2004. Climate change, land-use change and the "desertification" of Mediterranean Europe. In S. Mazzoleni, G. Di Pasquale, M. Mulligan, P. Di Martone & F. Rego (eds.), *Recent dynamics of the Mediterranean vegetation and landscape* (pp. 259–280). Chichester: Wiley & Sons.

Noy-Meir, I. 1998. Effects of grazing on Mediterranean grasslands: the community level. In Papanastasis, V.P. & Peter, D. (eds.), *Ecological basis of livestock grazing in Mediterranean ecosystems* (pp. 27–39). Proceedings of the International Workshop held in Thessaloniki (Greece) on October 23–25, 1997. Luxembourg: Office for Official Publications of the European Communities.

Oba, G., Post, E., Stenseth, N.C. & Lusigi, W.J. 2000. The role of small ruminants in arid zone environments: a review of research perspectives. *Annals of arid zone* 39(3): 305–332.

Papanastasis, V.P. 1998. Livestock grazing in Mediterranean ecosystems: an historical and policy perspective. In Papanastasis, V.P. & Peter, D. (eds.), *Ecological basis of livestock grazing in Mediterranean ecosystems* (pp. 5–10). Proceedings of the International Workshop held in Thessaloniki (Greece) on October 23–25, 1997. Luxembourg: Office for Official Publications of the European Communities.

Papoulia, S., Kazantzidis, S. & Tsiourlis, G. 2003. *Distribution of the bird fauna in shrub lands and forests of Lagadas area* (Thessaloniki, Greece). NAGREF – Forest Research Institute, GeoRange project.

Pickup, G. & Chewings, V.H. 1988. Estimating the distribution of grazing and patterns of cattle movement in a large arid zone Paddock: an approach using animal distribution models and Landsat imagery. *International Journal of Remote Sensing* 9(9): 1469–1490.

Pickup, G. & Chewings, V.H. 1994. A grazing gradient approach to land degradation assessment in arid areas from remotely-sensed data. *International Journal of Remote Sensing* 15(2): 597–617.

Pickup, G., Bastin, G.N. & Chewings, V.H. 1998. Identifying trends in land degradation in non-equilibrium rangelands. *Journal of Applied Ecology* 35(1): 365–377.

Roberts, D.A., Gardner, M., Church, R., Ustin, S., Scheer, G. & Green, R.O. 1998. Mapping chaparral in the Santa Monica Mountains using multiple endmember spectral mixture models. *Remote Sensing of Environment* 65: 267–279.

Röder, A. 2005. *A Remote Sensing Based Framework for Monitoring and Assessing Mediterranean Rangelands. Case Studies from Two Test Sites in Spain and Greece.* PhD Thesis, Trier (http://ubt.opus. hbz-nrw.de/volltexte/2006/350/).

Röder, A., Kuemmerle, T., Hill, J., Papanastasis, V.P. & Tsiourlis, G.M. 2007. Adaptation of a grazing gradient concept to heterogeneous Mediterranean rangelands using cost surface modelling. *Ecological Modelling*, 204: 387–398.

Röder, A., Udelhoven, Th., Hill, J., Del Barrio, G. & Tsiourlis, G.M. 2008. Trend analysis of Landsat-TM and –ETM+ imagery to monitor grazing impact in a rangeland ecosystem in Northern Greece. *Remote Sensing of Environment* (doi:10.1016/j.rse.2008.01.08).

Rowntree, K., Duma, M., Kakembo, V. & Thornes, J.B. 2004. Debunking the myth of overgrazing and soil erosion. *Land degradation & development* 15: 203–214.

298

Schlittgen, R. & Streitberg, B.H.J., 1999. Zeitreihenanalyse. München, Wien: Oldenburg Verlag.

Shoshany, M. & Svoray, T. 2002. Multidate adavptive unmixing and its application to analysis of ecosystem transition along a climatic gradient. *Remote Sensing of Environment* 82, 5–20.

Small, C. 2004: The Landsat ETM+ spectral mixing space. *Remote Sensing of Environment* 93: 1–17.

Smith, M.O., Ustin, S.L., Adams, J.B. & Gillespie, A.R. 1990. Vegetation in deserts: I. A regional measure of abundance from multispectral images. *Remote Sensing of Environment* 31: 1–26.

Sneath, P.H. & Sokal, R.R. 1973. *Numerical taxonomy*. San Francisco: Freeman.

Sparrow. A.D., Friedel, M.H. & Stafford-Smith, D.M. 1997. A landscape-scale model of shrub and herbage dynamics in Central Australia, validated by satellite data. *Ecological Modelling* 97: 197–216.

Stafford-Smith, D. M. & Reynolds, J.F. 2002. The Dahlem desertification paradigm: a new approach to an old problem. In J.F. Reynolds, & M.D. Stafford-Smith (eds.), *Global desertification. Do humans cause deserts?* (pp. 403–424). Dahlem workshop report 88. Berlin: Dahlem University Press.

Stellmes, M., Hill, J. & Röder, A. 2007. Maps of land-use/cover change and land degradation status. Lagadas - part 2. In DeSurvey-IP (Ed.), A Surveillance System for Assessing and Monitoring Desertification (p. 22). Brussels: European Commission, DG Research

Tanré, D., Deroo, C., Duhaut, P., Herman, J.J., Perbos, J. & Deschamps, P.Y. 1990. Description of a computer code to simulate the signal in the solar spectrum—the 5S code. *International Journal of Remote Sensing* 11(4): 659–668.

Teillet, P.M. & Fedosojevs, G. 1995. On the dark target approach to atmospheric correction of remotely sensed data. *Canadian Journal of Remote Sensing* 21: 374–387.

Teillet, P.M., Barker, J.L., Markham, B.L., Irish, R.R., Fedosejevs, G. & Storey, J.C. 2001. Radiometric cross-calibration of the Landsat-7 ETM+ and Landsat-5 TM sensors based on tandem data sets. *Remote Sensing of Environment* 78: 39–54.

Thome, K.J., Markham, B., Barker, J., Slater, P. & Biggar, S. 1997a. Radiometric calibration of Landsat. *Photogrammetric Engineering & Remote Sensing* 63(7): 853–858.

Thome, K.J., Crowther, B.G. & Biggar, S.F. 1997b. Reflectance- and irradiance-based calibration of Landsat-5 Thematic Mapper. *Canadian Journal of Remote Sensing* 23(4): 309–317.

Thornes, J.B. 1985. The ecology of erosion. *Geography*, 70(3): 222–236.

Thornes, J.B. 1990. The interaction of erosional and vegetational dynamics in land degradation: spatial outcomes. In J.B. Thornes (Ed.), *Vegetation and erosion: Processes and environments* (pp. 41–54). Chichester, New York, Brisbane, Toronto, Singapore: Wiley & Sons.

Thrash, I. & Derry, J.F., 1999: The nature and modeling of biospheres: A review. *Koedoe*, 42(2): 73–94.

Tomlin, C.D. 1990. *Geographic information systems and cartographic modelling*. New Jersey: Prentice Hall.

Tsiourlis, G.M. 1998. Evolution of biomass and productivity of grazed and ungrazed kermer oak shrubs in an insular phryganic ecosystem of Naxos, Greece. In Papanastasis, V.P. & Peter, D. (eds.), *Ecological basis of livestock grazing in Mediterranean ecosystems* (pp. 86–89). Proceedings of the International Workshop held in Thessaloniki (Greece) on October 23–25, 1997. Luxembourg: Office for Official Publications of the European Communities.

Tsiouvaras, C.N., Koukoura, Z., Platis, P. & Ainalis, A. 1998. Yearly changes in vegetation of a semi-arid grassland under various stocking rates and grazing systems. In Papanastasis, V.P., V.P. & Peter, D. (eds.), *Ecological basis of livestock grazing in Mediterranean ecosystems* (pp. 58–61). Proceedings of the International Workshop held in Thessaloniki (Greece) on October 23–25, 1997. Luxembourg: Office for Official Publications of the European Communities.

Yiakoulaki, M.D., Zarovali, M.P., Ispikoudis, I. & Papanastasis, V.P. 2002. Evaluation of small ruminant production systems in the area of Lagadas County. *Proceedings of 3rd National Rangeland Congress*. Karpenisi, Greece, 4–6 September 2002, (in press).

Recent Advances in Remote Sensing and Geoinformation Processing
for Land Degradation Assessment – Röder & Hill (eds)
© 2009 Taylor & Francis Group, London, ISBN 978-0-415-39769-8

Integrating GPS technologies in dynamic spatio-temporal models to monitor grazing habits in dry rangelands

T. Svoray & R. Shafran-Nathan
Ben-Gurion University of the Negev, Beer-Sheva, Israel

E.D. Ungar, A. Arnon & A. Perevolotsky
Agricultural Research Organization, Bet-Dagan, Israel

ABSTRACT: Land degradation in drylands occurs due to a combination of drought and misman-agement of the land while areas of rough topography or undeveloped soils are under greater threat. The role of habitat preference by grazers in these processes is difficult to study because of the complexity involved with animal tracking and mapping of spatio-temporal variation in vegetation status. We addressed this problem by integrating GIS and GPS technologies. In this chapter we assess the time spent by sheep and goat herds in plant habitats in a semi-arid environment. We used a predictive habitat distribution model which is spatially and temporally explicit, based on rules formulated using fuzzy logic. The results show that biomass availability together with a number of abiotic factors (slope decline and orientation, distance from the corral and topographic sub-slope units) could not fully explain the pathway of herds in the study area. However, the effect of these variables on the time spent by the herd in the different habitats was significant. Specifically, the time spent by the herd at a given location (grid cell) along the herding route was positively correlated with the biomass availability in the cell. Furthermore, the time spent by the herd in a cell responded non-linearly to the distance from the corral, being relatively low at near and far distances, and high at intermediate distances. The results imply that additional factors in these systems have a strong influence on the grazing habits.

1 INTRODUCTION

It has been widely agreed that desertification and land degradation in drylands occur due to a combination of drought, i.e., worsening in climate conditions (Le Houerou 1996), and human mis-management of the land (Foley et al. 2005). Other physical properties such as rough topography or undeveloped soils may also contribute to land degradation and such areas are therefore under greater threat. Desertification may cause economic and societal damages as well as environmental damages due to changes in ecosystem functioning (Ungar et al. 1999). In the early days of state-hood, fifty years ago, desertification was not a major problem in the semi-arid part of Israel and therefore was not of major public concern. However, in recent decades, high intensity gully erosion both in agricultural areas and rangelands, mainly in the loess areas, may be a sign of emerging desertification and potential future risks (Portnov & Safriel 2004). Therefore, there is a need for reliable indicators as well as efficient monitoring systems to measure the magnitude and rate of desertification processes in this dramatically fluctuating climatic region. Among the major factors that can cause accelerated land degradation processes is overgrazing which is, in many cases, very difficult to assess. Geographic information technologies provide an excellent infrastructure for the integration of: 1) remote sensing data on vegetation and soil characteristics; 2) GPS data on herd movements; and 3) dynamic spatio-temporal ecological and environmental modeling. The use of this framework for studies of ecosystem functioning as an indicator for desertification is particu-larly useful due to the dynamic nature of vegetation and the immediate response of ecosystems to land degradation and desertification.

Annual production of herbaceous vegetation plays a key role in understanding ecosystem functioning (Cao et al. 2004, Jobbagy et al. 2002). In semi-arid regions this production responds significantly to spatial and temporal changes in rainfall amounts (Noy-Meir 1973), soil moisture distribution and variation in grazing pressures (Adler et al. 2001). As a result, semi-arid regions are characterized by a patchy spatial pattern of primary production. However, this pattern has not yet been fully analyzed at the slope and watershed scales. In this study we integrate remote sensing images, geographic information techniques, ecological modeling and the tracking of animals by GPS, in a geo-computational framework, to study the effect of available biomass and abiotic factors on the grazing habits of small ruminant herds.

2 THE STUDY AREA

The study was conducted at the Lehavim Bedouin demonstration farm (31020'N, 34045'E), which is a Long-Term Ecological Research (LTER) station located in the northern Negev region of Israel (Fig. 1).

Average rainfall at the site is approximately 280 mm per annum, with most of the rainfall occurring in the period November to April. The mean annual temperature is 20.5°C, with a maximum of 27.5°C and a minimum of 12.5°C (The Meteorological Survey of Israel). The terrain is hilly and the dominant rock formations are Eocenean limestone and chalk with patches of calcrete. Soils are brown lithosols and arid brown loess partly covered by a microphytic crust. Alluvial soils predominate at the wadi shoulders. The phytogeography of the region is Irano-Turanian, but Mediterranean and Saharo-Arabian species can also be found. Vegetation at the site is characterised by scattered dwarf shrubs and patches of herbaceous vegetation, mostly annual, which spread between

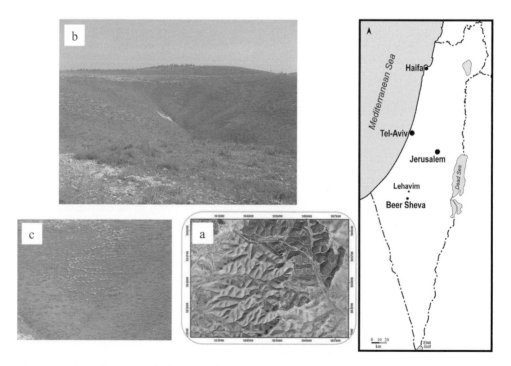

Figure 1.   The study area—Lehavim LTER site. (a) airphoto representing the hilly nature of the topography in the Lehavim LTER area; (b) a typical wadi shoulder habitat that serves as a productive grazing area; (c) a zoom-in of 1b with presence of sheep herd.

rocks and dwarf shrubs. The dwarf-shrub community develops a steppe-like landscape with diffuse vegetation. The dwarf shrub community is dominated by "*Sarcopoterium spinosum (L.) Spach, Corydothymus capitatus (L.) Reichenb.* [. . .] *Thymelaea hirsuta (L.) Endl.*" *Sarcopoterium spinosum (L.) Spach, Corydothymus capitatus (L.) Reichenb.* and *Thymelaea hirsuta* (L.) Endl. The herbaceous vegetation appears shortly after the first rains and persists as green forage for 3–4 months, depending on the amount and distribution of rainfall. The herbaceous vegetation is highly diverse, mostly composed of annual species that represent 56% of the regional flora. Forest plantations, natural reserves, and wheat fields surround the Lehavim site.

3 METHODS

The geoinformatics scheme that we employed (Fig. 2) included the following four components: a) remote sensing to determine rock coverage from aerial photographs and for validation of model predictions using NDVI layers extracted from an IKONOS image; b) process-based ecological modeling using fuzzy logic to predict herbaceous vegetation production; c) an environmental GIS database to apply explicitly in space the fuzzy rules of the ecological model; and d) the speed of movement in each grid cell of goat and sheep herds, monitored by animal-borne GPS. The following sections detail the role of each component of the scheme.

3.1 *Remote sensing*

Color (VIS) aerial photos with 1.25×1.25 m^2 ground resolution, acquired at 1336 h on 31 Dec. 2004 were used to classify three land cover categories: rock and stone, bare soil, and vegetation. We used maximum likelihood classification and, similarly to previous studies (Svoray et al. 2007), confusion matrices (n = 300) showed a very good correlation (>90% overall accuracy) between image processing results and field observations. Sampling scheme of validation data was based on 100 random sites for each of the land covers. Each site was visually identified and, in places of doubts, the sites were visited in the field. After classification accuracies were deemed satisfactory, the frequencies of rock class were calculated for each cell of a 25×25 m fishnet that covered the entire study area.

In addition to the use of airphotos for mapping rock coverage, one tile of IKONOS image from winter 2003, at a resolution of 4 meters, was used to calculate the recent version of the Normalized Difference Vegetation Index based on Ünsalan & Boyer (2004). The four IKONOS spectral

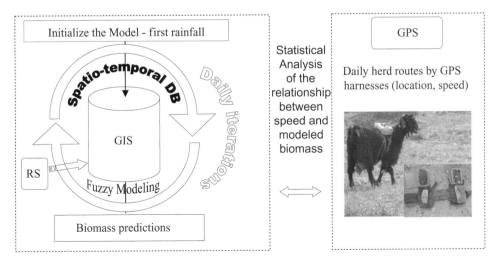

Figure 2. The geoinformatics scheme.

bands were calibrated against reflectance units according to the Empirical Line technique (Smith and Milton 1999). File coordinates of the two images were transformed to the Israel Grid, on the basis of more than 50 differential GPS (DGPS) measurements, through a first-order transformation (RMS < 1 pixel), and the image pixels were re-sampled by means of the nearest-neighbor technique. After preprocessing, NDVI values were calculated. NDVI has been used as a surrogate of biomass production in semiarid areas by other authors (e.g., Paruelo et al. 1997), who found good correlations between these two variables in grassland ecosystems. That is because NDVI is sensitive to green biomass, not to the woody components that in many ecosystems form the largest biomass contributor. Thus, the use of NDVI to predict biomass is especially appropriate for herbaceous vegetation. We compared NDVI and biomass of annuals in several plots in our study area, and found that the values were strongly correlated ($R2 = 0.81$, $P < 0.001$). Other studies reported that NDVI is of limited use in areas of sparse vegetation (e.g., Rondeaux et al. 1996, Elmore et al. 2000) and yet our data show that in the study area the NDVI is the most efficient surrogate to vegetation cover and biomass.

3.2 Ecological modeling and GIS database

The ecological model used in this study is described in detail in Shafran-Nathan (2006) and in Svoray et al. (Submitted) and therefore only the key principles of the model are presented here.

The model aim is to predict habitats of herbaceous vegetation in dry environments while the quality of habitats is assessed on the basis of their potential to produce herbaceous vegetation biomass. This potential may vary with time—due to variation in climatic conditions—and with space—due to variation in environmental and biotic conditions. We assumed that in Mediterranean, semi-arid and other water-constrained environments, water availability and temperature play a crucial role in determining herbaceous vegetation productivity. Therefore, the model was developed based on the hypothesized water requirements of herbaceous vegetation in the study area. The model created maps of the spatial variation in four variables that indirectly influence water availability: rock cover, radiation, runoff, and sub-slope units. Temporal variation in water availability was expressed by meteorological data.

As was described in section 3.1, rock cover was mapped using high resolution airphotos. The radiation, sub-slope units and runoff layers of the GIS database were predicted using digital elevation data with 25×25 m cell resolution. In addition, the runoff model (Dynamic TOPMODEL—Beven & Freer 2001) took advantage of data on soil characteristics extracted from a field survey. Mean catchment deficit was calculated by estimating the difference between saturated soil moisture content and mean gravimetric soil moisture content, both measured by us in the field ($n = 24$). Upslope contributing area and local slope angle were obtained from the DEM on a cell basis. The shaping parameter (m) for the form of exponential decline in conductivity with depth was extracted according to Beven (1984) from a regression model between empirical measurements of soil texture—taken by us in the field—that were converted to hydraulic conductivity by using the equations of Clapp and Hornberger (1978), and soil depth measurements taken by Kyumedjinski et al. (1988). Rock and soil surfaces differ significantly in their response to rainfall, and therefore the dynamics of the saturated zone cannot always be approximated by a quasi-steady-state representation. Therefore, to apply the soil data to the entire study area, we categorized the study area into eight sub-slope units and assigned soil properties to each of them based on data from soil surveys (with three replicas for each sub-slope unit and 24 sites overall). To calculate the soil water deficit, the cells in the study area were divided into sink and source units. Solar radiation in the study area was predicted on the basis of three DEM-based variables: slope orientation and decline and hillshade, as well as the sun's altitude and azimuth. The recently developed method of Park et al. (2001) and the DEM of Hall and Cleave (1988), with 25 m and 10 m vertical and horizontal resolutions respectively, were used for mapping six sub-slope units: interfluve, shoulder (seepage slope and convex creep slope), backslope (free face and transportational midslope), footslope (colluvial footslope), toeslope (alluvial toeslope), and channel (Fig. 3). The model of Park et al. (2001) predicts the occurrence of soils over a hilly landscape by separating the sub-slope units

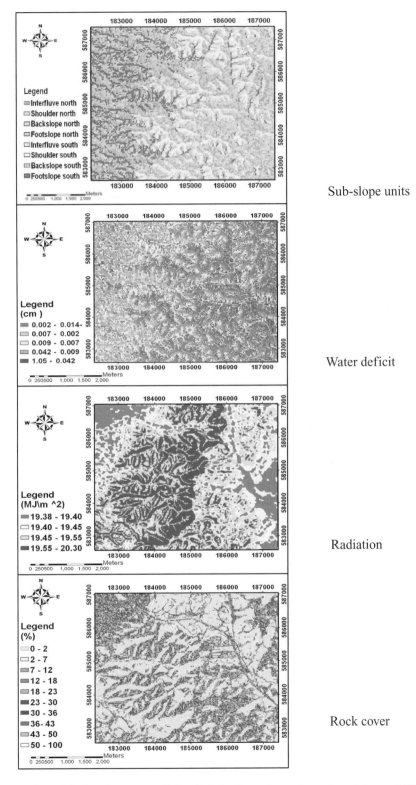

Figure 3. The GIS database layers as applied to the Lehavim study area (see colour plate page 398).

305

according to similarity in surface processes. The model is based on the idea of possible links between physiographic units and soil formations as presented by Conacher and Dalrymple (1977) who defined the slope as a three-dimensional pedogeomorphological body extending from the inter-fluve to the channel bed. Each point on a slope is allocated to one of nine units, and has distinct pedogenic characteristics reflecting the influence of soil–water–gravity interrelationships governed by surface forms.

The model used daily values of three climatic variables: temperature, evaporation and rainfall amount. These were obtained from meteorological stations of the Israeli Meteorological Survey.

The database was integrated to apply the rules of water requirements through the use of fuzzy logic. Fuzzy logic is a theory in formal mathematics that enables the computation of a definitive solution for problems that are complex, uncertain and unstructured (Burrough et al. 1992). There-fore, it is a most suitable modeling platform for hydrological and ecological phenomena (Svoray et al. 2004). Our fuzzy rule-based production model predicts two processes in the life cycle of the herbaceous vegetation: 1) germination—modeled in daily iterations until the conditions are satisfied; 2) primary production—modeled from germination until the end of the rainy season. In general, fuzzy formulation is composed of three primary elements: fuzzy sets, fuzzy membership functions and fuzzy logic joint membership functions (Zhu et al. 1996). In our model, the set A in equation 1 represents a group of cells, i.e., real world unit areas of 25×25 m, with optimal conditions for biomass production of annual herbaceous vegetation:

$$A = \{x, \mu_A(x)\} \quad \text{for each } x \in X \tag{1}$$

where: x is the variable and $A(x)$ is the membership function of x in A. The scientific literature provides numerous membership functions, including, for example, linear, trapezoidal, sigmoid and cosine (Robinson 2003). The membership functions (MFs) define the degree of membership for each cell or attribute in the set (A). The membership value is subjective and should reflect the way in which the researcher grasps the phenomenon. The degree of membership is limited in the 0–1 range, where 1 means complete membership. The set A was defined in our study as the set with optimal conditions for vegetation production. For example, as rainfall amount increases, more water is available to the plant and production is expected to be more intense. Therefore, the rainfall membership function we used is linear positive:

$$MF = \frac{x - \alpha}{\beta - a} \tag{2}$$

where X is the grid cell value or the record in the attribute cell (fuzzy set); α is the minimum value in the set, and β is the maximum. For the other variables, we used different membership functions (Shafran-Nathan 2006).

To integrate the effect of all membership values (spatial and temporal), we used the joint mem-bership function (JMF). There are several ways to perform a JMF and we have used the convex combination operation. Equation 3 shows the operation where every MF for each fuzzy set A is the membership function $\mu(X1 \cdots Xn)$, and the weight λ $(1 \cdots 7)$.

$$JMF = \frac{\lambda_1 mf_1 + \lambda_2 mf_2 + \cdots + \lambda_n mf_n}{\sum_{i=1}^{n} \lambda_i} \tag{3}$$

The weights (λ) determine the extent to which each membership function contributes to the final set (daily germination or production prediction). Consequently, they represent a hierarchy of the variable's contributions to the JMF and, hence, each variable's value in the final predictive model. In our model, the membership scores accumulate from the first rainfall event (day one). Germination conditions (JMFgermination) are calculated on a daily basis, until conditions for germination are fulfilled based on a threshold value that was set to 1. If a cell has germinated, it starts to accumulate daily conditions for productions (JMFproduction) until the end of the season.

To validate the production model we used ground biomass data that was measured in a group of 12 randomly selected plots that were sampled during six field campaigns between 2004 and 2006. Each plot represents an area of >9 model cells and represents a specific habitat in terms of botanical characteristics. Biomass estimations were determined in these plots using the harvest method and DGPS readings of each plot's center were taken to allow assessment with reference to the model data.

3.3 Herd movement

The use of GPS for tracking the movement of foraging animals and for testing their effects on vegetative landscapes is an emerging field (Forester et al. 2007, Clark et al. 2006, Ungar et al. 2005, Rodgers 2001). Nevertheless, we could not find in the literature spatially explicit studies of the effects of herds of sheep and goats on herbaceous vegetation production in semi-arid areas and thus on desertification and land degradation. The GPS analyses conducted here sought to represent spatial variation in grazing pressures.

A Bedouin household is resident at the study site, and has sole grazing rights. He owns two herds of animals: one of 400 sheep and another of 200 goats, which are reared in the traditional Bedouin manner. This involves a yearly management cycle: from January or February to May, the animals graze the green rangeland quite heavily, until the fields of wheat and barley in the region are harvested. During the summer transhumance period (June to September), the herds graze on the harvested fields and other agricultural aftermaths. The herds then return to the rangeland, and during the fall and early winter (October to January), graze the remaining dry vegetation, while drawing a significant portion of their nutritional requirements from supplemented feeds such as maize and barley grains (Ginguld et al. 1997). During the grazing periods on the station, the herds leave, and return to, a fixed corral and watering point, on a daily grazing excursion across the landscape. A typical excursion lasts about 7 hours, and a shepherd accompanies each herd. To track the routes of the herds, we used two off-the-shelf Trimble GEII Explorer GPS rover units and purpose-designed harnesses which were strapped to the back of one volunteer animal in each herd (see Figure 2). The harnesses had two pouches, one for the GPS unit and the other for an external battery, which enabled approximately eight hours of continuous data collection. An external GPS antenna was situated between the two pouches, facing straight up to the sky. The harness did not seem to disturb or influence the behavior of the individual carrying it in any way. The herding routes of the sheep and goat herds were tracked on 78 days in the green season of 2003 (February to May). The Trimble GEII Explorer is a six-channel device with a relatively large memory capacity which enables it to store over 10,000 locations under standard settings. Hence we were able to operate the device with a fix interval of just 30 seconds and download the stored data once every few days. For our purposes, there was no need to conduct differential correction of the data; tests we conducted with the stationary device showed that 50%, 90% and 95% of the values were within 2.8, 7.0 and 8.6 meters of the mean, respectively. (Note that this study was conducted after Selective Availability was turned off.)

3.4 Statistical analysis

For the purposes of the statistical analysis, the herding routes were overlaid on GIS raster layers containing data on vegetation production at a resolution of 25×25 m per cell. For the vegetation map of early season phase, we took all the herding routes gathered between 25 Feb. and 5 Mar. 2003. This yielded a total of 5993 GPS locations. For the vegetation map of late season phase, we took all the herding routes gathered between 25 Mar. and 10 Apr. 2003. This yielded a total of 8254 GPS locations. Each GPS location was joined with characteristics of the grid cell within which it was located. These characteristics included Aspect (north, east, south and west), Slope ($0–22°$), Catena (eight sub-slope units), the modeled primary production potential ($0–1$), and distance to the central corral. The speed of travel along the herding route was computed from adjacent GPS locations and the time interval. The speed of the herd movement in the cells represents the herd initiative to spend time at the cells.

We used multiple regression to analyze the speed of herd movement as a function of available biomass, distance to the corral, sub-slope units, and slope decline and orientation. Linear and quadratic terms were included in the model for the continuous variables. Data was analyzed using the software package JMP version 5.1 (SAS Institute Inc., Cary, NC).

4 RESULTS

The production model was executed and tested against field measurements (using the harvest method) acquired during three years 2003–2005. The results showed good agreement and the coefficients of determination of the correlations between predicted and measured values varied between 0.5 to 0.74, with P < 0.001 in all cases. In addition, the model was tested against satellite data and the regression analyses between the production model output and the biomass estimated from the IKONOS images showed even stronger correlations ($R2 = 0.81$, P < 0.0001).

The total area of the Lehavim Bedouin demonstration farm is approximately 5 km². Within this area the mean daily herd pathway length was 6.0 and 5.2 km for goats and sheep, respectively. Figure 4 shows the pathways of sheep and goat herds in the study area during the growing season of 2003.

The figure shows that both goats and sheep herds covered large parts of the study area and therefore most cells of the study area were visited by either sheep or goats. However the question

Figure 4. Annual routes of sheep (white dots) and goats (black dots) annotated on the background of the biomass model output from March 2003 (white colours represent low biomass potential while black colours represent cells of high biomass). For visual interpretation purposes, the biomass layer is overlaid, with 30% transparency, on a topographic map.

that arises is how much time did the herd spend in each cell (and thus what characterizes the cells that experience higher grazing pressures). The overall mean speed of herd movement was 0.26 metres per second. However, this differed according to species of animal and stage in season; goats tended to travel faster than sheep (0.297 vs. 0.250 m s^{-1}, respectively), and speed was greater in phase 1 of the season than phase 2 (0.300 vs. 0.246 m s^{-1}). We tested statistically the relationship between available biomass as predicted by the model for the period 25 Feb.–05 Mar. 2003 and speed of the goat or sheep herd in each cell. Table 1 shows that the effect of the abiotic factors on the speed of the herd was found significant or near significant in all cases.

When the number of variables in the model was reduced to available biomass and distance to corral, the level of significance of available biomass increased (Table 2).

For the goat herd, a negative coefficient was obtained for Available biomass, meaning that the herd moved more slowly in cells in habitats of higher quality, i.e., the herd spent significantly more time in areas of high available biomass and vice versa. For the sheep herd, however, the corresponding coefficient was positive, albeit much closer to zero. The response to Distance to corral for both herds was curvilinear: speed of herd movement declined with increasing Distance to a minimum at approximately 1000 m for goats and 900 m for sheep, and increased thereafter.

A similar analysis was conducted for the second phase using the herding routes for the period 25 Mar. to 10 Apr. 2003. Here too, the linear and quadratic terms for distance to the corral were highly significant, indicating a minimum speed at approximately 900 m for goats and 770 m for

Table 1. Levels of significance of the six environmental variables in a multiple regression analysis of speed of herd movement in the cells of the study area. The available biomass is the model prediction for mid winter 2003.

Variable	Goats	Sheep
Available biomass	0.0616	0.0682
Distance to corral (linear)	<0.0001	<0.0001
Distance to corral (quadratic)	<0.0001	<0.0001
Aspect	<0.0001	0.0014
Slope	0.0609	0.0396
Sub-slope units	0.0248	<0.0001

Table 2. Levels of significance and regression coefficients of the three independent variables in a multiple regression analysis of the speed of herd movement. The available biomass is the model prediction for mid winter 2003.

Variable	Goats		Sheep	
	P value	Coefficient	P value	Coefficient
Available biomass	0.0003	−0.006004	0.0114	0.001935
Distance to corral (linear)	<0.0001	−0.000195	<0.0001	−8.9e-5
Distance to corral (quadratic)	<0.0001	9.41e-7	<0.0001	2.00e-7

Table 3. Levels of significance and regression coefficients of the three independent variables in a multiple regression analysis of the speed of herd movement. The available biomass is the model prediction for spring 2003.

Variable	Goats		Sheep	
	P value	Coefficient	P value	Coefficient
Available biomass	0.0014	0.0021	0.0036	0.0020
Distance to corral (linear)	<0.0001	−6.5e-5	<0.0001	−2.14e-4
Distance to corral (quadratic)	<0.0001	2.92e-7	<0.0001	5.58e-7

sheep. Available biomass was significant for both animal species, and the coefficient was positive, suggesting increasing herd speed with increasing biomass (see Table 3). The change in direction of response of speed of travel to biomass availability between the two phenological phases may be due to an overall increase in herbage as the season progressed, combined with the fact that there tends to be a negative relationship between biomass and quality in the latter part of the season. This relationship derives from the effect of defoliation on plant maturation and hence quality (McNaughton 1984, Fryxell 1991, Milchunas et al. 1995). Thus, in the early part of the season, quality is generally high and the speed of the herd is inversely related to biomass. Later in the season, overall biomass availability is relatively high, and quality tends to be the more important factor controlling speed. Since this is negatively related to biomass, speed of travel is lower at lower biomass.

5   CONCLUSIONS

The integration of remote sensing, ecological modeling, GIS and GPS technologies in a geo-informatics scheme allows the grazing pressure imposed by herds of goats and sheep during the grazing season to be mapped and analyzed. The use of the geo-informatics scheme supported the hypothesis that biomass availability, distance from the corral, and slope decline and orientation all significantly affect the time spent by a herd in each cell of a semi-arid study area. The effect of biomass availability on herd residence time is not unidirectional and may depend on the relation between availability and forage quality at a given stage of the season. The distance from the corral was also found to be an important factor: both in the vicinity of the corral and in the remoter areas the herd tended to spend less time in each cell along its herding route (higher speed), while there was a tendency for greater residence times (lower speed) in cells located at intermediate distances from the corral.

The use of NDVI derived from IKONOS images, that were radiometrically corrected, was found to be efficient and the outcomes seem reliable against field measurements. Yet other studies (e.g., Volcani et al. 2005) have shown the limitations of the use of NDVI in the heterogeneous Mediter-ranean and semiarid environments. High spatial fragmentation in soil and vegetation formations in these environments causes mixture at spatial resolutions greater than a few meters. Differentiation between elementary vegetation and soil cover fractions that was achieved using the multispectral unmixing technique (e.g., Shoshany and Svoray 2002) could help to better represent the vegetation status in these areas.

The variation in vegetation production that was not explained by the ecological model may be attributed to sources of stress that were not expressed in our model. These include nutrient availability, fire regimes, and facilitation and competition processes. These could be explored in future research although the model needs to be further studied to achieve better understanding on uncertainty and error propagation through the layers that may also cause disagreement between model predictions and field measurements. Particularly we recommend further study of the effect of location error in all layers and classification and DEM error effect on the final predictions.

From the applied point of view, our study can serve as an example of a framework for studying the effect of grazers on the ecosystem and on land degradation that can be used in other semiarid areas. Yet application of this framework to areas grazed by other animals, shepherd habits, different vegetation types as well as environmental conditions is still needed to test the validity of our findings. Also quantification of the effect of grazers on vegetation status in a regional scale is much sought (Karnieli et al. This Book) and should be further studied over wide regions using Landsat TM or Spot images.

Future development of the geo-informatic framework shown here will allow us to locate cells that experience higher grazing pressures and to estimate if they suffer deterioration in primary production. Such an analysis will help to locate areas of higher desertification risks. Further study will be needed on the effect of the shepherd on the time spent by the herd in each cell. We suspect that a be large part of the variation that could not be explained by the variables that we studied may be attributable to the preferences of the shepherd, which will be studied in future research.

ACKNOWLEDGMENTS

We thank the Israel Science Foundation (ISF) grant no. 692/06 for financial support. We also thank Rafi Yonatan, Dani Barkai, Ezra Ben-Moshe, Hagit Baram and Shimon Brener for their help with field work and the people from the Nasasra tribe for their cooperation and help in tracking the herds.

REFERENCES

Adler, P.B., Raff, D.A. & Lauenroth, W.K. 2001. The effect of grazing on the spatial heterogeneity of vegetation. *Oecologia* 128: 465–479.

Beven, K. 1984. Infiltration into a class of vertically non-uniform soils. *Hydrological Science Bulletin* 29: 425–434.

Beven, K. & Freer, J. 2001. A dynamic topmodel. *Hydrological Processes* 15: 1993–2011.

Burrough, P.A., Macmillan, R.A. & Van Deursen, W. 1992. Fuzzy classification methods for determining land suitability from soil profile observation and topography. *Journal of Soil Science* 43: 193–210.

Cao, M., Stephen, D.P. Small, J. & Goetz, S.J. 2004. Remotely sensed interannual variations and trends in terrestrial net primary productivity 1981–2000. *Ecosystems* 7: 233–242.

Clapp, R.B. & Hornberger, G.M. 1978. Empirical equations for some soil hydraulic properties. *Water Resources Research* 14: 601–604.

Clark, P.E., Johnson, D.E., Kniep, M.A., Jermann, P., Huttash, B., Wood, A., Johnson, M., Mcgillivan, C. & Titus, K. 2006. An advanced low-cost GPS-Based animal tracking system. *Rangeland Ecology & Management* 59 (3): 334–340.

Conacher, A.J. & Dalrymple, J.B. 1977. The nine unit landsurface model: an approach to pedogeomorphic research. *Geoderma* 18: 1–153.

Elmore, A.J., Mustard, J.F., Manning, S.J. & Lobell, D.B. 2000. Quantifying Vegetation Change in Semi-arid Environments: Precision and Accuracy of Spectral Mixture Analysis and the Normalized Difference Vegetation Index. *Remote Sensing of Environment* 73: 87–102.

Foley, J.A., Defries, R., Asner, G.P., Barford, C., Bonan, G., Carpenter, S.R., Chapin, F.S., Coe, M.T., Daily, G.C., Gibbs, H.K., Helkowski, J.H., Holloway, T., Howard, E.A., Kucharik, C.J., Monfreda, C., Patz, J.A., Prentice, I.C., Ramankutty, N. & Snyder, P.K. 2005. Global consequences of land use. *Science* 309: 570–574.

Forester, J.D., Lves, A.R., Turner, M.G., Anderson, D.P., Fortin, D., Beyer, H.L., Smith, D.W. & Boyce, M.S. 2007. State-space models link elk movement patterns to landscape characteristics in Yellowstone National Park. *Ecological Monographs* 77: 285–299.

Fryxell, J.M. 1991. Forage quality and aggregation by large herbivores. *The American Naturalist* 138: 478–498.

Ginguld, M., Perevolotsky, A. & Ungar, E.D. 1997. Living on the margins: livelihood strategies of Bedouin herd-owners in the Northern Negev, Israel. *Human Ecology* 25: 567–591.

Hall, J.K. & Cleave, R.L. 1988. The DTM project. *Geological Survey of Israel* 6: 1–7.

Jobbagy, E.G., Sala, O.E. & Paruelo, J.M. 2002. Patterns and controls of primary production in the Patagonian steppe, A remote sensing approach. *Ecology* 83: 307–319.

Karnieli, A., Gilad, U. & Svoray, T. 2008. Satellite Image Processing and Geo-statistical Methods for Assessing Land Degradation around Watering Points in the Ust-Urt Plateau, Kazakhstan. In Röder, A. & Hill, J. (eds.), Recent Advances in Remote Sensing and Geoinformation Processing for Land Degradation Assessment: this book. Leiden: Taylor and Francis.

Kyumedjinski, H., Dan, Y., Soriano, S. & Nisim, S. 1988. Selected soil transects from the land of Israel, *Agricultural Research Organization, the Volcani Center*, Beit Dagan, 6–239.

Le Houerou, H.N. 1996. Climate change, drought and desertification. *Journal of Arid Environments* 34: 133–185.

McNaughton, S.J. 1984. Grazing lawns: animals in herds, plant form, and coevolution. *The American Naturalist* 124: 863–886.

Milchunas, D.G., Varnamkhasti, A.S., Lauenroth, W.K. & Goetz, H. 1995. Forage quality in relation to long-term grazing history, current-year defoliation, and water resource. *Oecologia* 101: 366–374.

Noy-Meir, I. 1973. Desert Ecosystems: Environment and producers. *Annual Review Ecology and Systematics* 4: 25–51.

Park, S.J., Mcsweeney, K. & Lowery, B. 2001. Identification of the spatial distribution of soils using a process-based terrain characterization. *Geoderma* 103: 249–272.

Paruelo, J.M., Epstein, H.E., Lauenroth, W.K. & Burke, I.C. 1997. ANPP estimates from NDVI for the central grassland region of the United States. *Ecology* 78: 953–958.

Portnov, B.A. & Safriel, U.N. 2004. Combating desertification in the Negev: Dryland agriculture vs. dryland urbanization. *Journal of Arid Environments* 56: 659–680.

Robinson, V.B. 2003. A perspective on the fundamentals of fuzzy sets and their use in Geographic Information Systems. *Transactions in GIS* 7: 3–30.

Rodgers, A.R. 2001. Tracking animals with GPS: The first 10 years. In: Tracking animals with GPS: An International Conference held at the Macaulay Land Use Research Institute, Aberdeen, 12–13 march 2001: 1–11. Aberdeen, UK.

Rondeaux, G., Steven, M. & Baret, F. 1996. Optimization of soil-adjusted vegetation indices. *Remote Sensing of Environment* 55 (2): 95–107.

Shafran-Nathan, R. 2006. *Prediction model for herbaceous vegetation production in a semi-arid environment: Embedding fuzzy logic in GIS*. MA Thesis, Ben-Gurion University of the Negev, Beer Sheva, Israel.

Shoshany, M. & Svoray, T. 2002. Multidate adaptive unmixing and its application to analysis of ecosystem transitions along a climatic gradient. *Remote Sensing Of Environment* 82 (1): 5–20.

Smith, G.M. & Milton, E.J. 1999. The use of the empirical line method to calibrate remotely sensed data to reflectance. *International Journal of Remote Sensing* 20: 2653–2662.

Svoray, T., Mazor, S. & Bar, P. 2007. How is shrub cover related to soil moisture and patch geometry in the fragmented landscape of the Northern Negev desert? *Landscape Ecology* 22: 105–116.

Svoray, T., Bar-Yamin, G., Henkin, Z. & Gutman, M. 2004. Assessment of herbaceous plant habitats in water-constrained environments: Predicting indirect effects with fuzzy logic. *Ecological Modelling* 180: 537–556.

Svoray, T., Shafran Natan, R., Henkin, Z. & Perevolotsky, A. 2008. Dynamic modeling of the effect of water and temperature on herbaceous vegetation production (Submitted to Ecological Modeling).

Ungar, E.D., Henkin, Z., Gutman, M., Dolev, A., Genizi, A. & Ganskopp, D. 2005. Inference of animal activity from GPS collar data on free-ranging cattle. *Rangeland Ecology and Management* 58: 256–266.

Ungar, E.D., Perevolotsky, A., Yonatan, R., Barkai, D., Hefetz Y. & Baram, H. 1999. Primary production of natural pastures of the hilly northern Negev—contributory factors and management implications. *Ecology & Environment* 5: 130–140.

Ünsalan, C. & Boyer, K.L. 2004. Linearized Vegetation Indices Based on a Formal Statistical Framework. *IEEE Transactions on Geoscience and Remote Sensing* 42 (7): 1575–1585.

Volcani, A., Karnieli, A & Svoray, T. 2005. The use of remote sensing and GIS for spatio-temporal analysis of the physiological state of a semi-arid forest with respect to drought years. *Forest Ecology and Management* 215: 239–250.

Zhu, A.X., Band, L.E., Dutton, B. & Nimlos, T.J. 1996. Automated soil inference under fuzzy logic. *Ecological Modelling* 90: 123–145.

Recent Advances in Remote Sensing and Geoinformation Processing
for Land Degradation Assessment – Röder & Hill (eds)
© 2009 Taylor & Francis Group, London, ISBN 978-0-415-39769-8

Satellite image processing and geo-statistical methods for assessing land degradation around watering points in the Ust-Urt Plateau, Kazakhstan

A. Karnieli & U. Gilad

The Remote Sensing Laboratory, Jacob Blaustein Institutes for Desert Research,
Ben Gurion University of the Negev, Israel

T. Svoray

Department of Geography and Environmental Development, Ben Gurion University of the Negev, Israel

ABSTRACT: Land degradation around watering points has been well observed by satellite images in many drylands around the world. It can be recognized as radial brightness belts fading as a function of the distance from the wells. The primary goal of this study was to characterize the spatial and temporal land degradation/rehabilitation in Ust-Urt plateau, Kazakhstan, in terms of vegetation and soil patterns, during different key historic periods, with respect to socio-economic changes before and after the collapse of the Soviet Union.

Landsat-derived Tasseled-Cap's Brightness Index was found as the best spectral transformation for enhancing the contrast between the bright-degraded areas close to the water wells and the darker surrounding areas far and in-between these wells. Empirical variograms were computed for each of the images and the exponential models were fitted. The kriging geo-statistical technique utilized the variograms for creating brightness maps. The maps demonstrate the grazing gradient as levels of degrading belts around the wells. Change detection analysis, based on the kriging maps, reveals some land rehabilitation between the 1975 and the 1987 images. However, land degradation was observed between the 1987 and the 2000 images due to recent exploration and exploitation of the gas and oil reserves in the region following independence of the Kazakhstan in 1991.

1 INTRODUCTION

The term *overgrazing* refers to "grazing by a number of animals exceeding the carrying capacity of a given parcel of land" (www.wiley.com/college/geog/cutter018104/ resources/glossary.htm). As a result, overgrazing by different types of livestock is perhaps the most significant anthropogenic activity that degrades rangelands in terms of plant density, plant chemical content, community structure, and soil erosion (Manzano & Návar 2000). In arid and semi-arid environments, land (soil and vegetation) degradation is mainly related to area surrounding point-sources of water, either natural or artificial, such as wells or boreholes (Lange 1969). Pickup & Chewings (1994) were the first who defined the term '*grazing gradient*' as "spatial patterns in soil or vegetation characteristics resulting from grazing activities and which are symptomatic of land degradation".

Domestic animals (sheep, goats, cattle, camels, yaks, and horses) prefer to graze near a watering point. When food is depleted in this area, they move away from the source of water but return regularly for drinking. Consequently, larger number of individuals frequently concentrate around watering points; a density that decreases gradually with increasing distance from water (Pickup et al. 1993, Friedel 1997). The livestock grazing distance is limited, depends on the water demand of different animals, season and weather conditions, and quality of the forage. Typical distance is 4–6 km, which can increase to 10 or even 20 km under extreme conditions (Hodder & Low 1978).

Since the radial pattern around watering points is well observed from space, most of the recent studies are based on the interpretation and modelling of remotely sensed data, which can be analyzed in a semi-automated and repeatable way over vast and remote areas. Various remote sensing based models have been developed to estimate the spatial distribution of the different variables around the watering points, these are the Perpendicular Distance (PD54) based on Landsat-MSS spectral bands (Pickup et al. 1993), the Normalized Difference Vegetation Index (NDVI) calculated from the Advance Very High resolution Radiometer (AVHRR) (Hanan et al. 1991), or a probabilistic linear spectral mixture model, called AutoMCU, based on Monte Carlo analysis (Harrise & Asner 2003).

The above-described ground and spaceborne observations have demonstrated not only that the grazing gradients are characterized by concentric circles around the watering points, but also the spatial nature of the measured biotic, abiotic, and environmental variables that are distributed in a common fashion. Each variable (e.g. vegetation cover, grass and annuals production, bush encroachment, soil pH, organic content, phosphate, and nitrate, soil nutrient concentrations, particularly potassium and phosphorus, and track density) has low (or high) value near the centre and changes continuously as the distance increases. Moreover, most of the observations show that the rate of improvement (or decline) of each variable does not change after several kilometres (usually 5 km as mentioned before) from the watering point.

The current paper presents another approach for assessing and mapping the grazing impacts around watering points. The primary goal of the study was to characterize the spatial and temporal land degradation process in Ust-Urt Plateau, Kazakhstan, in terms of vegetation and soil patterns, during different periods, with respect to the socio-economic changes before and after the collapse of the Soviet Union. More specific objectives of the study were: (1) to develop a geo-statistical model, based on the kriging technique and using satellite image processing in order to assess spatial and temporal land cover patterns in three key different years (1975, 1987 and 2000); (2) to conduct a change detection analysis based on the geo-statistical products in order to assess the direction and intensity of changes between the study periods; and (3) to link the previous findings to the socio-economic situations that influenced the grazing gradients and hence the land-use/land-cover state of the study sites before and after the collapse of the Soviet Union.

2 METHODOLOGY

2.1 Study site

The study site is located in the Ust-Urt desert plateau, ca. 160,600 km^2, between the Caspian and Aral seas in Central Asia and occupies the southern part of Kazakhstan, the northern part of the Karakalpak Republic, and Turkmenistan (Fig. 1). It rises to between ca. 150 and 300 m above mean sea level. It consists primarily of stony desert with grey-brown soils. The area has a semiarid continental climate with hot summers and cold windy winters. Vegetation consists on *Artemisia terrae-albae*, *Anabasis*, and *Salsae*. Its semi-nomadic population raises sheep, goats, and camels. The North Caspian basin is a petroleum-rich region with large oil and gas reserves but hardly explored yet (Ulmishek 2001).

2.2 Image processing

Three Landsat images of the study area were used. They were acquired in April 1975, June 1987, and June 2000 respectively by different sensors—MSS, TM, and ETM+. From the original images, subsets of the study sites were made, bounding an area of 1,184 km^{-2}. Image processing started by converting digital numbers to reflectance values (Markham & Barker, 1985). Reflectance values were used to calculate several vegetation indices for each image subset. These include: NDVI (Rouse et al. 1974), SAVI (Huete 1988), MSAVI (Qi et al. 1994), PVI (Richardson & Wiegand 1977), and the Tasseled Cap-derived Greenness and Brightness indices (Kauth & Thomas 1976). Performance analysis, in terms of standard deviation and stretch of the index values within the dynamic range, was applied to all indices. These analyses revealed that the Tasseled Cap-derived

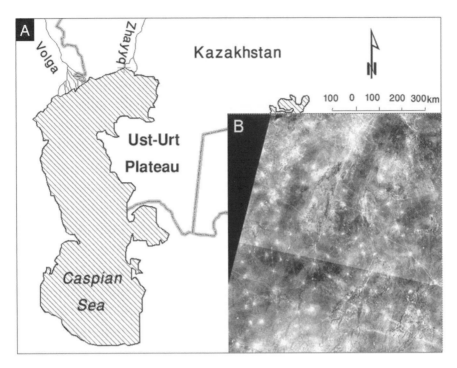

Figure 1. (A) Location map of the study area in the Ust-Urt plateau, Kazakhstan. (B) The specific study site on Landsat-TM image, bright spots indicate watering points.

Brightness Index (BI) had produced the best contrast and consequently had been selected for further analysis. The BI has the general form of

$$BI = \alpha 1(B1) + \alpha 2(B2) + \cdots\cdots + \alpha n(Bn) \qquad (1)$$

where Bn is the spectral band number and αn is the appropriate BI coefficients of each sensor—MSS, TM, and ETM+ (Crist et al. 1984, Kauth & Thomas 1976, Huang et al. 2002, respectively). Besides the statistical significant results showing the superior of the BI in terms of separability and that this index was originally designated to examine soil properties, its advantage is in the ability to compare between the different sensors that have different spectral bands, as it reduces the different spectral bands to one normalized layer of BI values.

Spatial resolution of the MSS is 80 m and that of the TM and ETM+ is 30 m. The resolution of the images was reduced by a factor of 3 and 6, for the MSS and TM/ETM+ images, respectively. The resulting pixel size of 171 m was processed easily by the used software (ARC-GIS and Surfer), as the sub-scene size was reduced from approximately 1.5 million to 40,000 pixels.

2.3 Geostatistical analysis

The ordinary kriging interpolation was used to model and map the spatial variation of surface brightness values in the study sites. The key concept of geostatistics is that of the regionalized variable (Matheron 1971), defined as a variable that can be characterized from a number of measurements that identify spatial structure. The basic assumption underlying this theory is that when assuming spatial continuity, samples that are located adjacent to one another tend to be more similar than samples located in remote areas. This spatially dependent variation may be treated statistically and described through a number of parameters derived from a semi-variogram that is the function relating the semi-variance to the directional distance between two samples. The semi-variance is defined as half the mean squared difference between two samples in a given direction and distance apart (Eq. 2). The direction and distance are defined by the vector h that is commonly referred as the lag:

$$\gamma(h) = \frac{1}{2N(h)} \sum_{i=1}^{N(h)} (z_{xi} - z_{xi+h})^2 \qquad (2)$$

where $\gamma(h)$ is the semi-variance at lag h, $N(h)$ is the number of sample-pairs a distance h apart, and Zi is the value of the regionalized variable at location i. In addition to the lag, the variogram is characterized by three other parameters—the nugget, range, and sill. The nugget is variability at zero distance and represents sampling and analytical errors. The range of influence designates the extent, say a distance a, beyond which autocorrelation between sampling sites is negligible. The sill represents the variability of spatially independent samples. An empirical semi-variogram can be calculated from the given set of observations and then fitted to several common theoretical models (Delhomme 1979). Once the theoretical semi-variogram is chosen, several criteria should be applied to determine the correctness of the model and to adjust its parameters. The current study uses four of these criteria:

1. Quantile-Quantile (Q-Q) plot—this is a visual comparison of two distributions while the quantiles from the two distributions are plotted versus one another. Thus a Q-Q plot of two identical distributions will result in a straight line while departure from this line reveals where they differ. To test the normality of the BI values, they were plotted versus standard normal distribution data (provided by ArcGIS Geostatistical Analyst tool).
2. Cross-validation scatter plot—in this plot, the measured and predicted data are regressed and the cloud of points is compared to a 1:1 line and to a line of best fit.
3. Mean kriged estimation error:

$$\frac{1}{n} \sum_{i=1}^{n} (Z_{xi} - Z_{xi}^*) = \frac{1}{n} \sum_{i=1}^{n} \varepsilon_i \approx 0 \qquad (3)$$

where ε is the difference between the kriged and the known point value (this term should approach 0).
4. Mean standardized squared estimation error:

$$\frac{1}{n} \sum_{i=1}^{n} [(Z_{xi} - Z_{xi}^*)/s_i^*]^2 = \frac{1}{n} \sum_{i=1}^{n} (\varepsilon_i/s_i)^2 \approx 1 \qquad (4)$$

where s_i^* is the estimation standard deviation (this term should approach 1).

The rationale for choosing the kriging technique was the similarity in spatial structure of most of the above-mentioned variables, gradually increasing (or decreasing) as a function of the increasing distance from the watering point until reaching the limit of no grazing effects, and the typical shape of the variogram. Curran (1988) and Woodcock et al. (1988a, 1988b) introduced the semi-variogram to remote sensing and discovered that the parameters of the variogram can be directly related to a feature in an image. The kriging technique has recently become very common for analyzing spaceborne data (Oliver et al. 2000).

2.4 Image differencing change detection analysis

Post-processing change detection method, namely BI Differencing (Yuan et al. 1998), was implemented to compute the degree and direction of the changes in each site and between the imaging periods. The general function of the change can be considered as:

$$change = \begin{cases} 0 & \text{if } |BI_t - BI_{t+1}| \leq T \\ 1 & \text{if } |BI_t - BI_{t+1}| > T \end{cases} \tag{5}$$

where t and $t + 1$ represent the two time periods and T is the threshold value. A common way to assess changes is based on determination of thresholds in terms of standard deviation levels below and above the mean of the difference between the BI values $(\overline{\Delta BI})$ of the images under study. In this manner, one can distinguish between changed and unchanged pixels as well as between negative and positive changes (Jensen 1986). In the current study, one standard deviation from the $\overline{\Delta BI}$ was defined as the threshold and steps of one standard deviation beyond this threshold determined the magnitude and direction of the change.

2.5 Semivariance for change detection

Addink (2001) introduced the Absolute Normalized Difference at lag h, $AND(h)$, to calculate differences between two sets of semivariances, $\gamma(h)$, from different dates, t and $t + 1$:

$$AND(h) = \frac{\frac{1}{2}|\gamma_{t+1}(h) - \gamma_t(h)|}{\frac{1}{2}(\gamma_{t+1}(h) + \gamma_t(h))} \tag{6}$$

The $AND(h)$ algorithm can produce values ranging from 0 to 1, indicating no and total change, respectively.

3 RESULTS AND DISCUSSION

Subsets of the study site were extracted from Landsat MSS, TM, and ETM+ of the years 1975, 1987, and 2000, respectively. Figure 2 represents the BI products as calculated by Equation 1 with the appropriate BI coefficients. Watering points can be recognized as bright spots spread over mostly the MSS and TM images, but the ETM+ one (Fig. 2C). In the latter image many of the watering points disappeared, however, a wide bright area exists in the center of the image. The brightness levels were equally stretched for the different sensors and ranging between 0.65 and 0.96. High values correspond to bare soil while low values indicate vegetation.

The three BI images were used for the kriging analysis. The first step in this course of action was to establish the empirical semi-variogram based on ca. 40,000 pixels in each image, for the three periods. Subsequently, several theoretical models were examined and the exponential model

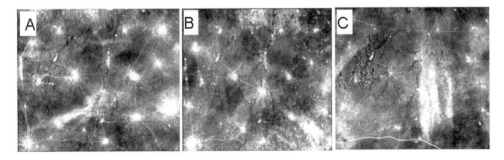

Figure 2. Brightness index products of sub-images used for the geostatistical analysis. (A) Landsat-MSS (1975) image; (B) Landsat-TM (1987) image; (C) Landsat-ETM+ (2000) image. Bright areas indicate degraded regions due to grazing and/or technological desertification. Dimension of each panel is about 30 × 30 km.

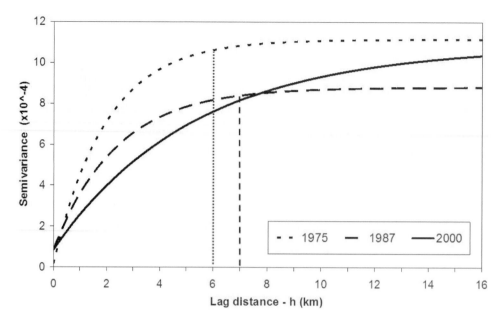

Figure 3.　Modeled variograms for the Brightness Index values of the Ust-Urt Plateau in 1975; 1987; and 2000.

was selected due to the best cross validation results. Thus, all empirical models were fitted with an equation of the form:

$$\gamma(h) = C_0 + C_1\left(1 - \exp\left(-\frac{|h|}{a}\right)\right) \qquad (7)$$

where a is the range, h is the lag, C_0 is the nugget, and $C_0 + C_1$ equals the sill. The fitted exponential models are illustrated in Figure 3. All variograms were processed with 16 lags of 1,000 m each. Visually, the 1975 and 1987 variograms look quite similar, having a typical variogram shape. Note that when the shape of the variogram is more round it can be referred to more symmetrical features in the image. Contrary, the 2000 variogram reaching about the same sill level only after 16,000 km and its shape is more linear than round.

The Q-Q plot (Fig. 4A) shows that in principle the distribution of the brightness values does not deviate much from normal distribution. A small deviation is observed in the region of the upper quantile and in the lower regions. However, these deviations are considered to be relatively small and thus can be ignored. The cross validation graph (Fig. 4B) shows a very good correlation between the measured and predicted values ($R^2 = 0.92$). The slope coefficient is very close to unity and the intercept coefficient is very close to zero. This means that the point distribution best fit is very close to the desired 1:1 line.

In the next step, the kriging interpolation maps were performed based on exponential models. Since the grazing impact was assumed to be an isotropic feature and the direction has no influence on the spatial variation, the maps were derived from the omni-directional variograms. Figure 5 depicts the final products for the distribution of the BI values for the three periods. In the 1975 and 1987 maps, one can observe the belts around the watering points, indicating progressive land degradation radiating from the wells, i.e. the grazing gradient. The dark-red areas in the images are related to zones where the grazing impact is the predominant feature that has a strong effect on the spatial variation. These areas can be considered as the centre of the grazing impact, denoted as the 'sacrifice zone' by Perkins & Thomas (1993). The surrounding light red and yellow belts represent a mixed zone where grazing impact and natural variability overlay each other or create a stable balance. This zone can be compared to an edge zone of the grazing impact and highlights

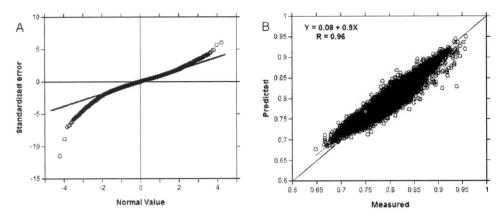

Figure 4. Cross validation of the exponential model semi-variogram fittings. (A) Q-Q plot; (B) Predicted vs. measured plot.

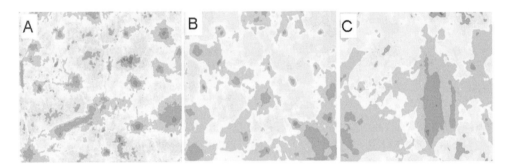

Figure 5. Kriging interpolation maps for images shown in Figure 2 based on the Brightness Index values. Dark tones indicate land degradation due to grazing and/or technological desertification. (Colour figure at the end of the volume) (see colour plate page 399).

the principal migration routes of livestock. The zone coloured by blue tones is considered to be an area where natural variability overbalances the grazing impact, denoted as '*grazing reserve*'. The radial pattern, related to the grazing gradient, is not seen in the 2000 map. Instead, the dominant feature in the middle of the scene is coloured in red tones.

Image differencing change detection analysis, based on the *BI* values, was performed on the two pairs of kriging maps—1987 vs. 1975, and 2000 vs. 1987. Results are illustrated in Figure 6. Figures 6A and 6C are the change maps while Figure 6B and 6D are the respective frequency histograms of the change categories. The SD lines are also presented. The difference map computed from 1987 and 1975 shows that despite of the area that was considered as 'no change' (64.0%), more area underwent rehabilitation process than degradation (22.1% vs. 13.8%). As opposed to this favourable land cover change, degradation process characterizes considerable portion of the area (30.7%) during the second period (2000 vs. 1987), while rehabilitation occurred only in minor part (12.5%).

Two different processes govern the land-use and land-cover changes in the Ust-Urt Plateau. On the one hand degradation processes have developed due to recent exploration and exploitation of the gas and oil reserves in the region (Ulmishek 2001). Consequently, large areas went through intensive technological desertification that means utilizing large amount of heavy-duty equipments, large-scale plants, and vehicles that damage the soil surface. On the other hand, following independence of the former Soviet states in 1991, and the imposition of difficult economic conditions with transition reforms, several major socio-economic changes occurred. Strong centralized government subsidy

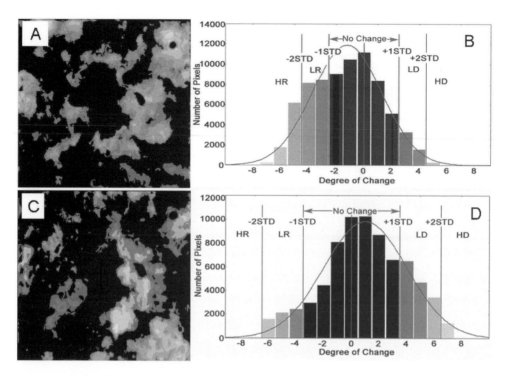

Figure 6. Change detection results shown as maps and histograms. HR = high rehabilitation; LR = low rehabilitation; HD = high degradation; LD = low degradation. (A, B) difference between the 1987 and 1975 images; (C, D) difference between the 2000 and 1987 images (see colour plate page 399).

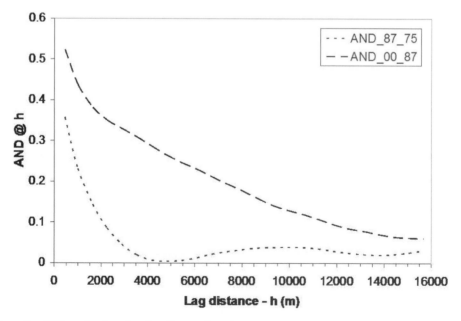

Figure 7. AND results plotted against distance from the water source.

programs terminated, including the practice of guaranteed supplemental forage in cold winters and drought years. Farmers could no longer feed the livestock during the harsh winters, water wells were demolished, pumps were stolen or broken, and there were no longer transportation means to convey the animals to the markets in the central cities (Antonchikov et al. 2002). Consequently, drastic decline in livestock populations were observed after 1991 that resulted in much less grazing pressure and hence recovery of the natural vegetation and rehabilitation of the land.

The AND analysis that is illustrated in Figure 7 reveals that during the first period most of the change between 1987 and 1975 occurred within 5 km, more likely from the watering points, with minimal change in further lag. These results support the grazing gradient approach related to the average grazing distance of livestock from the drinking source. However, during the second period, this pattern is not evident. The change decreases gradually as a function of the lag distance up to approximately 16 km.

4 CONCLUSIONS

The Tasseled Cap-derived Brightness Index (BI) was selected to describe the spatial surface patterns since (1) it produced the best contrast (separability); (2) it was originally designed to examine soil properties; and (3) it is able to compare between the different sensors with different spectral bands, as it reduces their different spectral bands to one normalized layer of BI values. Geostatistical analysis, based on the ordinary kriging interpolation technique, was found to be a suitable method for modelling the spatial patterns of land-use/land-cover, especially around watering points in arid and semi-arid regions. The reason is the similarity between the shape of the variogram and the directional change of many biotic, abiotic, and environmental variables along the grazing gradient. Temporal changes were effectively conducted by the index differencing technique. The study demonstrates the ability of spaceborne image analysis to follow after land-use/land-cover changes caused by dramatic socio-economic changes as occurred in Kazakhstan after its separation from Russia in 1991. In the Ust-Urt plateau, the area was degraded due to recent exploration and exploitation of the gas and oil reserves.

REFERENCES

Addink, E. 2001. *Change Detection with Remote Sensing. Relating NOAA-AVHRR to Environmental Impact of Agriculture in Europe*. Wageningen University: PhD Thesis: 133.

Antonchikov, A.N., Bakinova, T.I., Dushkov, V.Y., Zalibekov, Z.G., Levykin, S.V., Neronov, V.M., Okolelova, A.A., Pazhenkov, A.S., Cherniakhovsky, D.A., Chibilev, A.A. & Yunusbaev, U.B. 2002: Desertification and Ecological Problems of Pasture Stockbreeding in the Steppe Regions of Southern Russia. Moscow: IUCN—World Conservation Union,

Crist, E.P. & Cicone, R.C. 1984. A physically-based transformation of Thematic Mapper data—the TM Tasseled Cap. *IEEE Transactions on Geoscience and Remote Sensing* 22: 256–263.

Curran, P.J. 1988. The semi-variogram in remote sensing: an introduction. *Remote Sensing of Environment* 24: 493–507.

Delhomme, J.P. 1979. Spatial variability and uncertainty in groundwater flow parameters: a geostatistical approach. *Water Resources Research* 15: 269–280.

Friedel, M.H. 1997. Discontinuous change in arid woodland and grassland vegetation along gradients of cattle grazing in central Australia. *Journal of Arid Environments* 37: 145–164.

Hanan, N.P., Prevost, Y., Diouf, A. & Diallo, O. 1991: Assessment of desertification around deep wells in the Sahel using satellite imagery. *Journal of Applied Ecology* 28: 173–186.

Harrise, A.T. & Asner, G.P. 2003. Grazing gradient detection with airborne imaging spectroscopy on a semi-arid rangeland. *Remote Sensing of Environment* 55: 391–404.

Hodder, R.M. & Low, W.A. 1978. Grazing distribution of free-ranging cattle at three sites in the Alice spring district, Central Australia. *Australian Rangeland Journal* 1: 95–105.

Huang, C., Wylie, B., Yang, L., Homer, C. & Zylstra, G. 2002. Derivation of a tasseled cap transformation based on Landsat 7 at-satellite reflectance. *International Journal of Remote Sensing* 23: 1741–1748.

Huete, A.R. 1988. A Soil-Adjusted Vegetation Index (SAVI). *Remote Sensing of Environment* 25: 295–309.

Jensen, J.R. 1986. *Introductory Digital Image Processing, A Remote Sensing Perspective.* New Jersey: Prentice-Hall.

Kauth, J. & Thomas, G.S. 1976. The tasseled Cap—A Graphic Description of the Spectral-Temporal Development of Agricultural Crops as Seen by LANDSAT. *Proceedings of the Symposium on Machine Processing of Remotely Sensed Data*: 4B-41—4B-51. Purdue University of West Lafayette, Indiana.

Lange, R.T. 1969. The piosphere: sheep track and dung patterns. *Journal of Range Management* 22: 396–400.

Manzano, M.G. & Návar, J. 2000. Process of desertification by goats overgrazing in the Tamaulipan thornscub in north-eastern Mexico. *Journal of Arid Environment* 44: 1–17.

Markham, B.L. & Barker, J.L. 1985. Spectral characterization of the LANDSAT Thematic Mapper sensors. *International Journal of Remote Sensing* 6: 697–716.

Matheron, G. 1971: *The Theory of Regionalized Variables and its Applications,* Cahiers du Centre de Morphologie Mathematique de Fontainebleau, No. 5, Ecole Superieure des Mines, Fontainebleau.

Oliver, M.A., Webster, R. & Slocum, K. 2000. Filtering SPOT imagery by kriging analysis. *International Journal of Remote Sensing* 21: 735–752.

Perkins, J.S. & Thomas, D.S.G. 1993. Spreading deserts or spatially confined environmental impacts—land degradation and cattle ranching in the Kalahari desert of Botswana. *Land Degradation and Rehabilitation* 4: 179–194.

Pickup, G. & Chewings, V.H. 1994. A grazing gradient approach to land degradation assessment in arid areas from remotely sensed data. *International Journal of Remote Sensing* 15: 597–617.

Pickup, G., Chewings, V.H. & Nelson, D.J. 1993. Estimating changes in vegetation cover over time in arid rangelands using Landsat MSS data. *Remote Sensing of Environments* 43: 243–263.

Qi, J., Chehbouni, A., Huete, A.R., Kerr, Y.H. & Sorooshian, S. 1994. A modified soil adjusted vegetation index. *Remote Sensing of Environments* 48: 119–126.

Richardson, A.J. & Wiegand, C.L. 1977. Distinguishing vegetation from soil background information. *Photogrammetric Engineering and Remote Sensing* 43: 1541–1552.

Rouse, J.W., Haas, R.W., Schell, J.A., Deering, D.W. & Harlan, J.C. 1974. Monitoring the vernal advancement and retrogradation (Greenwave effect) of natural vegetation. NASA/GSFCT Type III Final Report, Greenbelt, MD, USA.

Ulmishek, G.F. 2001. Petroleum Geology and Resources of the North Caspian Basin, Kazakhstan and Russia. Denver: U.S. Geological Survey Bulletin 2201-B. USGS.

Woodcock, C.E., Strahler, A.H. & Jupp, D.L.B. 1988a. The use of the variogram in remote sensing: I. Scene models and simulated images. *Remote Sensing of Environment* 25: 323–348.

Woodcock, C.E., Strahler, A.H. & Jupp, D.L.B. 1988b. The use of the variogram in remote sensing: II. Real digital images. *Remote Sensing of Environment* 25: 349–379.

Yuan, D., Elvidge, C.D. & Lunetta, R.S. 1998. Survey of multispectral methods for land cover change analysis. In R.S. Lunetta & C.D. Elvidge (eds), *Remote Sensing Change Detection, Environmental Monitoring Methods and Applications* (pp. 21–39). Michigan: Ann Arbor Press.

Recent Advances in Remote Sensing and Geoinformation Processing
for Land Degradation Assessment – Röder & Hill (eds)
© 2009 Taylor & Francis Group, London, ISBN 978-0-415-39769-8

Landscape analysis using multi-scale segmentation and object-oriented classification

B.J.F. Clark & P.K.E. Pellikka

Department of Geography, Faculty of Science, University of Helsinki, Finland

ABSTRACT: In pressured environmentally sensitive and ecologically important areas, such as the Taita Hills study area in Kenya, there is a continuing need for accurate up-to-date and historical land cover mapping to be derived from remotely sensed data, that can be used in change detection studies and for developing sustainable land use policies. However, traditional classification techniques based solely on the spectral response of individual pixels achieve only limited success in complex heterogeneous environments. In an attempt to improve on this situation, multispectral SPOT data from 1987, 1992 and 2003 was subject to an object-oriented classification approach to identify 11 land use/land cover classes derived using the Land Cover Classification System (LCCS) protocol. Ground reference test data was collected to enable accuracy assessment and a comparison with the standard maximum-likelihood technique. The derived maps were used to identify major landscape changes that have occurred in the Taita Hills over the period 1987 to 2003.

1 INTRODUCTION

Land degradation is a composite term which describes the negative impacts of a large number of naturally occurring but human influenced processes on an environment (Stocking and Murnaghan, 2000). It implies a reduction in the productive capacity of an ecosystem and in its value as an economic resource (UNEP 1992). At a landscape level there is a strong cyclical connection between land use and degradation. Changes in land use over time, such as the expansion of agriculture, influence degradation processes and this in turn affects the future utilization of the land. Whilst specific erosion processes and resultant landforms, such as gully formation through soil erosion by water, can be usefully studied at a micro level, land degradation cannot be directly assessed at a macro level through any single measure. Rather, use must be made of indicator variables which demarcate the likely occurrence of degradation. At a landscape level, these indicators are changes in land cover patterns over time such as loss of vegetation cover, including deforestation.

A crucial first step in an understanding of landscape changes is, then, to accurately quantify and map land use and land cover (LULC) over time. Remote sensing offers the most efficient methodology for routinely monitoring at a landscape level over regional, as well as national and continental, scales. However, as is noted by Burnett and Blaschke (2003), in common with all observation of reality, remotely sensed images are an imperfect capturing of patterns, which are themselves an imperfect mirror of ecosystem processes. The ground instantaneous field of view (GIFOV) of a sensor, which is realized as the pixel resolution of the imagery, is actually a complex phenomenon determined by technical constraints as much as by mapping requirements. Moreover, there is not one individual scale which is appropriate for mapping a landscape if we accept that reality is formed of a mosaic of process continuums.

Wu and Loucks (1995) and Wu (1999) argue that by breaking down ecological complexity through a hierarchical scaling strategy, so called hierarchical patch dynamics (HPD) provides a conceptual framework within which the interaction between ecological processes operating at different scales can be understood. This multi-scale analysis perceives a landscape as a spatially nested patch hierarchy where larger patches are formed of smaller, functional patches. These systems exhibit instability at lower levels, but possess meta-stability at higher levels as, in general, small scale processes tend to be more stochastic. The hierarchical structuring of a landscape is defined at various critical levels of organization where interactions are stronger within levels than between them, and where each level operates at specific spatial and temporal scales (Hay and Marceau 2004). Stronger gradients in these flux rates results in more apparent boundaries, or local heterogeneity (Burnett and Blaschke 2003).

The concept of scale then becomes central to an understanding and analysis of landscapes (Levin 1992). Scale represents the 'window of perception', the filter, or measuring tool, with which a system is viewed and quantified; consequently real-world objects only exist as meaningful entities over a specific range of scales (Hay et al. 2002). Landscape ecologists define scale as having grain and extent, where grain refers to the smallest intervals in an observation set and extent refers to the range over which observations at a specific grain are made (O'Neill and King 1998). In remote sensing, grain is equivalent to the spatial, spectral and temporal resolutions of the image pixels, whilst the extent represents the geographic area, combined spectral bandwidths and temporal duration covered by an image as a whole (Hay et al. 2001). Scale may be measured in absolute units or relative to the phenomenon of interest; the 'focal scale' as it is known in hierarchy theory.

Standard image classification procedures, such as the maximum-likelihood (ML) classifier, work on a uni-scale pixel-by-pixel basis and therefore ignore both useful spatial information surrounding the pixel and multi-scale information within the image. Class assignment is based solely on the principle that pixels of the same land cover type will be close in multi-spectral feature space. As Burnett and Blaschke (2003) themselves note, this does not hold true for complex environments. Rather, Burnett and Blaschke (2003) propose a multi-scale segmentation/object relationship modelling (MSS/ORM) methodology for landscape analysis based on HPD theory and suggest that more accurate analysis can be derived through the application of this technique, especially for heterogeneous landscapes. Central to this methodology is the generation of meaningful image objects relating to landscape patches by multi-scale segmentation, where a search is made for the gradient of flux zones within and between patches. Critically, because MSS/ORM is a move away from pixel-based to object-based analysis, it is possible to explore multiple scales of objects within the *same* image as well as within a GIS database formed from imagery at different resolutions. Once captured, a model of the hierarchical relationships between the image objects is built up using both directly calculable properties, such as the mean spectral values or the number of sub-objects, and by the derivation of semantic rules requiring the input of a human expert on the landscape in question. This can be considered as a training phase in an object-oriented (OO) LULC classification methodology.

2   MULTI-SCALE SEGMENTATION

Segmentation is the division of remotely sensed images into discrete regions or objects formed of aggregations of pixels that are homogenous with regard to spectral and/or spatial characteristics. Homogenous in this instance refers to the fact that the within-object variance is less than the between-object variance. Research into segmentation techniques is not new, see for example Haralick et al. (1973), and there are a large number of possible methodologies, but the availability of operational software is a relatively recent development. In this study, all MSS/ORM work was implemented with the eCognition software, which utilizes a fractal net evolution approach (FNEA) to multi-scale segmentation. The full details of the FNEA methodology and of the workings of eCognition's object-oriented fuzzy analysis and classification are covered in depth elsewhere (see Baatz and Schäpe 2000, Benz et al. 2004), so it is only useful here to give a general description of this segmentation process and highlight why its use is particularly appropriate in this instance.

In FNEA, image information is considered to be fractal in nature; that is to say there is self-similarity across scales—structures appear at different scales within an image simultaneously. As image object attributes are scale dependent, to derive meaningful image objects it is necessary to focus on different scale levels of analysis. This is different from other segmentation techniques, such as watershed algorithms or Markov random fields, where the focus is not on specific scales but rather certain heterogeneity criteria (Hay et al. 2003). In FNEA, definition of a specific level of analysis is freely adaptable and leads to a derivation of objects of a similar size and comparable scale, facilitating multi-scale segmentation and enabling the implementation of the MSS/ORM conceptual approach to landscape analysis. FNEA is a bottom-up region merging technique, commencing with single pixel objects. In subsequent pairwise clustering iterations, smaller image objects are merged into larger ones based on the user defined scale, spectral (termed 'color') and shape parameters, and the neighbourhood function, which dictate the growth in heterogeneity between adjacent image segments.

The spectral heterogeneity (h_{color}) of an image object is computed as the sum of the standard deviations for each of k image bands (σ_k) multiplied by the band weights (w_k):

$$h_{color} = \sum_k w_k \cdot \sigma_k \tag{1}$$

To enable the inclusion of image textural features, and because the use of spectral segmentation alone leads to object borders with fractal geometry, a spatial heterogeneity criterion (h_{shape}) composed of two well known landscape ecology metrics is also incorporated into the FNEA segmentation process (Baatz and Schäpe 2000). Firstly, the deviation from a compact shape, the *compactness* (*cpt*), defined as the ratio of the object fractal edge length (l) and the square root of the number of pixels (n) forming the object;

$$cpt = \frac{l}{\sqrt{n}} \tag{2}$$

and, secondly, the deviation from the shortest possible edge length determined by a bounding box, the so-called *smoothness* (*smooth*), defined as the ratio of the object fractal edge length (l) and the border length (b) of a box bounding the object:

$$smooth = \frac{l}{b} \tag{3}$$

The shape heterogeneity criterion is a weighted combination of the two measures as follows (Benz et al. 2004):

$$h_{shape} = w_{cpt} \cdot h_{cpt} + (1 - w_{cpt}) \cdot h_{smooth} \tag{4}$$

Given these measures of image object heterogeneity, the heuristics to decide which adjacent objects to merge can be defined. Global mutual best fitting is the most common solution to such an optimization problem but has the disadvantage that it builds initial segments in regions of low spectral variance, resulting in uneven growth of image objects across the scene (Hay et al. 2003). To counter this, FNEA incorporates local mutual best fitting which always undertakes the most homogeneous merge in the local vicinity, following the gradient of best fitting. An arbitrary starting point is required and, to ensure simultaneous growth of similar sized objects, it is necessary that each object is treated once per iteration and that subsequent merges are distributed as far away as possible from each other over the whole scene. The application of a random sequence here is sub-optimal because clustering can occur. Consequently, a distributed treatment order derived from a dither matrix generated by a binary counter, which systematically takes points with a maximum distance to all other points treated previously, is implemented in FNEA.

Starting with single pixel objects, pairwise merging will evidently increase the heterogeneity. The aim of the optimization procedure is to minimize the incorporated heterogeneity at each single merge. An image object should, therefore, be merged with the adjacent object that incorporates the minimum increase in defined heterogeneity. To assess this, the 'merging cost' representing the 'degree of fitting' for every possible pair of adjacent objects is described by the change in heterogeneity (Δh) before and after a virtual merge (*mg*). The spectral criterion (Δh_{color}) is the change in heterogeneity of the band weighted standard deviations, weighted by the object's size in pixels (n); defined as (Benz et al. 2004):

$$\Delta h_{\text{color}} = \sum_{k} w_k (n_{mg} \cdot \sigma_k^{mg} - [n_{ob1} \cdot \sigma_k^{ob1} + n_{ob2} \cdot \sigma_k^{ob2}]) \tag{5}$$

The change in shape heterogeneity caused by a virtual merge is calculated for the compactness as (Benz et al. 2004):

$$\Delta h_{cpt} = n_{mg} \cdot \frac{l_{mg}}{\sqrt{n_{mg}}} - \left(n_{ob1} \cdot \frac{l_{ob1}}{\sqrt{n_{ob1}}} + n_{ob2} \cdot \frac{l_{ob2}}{\sqrt{n_{ob2}}} \right) \tag{6}$$

and for smoothness as (Benz et al., 2004):

$$\Delta h_{\text{smooth}} = n_{mg} \cdot \frac{l_{mg}}{b_{mg}} - \left(n_{ob1} \cdot \frac{l_{ob1}}{b_{ob1}} + n_{ob2} \cdot \frac{l_{ob2}}{b_{ob2}} \right) \tag{7}$$

The merging process stops when the smallest possible growth exceeds a user defined scale threshold, termed the scale parameter. A larger scale parameter results in larger image objects, thus enabling a mechanism for multi-scale segmentation (Benz et al. 2004). The general FNEA segmentation function (S_f) is defined as:

$$S_f = w_{\text{color}} \cdot h_{\text{color}} + (1 - w_{\text{color}}) \cdot h_{\text{shape}}, \quad w_{\text{color}} \in [0, 1], \quad w_{\text{shape}} \in [0, 1], \quad w_{\text{color}} + w_{\text{shape}} = 1 \tag{8}$$

FNEA is the most appropriate segmentation methodology for enabling the implementation of the desired MSS/ORM conceptual approach to landscape analysis. Additionally, a major advantage of FNEA is that the heuristics do not evaluate the absolute heterogeneity of a region, but rather evaluate the change in heterogeneity over a merge. This has the desirable effect of enabling relatively homogeneous image segments to remain separate, even if the mean values of adjacent regions are similar. This is important in terms of deriving ecologically meaningful image segments from medium resolution data, such as SPOT imagery, where there are large numbers of pixels formed of mixed land cover types and general spectral overlaps, i.e. there is low spectral contrast. The segmentation parameters that were used to derive 'image object primitives', that is to say objects which do not yet posses real world meaning, in the SPOT data used in this study are reported in Section 3.2.1 below.

3 CASE STUDY—THE TAITA HILLS, KENYA

The MSS/ORM approach to landscape analysis was applied to the pressured environmentally sensitive and ecologically important Taita Hills study area located in the Taita-Taveta District of south-east Kenya at latitude 3°25′S, longitude 38°20′E; see Figure 1. The hills cover an area of approximately 1000 square kilometers and are surrounded by semi-arid Acacia/Commiphora shrubland and dry savannah, some of which falls within sections of Tsavo National Park. The actual area of mapping covers 89,220 ha from 3°31′27″ S to 3°16′46.5″ S and from 38°14′21.6″ E to 38°22′11″ E. Whilst the surrounding plains are at an elevation of 600–700 m a.s.l., the Taita Hills rise abruptly in a series of ridges with the highest peak of Vuria at 2208 m, although the average elevation of the hills is 1500 m. The climate of this region is influenced by the Inter-Tropical Convergence Zone (ITCZ) which leads to a bi-modal rainfall incidence, with a longer rainy season during March–May/June and short rains in October–December, although the annual variability of precipitation is high, especially in the semi-arid shrubland surrounding the hills. Despite that the Taita Hills lie approximately 150 km inland from the coast, orographic rainfall plays an important

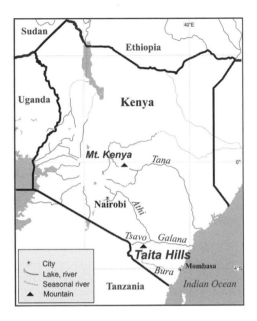

Figure 1. The location of the Taita Hills study area.

part in the local climate as the hills form the first significant barrier which moisture laden air from the Indian Ocean encounters. Mist and cloud precipitation usually occur throughout the year; consequently, whilst the annual rainfall is circa 600 mm on the plains it is over 1200 mm in the hills (Beentje and Ndiang'ui 1988). A 'rain shadow' effect is discernable on the north western side of the hills, with the distinctive *Euphorbia candelabrum* and more commonly *Euphorbia bussei* var. *kibwezensis* growing in the drier conditions.

The Taita Hills form the northernmost part of Africa's Eastern Arc Mountains, which have been identified by Conservation International as one of the top ten biodiversity hotspots in the world. Of particular scientific and conservation interest are the indigenous forest patches which are home to many rare or endangered endemic animals and plants. Today, only a small amount of native forest remains, occurring in a scatter of three larger hilltop remnants; Mbololo (c. 179 ha), Ngangao (c. 136 ha) and Chawia (c. 94 ha) (as reported by Lens et al. 2002, from field survey), and further much smaller fragments embedded in a mosaic of human settlements, small-holder cultivation plots (known locally as 'shambas'), and plantations of exotic tree species such as *Cupressus lusitanica*, *Pinus* spp., *Eucalyptus* spp., and *Grevillea robusta*. The indigenous forest cover has been termed upland moist or mist forest by Beentje and Ndiang'ui (1988), but is also referred to as montane forest or cloud forest by other workers. The characteristic tree species include *Newtonia buchananii*, *Tabernaemontana stapfiana*, *Macaranga conglomerata*, *Albizia gummifera*, *Phoenix reclinata*, *Strombosia scheffleri*, *Cola greenwayi*, *Podocarpus* spp., *Ochna holstii*, and *Millettia oblate* (Beentje and Ndiang'ui 1988). These indigenous forest patches also play an important role in both capturing additional moisture and storing the precipitation on the hilltops.

The population of the whole Taita-Taveta district has grown from 90 000 (1962) persons to over 300 000, consequently this has been a driving factor behind rising environmental pressure on the Taita Hills. There has been an increase in the area under cultivation and due to poor agricultural management, erodible soils and the large relative height differences in the hills, the foothills especially are subject to land degradation and accelerated soil erosion (KARI 2005). Identified threats to the remnant forest patches include encroachment (for settlement, agriculture and livestock grazing), over extraction of firewood and building materials, charcoal burning, poor enforcement of government policies and regulations, illegal logging, lack of awareness among the communities living adjacent to forests, fires (both deliberate and naturally occurring) and colonization by

327

suppressive and fast growing exotic tree species (EAWLS 2005). There is, therefore, a pressing requirement to map and assess LULC changes in the Taita Hills to allow planning for the sustainable use of natural resources. The production of a digital geographic database of the Taita Hills, and the analysis of LULC changes over time, have been major aims of the TAITA research project undertaken at the University of Helsinki.

3.1 Materials and methods

3.1.1 Data and software

The TAITA project has available SPOT XS data for the years 1987, 1992 and 2003, see Table 1, with a 20 m pixel resolution and green, red and near infrared (NIR) spectral bands. All images represent dry conditions, despite the non-anniversary dates, and share very similar off-nadir sensor view angles. Nevertheless, mapping the Taita Hills landscape from this imagery represents a challenge because of the limited spectral information and because of the complex heterogeneous nature of the land cover relative to the sensor GIFOV; both in terms of small-scale cultivation areas with mixed cropping and extensive use of agroforestry, and in terms of the bare soil and low density dry vegetation mix of the shrubland areas. Consequently, the majority of the imagery is formed from pixels of mixed land cover types, with only areas of closed shrublands and forests giving very strong homogeneous spectral responses. There is also some spectral overlap between different LULC classes. It was therefore hoped that the utilization of an MSS/ORM approach would improve classification accuracies over the pixel-centered ML technique. All MSS/ORM work was implemented in eCognition, image processing and ML classifications were carried out in ERDAS IMAGINE, and LULC change detection utilized IDRISI Andes and the ArcGIS vLATE extension. Before the imagery was segmented and classified, however, it needed to be preprocessed.

3.1.2 Image preprocessing

The first step in processing the SPOT imagery was to make the multi-temporal scenes spatially comparable through geometric correction. Because of the rugged terrain in the Taita Hills, it was necessary to orthorectify the imagery utilizing a 20 m planimetric resolution digital elevation model (DEM) interpolated from 50-feet interval contours captured from 1:50,000 scale topographic maps. The 2003 image was orthorectified first using the scan-maps. The 1987 and 1992 scenes were then orthorectified to this geometric master scene to ensure the best possible alignment (inter-scene agreement 0.45 pixels RMSE) and, finally, the two adjacent 1992 scenes were mosaiced together. A nearest-neighbour resampling technique was employed to ensure that the original pixel values were preserved.

Accurate LULC classification and change detection in a set of multi-temporal SPOT data is dependent on the ability to successfully relate differences in corrected reflectance measurements to actual changes in vegetative state or land cover on the ground. This requires both absolute radiometric calibration and topographic corrections to be applied. As a first step in radiometric processing, the raw digital numbers (DNs) were divided, on a per channel basis, by the supplied gain values to derive at-sensor radiance (L_{SAT}) in $\text{W m}^{-2}\,\text{sr}^{-1}\,\mu\text{m}^{-1}$. From there it was necessary to correct for variations in the solar zenith angle, Earth-Sun distance, and atmospheric scattering and

Table 1. SPOT data utilized in this study.

Image date	Path and Row	SPOT sensor	Sensor view angle
1987-07-01	143-357	SPOT 1 HRV 1	R 10.35°
1992-03-25	142-357*	SPOT 2 HRV 1	R 13.8°
1992-03-25	143-357*	SPOT 2 HRV 2	R 9.3°
2003-10-15	143-357	SPOT 4 HRVIR 1	R 10.4°

* Adjacent scenes captured simultaneously.

absorption between image dates. However, no detailed overpass concurrent atmospheric measures were available that could be used as inputs into radiative transfer models (RTM), such as 6S. Additionally, with only three broad spectral bands available in the visible/NIR, the estimation of the necessary atmospheric optical properties for an RTM correction from the imagery itself was not possible. During field work in 2005, surface reflectance (P_S) measurements were made at a limited number of spectrally pseudo-invariant sites in the Taita Hills. This enabled the verification and comparison of an historical empirical line method (HELM), developed to account for the circumstances and limitations of the TAITA project (Clark and Pellikka 2005), with alternative image based correction approaches, such as the Cosine of Solar Zenith Angle (COST) method (Chavez 1996).

The empirical line method (ELM) corrects L_{SAT} data to P_S measurements, made at a number of spectrally stable calibration sites, utilizing a standard linear regression equation in the form $y = ax + b$; where a is the slope of the regression line, representing the atmospheric attenuation, and b is the intercept with the x-axis, representing the atmospheric path radiance. A separate correction is derived for each spectral band. The main assumptions are that the atmosphere is approximately homogenous throughout the image area and that there is a linear relationship between L_{SAT} and P_S. As Moran et al. (1990) note, although this relationship is quadratic for the full range of reflectance (0–100%), it is sufficiently linear over the range 0–70% to allow interpolation with negligible error (all surface reflectances in the Taita Hills are <70%).

Previous researchers have successfully retrieved P_S from remotely sensed data utilizing ELM (e.g. Karpouzli & Malthus 2003). The main problem applying this method to SPOT data is to identify ground targets that are large enough to counter the contaminating effects of the point spread function (PSF) on the GIFOV of the sensor. As Karpouzli and Malthus (2003) note, the calibration and validation targets need to be at least three times the pixel size (60 × 60 m for 20 m resolution SPOT 1–4 data) to derive representative L_{SAT} values. In outlining their Refined Empirical Line (REL) method for Landsat data, Moran et al. (2001) showed that, because of the near-linear relationship between L_{SAT} and P_S, an accurate estimation of the correction line can be obtained using detailed field measurements of only *one* appropriate within-scene bright calibration target, and a "reasonable" estimate of path radiance for $P_S = 0$ derived using an RTM. HELM is based on the REL method, but derives the path radiance estimate directly from the imagery assuming 1% reflectance for so called 'dark objects', such as areas of complete topographic shadow, rather than using an RTM. If the calibration target is truly spectrally pseudo-invariant over time then P_S measurement need not coincide with the image acquisition. The objective of HELM is, therefore, to (re)construct the historical linear relationship between L_{SAT}, as recorded by the multi-temporal imagery, and P_S for the pseudo-invariant pixels (PIPs) as measured in the field.

The chosen bright calibration site was a roadside quarry where half-day long measurements were made in an attempt to determine changes in P_S with the solar zenith angle, as suggested by Moran et al. (2001). 15 sets of P_S measurements, each with a sample average of 15, were taken every 10 minutes using an ASD FieldSpec® Handheld VNIR (325–1075 nm, 3.5 nm spectral resolution) spectroradiometer calibrated to a Spectralon® BaSO$_4$ 99% reflectance panel before each measurement set. The device was handheld at ~1.2 m height, with a 25° bare-head optic giving an at-nadir ground view of 53 cm in diameter. Upscaling to match the SPOT GIFOV was achieved by taking multiple measurement points within the site and averaging spatially during the data processing phase. In the event, it was found that the noise level of the handheld spectrometer measurements exceeded the signal of variation in P_S with the solar zenith angle, so it was not possible to quantify this relationship. However, it was inferred from this that these variations must be relatively small and therefore that the calibration target exhibited near-Lambertian reflectance behaviour. It was thus considered that the average nadir reflectance characteristics of the target had been captured and that the SPOT imagery with varying sensor view angles could be normalized to this data with minimal error, given the measurement noise. In order to increase the sample size to include validation data, multiple nadir P_S measurements were made of a sandy school playground, an area of compacted red soil, and a tarmac road.

The spectrometer derived P_S data were processed to synthesis the SPOT response for the PIPs at each date based on the specific spectral sensitivities of each band for the SPOT sensor involved. The L_{SAT} values for the darkest in-scene object and the bright calibration site were determined for each spectral band and regressed to the synthesized P_S to derive a correction equation which was then applied to the whole scene. Based on a comparison of the actual and predicted P_S of the verification sites, HELM corrected the SPOT data with an average RMSE better than 2% reflectance for all bands and all dates. This was more accurate than the tested alternative image based COST and DOS (Dark Object Subtraction) methods, which predicted the dark tarmac reflectance very accurately but had significant bias (~10% reflectance) in underestimating the bright targets (a known problem with such techniques). HELM also reduced the average difference in mean P_S of all bands for all dates to 2.2%, compared to 3.4% for the uncorrected top-of-atmosphere reflectance data. A disadvantage of applying HELM, however, was that it was not possible to calibrate the 2003 band 4 SWIR data, as the utilized spectrometer was limited to the 325–1075 nm range, but this channel was unavailable for 1987 and 1992 in any case; see Clark and Pellikka (2005) for full details on HELM.

Topographic correction of the satellite imagery over the rugged mountainous terrain of the Taita Hills is at least as important as atmospheric correction, if comparable surface reflectance values are to be taken throughout the area. This is critical both for traditional classification techniques and for image segmentation procedures, where changes in reflectance should relate solely to differences in land cover types and not to variation in illumination conditions. Illumination can be defined as the cosine of the solar incidence angle (cos i), representing the proportion of direct solar radiation hitting a pixel within an image. The amount of illumination is therefore dependent on the relative orientation of the pixel toward the Sun's actual position during image acquisition, as determined from a DEM of the area. Cos i was calculated as:

$$\cos(i) = \cos S \cos \theta_z + \sin S \sin \theta_z \cos (\phi_s - \phi_n) \tag{9}$$

where S is the slope of the pixel, θ_Z is the solar zenith angle, θ_S is the solar azimuth angle, and θ_n is the azimuth angle of the pixel (i.e. the aspect). If the surface is flat the aspect is undefined and i is simply θ_Z. Removal of slope-aspect effects from the HELM corrected imagery utilized a method based on the cosine function, similar to that originally proposed by Teillet et al. (1982), with band specific 'c' correction factors calculated for identified general vegetation classes:

$$P_H = P_T \frac{\cos \theta_z + c}{\cos i + c}$$
$$c = \frac{b}{m} \tag{10}$$

where P_H is the surface reflectance of a corrected pixel, P_T is the surface reflectance of an uncorrected pixel, and b is the y-intercept and m is the gradient of the linear regression line of cos i against P_T for a specific spectral band and vegetation cover type combination. To identify general vegetation classes *before* a topographic correction and classification has been applied, Normalized Difference Vegetation Indexes (NDVI) were derived for each image and ten cluster classes were identified within them using the automated ISODATA algorithm in ERDAS IMAGINE. NDVIs were used because they are a ratio between the red and NIR bands and are consequently relatively unaffected by topographic effects, i.e. both bands respond to the variation in illumination in a similar way. The ISODATA algorithm is more usually utilized to automatically identify data clusters in multi-spectral feature space, but here it is used in a one-dimensional manner to capture naturally occurring frequency clusters in the NDVI. The number of classes was set at 10 after experiments with the data to identify a generally applicable value.

For each image date, and for each spectral band and vegetation class combination, the reflectance values for pixels with a slope greater than 5° were regressed against their cos i values using standard linear regression. Where there was not a significant relationship, which occurred with some of the least vegetated NDVI clusters, the classes were merged and re-regressed until groupings with a usable stronger coefficient of determination were generated. One c correction factor for each utilized spectral band and vegetation class combination was then calculated by dividing the intercept of the finalized regression line by its slope. 5° was taken as the cutoff for modelling as it was considered there was no quantifiable relationship between $\cos i$ and P_T for shallower slopes. Similarly, no c factors were applied to slopes less than 5° in the implementation of the correction, as the cosine correction here is very slight in any case, but band specific masks for the various c factors were used elsewhere. It is considered that use of c factors accounts for both diffuse irradiance and the non-Lambertian reflectance behaviour of the vegetation within each generalized group, and also has the effect of limiting the overcorrection of weakly illuminated pixels. This model does have limitations though; for example, it does not account for irradiance reflected from surrounding terrain. However, several workers have found that good correction results can be obtained using the c-correction model (see, for example, Riano et al. 2003). Visual inspection and re-regression of the corrected reflectance values for each class area against the $\cos i$ values, which derived no relationships, demonstrated that topographic effects had been successfully removed from the SPOT imagery.

3.1.3 *Classification nomenclature*

Standardization of LULC classification schemes is an important issue if greater use and understanding of digital mapping products is to be facilitated. The software and protocol of the Land Cover Classification System (LCCS) developed by the Food and Agriculture Organization of the United Nations (FAO) and the United Nations Environment Programme (UNEP) has emerged as a widely accepted format and was consequently utilized to derive the LULC classes for the TAITA project, as shown in Table 2. The LCCS is a comprehensive, standardized *a priori* classification system which can be used for any mapping exercise regardless of the methodology, scale, source material and geographic location (FAO 2005). As well as logically guiding the user through the derivation of appropriate mutually exclusive LULC classes, as is shown in Table 2, the LCCS software generates unique codes (third column in Table 2) and Boolean formulas (fourth column in Table 2) for each class which allows other users to precisely reconstruct the detailed definitions utilized. This is very useful as previously it would not necessarily be certain what was meant when a map contained a LULC class name, such as 'Thicket'.

Table 2.   TAITA project LCCS nomenclature adopted for SPOT imagery LULC mapping.

ID	User land cover name	LCCS code	LCCS Boolean formula (classifiers)
1	Cropland	11251 – 12699	A3B2XXC2D1 – C4C10C17C13C17
2	Shrubland (20% to 70% cover)	20373	A4A11B3XXXXXXF1
3	Thicket (Closed Shrubland >70% cover with emergent trees)	20354 – 13554	A4A10B3XXXXXXF2F5F10G2F1 – B9G7
4	Woodland	20013	A3A11
5	Plantation Forest	10001 – S1002S1003W7	A1-S1002S1003W7
6	Broadleaved Closed Canopy Forest	20088 – 13152	A3A10B2XXD1 – B5
7	Grassland with scattered shrubs and trees	20412 – 104774	A2A10B4XXXXXXF2F5F10G2F2F6F 10G3 – B12G7G9
8	Bare Soil & Other unconsolidated material	6005	A5
9	Built-up Area	5001	A1
10	Bare Rock	6002-1	A3 – A7
11	Water	8002-5	A1B1 – A5
12	Burned Area	*Not available*	–
13	Cloud/Cloud shadow	*Not available*	–

In LCCS terminology, a 'classifier' is one of many measurable diagnostic characteristics that are used in the definition of a land cover class, such as vegetation cover and height. Additionally, a 'modifier' is a further optional refinement to a classifier which helps specify the exact properties of an LULC class. Thanks to the code and Boolean formula in LCCS, the user is free to call the class any general or colloquial name that he/she desires and retain interoperability. The classes listed in Table 2 were developed based on extensive inspection of the SPOT imagery and fieldwork knowledge to determine what was feasible to map. Note that burned areas, and clouds and cloud shadows are not included as possible classes in the LCCS software but they occur in the imagery and are consequently specified in the TAITA map legend.

3.2 Results and discussion

3.2.1 Segmentation and object oriented classification

As discussed above, initial segmentation of 'image object primitives' with FNEA is based on user-defined parameters for the scale, for the weighting of the shape versus spectral information, for the balance in smoothness and compactness parameters that make up the shape factor, for weighting the image bands, and for determining the neighbourhood function (Benz et al. 2004). Each of the three spectral bands was given an equivalent weighting of 1, as all were considered to contribute equally useful information. The most appropriate segmentations were derived when 'color' was weighted at 0.8 (shape criterion 0.2), as spectral information in the imagery was stronger than the spatial information. Within the shape setting, the smoothness was set to 0.8 (compactness 0.2) as the objects of interest discernable in the SPOT data are mostly of variable shape, better described by smoothness. Because the features of interest are close to the pixel scale, diagonal segmentation was used to allow objects to grow in NE-SE-SW-NW trajectories (in addition to the N-E-S-W orientations of the simpler plane 4 neighbourhood), following the complex patterns of agricultural terracing and other irregular features on the ground.

The scale is a unitless parameter which determines the average size of the segments, enabling multi-scale segmentation. The first segmentation is critical because the borders defined at this stage will be adhered to by any subsequent segmentations, either subdividing the image object primitives or combining them into larger objects. As the patterns of small-scale cultivation areas in the Taita Hills are close to the SPOT GIFOV, a very detailed initial segmentation, with a scale parameter of 2, was required to successfully capture them. As is illustrated in Figure 2, fields and patches of shrublands occurring in the foothills and lowlands areas, however, were best described by an aggregation with a scale parameter of 4. As the actual final 'mapping level' was to be level 2,

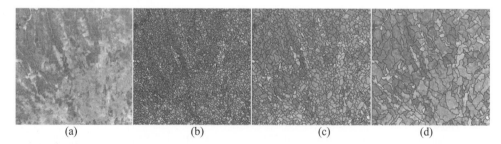

| (a) | (b) | (c) | (d) |

Figure 2.   Detail of multi-scale segmentation of the 2003 image for a circa 6 km by 6 km lowlands area directly to the north west of the Taita Hills: (a) 2003 SPOT image red band. Darker areas are shrubland, whilst lighter patches are encroaching croplands. (b) Image objects derived from a level 1 segmentation with a scale parameter of 2 do not coincide with the focal scale of cropland patches desired for this lowland area; this is an over-segmentation. (c) A level 2 segmentation with a scale parameter of 4 successfully captures the field boundaries deriving ecologically meaningful landscape patches for the croplands. (d) A level 3 segmentation with a scale parameter of 10 delineates general areas of cultivation but has amalgamated the smaller within- and between-field shrubland patches into the croplands polygons.

in order to derive ecologically meaningful objects throughout, it was necessary to perform a merge of these segmentation levels for highland and lowland areas.

The experimentally derived finally utilized parameters and the segmented image object statistics are summarized in Table 3. As can be seen, there was a difference in the number of objects at all levels between dates, but with a small coefficient of variation (CV). This was due to varying amounts of clouds (minimal in 2003; see Table 6 below), which added segments, especially at levels 3 and 4 where they formed small objects with high spectral contrast, increasing the CV. An expansion in homogenous croplands in 2003 also led to less objects. As might be expected due to the workings of the scale parameter in FNEA, the average size of the objects at each level between dates was very similar, with a rising CV at higher hierarchy levels reflecting greater variance. For all segmentation levels in 1987 and 1992, the smallest objects were extreme outliers, as can be seen from the mean size and small CV, related to clouds and their translucent edges. In 2003, where there was little cloud, the smallest object at level 4 was a patch of riverine forest, but a small patch of very bright high contrast bare soil for the other levels. At all dates and all segmentation levels, the largest objects related to spectrally homogenous patches of crops or grassland, where segment growth was only constrained by the scale parameter. Overall, it is noted that the FNEA multi-scale segmentation approach was very successful at capturing image objects visually identifiable in the SPOT data. This is important, as the quality of the segmentation directly affects the quality of the classification, as objects and their derived properties should relate to meaningful and coherent landscape features.

The same segmentation parameters and OO classification procedure were applied to all three image dates, as was a ML classification, starting with 2003. For this date, ML classification was undertaken based on spectral training areas identified during fieldwork for each of the LULC classes, except Burned Areas where the training areas were derived directly from the imagery, and the Built-up Area and Bare Rock classes which where not spectrally separable and could not therefore be mapped using this approach. For the OO classification, segmented image objects covering the same sample areas were used as training polygons for the fuzzy nearest neighbour classifier implemented in eCognition. For the 1992 and 1987 scenes, the training areas were modified to remove those that had a different land cover from 2003 on the basis of detailed inspection of the imagery. In the OO classification it also proved necessary to make some edits to the spectral values of the fuzzy classifier rules when transferring the protocol to the 1992 and 1987 scenes,

Table 3. Segmentation scale parameters and object statistics for the TAITA SPOT data.

| Level | Scale | Shape | Smoothness | Number of objects | | | *Av. No. Ob. (SD) [CV] | | |
				1987	1992	2003			
4	100	0.2	0.8	92	104	79	92	(10.2)	[11.1]
3	10	0.2	0.8	8,229	9,015	7,805	8,350	(500.9)	[6.0]
2	4	0.2	0.8	47,714	50,088	44,962	47,588	(2,094.6)	[4.4]
1	2	0.2	0.8	169,187	170,290	154,850	164,776	(7,032.9)	[4.3]

| Level | Scale | **Av. Ob. Size Pxls (SD) [CV] | | | Range of object size in Pixels (Min–Max) | | |
					1987	1992	2003
4	100	24,642	(2,785)	[11.3]	277–102,489	432–101,580	2,181–146,379
3	10	268.1	(15.8)	[5.9]	7–2,049	6–2,386	2–3,094
2	4	46.9	(2.1)	[4.5]	1–569	2–559	1–593
1	2	13.6	(0.6)	[4.4]	1–184	1–185	1–255

* Av. No. Ob. (SD) [CV] = Average Number of Objects (Standard Deviation) [Coefficient of Variation].
** Av. Ob. Size Pxls (SD) [CV] = Average Object Size in Pixels (Standard Deviation) [Coefficient of Variation].

despite the calibration of the images. These reflectance variations are most likely due to actual phenological and moisture differences between the scenes. For the 2003 scene, the average transformed divergence separability of the training areas was 1989, and the value was above 1900 for all class pairs (indicating good spectral separability), with the exception of Shrubland and Thicket (1845), and Crops and Shrubland which had a value of 1266, indicating spectral confusion. As would be expected from this, Shrubland proved to be the problematic class, both in the ML and the OO classifications, as is reflected in the accuracy assessment shown in Table 4 below.

The other segmentation levels 3 and 4, as detailed in Table 3, formed part of the OO classification process. Clouds and cloud shadows could be mapped semi-automatically at level 3, where the segment boundaries accurately captured their occurrence. The clouds themselves where identified by image derived spectral training areas and a threshold rule of high reflectance in the green band. The cloud shadows were identified as being image segments with a low mean reflectance in the NIR that also had adjacency to cloud segments. This was enabled by a two-step 'spectral-only' followed by a rules based classification, as the cloud objects must first be mapped before their shadows can be identified. The majority of the clouds and shadows were recognized in this manner and only minor manual edits were required to capture objects describing smaller and more translucent clouds which did not meet the class rules. The existence of class Cloud/Cloud Shadow super-objects with a feature distance of 1 (i.e. from the next level up in the segmentation hierarchy) was then used to classify level 2 patches that were clouds. In the ML classification the clouds were manually digitized and masked out from the imagery before processing.

The level 4 segmentation was used to identify large landscape objects which described general grassland and shrubland areas, croplands, and closed canopy forest patches. Because a diagonal neigbourhood function was utilized, elongated segments following riverine woodland were also derived. To assist in the classification of level 2 landscape patches, these objects were categorized into three classes: 'No Agriculture', 'Agriculture Possible' and 'Rivierine Woodland'; the first two being identified by spectral criteria, the third by the length/width ratio of the elongated object shape. 'No Agriculture' areas were segments consisting entirely of shrubland or other natural vegetation types. Fuzzy logic rules then utilized these super-objects in a mask like manner to disallow the occurrence of Cropland and Built-up Area as sub-object classes within identified 'No Agriculture' and 'Riverine Woodland' areas (by allocating a fuzzy membership value of zero). These objects then became classified to the next most likely LULC group which was usually the correct class. The level 4 delineation between pure shrubland and areas of cultivation was not perfect as some segments identified as 'No Agriculture' did contain a small number of fields. In comparison to the ML approach, however, this step had a major effect in reducing erroneous allocations of the

Table 4. 2003 classification accuracy assessment.

LULC class	Maximum-likelihood		Object-oriented		OO manually edited	
	Producer	User	Producer	User	Producer	User
Cropland	62.5%	61.0%	67.5%	67.5%	95.9%	81.8%
Shrubland	41.2%	54.9%	47.1%	65.3%	64.2%	82.7%
Thicket	88.6%	68.9%	82.9%	72.5%	92.9%	87.8%
Woodland	71.7%	62.3%	86.8%	70.8%	91.5%	91.5%
Plantation Forest	47.1%	96.0%	70.6%	85.7%	97.2%	94.6%
Broadleaved Forest	87.8%	81.2%	85.7%	85.7%	97.0%	100.0%
Grassland	58.1%	66.7%	71.0%	45.8%	71.0%	95.7%
Bare Soil	85.5%	53.6%	80.7%	68.5%	84.7%	90.9%
Built-up Area*	–	–	12.0%	100.0%	96.0%	92.3%
Water	95.0%	100.0%	90.0%	100.0%	100.0%	100.0%
Overall accuracy		65.6%		73.5%		89.0%
Overall KIA		0.60		0.66		0.87

* This class was not included in the overall accuracy calculations for the ML and OO classifications.

Cropland and Built-up Area classes caused by spectral overlap between irrigated crop areas and riverine woodland and between built-up areas and dry grassland.

The biggest disappointment when investigating possible classifiers for shrubland patches was that no textural measures, including the grey-level co-occurrence matrix (GLCM) calculations developed by Haralick et al. (1973), were able to successfully differentiate between cropland and shrubland segments. This was surprising because a visual identification is based on the typical 'speckle' of shrubland, particularly from the red band, caused by pixels with continuously varying proportions of bare soil and shrubs, compared to the more homogenous texture of a field. The texture measures were applied at all the segmentation levels to check if the smaller number of pixels forming segments at level 2 may have been reducing the effectiveness of the calculations. However, at no segmentation level was a texture based differentiation possible.

Rather than detailing the various individual rules used to classify each of the 13 LULC classes, it is more instructive to note that the fuzzy nearest neighbour classifier was utilized in the rule set of every class, as it remained the most effective method for establishing general class membership which could then be refined by the application of other fuzzy logic rules. Based on the sample objects collected for the LULC classes, an automated feature space optimization was undertaken to identify which combination of available spectral properties derived the best minimum distance to mean separability. That is to say, the combination with the greatest linear separation distance between the samples in multi-dimensional feature space, and therefore most appropriate for utilization in the implemented nearest neighbour classifier. To allow the comparison of various ranges, the data were normalized by the standard deviation of the feature. A combination of five parameters was identified: an object's mean reflectance for each of the three available bands (green, red, NIR), and the ratio property for the red and for the NIR band. Here, the ratio property is the object's mean reflectance in the specified band divided by the sum of the object's mean reflectance for all the available spectral bands. It should also be noted that, due to heavy spectral overlap with Built-up Areas, it was not possible to automatically map Bare Rock with either the ML or the OO methodology, so this class was manually captured. Similarly, due to spectral overlap with dry grassland areas, it was also not possible to map Built-up Areas using the ML approach. Furthermore, once the automated OO classifications had been generated, a manual editing pass was made through the mapped area to correct errors identifiable relative to the original SPOT imagery. Note that the unedited 2003 OO map was used as the basis for the comparison with 2003 ML classification, but the manually edited OO maps were used in the change detection exercise.

3.2.2 Classification accuracy assessment

To enable accuracy assessment of the 2003 classifications, ground reference test data were collected during field visits to the Taita Hills in January 2005 and 2006 using stratified random road sampling (points falling in areas visually identified to have changed land cover relative to the 2003 image were discarded and regenerated), and from 0.5 m resolution true-colour digital aerial photography flown in January 2004 (3 months after the SPOT acquisition) using stratified random sampling. The photography is limited to 8 mosaic areas covering 12% of the Taita Hills, mostly in the highlands, and whilst the road sampling extended into the lowlands, less reference points were collected in the field because of logistical and financial constraints. A minimum statistically valid class sample size of 60 was calculated based on the multinomial distribution approach outlined by Plourde and Congalton (2003). Lesser numbers of points were collected for the spatially limited classes, such as Water, and more for the spatially extensive classes, like Cropland. The ephemeral Burned Area and Cloud/Shadow classes could not be sampled, and Bare Rock was not assessed as it could not be automatically mapped. A comparison of the 2003 ML and OO classifications is detailed in Table 4. Because of a lack of timely ground reference test data or aerial photography, the 1987 and 1992 classifications could not be assessed. However, the manually edited 1987 and 1992 OO classifications used in the change detection study are assumed to have an accuracy comparable to the edited 2003 OO classification.

Table 4 indicates that utilizing an OO classification approach, as opposed to the ML technique, derived an improvement in the overall accuracy from 65.6% to 73.5%, and in the Kappa index of agreement (KIA) from 0.6 to 0.66, with variable class specific commission errors (user's accuracy)

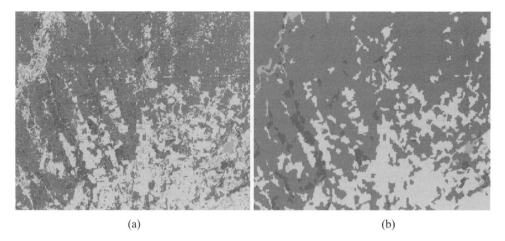

(a)   (b)

Figure 3.   Detail of (a) ML and (b) OO classification of the 2003 image for a lowlands area directly to the north west of the Taita Hills. The darkest tone is Thicket, the lightest Croplands which are encroaching into Shrubland. The presence of 'salt-and-pepper' noise in the ML classification is clearly noticeable.

and omission errors (producer's accuracy). In fact, the OO classification is actually much more accurate than the ML map than is suggested by the results in Table 4, as significant areas of lowlands around the hills were not sampled. Here, large amounts of error in misclassifying shrubland and grassland as Cropland is known to exist in the ML classification, but is not present in the OO map because of the level 4 segmentation, as described above. Also not reflected in this assessment is the visual superiority of the OO results in generating coherent landscape patches and in completely removing the 'salt-and-pepper' effect typical of ML results, as is shown in Figure 3. These improvements are due to the spectral properties of a landscape patch being more stable and separable than an individual pixel, and the inclusion of an object hierarchy and contextual relationships into the classification process. Manually editing the visible errors further improved the overall accuracy of the OO classification to 89% (KIA 0.87), which is equivalent to many multi-class land cover maps in use and is suitable for use in studying land cover change. As can be seen in Table 4, the main problem class was Built-up Area, where all but the most developed sites were misclassified as Bare Soil or Grassland in the OO approach, and the class could not be mapped at all using the ML technique. Manual editing improved the situation, but even visual interpretation proved difficult. Utilizing the OO approach, Cropland, Shrubland and Woodland showed a reduction in both omission and commission error. Shrubland, however, remained a problematic class with a low producer's accuracy and misclassification with Cropland and Grassland. Unsurprisingly, perhaps, given its variability from the dry lowlands to the verdant hills, the Cropland class had the widest range of commission errors with other classes (although not the lowest user's accuracy which fell to Grassland). In contrast, Water, being the most easily spectrally separable class, was well mapped by both methods, although a few edge-segments formed from 'mixels' were wrongly identified as Burned Areas.

3.2.3   Change detection

Once delineated from the remote sensing data, patch objects can be explored using landscape metrics (McCarigal and Marks 1995) inferring ecological function from the structure as proposed by Levin (1992). However, this implies calibrating the applied metrics to species distribution and persistence data, otherwise the ecological meaning is effectively unknown (Opdam et al. 2003). McGarigal and Marks (1995) themselves warn against the non-selective and uncritical application of metrics, and understanding the relationships between landscape metrics and ecological processes remains a research area. Within the Taita Hills, the most ecologically important habitat is the indigenous forest patches. Consequently, special emphasis was placed on their study utilizing the Vector-based Landscape Analysis Tools Extension (vLATE) for ArcGIS, developed at the

Table 5. Selected landscape structural metrics calculated for Broadleaved Closed Canopy Forest.

	1987	1992	2003	Change 1987–2003	
				Amount	% Change
Area related					
Number of patches	90	97	84	−6	−6.7
Mean patch size (ha)	8.6	7.6	8.3	−0.3 ha	−3.5
Patch size standard deviation (ha)	28.1	26.2	27.6		
Taita Thrush Inhabited Area**	450.3	431.3	431.6	−18.7 ha	−4.2
Core Area*					
Number of remaining Core Areas	37	30	23	−14	−37.8
Total Core Area (ha)	297.9	280.1	280.7	−17.2 ha	−5.8
Taita Thrush Inhabited Core Area**	272.2	261.8	261.2	−11.0 ha	−4.0

* Based on *Turdus helleri*; see discussion for details.
** Formed of three forests: Mbololo, Ngangao and Chawia.

Table 6. Summary of LULC changes in the Taita Hills for 1987, 1992 and 2003.

	1987		1992		2003		Change amounts	
LULC Class	ha	%	ha	%	ha	%	87-03 ha	87-03%
1. Cropland	25980.5	29.1	27132.6	30.4	36458.8	40.9	10478.4	40.3
2. Shrubland	26411.9	29.6	22003.6	24.7	20108.0	22.5	−6303.9	−23.9
3. Thicket	25640.3	28.7	26041.2	29.2	21070.2	23.6	−4570.1	−17.8
4. Woodland	4648.2	5.2	5774.0	6.5	5062.3	5.7	414.1	8.9
5. Plantation Forest	1991.8	2.2	1827.8	2.0	2024.4	2.3	32.6	1.6
6. Broadleaved Forest	773.6	0.9	740.9	0.8	693.6	0.8	−79.9	−10.3
7. Grassland	1477.8	1.7	1504.5	1.7	1853.1	2.1	375.4	25.4
8. Bare soil	378.4	0.4	680.7	0.8	924.8	1.0	546.4	144.4
9. Built-up Area	75.2	0.1	90.5	0.1	100.5	0.1	25.4	33.7
10. Bare Rock	228.4	0.3	225.2	0.3	188.9	0.2	−39.5	−17.3
11. Water	84.0	0.1	50.4	0.1	19.5	0.0	−64.6	−76.8
12. Burned Area	58.6	0.1	261.4	0.3	642.8	0.7	584.2	997.7
13. Cloud	1471.5	1.6	2887.2	3.2	73.1	0.1	−1398.4	−95.0
Total Area (ha)	89220.04		89220.04		89220.04			

University of Salzburg (Lang and Tiede, 2003). As can be seen from Table 6, there was a 10% reduction in the total area of indigenous (mapped as Broadleaved Closed Canopy) forest over the study period, dropping from 774 ha in 1987, to 741 ha in 1992 and 694 ha in 2003. In combination with this, as Table 5 shows, the number of patches increased slightly between 1987 and 1992 whilst the mean size reduced, indicating fragmentation, but then by 2003 the total number of patches fell while the mean size increased, reflecting a loss of smaller patches. Table 7 indicates that most of this loss (54 ha) was to the Plantation Forest and Woodland classes, with only 22 ha being mapped as converted to Croplands. It therefore appears that there has been little deforestation in the Taita Hills during this 15 year period, as both Woodland and Plantation Forest cover have increased (9% and 1.5% respectively). This contrasts with a recent study by Ward et al. (2004) based on an analysis of Landsat imagery from 1987 and 1999 utilizing unsupervised classification methods which reported a large 37% decrease of indigenous forest in the Taita Hills, but mapped for both 1987 and 1999 with an erroneous over-estimation of cover with nearly the entire upland areas of the hills depicted as forest. In reality the remnant patches present during this time period are very small and cover less than 1% of the total area, as indicated both from field measurements Lens et al. 2002 and from the results of this study.

Of the rare endemic animal and plant species living in the remnant forest patches, the Taita Thrush (*Turdus helleri*), a bird from the Turdidae family, is considered to be the most critically endangered due to its very limited occupied range, low population (estimated to be <1,350), habitat sensitivity and high levels of measured genetic stress (Bytebier 2001). Consequently, this species was taken as the basis for a core area analysis. As mentioned above, Mbololo, Ngangao and Chawia are the three main largest remaining forest fragments. The Taita Thrush is present in all three, but as it prefers well-shaded areas with a dense understorey, high litter-cover and little or no herbaceous cover, it is found at a greater density in Mbolobo, the least disturbed forest area, and is rarest in Chawia, which has a more open canopy and a very shrubby understorey (Birdlife International 2007). Very few inter-fragment movements have been recorded despite extensive research, although Taita Thrushes have been seen in a further very small heavily disturbed patch known as Yale (c. 2 ha). According to Lens et al. (2007), out of 1280 telemetry fixes for the species, 1 observation was 10 m from the edge, 2 were at 17 m, but all the rest were >50 m into the forests. An ecologically meaningful edge distance defining the core area for analysis was therefore taken to be 50 m. In addition to the mapped total core area, it is useful to consider the fragments actually inhabited by the Taita Thrush. As Yale is so small, this left the summation of Mbololo, Ngangao and Chawia core areas as the critical inhabitable area which, as can be seen from Table 5, reduced between 1987 and 1992 to 262 ha, but then remained stable between 1992 and 2003, as did the total core area. This is heartening and conservation efforts are currently underway both to safeguard these remaining fragments and to establish connectivity between the patches so that inter-fragment movements can occur.

When considering quantified LULC change analysis from digital maps it is important to keep in mind issues of error propagation from the classifications. As indicated in Table 4, the crucial indigenous forest was well mapped and given the estimated overall error of circa 10% for the edited OO maps, it is reasonable to suppose that identified major differences in all LULC classes are indicative of real changes on the ground. Furthermore, as discussed in the Introduction, strong patterns in LULC change at a landscape level can be used as indicators for the likely presence and extent of land degradation processes, at the very least allowing the derivation of target areas for further, more detailed, study. The variation in position and extent of clouds and cloud shadows in the imagery, which covered 1471 ha in 1987, 2887 ha in 1992 and 73 ha in 2003, also obfuscates true changes in LULC. In addition, the inherent limitations in a three-date 'snap-shot' change detection study, which is vulnerable to seasonal variations or unusual circumstances over long term changes, is recognized.

LULC change analysis was conducted on the OO edited classifications as raster maps in IDRISI Andes software. In addition to indigenous forest discussed above, Tables 6 and 7 suggest that the

Table 7. From-To landscape changes in the Taita Hills 1987 (read as rows*) to 2003.

LULC Class	1	2	3	4	5	6	7	8	9	10	11	12	13
1. Cropland	*20205.0*												
2. Shrubland	−6414.4	*13183.2*											
3. Thicket	−3299.6	−925.5	*13889.7*										
4. Woodland	−120.7	−47.7	302.3	*1796.9*									
5. Plantation F.	−29.5	−0.2	60.8	−73.2	*1159.4*								
6. Broadleaved F.	−21.8	0.1	0.5	−25.4	−28.8	*583.8*							
7. Grassland	−60.7	384.2	48.6	−5.8	0.3	0.2	*1190.8*						
8. Bare soil	163.5	262.5	55.6	11.2	2.2	0.6	19.9	*128.1*					
9. Built up Area	19.1	5.4	4.0	2.0	0.0	0.3	0.2	−8.7	*23.8*				
10. Bare Rock	1.9	−20.4	−8.8	−5.0	−7.2	3.1	−4.4	0.8	0.0	*94.4*			
11. Water	−19.2	−5.6	−0.5	−10.1	0.0	0.0	−10.6	−18.6	0.0	0.0	*17.8*		
12. Burned Area	71.3	330.4	135.4	36.6	10.9	0.0	−13.6	1.9	0.0	2.9	0.1	*0.0*	
13. Cloud	−768.2	−93.8	−253.0	−210.6	−52.2	0.3	−0.1	−6.3	−2.9	−3.3	0.0	−8.3	*0.0*

* All values in ha. A negative value indicates a loss in the class from 1987 to 2003, positive values a gain.

most important indicative landscape changes were in the Croplands, Bare Soil, Water and Burned Area classes, which can be linked to increasing environmental pressure on the Taita Hills. Areas of Bare Rock relate to a number of granitic bornhardts which are very large features, of extreme longevity, within the Taita Hills. The mapped 40 ha reduction in this class, though, probably does not represent the real world situation, because the class was mapped manually and has not been accuracy assessed. It is more likely that the area of Bare Rock has actually remained similar. Other mapped changes, however, are much more pronounced and strongly indicate landscape level processes occurring in the area. The Cropland class has expanded by 10,478 ha mainly into Shrubland (6,414 ha) and Thicket (3,300 ha) areas, especially on the lowlands and foothills. This is undoubtedly in direct response to local population growth, which has seen an increase in cultivated land from 29% to 41% of the total study area over the 15 year period. The lowlands have a semi-arid climate and poor soils for crowing crops which suggests they will be likely areas of land degradation; indeed gullies and sheet erosion features are present in these areas. The near 1000% increase in the Burned Area class reflects a continued expansion of agriculture, with many burned areas present along the edge boundaries of field areas. The standard procedure for establishing new fields in these shrubland areas around the fringes of the Taita Hills is to cut down the bushes and trees along the boundaries then burn the encircled area to clear the vegetation and add some nutrients to the soil.

Increasing environmental pressure on the Taita Hills is also implicated by other mapped landscape level changes. Although strong seasonal variations in water levels are acknowledged for this East African environment, there was a major 77% reduction in the extent of standing water from 84 to 19.5 ha, despite that reference to local meteorological data shows that all three images were acquired during similarly dry periods. This suggests that a greater proportion of available moisture is being utilized by intensified agriculture. Also, the size of the main water feature in the area, the Mwatate reservoir, was reduced from 30.6 ha to 14.6 ha in its northern extent, a decrease of 52%, because it is filling up with sediment. This is undoubtedly caused by increased sedimentation due to soil erosion up stream in the foothills. There were also a large 145%, 546 ha increase in the area of Bare Soil, mainly from Cropland and Shrubland areas (see Table 7), and a 34% increase in Built-up Areas. As all the indications are that environmental pressure on the Taita Hills is growing, it is heartening to report that the extent of the core area of the remnant indigenous forest patches appears to have remained stable since 1992. It should be noted, however, that in this study no attempt was made to assess the health of the forests from the SPOT imagery, which is an important factor for habitat quality sensitive species, such as the Taita Thrush.

4 CONCLUSIONS

A multi-scale segmentation/object relationship modelling (MSS/ORM) approach was applied to map land use/land cover (LULC) at a landscape level in the Taita Hills, Kenya, from multi-temporal SPOT XS satellite imagery. This object-oriented procedure was shown to derive improvements over a uni-scale maximum-likelihood technique in this complex area, both in terms of an increase in the assessed overall accuracy of the classification from 65.6% to 73.5%, and in a Kappa Index of Agreement from 0.6 to 0.66, but also more significantly in the derivation of visually superior land cover maps based on meaningful homogeneous landscape patches and free from the 'salt-and-pepper' classification noise effect typical of maximum-likelihood results. This is due to the theoretical advancements possible when conceptualizing a landscape and its depiction in a remotely sensed image as a spatially nested patch hierarchy definable at various critical levels of organization operating at specific spatial and temporal scales. Useful spatial information surrounding each pixel and multi-scale information within the image are incorporated into the classification process by the MSS/ORM approach, where a search is made for apparent boundaries in the gradient of flux zones within and between landscape patches identifiable through local heterogeneity. In particular, the fractal net evolution approach to multi-scale segmentation was successful at capturing image objects relating to ecologically meaningful landscape patches identifiable in the SPOT data. Segmentation studies are usually focused on so called 'high resolution' imagery, such as IKONOS data or digital aerial photography, so it is interesting to note the successful application of the MSS/ORM approach to data derived from a 'medium resolution' sensor such as SPOT.

Further manually editing work, by reference to the original imagery, was needed, however, to increase the accuracy of the LULC classifications to a level suitable for further utilization in various multi-disciplinary applications. The derived maps were used to identify major landscape changes that have occurred in the Taita Hills over the study period 1987 to 2003. It is acknowledged that the three-date 'snap-shot' study reported here forms a minor temporal sampling of a complex environmental system in constant flux. Nevertheless, in an area of the world where detailed accuracy assessed landscape level digital mapping and change analysis is sparse, this information derived using the standardized Land Cover Classification System nomenclature is useful in many applications, as well as an indicator for the likely presence and extent of land degradation processes occurring in the region.

REFERENCES

Baatz, M. & Schäpe, A. 2000. Multiresolution segmentation: an optimization approach for high quality multi-scale image segmentation. In J. Strobl, & T. Blaschke (eds), *Angewandte geographische Informationsverarbeitung* Vol. XII (pp. 12–23). Heidelberg, Germany: Wichmann.

Beentje, H.J. & Ndiang'ui, N. 1988. An ecological and floristic study of the forests of the Taita Hills, Kenya. *Utafiti* 1 (2): 23–66.

Benz, U.C., Hoffmann, P., Willhauck, G., Lingenfelder, I. & Heynen, M. 2004. Multi-resolution, object-oriented fuzzy analysis of remote sensing data for GIS-ready information. *ISPRS Journal of Photogrammetry and Remote Sensing* 58: 239–258.

BirdLife International 2007. Species factsheet: *Turdus helleri*. Downloaded from http://www.birdlife.org (accessed June, 2007).

Burnett, C. & Blaschke, T. 2003. A multi-scale segmentation/object relationship modelling methodology for landscape analysis. *Ecological Modelling* 168: 233–249.

Bytebier, B. 2001. Taita Hills Biodiversity Project Report. National Museums of Kenya, Nairobi.

Chavez, P.S. Jr. 1996. Image-Based Atmospheric Corrections—Revisited and Improved. *Photogrammetric Engineering and Remote Sensing* 62 (9): 1025–1036.

Clark, B. & Pellikka, P. 2005. The Development of a Land Use Change Detection Methodology for Mapping the Taita Hills, South-East Kenya: Radiometric Corrections. *Proceedings of the 31st International Symposium on Remote Sensing of Environment (ISRSE)*, 20–24 June, 2005, St Petersburg, Russian Federation. CD-Publication, no page numbers. (available from: http://www.helsinki.fi/geography/b-clark_p-pellikka-ISRSE-paper.pdf).

EAWLS 2005. *East African Wildlife Society: Stakeholders Workshop on the Conservation and Management of Taita Hills Forests*—Workshop Summary Report.

Food and Agriculture Organization of the United Nations (FAO) 2005. *Land Cover Classification System (LCCS), version 2: Classification Concepts and User Manual*. FAO Environment and Natural Resources Service Series No. 8. FAO, Rome.

Haralick, R., Shanmugan, K. & Dinstein, I. 1973. Textural features for image classification. *IEEE Transactions on Systems, Man and Cybernetics* 3 (1): 610–621.

Hay, G.J., Marceau, D.J., Dubé, P. & Bouchard, A. 2001. A multi-scale framework for landscape analysis: object-specific analysis and upscaling. *Landscape Ecology* 16 (6): 471–490.

Hay, G.J., Dubé, P., Bouchard, A. & Marceau, D.J. 2002. A scale-space primer for exploring and quantifying complex landscapes. *Ecological Modelling* 153 (1–2): 27–49.

Hay, G.J., Blaschke, T., Marceau, D.J. & Bouchard, A. 2003. A comparison of three image-object methods for the multi-scale analysis of landscape structure. *ISPRS Journal of Photogrammetry & Remote Sensing* 57: 327–345.

Hay, G.J. & Marceau, D.J. 2004. Multiscale Object-Specific Analysis (MOSA): An Integrative Approach for Multiscale Landscape Analysis. In de Jong, S. & van der Meer, F. (eds), *Remote Sensing Image Analysis: Including the Spatial Domain*: 71–92. Dordrecht, The Netherlands: Kluwer Academic Publishers.

KARI 2005. Assessment of land degradation and its impacts on land use sustainability in Taita Taveta catchment. Muya, E.M. & Gicheru, P.T. (eds), Kenya Agricultural Research Institute, Miscellaneous Paper No. 63.

Karpouzli, E. & Malthus, T. 2003. The empirical line method for the atmospheric correction of IKONOS imagery. *International Journal of Remote Sensing* 20 (13): 2653–2662.

Lang, S. & Tiede, D. 2003. vLATE for ArcGIS. Downloaded from http://www.geo.sbg.ac.at/larg/vlate.htm (accessed June, 2007).

Lens, L., Van Dongen, S., Norris, K., Githiru, M. & Matthysen, E. 2002. Avian Persistence in Fragmented Rainforest. *Science* 298: 1236–1238.

Lens, L., Lehouck, V. & Githiru, M. 2007. *Terrestrial Ecology Unit, University of Ghent, Belgium*. Personal e-mail communication.

Levin, S.A. 1992. The problem of pattern and scale in ecology. *Ecology* 73 (6): 1943–1967.

McGarigal, K. & Marks, B.J. 1995. FRAGSTATS: Spatial Analysis Program for Quantifying Landscape Structure. *USDA Forest Service General Technical Report*. PNW-GTR-351.

Moran, M.S., Bryant, R., Thome, K., Ni, W., Nouvellon, Y., Gonzalez-Dugo, M.P., Qi, J. & Clarke, T.R. 1990. Obtaining Surface Reflectance Factors from Atmospheric and View Angle Corrected SPOT-1 HRV Data. *Remote Sensing of Environment* 32:203–214.

Moran, M.S., Bryant, R., Thome, K., Ni, W., Nouvellon, Y., Gonzalez-Dugo, M.P., Qi, J. & Clarke, T.R. 2001. A refined empirical line approach for reflectance retrieval from Landsat-5 and Landsat-7 ETM+. *Remote Sensing of Environment* 78:71–82.

O'Neill, R.V. & King, A.W. 1998. Homage to St. Michael: Or why are there so many books on scale? In Peterson, D.L. & Parker, V.T. (eds), *Ecological Scale, Theory and Application*: 3–15. Irvington: Columbia University Press.

Opdam, P., Verboom, J. & Pouwels, R. 2003. Landscape cohesion: and index for the conservation potential of landscapes for biodiversity. *Landscape Ecology* 18: 113–126.

Plourde, L. & Congalton, R.G. 2003. Sampling Method and Sample Placement: How Do They Affect the Accuracy of Remotely Sensed Maps. *Photogrammetric Engineering and Remote Sensing* 69 (3): 289–297.

Riano, D., Chuvieco, E., Salas, J. & Aguado, I. 2003. Assessment of different topographic corrections in Landsat TM data for mapping vegetation types. *IEEE Transactions on Geoscience and Remote Sensing* 41 (5): 1056–1061.

Stocking, M. & Murnaghan, N. 2000. *Land Degradation—Guidelines for Field Assessment*. UNU/UNEP/PLEC Working Paper. Overseas Development Group, University of East Anglia, Norwich, (Available online at: www.unu.edu/env/plec/l-degrade/index-toc.html).

Teillet, P.M., Guindon, B. & Goodenough, D.G. 1982. On the Slope-aspect Correction of Multispectral Scanner Data. *Canadian Journal of Remote Sensing* 8 (2): 84–106.

UNEP. 1992. *World Atlas of Desertification*. London: Edward Arnold.

Ward, J., Dull, C., Hertel, G., Mwangi, J., Madoffe, S. & Douce, K. 2004. Monitoring for Sustainable Forestry and Biodiversity in the Eastern Arc Mountains of Tanzania and Kenya. *The International Archives of Photogrammetry, Remote Sensing and Spatial Information Sciences* 34, Part XXX. (see also: USDA Forest Service, 1999, Monitoring Changes in Forest Condition, Land Fragmentation and Conversion: Eastern Arc Mountains of Tanzania and Kenya http://www.easternarc.org/landchange/arcfinal1100_files/frame.htm [accessed June, 2007]).

Wu, J. 1999. Hierarchy and scaling: extrapolating information along a scale ladder. *Canadian Journal of Remote Sensing* 25: 367–380.

Wu, J. & Loucks, O.L. 1995. From balance-of-nature to hierarchical patch dynamics: a paradigm shift in ecology. *Quarterly Review of Biology* 70: 439–466.

Recent Advances in Remote Sensing and Geoinformation Processing
for Land Degradation Assessment – Röder & Hill (eds)
© 2009 Taylor & Francis Group, London, ISBN 978-0-415-39769-8

Land use and carbon stock capacity in slash-and-burn ecosystems in mountainous mainland of Laos

Y. Inoue
National Institute for Agro-Environmental Sciences, Tsukuba, Japan

J. Qi
Michigan State University, East Lansing, USA

Y. Kiyono & Y. Ochiai
Forestry and Forest Products Research Institute, Tsukuba, Japan

T. Horie, T. Shiraiwa, H. Asai & K. Saito
Kyoto University, Kyoto, Japan

L. Dounagsavanh
National Agriculture and Forest Research Institute, Vientiane, Lao PDR

A. Olioso
INRA-CSE, Avignon, France

ABSTRACT: In the mountainous region of northern Laos, land use is changing drastically as affected by the combined pressure of increasing population and governmental regulation on land-use. Land use is strongly related to the ecosystem carbon stock as well as to food security and sustainability of forest resources. However, quantitative geo-spatial land use information is still quite limited. The objective of our study was the regional assessment of land use and carbon stock capacity under the present and alternative land-use scenarios. We derived the chrono-sequential changes in land use and stand age (fallow length) through polygon-based classification of high-resolution and time-series satellite images with the support of *in situ* spectral measurements and field survey. The area of slash-and-burn land-use has been increasing consistently to date with the annual rate of 3–5% in the past decade. In average, 77% was abandoned after a single year cropping. The slash-and-burn area with fallow periods shorter than 4 years was approximately 64%. The ecosystem carbon stock was assessed based on synthesis of the land use and ground-based measurements of carbon in the soil and fallow vegetations. The chrono-sequential average of ecosystem carbon stock showed a large variation (up to 33 tC/ha per year) depending on the crop-fallow cycle patterns. Results suggested that the ecosystem carbon stock would continue decreasing under the present land-use condition with short fallow cycle. It was also suggested that the carbon sink capacity of ecosystem would be recovered or enhanced by alternative land-use/ecosystem managements that would allow longer fallow periods.

1 INTRODUCTION

The "slash-and-burn agriculture" ("shifting cultivation") has been an important food production system in South and Southeast Asia, and is still widely practiced in the mountainous regions of Laos, Bangladesh, and northern part of India (Rasul & Thapa 2003). It is a major agricultural land use especially in the northern part of Laos, which may be up to 80% of the soils used for agriculture if all the fallow land is included (Roder 2001). In this type of land use, a patch of vegetation is cleared by the slash-and-burn method, and used to grow crops for a few seasons, and then abandoned for regeneration of vegetation (Spencer 1966). The slash-and-burn agriculture is considered to be a suitable land use for the hilly regions, and sustainable as far as the fallow period is long enough to regenerate the vegetation cover and soil fertility (Moran 1979, Warner 1991). Although the productivity is low, it allows high returns to labor with low energy requirements. It is also recognized that weed and insect problems can be reduced by extended fallow periods (Roder 2001). It was the best land-use option for the rural population with low income and low access to inputs and infrastructures when the population density was low.

The fallow period used to be kept for 15–20 years until 1970s (Roder 1997). However, thereafter, increasing population pressure forced the shifting cultivators to expand the slash-and-burn land use. According to FAO (1985), 49 per cent of deforestation in South and Southeast Asia was caused by the slash-and-burn agriculture. The global awareness of negative off-site effects might have pressed indirectly or directly the governmental policies to abolish the slash-and-burn agriculture. For example, in 1990, the government of Laos planned to replace the slash-and-burn agriculture by permanent agriculture by the year 2000 (Fjisaka 1991). The government started a strong regulation on local land use to protect the forest by reducing the slash-and-burn land use, but the situation has been even worse to date (Pravongvienkham 2004). Consequently, the combined pressure of increasing population and governmental regulation of land use forced the shifting cultivators to shorten fallow periods.

The average fallow period was estimated roughly to be 5 years in 1990s (Roder 2001) or shorter in 2000s according to our preliminary interview with villagers. Slash-and-burn cultivation with short fallow periods is associated with deterioration of soil nutrient and physical conditions, and increase of weeds, pests and diseases, which results in lowered crop yield, less labor productivity as well as degradation of forest resources including non-timber forest products (NTFP) (Fjisaka 1991, Roder 2001). It is also suggested that natural vegetative succession back to forest is unlikely even with long fallow after prolonged slash-and-burn land use (Gomets-Pompa et al. 1972). Hence, the land use may be no longer sustainable, and the situation is often referred to as the vicious spiral of shortening of fallow periods, decreasing crop yield and increasing cropping labor (mainly weeding), increasing food insecurity and poverty, and expansion of slash-and-burn land use (e.g., Pravongvienkham 2004). It is obvious that forest resources are linked inevitably with the slash-and-burn land use and that crop productivity. It is strongly required to develop alternative cropping and ecosystem management scenarios for food, resource and environmental security. A number of assistance programs by international agencies (e.g. SIDA) have been implemented to protect forests and to reduce poverty mostly through socio-economic, forestry and agricultural activities. However, quantitative assessment of land use, especially at a geo-spatial basis, is still limited. Accurate statistics on land use are hardly available.

On the other hand, regional land-use change is strongly linked not only with crop productivity but also with the CO_2 flux by biomass burning or biomass stock capacity as found in such as Brazil (Cochrane et al. 1999, Czimczik et al. 2005). From the viewpoint of carbon cycle science, quantitative estimation of carbon exchange between the atmosphere and ecosystems as affected by the land-use change is also required since the site specific data in the region are still very limited (IPCC 2003). The ecosystem carbon stock is one of key variables in the assessment of alternative ecosystem management options in the contexts including global carbon sequestration (IPCC 2003). Nevertheless, scientific data and assessment of the long-term and wide-area changes of land use and carbon balance at the ecosystem scale in the region are very few for the region (Roder 2001). According to the preliminary analysis using the classified Landsat imagery produced by Michigan State University (Tropical Rain Forest Information Center), the forest area in the northern part of Laos (186,000 km^2) was more than 90% in both 1973 and 1985, but it decreased down to around 80% in 1992 (Inoue et al. 2005). It was presumed that the deceasing trend might have been accelerated after 1992, but recent change in the region has not been well quantified.

Thus, the objective of this study is to quantify the chrono-sequential change of land use, vegetation, and carbon stock capacity under present and alternative land-use scenarios based on satellite imagery and *in situ* measurements. The carbon stock in the soil, crops and fallow vegetations has been investigated in the region (Kiyono et al. 2007, Asai & Saito 2006). Geo-spatial data from remote sensing and GIS are synthesized with these *in situ* measurements and survey for regional

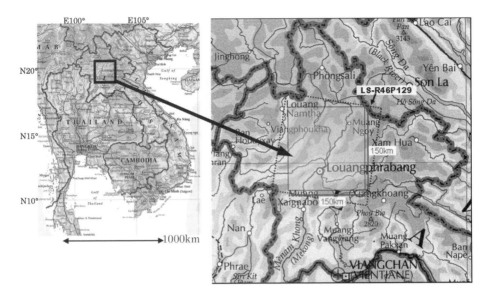

Figure 1. The Study area; northern part of Lao P.D.R. The dotted area indicates a full Landsat scene (Row46/Path129).

assessment of dynamic change in land use and carbon stock. Since the upland rice is the major crop in the region, we assume that alternative land-use scenarios would be based on rice-based cropping systems using high yielding cultivars (Saito et al. 2006a) as well as forage and cash crops (Saito et al. 2006b).

2 STUDY AREA

We selected the central part of northern Laos (150 km × 150 km, Fig. 1) as a study area since the slash-and-burn agriculture is the most important agricultural system there, and the area may be most typical of the similar ecosystems in mountainous mainland in Southeast Asia (Rasul & Thapa 2003).

The central position of the area is [102°03′48.9E, 20°13′12.8″N], and the whole area is covered by a single Landsat scene (Row46/Path129). It is reported that people in the area consists of more than sixty ethnic groups (Fujisaka 1991). The elevation ranges from 300 m to 2000 m, and the slope is from 40% to 100%. The most important crop in the area is upland rice, and some other crops such as job's tear, sesame, paper mulberry are grown. Major tree species include *Irvingia malayana and Castanopsis echinocarpa*. Teak plantation was started about 10–15 years ago. The mean annual rainfall for the area is about 1300 mm with the annual variability (SD) of 260 mm, but more than 90% of the rainfall is during the wet season from April to October. Within the large study area, we also set several areas for intensive field survey. The region is also typical for opium cultivation, which was found to be correlated with deficit in rice production (Roder 2001). This paper focuses on the results from the "HK-site" (15 km × 15 km) in Luang Prabang province.

3 DATA AND METHODS

3.1 *Data*

Multi-temporal Landsat images (Row46/Path129) have been collected for time-series analysis of land use in the study area. A high resolution image from QuickBird was also used for detailed land-use classification in 2003 when the ground-truth data were available. The intensive study area (HK-site) was covered by the QuickBird image. The list of satellite imagery used for this study

Table 1. Satellite images used for this study (LS = Landsat).

No.	Date (yyyy.mm.dd)	Sensor	No.	Date (yyyy.mm.dd)	Sensor	No.	Date (yyyy.mm.dd)	Sensor
1	1973.01.24	LS-MSS	10	1992.02.06	LS-TM	19	2000.03.07	LS-ETM
2	1975.12.31	LS-MSS	11	1993.02.08	LS-TM	20	2000.11.02	LS-ETM
3	1986.03.17	LS-MSS	12	1994.01.10	LS-TM	21	2001.02.06	LS-ETM
4	1988.01.26	LS-TM	13	1995.02.14	LS-TM	22	2001.11.21	LS-ETM
5	1989.03.25	LS-TM	14	1996.02.17	LS-TM	23	2002.02.09	LS-ETM
6	1989.12.14	LS-TM	15	1997.02.13	LS-TM	24	2002.11.08	LS-ETM
7	1990.01.31	LS-TM	16	1997.11.02	LS-TM	25	2003.10.18	Quickbird
8	1990.11.15	LS-TM	17	1999.01.24	LS-TM	26	2003.12.05	LS-TM
9	1991.04.08	LS-ETM	18	1999.11.16	LS-ETM	27	2004.04.11	LS-TM

is shown in Table 1. Twenty seven images cover the period of 18 years from 1986 to 2004. The other intensive sites were also covered by high resolution satellite images such as IKONOS and QuickBird.

Ground-based spectral measurements were made over a wide range of ecosystem surfaces, i.e. water bodies, bare soils, crops, slash/burn surfaces, and fallow vegetations. The reflectance spectra were obtained under clear sky conditions using a portable spectro-radiometer (FS-FR; ASD). Spectral range, resolution and the field of view of the sensor were 350–2500 nm, 2 nm, and 22°, respectively. Data were calibrated using a standard white reference (Spectralon; Labsphere). When ecosystem surfaces were not uniform, hundreds of measurements were taken moving over each surface, and averaged for spectral analysis.

Information on the history of land use and cropping conditions was obtained through field survey and interviews with the villagers. Several fallow plots (40 m × 40 m) with different canopy ages were fenced off for periodical measurement of fallow biomass without human disturbance. Fallow biomass was made by complete enumeration method. Biomass, litter, crop yield, soil carbon content, and soil CO_2 flux during both cropping and fallow periods were measured based on destructive sampling, chamber method, and chemical analysis in experimental plots in the intensive study area. Details of these measurements are reported in Kiyono et al. (2007) and Asai et al. (2006). Based on these measurements, all carbon pools, i.e. above-ground and below-ground biomass, soil, dead wood, litter, and soil organic matter were estimated as a function of fallow period (IPCC 2003, Kiyono et al. 2007, Asai et al. 2006). We also collected basic regional data/information such as topographic data, climatic data, soil map, and agricultural statistics.

3.2 Approaches

First, we analyzed the in situ measurements of reflectance spectra to elucidate the spectral characteristics of the ecosystem surfaces. Second, the recent land-use status was analyzed using the high resolution satellite imagery, i.e. QuickBird with the ground-truth data. Third, based on the results from previous two steps, the temporal change of slash-and-burn area was derived from the time-series Landsat images. The geospatial and dynamic change of slash-and-burn land use such as fallow periods was estimated at the scale of land-use patches. The land-use patches were generated from the satellite images using a segmentation technique. Fourth, the ecosystem carbon stock capacity was estimated from the regional estimates of fallow periods (i.e. stand age) and in situ measurements of the carbon cycle parameters. In this study, the stand age distribution was the key variable for synergizing the data on land use and carbon balance since the carbon pools were assessed as a function of stand age.

3.2.1 Processing of satellite imagery for land use analysis

All satellite images were registered to a standard image through image-to-image registration technique so that the rectification error should be less than 0.2 pixels. A Landsat-TM image taken on Nov. 16 in 1999 was used as the standard image and more than 50 GCPs were taken for the rectification. Image analysis was conducted by using Imagine 8.7 (ERDAS). Digital number was used without radiometric, atmospheric and topographic corrections.

Since the land use in slash-and-burn ecosystems is managed on a basis of land patches, it is reasonable to identify land-use areas on an image at a patch basis instead of a pixel basis. Through segmentation of a raster image, a group of pixels can be considered as a single unit (object) with a uniform land use. The image segmentation was performed semi-automatically using eCognition (Baatz et al. 2001). The image objects were generated through a bottom-up clustering process starting with one-pixel object calculating spectral and spatial heterogeneity. The segmentation was based on local optimization procedure to minimize the weighted heterogeneity of the whole image. The criteria of the heterogeneity were adjusted by "scale", "color" and "shape" parameters. A set of parameters was determined through visual inspection of the segmented images based on the ground-truth information. The scale parameter affects the average size of objects. All visible and infrared spectral bands were used for the procedures, i.e. four and six bands in QuickBird and Landsat images, respectively.

The segmented images were used for land-use classification. The classification was performed using eCognition which used the supervised classification-algorithm based on fuzzy logic (Baatz et al. 2001). Based on the field survey, a total of 9 land-use classes were considered; slash-and-burn fields (S/B), short fallow fields (1–3 years; F1), longer fallow fields (4 years or longer; F2), conservation forest (CF), Teak plantation (TP), paddy and other croplands (P), water surfaces (rivers/pond; W), and the area of road/bare soils/houses (RBH). Areas of cloud, shadow, and noise in each image were also classified. For the analysis of QuickBird Image in 2003, two sets of 199 polygons (S/B, 36; F1, 23; F2, 11; CF, 6; TP, 16; P, 12; W, 20; RBH, 33) were derived from ground-truth information, and used for training the fuzzy classifier and for validation of classification results, respectively. Clouds and shadow polygons were also identified on images through visual interpretation. The slash-and-burn patches were identified clearly on the B-G-R views compositing 550, 660, and 830 nm bands.

The chrono-sequential analysis of land use was conducted using the time-series Landsat images, where we focused mainly on extraction of the slash-and-burn patches in each year. The slash-and-burn patches were identified clearly by visual comparison of images using the 660, 830, 1650 and/or 2200 nm bands for three consecutive years. The robustness in extraction of the slash-and-burn patches was confirmed with the QuickBird and Landsat images in the same year based on ground-truth data as well as the relative change in vegetation indices. An image of two classes, i.e. slash-and-burn and the others, was created for each year from the classification results. The binary images for all available years were stacked into an image as time-series layers. The multi-layered image enabled to extract the history of slash-and-burning at a pixel basis. For example, the year of last burning could be inferred at a pixel basis. The number of consecutive years under the fallow or slash-and-burn conditions could be estimated for each pixel. Thus, the image allowed estimating the area and spatial distribution of various land-use patterns at the regional scale. Since the length of fallow period is equivalent to the stand age, the area and spatial distribution of stand age could be derived from the image.

3.2.2 *Use of satellite data for assessing ecosystem carbon stock*

The use of geospatial information from satellite is crucial for assessment of carbon dynamics in terrestrial ecosystems. In general, satellite imagery may be used in three different ways; (a) linking satellite-derived land-use/land-cover information with default values or semi-empirical equations derived from ground-based measurements on carbon balance, (b) linking satellite-derived land-use/land-cover information with a biophysical process-based model, and (c) linking biomass information derived directly from satellite signatures with biophysical process-based models. Nevertheless, the approach (b) requires a lot of input data and biological parameters for the process-based model, and the approach (c) further requires tedious tasks for radiometric, atmospheric and topographic (BRDF) corrections to derive physically-consistent spectral signatures. Use of the radiometric, atmospheric and topographic corrections for the satellite images is still a challenging task for the hilly regions in monsoon climate. These two approaches are attractive from the methodological point of view (Inoue 2003), but need a lot more data that are not available in this study area, especially for past periods. Hence, as the first step, we focus on the results from the approach (a), which is simple but may allow reasonable assessment under less data availability. The fallow length (stand age) derived from time-series images were linked with the data on carbon stock pools such as fallow biomass and soil organic matter since they were expressed as a function of fallow length. Then, all information was synthesized to assess the chrono-sequential change in ecosystem carbon stock at the regional scale under present and alternative land-use/ecosystem management scenarios.

4   RESULTS AND DISCUSSION

4.1   *Spectral characteristics of major ecosystem surfaces*

It is essential to understand the reflectance characteristics of ecosystem components for inter-pretation and quantitative analysis of satellite images. We investigated the *in situ* measurements of reflectance spectra over various land surfaces. Figure 2 depicts the averaged hyperspectra for typical ecosystem conditions and the simplified transition process of the reflectance spectra after crop harvesting through re-growth, fallow forest, slashing, to burning. Reflectance spectra for dense vegetation, slashed/senescent vegetation and burnt surfaces were largely different especially in visible and near infrared wavelength regions. The sparse vegetation showed the intermediate spectra between dense vegetation and the slashed/senescent vegetation in general. Reflectance at shortwave infrared wavelengths showed clear differences as affected by moisture and burning con-ditions, respectively. Note that the burnt surface showed lowest reflectance at 1650 nm while higher than dense vegetation in 2200 nm.

Normalized difference indices NDI[i, j] were derived from the hyperspectral data. The NDI[i, j] is defined as $[\rho i - \rho j]/[\rho i + \rho j]$ where ρi and ρj are reflectance values at wavelengths i and j. Therefore, NDI[830, 660] is the normalized difference vegetation index NDVI. Wavelengths

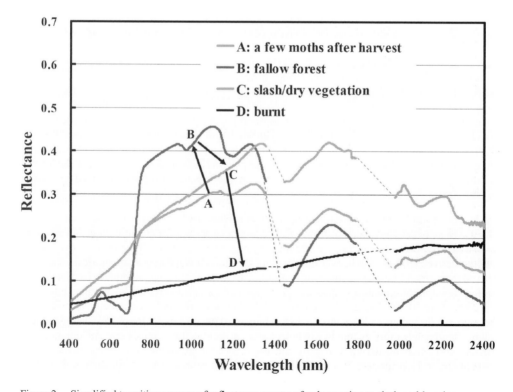

Figure 2.   Simplified transition process of reflectance spectra after harvesting to slash-and-burning.

348

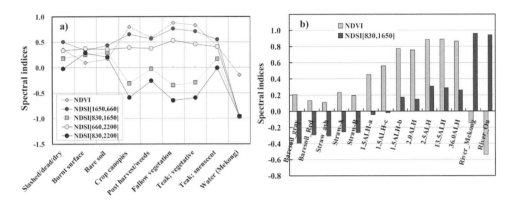

Figure 3. Response of vegetation indices to major land-surface targets in the study area.
Note: The number with ALH indicates the number of months after last harvesting.

were selected so that they are equivalent to the optical bands of Landsat-TM. Figure 3 shows the response of five vegetation indices, i.e. NDVI, NDI[830, 1650], NDI[830, 2220], NDI[660, 2200] and NDI[1650, 660] to the typical ecosystem surfaces.

From previous reports NDI[830, 1650], NDI[830, 2220], and NDI[1650, 660] were presumed to be sensitive to moisture, burning, and senescence, respectively (Wilson & Sader 2002). The NDI[660, 2200] showed little change for the wide range of land surfaces, so that it could not be suitable for change detection. The other four indices had similar information that was useful for distinguishing green vegetation, slashed/senescent vegetation, burnt surface, and bare soil surfaces, respectively. Nevertheless, the NDI[830, 2200] seemed to be useful specifically for distinguishing burnt surfaces, while NDI[830, 1650] may be effective to detect slashed/senescent vegetation areas due to higher sensitivity to moisture conditions. Regeneration of fallow vegetation may be best inferred from the NDVI since it was most sensitive to green biomass although it saturates at some vegetation coverage as shown in Figure 3b. The NDI[1650, 660] had similar response as NDVI, NDI[830, 1650] and NDI[830, 2220], but its dynamic range was narrower than the others. Therefore, when shortwave infrared wavelength is available as in Landsat-TM, it is most efficient to use NDI[830, 1650] and NDI[830, 2220] with NDVI. Our results agreed well with a previous report for savannah (Trigg 2001). In fact, slash-and-burn patches were visualized clearly on images with the combination of 660, 830, 1650 or 2200 nm Landsat bands. An example set of Landsat images with the band combination of 660, 830, and 1650 nm in three consecutive years (2000, 2001, and 2003) is shown in Figure 5 (section 4.3). Areas A, B and C, for example, were obviously used for slash-and-burn cropping for each single year in 2000, 2001, 2002, respectively. Visual interpretation of consecutive images allows robust identification of the year of slash-and-burning at the patch basis. When the wavelength bands are limited to visible and near-infrared regions as for IKONOS and QuickBird, the NDVI can be used for the same purpose. These results were quite useful for interpretation of satellite imagery and supported the analysis of satellite-derived spectral indices in sections 4.2 and 4.3.

4.2 Recent status of land use using high resolution satellite imagery

The land-use condition in 2003 in the intensive study area (HK-site, 320 km^2; Luang Prabang Province) was analyzed using high-resolution multi-spectral imagery (QuickBird, 18 Oct., 2003). A 3D view of the intensive study area is shown in Figure 4. The spatial resolution was 2.4 m for the multi-spectral image and 10 m for the digital elevation map, respectively. Since the green, red and near-infrared bands are assigned to B, G, and R colors, the bluish color represents the slash-and-burn area in the year, and reddish color indicates the densely vegetated areas.

10km

Figure 4.   A 3D view of the intensive study area near Luang Prabang (HK-site) (see colour plate page 400). Note: The spectral image is from QuickBird on Oct. 18, 2003. The spatial resolution is 2.4 m for spectral images and 10 m for digital elevation map, respectively. The G, R, and NIR bands are assigned to B, G, and R colors, so that the bluish color represents the slash/burn land-use during the year, and reddish color indicates the densely vegetated areas, respectively. The area was approximately 320 km².

Table 2.   Land use in 2003 in the intensive study area estimated from classification of high-resolution satellite image (QuickBird).

Land Use	Polygons	Area (ha)	Area (%)	Producer's accuracy (%)	User's accuracy (%)
1 Slash-and-burn (S/B)	4031	3952	12.9	100.0	86.0
2 Fallow [1–3 years] (F1)	5348	10656	34.8	91.3	80.8
3 Fallow [4< years] (F2)	3467	6010	19.6	63.6	70.0
4 Conservation Forest (CF)	4704	7759	25.4	83.3	70.0
5 Teak Plantation	824	1515	5.0	62.5	71.4
6 Paddy/Other cropland	177	202	0.7	83.3	90.9
7 Water (river/pond)	85	174	0.6	100.0	100.0
8 Road/Bare soil/House	603	326	1.1	81.8	100.0
Total	19239	30595	100	Overall accuracy = 89.5%	
SB+F1+F2	12846	20618	67.4	Kappa = 0,878	
SB+F1+F2+CF+Teak [potential forest area]	18374	29892	97.7		

Notes: Land-use polygons were derived from the QuickBird image shown in Figure 5 through segmentation analysis. Accuracy assessment was based on ground-truth samples. The total area was approximately 320 km². Polygons for clouds and shadow were excluded based on classification.

Table 2 shows that classification results for the region. The slash-and-burn patches, water surface, and the area for road/bare soils/houses were classified with high accuracy. The classification accuracy was less for fallow, conservation forest and teak plantation areas although the short fallow had the highest accuracy among them. Results suggest the difficulty in distinguishing long fallow, conservation forest, and teak plantation areas. Since these three classes have similar spectral characteristics, they are hardly discriminated on a single image. The slash-and-burn fields were identified clearly on each single image, which may be the most robust information on land use from a single image. The polygon-based classification is useful especially for distinguishing the slash-and-burn patches because spectral reflectance within each land-use patch is not uniform in semi-natural ecosystems and because a cropping area is shifted by the unit of land-use patch in the slash-and-burn ecosystems. The averaged spectra and/or indices for each patch may be more representative of its land-surface conditions.

The areas for slash-and-burn fields and short fallow (1–3 years) patches in 2003 were 12.9% and 34.8%, respectively (Table 2). The ratio of the latter to the former was about 2.7, which implies that the average land-use cycle in the region is approximately 4 years, i.e. 1 year for cropping and 3 years for fallow period. The potential area for slash-and-burn land use (SB+F1+F2) was estimated to be 67.4%, although the potential forest area (SB+F1+F2+CF+Teak) could be up to 97.7%. The area of paddy and other cropland was very limited to the narrow flat areas along streams.

4.3 Long-term assessment of slash-and-burn area and stand age

Figure 5 depicts an example image using the 660, 830 and 1650 nm bands for three consecutive years 2000–2001–2002. Sample areas A, B and C indicate the typical patches with the land-use pattern of slash/burn-fallow-fallow, fallow-slash/burn-fallow, and fallow-fallow-slash/burn, respectively. The land use was confirmed with the ground-truth information. Change of the vegetation indices NDVI, NDI[830, 1650], and NDI[830, 2200] in the consecutive years is shown in Figure 6.

These three NDIs proved to be useful for detecting the slash-and-burn patches through the analysis of the *in situ* hyperspectral measurements (section 4.1). All vegetation indices showed significant differences between the slash-and-burn and fallow conditions in all three land-use patterns (Figure 6a-c). Without any exception, all indices were significantly low in the year of slash-and-burn land use. The NDVI value was 0.56 ± 0.01 in slash-and-burn patches in all three years (60 patches), while it was 0.86 ± 0.02 in fallow patches irrespective of before and after the slash-and-burn year. It was notable that the NDVI values recovered quickly after abandoning to the similar level as those of fallow vegetations before slash-and-burning. As suggested by the *in situ* reflectance measurements, the index would be already saturated during the first year of fallow period. The NDI[830, 1650] was 0.30 ± 0.01 in slash-and-burn patches and 0.55 ± 0.02 in patches, while NDI[830, 2200] was 0.45 ± 0.02 in slash-and-burn patches and 0.74 ± 0.05 in fallow patches. These two indices also

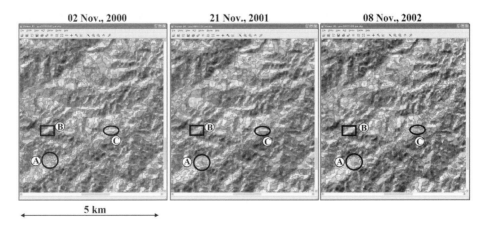

Figure 5.   Slash-and-burn patches identified on the composite image of 660 nm (B), 830 nm (G), and 1650 nm (R) in three consecutive years (see colour plate page 400).
Note: Sample areas A, B and C indicate typical patches of slash-and-burning in 2000, 2001 and 2002, respectively.

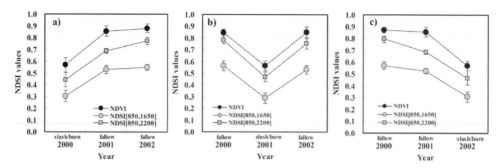

Figure 6. Normalized difference indices, NDVI, NDI[730, 1650] and NDI[830, 2200] in the slash-and-burn fallow patches in three consecutive years.
Note: Indices were calculated for three sets of 20 patches with different land-use patterns during the three consecutive years, i.e. a) slash/burn-fall-fallow, b) fallow-slash/burn-fallow, and c) fallow-fallow-slash/burn, respectively.

showed the similar values for the fallow patches before and after the slash-and-burning, respectively. The relative sensitivity to the slash-and-burn land use as defined by the difference of NDI between slash-and-burn and fallow conditions divided by the NDI in fallow conditions was 0.45, 0.39, and 0.35 for NDI[830, 1650], NDI[830, 2200] and NDVI, respectively. Hence, the NDI[830, 1650] may be the most useful index in distinguishing slash-and-burn patches, which support the common practice to utilize the 1650 nm band for finding patches of forest change during the post-classification editing (e.g. Woodcock et al. 2001). Unfortunately, it was obvious that none of these indices could be useful in quantifying vegetation biomass during the fallow periods since they showed little difference before and after the slash-and-burn land use in spite of large differences in biomass. Since direct estimation of forest biomass using optical and radar (SAR) images are not yet feasible (Salas et al. 2002), indirect approach using such as stand age may be more robust. Stand age is a simple descriptor of forest condition which may be used to summarize or capture the differences in tree size, height, closure and other attributes (Franklin et al. 2003). The stand age as well as canopy height or crown closure is one of major parameters in forest inventory, and many attempts have been made to estimate such parameters using satellite imagery (e.g. Wynne et al. 2000, Woodcock et al. 2001, Franklin et al. 2003). Spectral response of forest to age (Nilson & Peterson 1994, Jensen et al. 1999), canopy closure (Spanner et al. 1990), and height (Gemmell 1995) are reasonably well understood. However, considering the spectral analyses in previous sections, spectral discrimination of such parameters on a single satellite image may not be precise even in the areas of pure tree species (Franklin et al. 2003). Derivation of stand age distribution from a single image using vegetation indices may not be accurate enough for detailed land-use analysis (Zhang et al. 2004). Therefore, time-series analysis of classified images focusing on the slash-and-burn polygons may be the most straightforward approach to estimate the stand age distribution (Hayes & Sader 2001).

On the basis of the classified images, we estimated the slash-and-burn areas for all available time-series images. Figure 7 shows the temporal change of the slash-and-burn area from 1973 to 2004.

The area increased rapidly from around 5% in 70's to 12% in 00's. The increase during 90's was striking while it was small before mid-80's in the region. The relative increase in the past decade was approximately 40 % and the annual rate of increment was estimated to be 3.8%. Although the relative areas for several crops might have been changing in recent years (Pravongvienkham 2004), the total area for slash-and-burn land use has been increasing.

Next, we analyzed the structure of land-use patterns in the region. The multi-layered image of the classified time-series images was used to estimate the spatial distribution of various land-use patterns at a pixel basis. Figure 8 shows the relative areas for different consecutive years under slash-and-burn land use.

Results indicate that, in average, 77% was abandoned for fallow after a single year of slash-and-burn cropping, and areas under one or two-year consecutive use was 94%, while the area for four

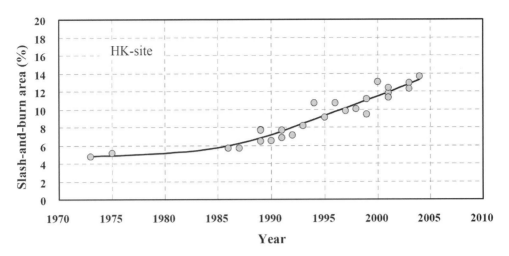

Figure 7. Temporal change of slash-and-burn area derived from the polygon-based classification of Landsat images.
Note: The intensive study area is 320 km2. Information on images is listed in Table 1.

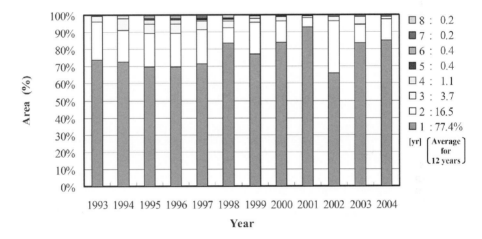

Figure 8. Relative areas for consecutive slash-and-burn land use during recent 12 years.
Note: The "2yr", for example, indicates that the land was used for slash-and-burn cropping for two consecutive years before shifting.

or longer consecutive years was negligible. The stand age was determined at a pixel basis as the number of years after last slash-and-burn land use in each year. The area distribution of stand age is shown in Figure 9 where values for years 2003 and 2004 are averaged.

Results indicate that approximately 11% was used for slash-and-burn cropping (stand age = 0). The gradual decrease in area with stand age indicates that some of fallow area in the previous year has been returned to slash-and-burn use. The stand age distribution of slash-and-burn areas (Figure 10) was derived from the decrements between the areas in consecutive stand ages in Figure 9. The 0-year means the consecutive slash-and-burn use from the previous year. Results indicate that the largest part (20%) of the slash-and-burn area was from 3-year fallow areas, second largest (18%) from 2-year fallow areas, and third (13%) from 4-year fallow areas. It was also confirmed that long-fallow or protected areas are also used, in part (8%), for slash-and-burn cropping.

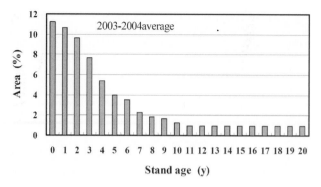

Figure 9.   Areas for different stand ages.
Note: Values for stand age longer than 11 years were assumed to be consistent due to limited availability of satellite images.

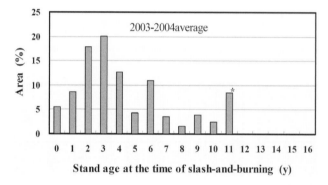

Figure 10.   Stand age distribution of slash-and-burn area in 2003–2004 periods.
Note: The value for the stand age of 11 years include the sum of longer years due to limited availability of satellite images.

4.4   Regional assessment of ecosystem carbon stock

We estimated ecosystem carbon stock at the regional scale based on the stand age distribution and carbon pools derived from *in situ* measurements of carbon exchange processes. The ecosystem carbon stock C_E can be expressed from the area A_k and carbon stock per unit area C_k for a specific land-use pattern k by Equation (1).

$$C_E = \sum A_k \cdot C_k \qquad (1)$$

The unit carbon stock C_k was estimated from the biomass of fallow vegetation C_{FB} and carbon in dead organic matter C_{DOM}, and soil organic carbon C_{soil} by Equation (2).

$$C_k = C_{FB} + C_{DOM} + C_{soil} \qquad (2)$$

The components C_{FB}, C_{DOM} and C_{soil} were estimated from stand age (fallow length) using semi-empirical equations (Asai & Saito 2006, Kiyono et al. 2007). The equation for $C_{FB} + C_{DOM}$ was a regression curve derived from yearly measurements of above-ground biomass at fallow plots of known stand ages covering 0–20 fallow periods. C_{soil} was estimated from a simple compartment model where plant litter, root decay, soil organic carbon, and its decomposition were calculated. Coefficients for root decay and decomposition of soil organic carbon were derived from measurements of soil carbon content, soil respiration, plant litter, and root mass. The stand age was used

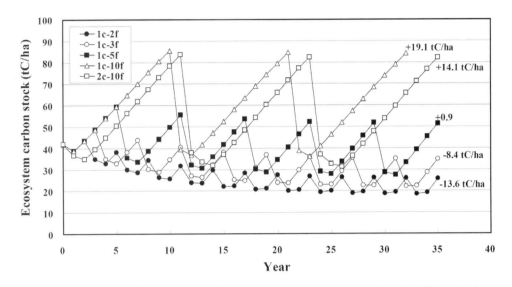

Figure 11. Simulation of chrono-sequential change in ecosystem carbon stock under different land-use patterns.

Note: "c" and "f" mean cropping and fallow, respectively, so that "1c-10f", for example, the land-use cycle of 1-year cropping and 10-years fallow. The numbers indicate the averaged increment from the initial value during the periods.

as a key parameter for synthesizing land use, fallow biomass, and soil carbon in calculation of ecosystem carbon stock. Results suggested that the slash-and-burn land use for one year caused a loss of about 6 tC/ha in the soil, while it would take about 12-year fallow period to recover the initial level of soil carbon stock. The $C_{FB} + C_{DOM}$ was expressed by a log function of stand age where it increased up to about 45 tC/ha after 10-year fallow.

First, we simulated the chrono-sequential change of ecosystem carbon stock C_E under various land-use patterns (Figure 11).

The average of ecosystem carbon stock for the land-use cycles over 30 years was kept near the similar level as the initial soil carbon stock under 1c-3f land-use cycle, but it was lower than the initial value when the fallow period was shorter than 3 years. The 1c-10f pattern allowed carbon stock increase of +25.6 tC/ha than the initial condition. The 10-year fallow may allow an increment of +20.1 tC/ha even after cropping for two consecutive years (2c-10f). Consequently, the chrono-sequential average of ecosystem carbon stock showed a large variation (up to 33 tC/ha per year) depending on the crop-fallow cycle patterns. It was suggested that the ecosystem carbon stock would continue decreasing from the initial level if the fallow period is shorter than 3 years. There would be a serious risk that the ecosystem would never recover the initial level of soil carbon stock as well as the soil fertility or vegetation even after long fallow (Gomets-Pompa et al. 1972).

Next, the change of ecosystem carbon stock was estimated at the regional scale using the stand age distribution (Figure 12).

Figure 12 shows the relative contribution of various land-use patterns to the carbon stock in the region. The overall average for all land-use patterns in the region was 5.96 tC/ha, which suggests that the ecosystem carbon stock could not be much larger than the initial level of soil carbon stock as far as the present land use would be continued. This drastic decrease in regional carbon stock implies that the large carbon emission from ecosystem into the atmosphere as well as the decreasing crop productivity and forest resources. The 1c–2f land use has the largest negative impact on the regional change while all land-use patterns with short fallow (1–3 years) have large negative effects. In this simulation, we assumed that the crop productivity and growth of fallow vegetation were unaffected by land-use or the level of carbon stock, but it has been pointed that repeated cycles of short-term slash-and-burn cropping would have negative effects on soil productivity and weeding.

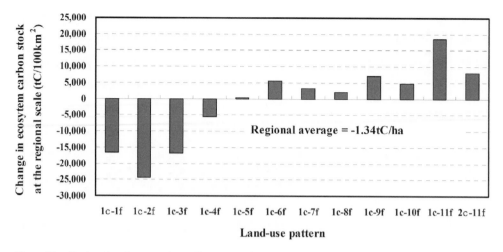

Figure 12. Regional-scale comparison of long-term change in carbon stock under different land-use patterns. Note: The relative increment from the initial carbon stock was averaged for 35 years for each land-use patterns. The "Regional average", 5.96 tC/ha, was overall average for all land-use patterns in the region. Simulation was performed assuming the present land-use situation.

Weeding is the most tedious management practices in rice cropping in slash-and-burn cropping (Roder 2001). If these negative effects were considered, the results would be more serious than the present simulation.

Thus, it is essential to extend the fallow period to reduce the negative impact. Development of alternative ecosystem management options is strongly required to allow longer fallow periods although few technologies are available to date. Nevertheless, it may not be suitable to apply modern agricultural technologies such as advanced machinery and agro-chemicals, considering the possible economic inputs, environmental impact, technological background, and the topographical conditions in the region. In this context, rice-based cropping systems may be most appropriate options. The food security is of the highest priority in the region. Alternative cropping systems using the high yielding rice cultivar (Saito et al. 2006a), or legume forage (Saito et al. 2006b) and other cash crops (paper mulberry, etc.) may be useful for reducing the slash-and-burn area and extending the fallow period through improvement of yield and income for villagers. Further research works are needed to propose high-yielding and sustainable production systems.

5 CONCLUSIONS

Polygon-based classification of high-resolution and time-series satellite images based on spectral investigations enabled the robust and accurate estimation of land-use patterns and stand age (fallow length) distributions. The combined use of segmentation approach and time-series satellite images proved quite useful especially in the chrono-sequential analysis of land use in slash-and-burn ecosystems. The area of slash-and-burn land-use has increasing consistently to date with the annual rate of 3–5% in the past decade. In average, 77% was abandoned after a single year cropping. The slash-and-burn area with fallow period of 4 years or shorter was approximately 64%.

We could assess the chrono-sequential change of ecosystem carbon stock from the results of land use, fallow biomass, and soil carbon by using the stand age as a key parameter to synthesize them. The chrono-sequential average of ecosystem carbon stock showed a large variation (up to 33 tC/ha per year) depending on the crop-fallow cycle patterns. Results suggested that ecosystem carbon stock would continue decreasing under the current land-use and cropping conditions with short fallow cycle. Since crop productivity is strongly linked with soil carbon content and land use is directly related to the fallow period, this trend may affect food security and sustainability of forest resources. Such impact may be large since it was obvious that the slash-and-burn ecosystems are common throughout the study area irrespective of ethnic group, elevation and slope in the northern part of Laos, which are also common for the surrounding regions in Thailand, Myanmar, Vietnam and China with similar ethno-environmental conditions.

Since it is hardly realistic to abolish the slash-and-burn land use from the region, alternative land use and ecosystem management strategies are needed to allow longer fallow cycle. Considering the highest priority of food security for the region, rice-based cropping systems with longer fallow periods should be most appropriate options. Agronomic research works on site are required to develop more productive and sustainable cropping systems for the region.

The data, knowledge and analysis in the present study are highly original but still preliminary. Further collection of *in situ* data and improvement of models are needed for more precise and reliable assessment of the ecosystem carbon dynamics. The synergistic use of remote sensing, GIS, modeling, and *in situ* measurements is an indispensable approach for the ecosystem and environmental issues at the regional scale.

ACKNOWLEDGEMENTS

This work was supported partly by the Global Environment Research Fund, Ministry of Environment in Japan.

REFERENCES

Asai, H. & Saito K. 2006. Carbon dynamics and sustainability under shifting cultivation. *International Workshop on Indigenous Eco-knowledge and Development in Northern Laos*, 15–16 March, 2006, at Oudomxay Province, Lao PDR: RIHN, CSEAS & NAFRI.
Baatz, M., Benz, U., Dehghani, S., Heynen, M., Höltje, A., Hofmann, P., Lingenfelder, I., Mimler, M., Sohlbach, M., Weber, M. & Willhauck, G. 2001. *eCognition Users Guide*. München: Definiens Imaging.
Cochrane, M.A., Alencar, A., Schulze, M.D., Souza Jr., C.M., Nepstad, D.C., Lefebvre, P. & Davidson, E.A. 1999. Positive feedbacks in the fire dynamic of closed canopy tropical rain forests. *Science* 284: 1832–1835.
Czimczik, C.I., Mund, M., Shulze, E.D. & Wirth, C. 2005. Effects of reforestation, deforestation, and afforestation on carbon storage in soils. In H. Griffiths and P.G. Jarvis (eds.), *The Carbon Balance of Forest Biomes*: 319–330. Taylor & Francis.
Franklin, S.E., Hall, R.J., Smith, L. & Gerylo, G.R. 2003. Discrimination of conifer height, age and crown closure classes using Landsat-5 TM imagery in the Canadian Northwest Territories. *International Journal of Remote Sensing* 24: 1823–1834.
Gemmell, F.M. 1995. Effects of forest cover, terrain, and scale on timber volume estimation with Thematic Mapper data in a Rocky Mountain site. *Remote Sensing of Environment* 51: 291–305.
Gomets-Pompa, A., Vasquez-Yanes, C. & Guevara, S. 1972. The tropical rainforest: non-renewable resource. *Science* 177: 762–765.
Hayes, D.J. & Sader, S.A. 2001. Comparison of change-detection techniques for monitoring tropical forest clearing and vegetation regrowth in a time series. *Photogrammetric Engineering and Remote sensing* 67: 1067–1075.
Inoue, Y. 2003. Synergy of remote sensing and modeling for estimating ecophysiological processes in plant production. *Plant Production Science* 6: 3–16.
Inoue, Y., Horie, T., Kiyono, Y., Ochiai, Y., Saito, K., Asai, H., Qi, J. & Shiraiwa, T. 2005. Assessing the change of land-use and carbon sink capacity in the shifting cultivation region of northern Laos. *Proc. International Geoscience and Remote Sensing Symposium IGARSS2005*: 3098–3100, IEEE.
IPCC 2003. *Good Practice Guidance for Land Use, Land-Use Change and Forestry*. 1–275. Hayama: IGES.
Jensen, J.R., Qui, F. & Minhe, J. 1999. Predictive modeling of coniferous forest age using statistical and artificial neural network approaches applied to remote sensor data. *International Journal of Remote Sensing* 20: 2805–2822.

Kiyono, Y., Ochiai, Y., Chiba, Y., Asai, H., Shiraiwa, T., Horie, T., Songnoukhai, V., Navongxai, V. & Inoue, Y. 2007. Predicting chronosequential changes in carbon stocks of pachymorph bamboo communities in slash-and-burn agricultural fallow, northern Lao PDR. Journal of Forest Research 12 (in press).

Moran, E.F. 1979. *Human adaptability: An Introduction to Ecological Anthropology*: 1–300. North Scituate: Duxbury Press.

Nilson, T. & Peterson, U. 1994. Age dependence of forest reflectance: analysis of main driving factors. *Remote sensing of Environment* 48: 319–331.

Pravongvienkham, P. 2004. Upland natural resources management strategies and policy in the Lao PDR. In H. Furukawa (ed.), *Ecological destruction, Health, and Development: Advancing Asian Paradigms* 8: 481–501. Kyoto: Center for Southeast Asian Studies.

Rasul, G. & Thapa, G.B. 2003. Shifting cultivation in the mountains of south and Southeast Asia: Regional patterns and factors influencing the change. *Land Degradation & Development* 14: 495–508.

Roder, W. 2001. *Slash-and-burn rice systems in the hills of northern Lao PDR: Description, challenges and opportunities*: 1–201. Los Banos: International Rice Research Institute.

Saito, K., Linquist, B., Atlin, G.N., Phanthaboon, K., Shiraiwa, T. & Horie, T. 2006a. Response of traditional and improved upland rice cultivars to N and P fertilizer in northern Laos. *Field Crops Research* 96: 216–223.

Saito, K., Linquist, B., Keobualapha, B., Phanthaboon, K., Shirakawa, T. & Horie, T. 2006b. *Stylosanthes guianensis* as a short-term fallow crop for improving upland rice productivity in northern Laos. *Field Crops Research* 96: 438–447.

Salas, W.A., Ducey, M.J., Rignot, E. & Skole, D. 2002. Assessment of JERS-1 SAR for monitoring secondary vegetation in Amazonia: I. Spatial and temporal variability in backscatter across a chrono-sequence of secondary vegetation stands in Rondonia. *International Journal of Remote Sensing* 23: 1357–1379.

Spanner, M.A., Pierce, L.L., Running, S.W. & Peterson, D.L. 1990. Remote sensing of temperate coniferous forest leaf area index: the influence of canopy closure, understory vegetation, and background spectral response. *International Journal of Remote Sensing* 11: 95–111.

Trigg, S. & Flasse, S. 2001. An evaluation of different bi-spectral spaces for discriminating burned shrub-savannah. *International Journal of Remote Sensing* 22: 2641–2647.

Warner, K. 1991. Shifting cultivators: local technical knowledge ad natural resource management in the humid tropics. *Community Forestry Note* 8: 1–20. Rome: FAO.

Wilson, E.H. & Sader, S.A. 2002. Detection of forest harvest type using multiple dates of Landsat TM imagery. *Remote Sensing of Environment* 80: 385–396.

Woodcock, C.E., Macomber, S.A., Pax-Lenney, M. & Cohen, W.B. 2001. Monitoring large areas for forest change using Landsat: Generalization across space, time and Landsat sensors. *Remote Sensing of Environment* 78: 194–203.

Wynne, R.H., Oderwald, R.G., Reams, G.A. & Scrivani, J.A. 2000. Optical remote sensing for forest area estimation. *Journal of Forestry* 98: 31–36.

Zhang, Q., Pavlic, G., Chen, W., Latifovic, R., Fraser, R. & Cihlar, J. 2004. Deriving stand age distribution in boreal forests using SPOT VEGETATION and NOAA AVHRR imagery. *Remote Sensing of Environment* 91: 405–418.

*Recent Advances in Remote Sensing and Geoinformation Processing
for Land Degradation Assessment – Röder & Hill (eds)
© 2009 Taylor & Francis Group, London, ISBN 978-0-415-39769-8*

Author index

Subject index

ISPRS Book Series

1. Advances in Spatial Analysis and Decision Making (2004)
   Edited by Z. Li, Q. Zhou & W. Kainz
   ISBN: 978-90-5809-652-4 (HB)

2. Post-Launch Calibration of Satellite Sensors (2004)
   Stanley A. Morain & Amelia M. Budge
   ISBN: 978-90-5809-693-7 (HB)

3. Next Generation Geospatial Information: From Digital Image Analysis to Spatiotemporal Databases (2005)
   Peggy Agouris & Arie Croituru
   ISBN: 978-0-415-38049-2 (HB)

4. Advances in Mobile Mapping Technology (2007)
   Edited by C. Vincent Tao & Jonathan Li
   ISBN: 978-0-415-42723-4 (HB)
   ISBN: 978-0-203-96187-2 (E-book)

5. Advances in Spatio-Temporal Analysis (2007)
   Edited by Xinming Tang, Yaolin Liu, Jixian Zhang & Wolfgang Kainz
   ISBN: 978-0-415-40630-7 (HB)
   ISBN: 978-0-203-93755-6 (E-book)

6. Geospatial Information Technology for Emergency Response (2008)
   Edited by Sisi Zlatanova & Jonathan Li
   ISBN: 978-0-415-42247-5 (HB)
   ISBN: 978-0-203-92881-3 (E-book)

7. Advances in Photogrammetry, Remote Sensing and Spatial Information Sciences. 2008 ISPRS Congress Book
   Edited by Zhilin Li, Jun Chen & Emmanuel Baltsavias
   ISBN: 978-0-415-47805-2 (HB)
   ISBN: 978-0-203-88844-5 (E-book)

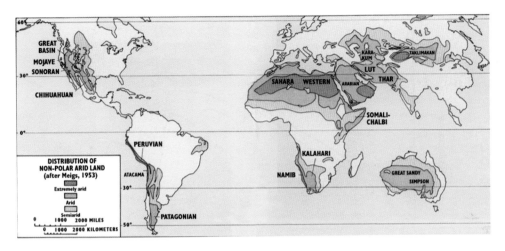

Figure 1.   Global distribution of arid ecosystems (MEA 2005) (S.L. Ustin et al., see page 16).

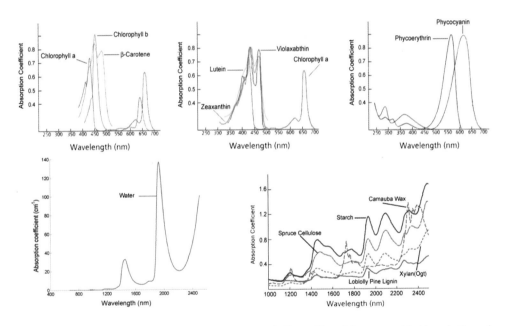

Figure 4.   Specific absorption coefficients for a) chlorophyll a and b and β-carotene, b) chlorophyll a and xanthophyll pigments: leutin, violaxanthin and zeathanthin, c) phycoerythrin and phycocyanin, d) water (after Kou et al., 1993) and e) five common biochemical compounds found in leaves (S.L. Ustin et al., see page 19).

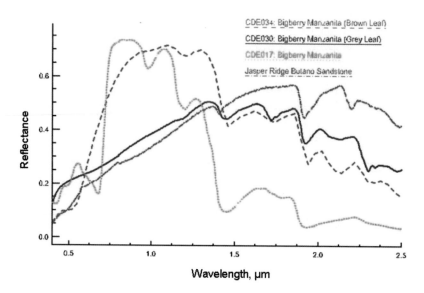

Figure 5. Typical spectra of fresh green leaf from bigberry Manzanita (¦), two stages of dry leaf weathering (¦, ¦), and bare soil (- - - - bulano sandstone) from Jasper Ridge Biological Preserve, California, showing spectral differences that can be used to differentiate them. Note that the Quercus pubescens leaf in Figure 3b was dried rapidly since a strong chlorophyll absorption remains in the 400–700 nm region. The dry grass residue here lacks a red-edge feature (S.L. Ustin et al., see page 21).

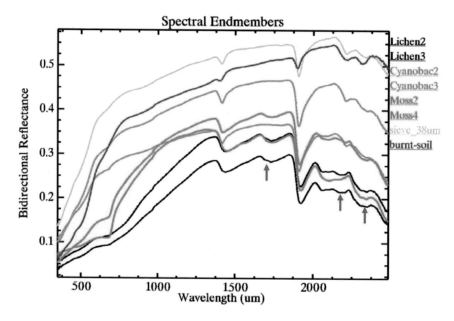

Figure 6. Components of biological soil crust from a Mojave desert habitat at the Nevada Desert Test Site, USA. Sieved and burnt (SOM exhausted) soil has higher reflectance than lichen, cyanobacteria, or mosses (P. Valko, unpublished results). Red arrows denote wavelengths where organic matter of biological curst are reported to show absorption features (S.L. Ustin et al., see page 22).

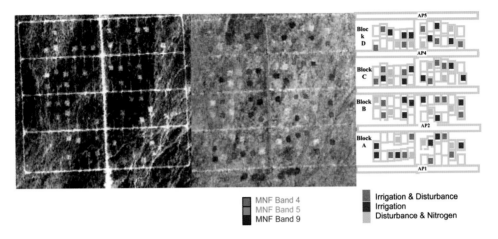

■ MNF Band 4	■ Irrigation & Disturbance
■ MNF Band 5	■ Irrigation
■ MNF Band 9	■ Disturbance & Nitrogen

Figure 7. False color Airborne Advanced Visible Infrared Imaging Spectrometer (AVIRIS) image (left), and minimum noise fraction image (MNF) acquired 9 July 2003 at ~3 m spatial resolution over the Mojave Global Change Facility experimental plots, Nevada, USA (P. Valko, unpublished results). Each of the 96 plots is 14 m × 14 m. Treatments include added monthly summer irrigation equivalent to three extra monsoon storm events, two levels of nitrogen enrichment simulating dry deposition (added in the fall before winter precipitation period, and crust disturbance simulating effect of over-grazing, performed annually. Treatments began summer 2001 (S.L. Ustin et al., see page 23).

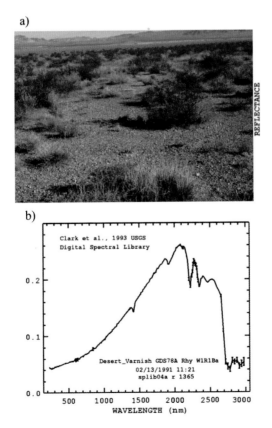

Figure 8. a) Desert pavement in Mojave Global Change Facility site, near Mercury, Nevada, USA. b) Spectrum of coating on desert pavement composed of rhyolite and quartz cobbles (desert varnish.gds78A) http://speclab.cr.usgs.gov (S.L. Ustin et al., see page 24).

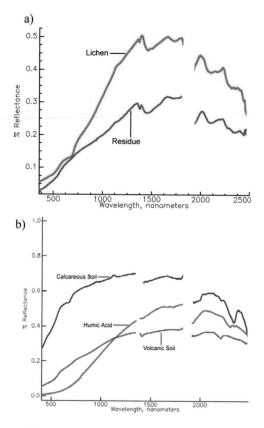

Figure 9.   a) Contrasting spectral shapes between biologically active lichen and dry plant residues. b) Respiration resistant humic acid soil component and two contrasting soils, derived from calcareous and volcanic parent material (S.L. Ustin et al., see page 26).

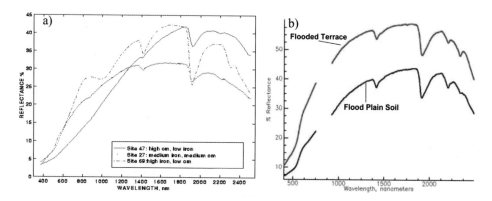

Figure 11.   a) Three soils showing effect of organic matter on soil reflectance. b) Soil spectra from an eroded calcareous terrace and an alluvial floodplain (S.L. Ustin et al., see page 28).

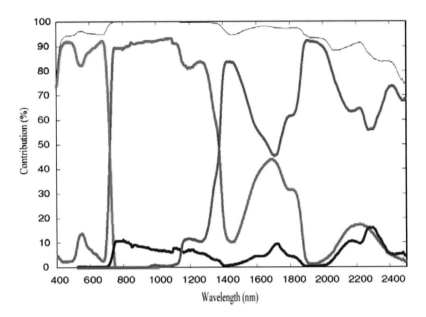

Figure 13.   Contribution to leaf transmittance simulated by PROSPECT for chlorophyll concentration Cab (green), water content Cw (blue), dry matter content Cm (brown) and the structure parameter N (red) (Pavan, unpublished) (S.L. Ustin et al., see page 33).

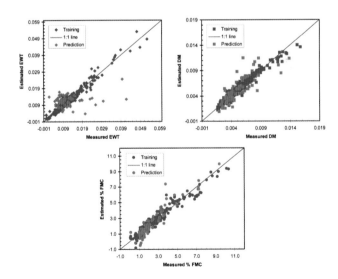

Figure 14.   a) Prediction of leaf water content from fresh leaves b) leaf dry matter from dry leaves, and c) leaf dry matter from fresh leaves using a generic algorithm-partial least square regression, GA-PLS (red dots) on LOPEX leaf data for calibration and validation (blue dots). Leaf reflectance and transmission data and biochemistry from the LOPEX dataset (Hosgood et al., 1994). The mean reflectance of the samples at each spectral band is subtracted before running GA-PLS (S.L. Ustin et al., see page 34).

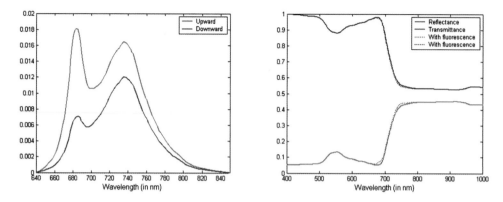

Figure 15. Simulation of the adaxial and abaxial chlorophyll fluorescence (on the left) and of the reflectance and transmittance with and without fluorescence (on the right) using fluormodleaf. Structure parameter: N = 1.5, chlorophyll a + b content: Cab = 33 μg cm^{-2}, equivalent water thickness: Cw = 0.025 cm, dry matter content: Cm = 0.01 g cm^{-2}, fluorescence quantum yield: Φ = 0.04, temperature: T = 20°C, species: green bean, PSII/SPI ratio: Sto = 2.0 (S.L. Ustin et al., see page 35).

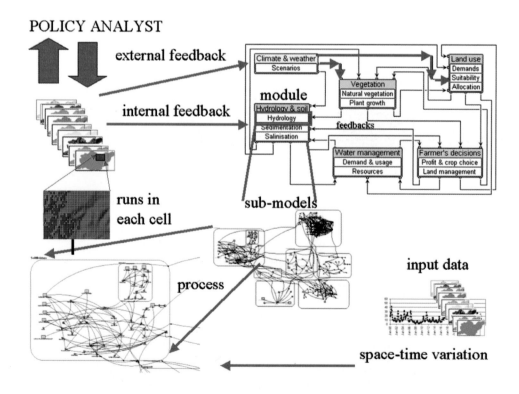

Figure 1. A Basic outline of the structure of the MedAction PSS (M. Mulligan, see page 63).

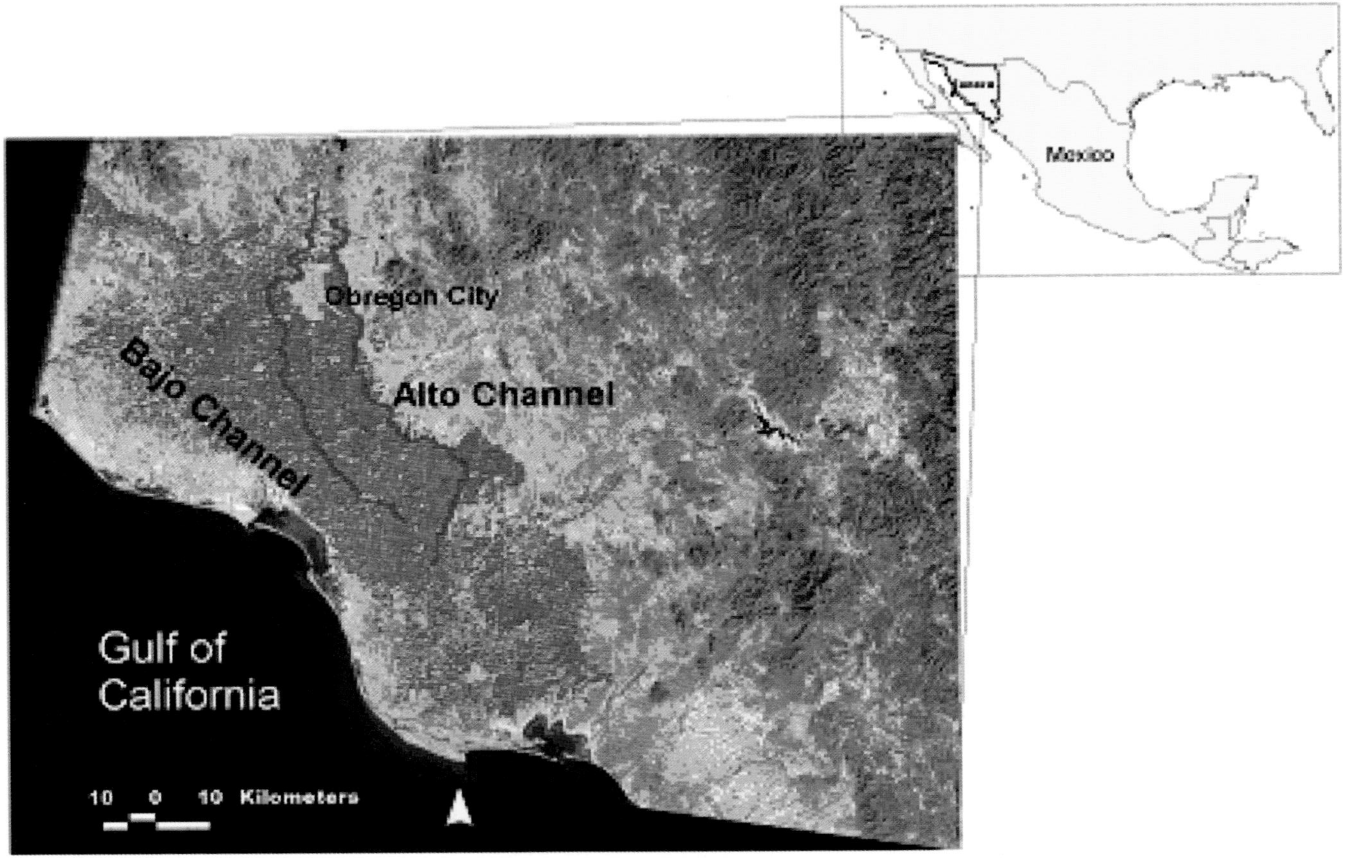

Figure 1. The Yaqui valley (Northwest Mexico) as described by Landsat ETM+ (false colour composite, 4, 3, 2 channels) February 26, 2000 (A. Chehbouni et al., see page 77).

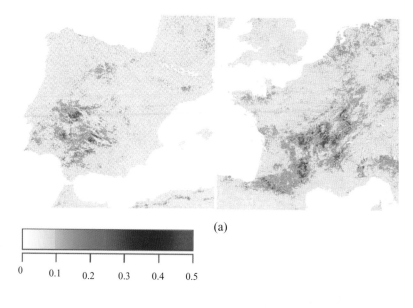

(a)

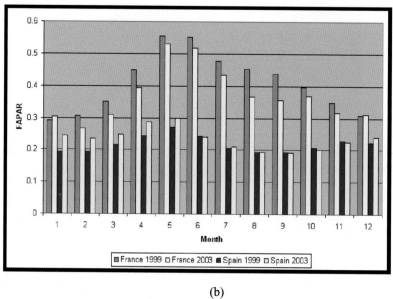

(b)

Figure 2.   (a) FAPAR anomalies over Spain, April 1999 (left), and France, August 2003 (right). (b) The seasonal cycle of nationally averaged FAPAR for France and Spain for 1999 and 2003 (N. Gobron et al., see page 94).

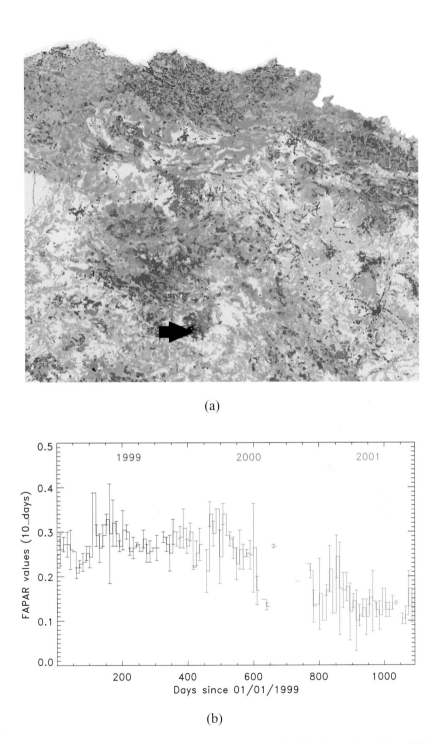

(a)

(b)

Figure 3. (a) Corine Land Cover 2000 over northern Spain. The full land cover legend is available from the source web site. In general, yellow, green and red colours stand for agricultural, forest and urban areas, respectively. (b) The seasonal cycle of monthly FAPAR products over a 'burnt area' (black spot on the map), located at [2.65° W, 41.545° N], is shown from January 1999 to December 2001. The FAPAR data are from the JRC-FAPAR product (N. Gobron et al., see page 95).

(a) FAPAR inAugust 2002

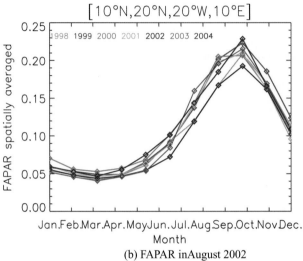

(b) FAPAR inAugust 2002

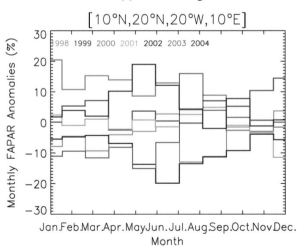

Figure 4. (a) Map of FAPAR for the Sahel region in August 2002. The colour scale is the same as for Figure 5. Black pixels correspond to missing values, either due to instrumental issues or the persistent presence of clouds. Time series of FAPAR (b) and FAPAR anomalies (c) averaged over the Sahel region for 1998 to 2004 (N. Gobron et al., see page 96).

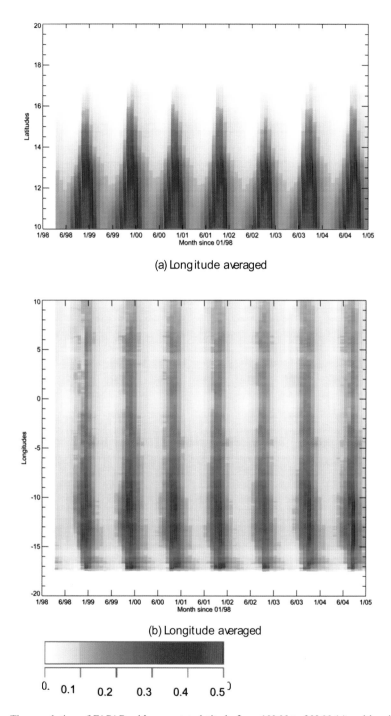

Figure 5. Time evolution of FAPAR with respect to latitude from 10° N to 20° N (a) and longitude from 20° W to 10° E (b) for the period 01/1998 to 12/2004 (N. Gobron et al., see page 97).

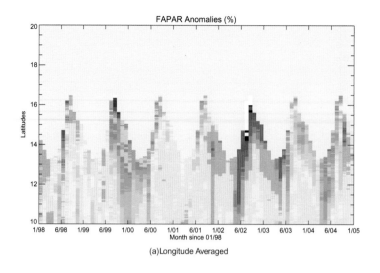

(a)Longitude Averaged

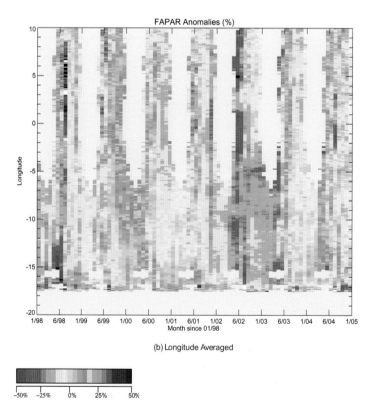

(b) Longitude Averaged

Figure 6. Time evolution of FAPAR anomalies with respect to the average value during the record period, displayed as a function of latitude from 10° N to 20° N (a) and longitude from 20° W to 10° E (b) for the period from 01/1998 to 12/2004 (N. Gobron et al., see page 98).

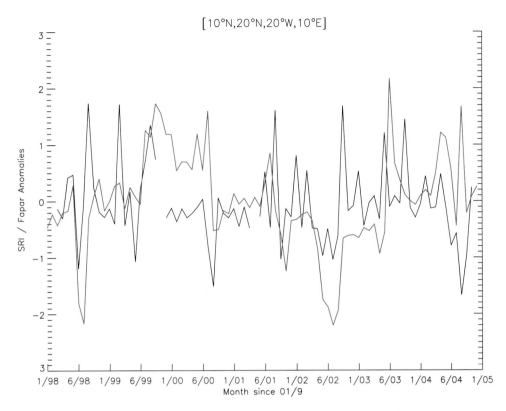

Figure 7. FAPAR anomalies (red line) and Sahel rainfall index (black line) over the Sahel region from 01/1998 until 12/2004 (N. Gobron et al., see page 99).

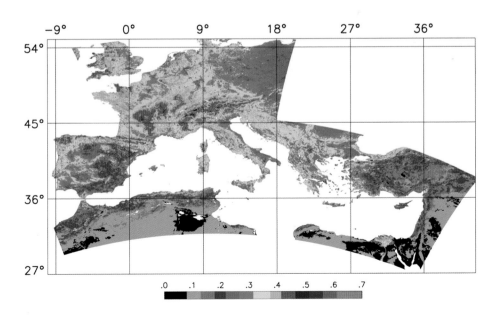

Figure 2. Sample of a MEDOKADS derived product. Annual mean of the period 1989 to 2004 showing the region covered with data (K. Friedrich & D. Koslowsky, see page 106).

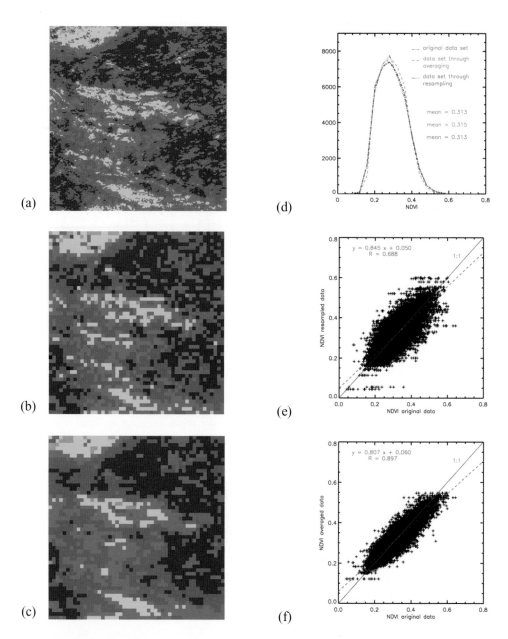

Figure 5. Effect due to re-sampling (b) and averaging (c) of original image (a), histogram for data sets (d), linear regression between original and re-sampled data (e) and averaged data (f) for an NDVI image (K. Friedrich & D. Koslowsky, see page 109).

382

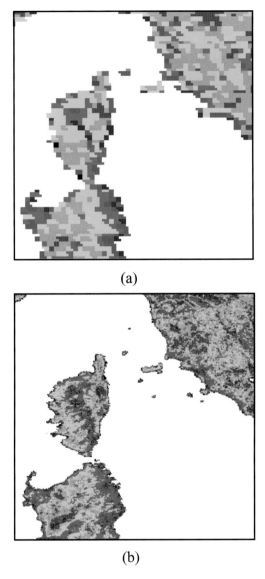

(a)

(b)

Figure 6.   Example for MEDOKADS (a) and PAL (b) data (last decade of month May 1995, region around Corsica) (K. Friedrich & D. Koslowsky, see page 110).

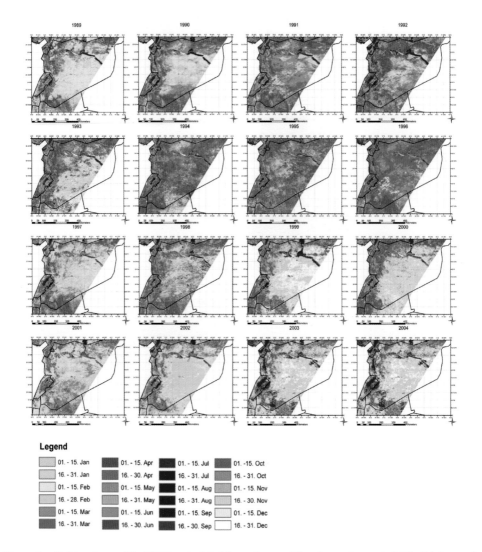

Figure 3.   Peaking times of the NDVI, calculated for each year of the observation period (Th. Udelhoven & J. Hill, see page 123).

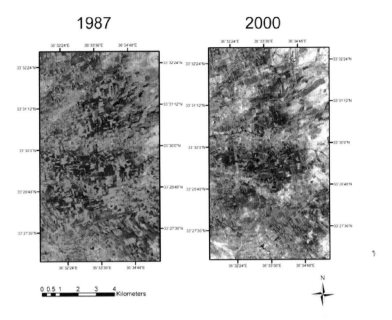

Figure 7.   Landsat TM sub-scenes (26.5.1987 and 21.5.2000) showing the Ghouta area of Damascus (RGB channels: 5-4-3) (Th. Udelhoven & J. Hill, see page 126).

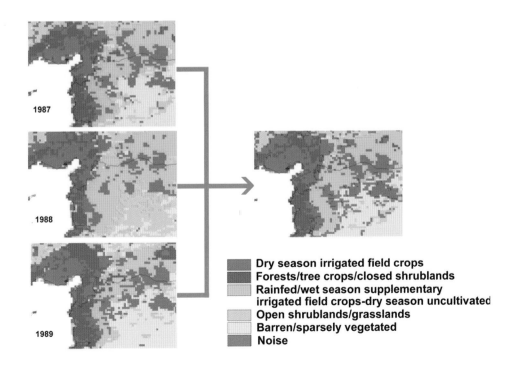

Figure 5.   Majority land cover classes for different years, subset northwest Syria (on the left the land cover classes for the years 1987, 1988, 1989; on the right the majority land cover class) (D. Celis & E. De Pauw, see page 138).

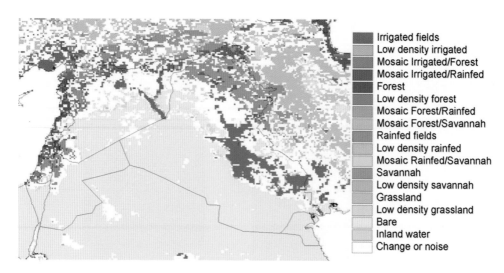

Irrigated fields
Low density irrigated
Mosaic Irrigated/Forest
Mosaic Irrigated/Rainfed
Forest
Low density forest
Mosaic Forest/Rainfed
Mosaic Forest/Savannah
Rainfed fields
Low density rainfed
Mosaic Rainfed/Savannah
Savannah
Low density savannah
Grassland
Low density grassland
Bare
Inland water
Change or noise

Figure 6.   Stable classes for parts of West Asia (D. Celis & E. De Pauw, see page 139).

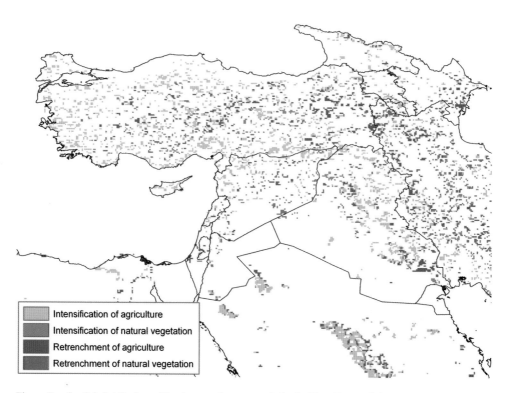

Intensification of agriculture
Intensification of natural vegetation
Retrenchment of agriculture
Retrenchment of natural vegetation

Figure 7.   Spatial distribution of land cover change trends in the Near East and the Caucasus (D. Celis & E. De Pauw, see page 140).

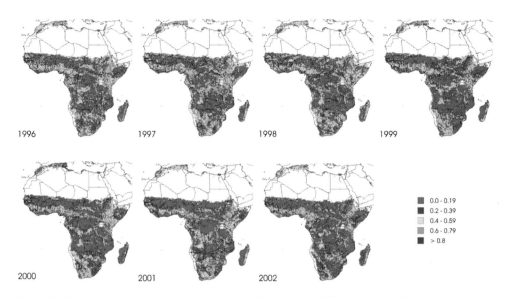

Figure 3.　Maps of *AI* showing inter-annual variation for the month of October during the seven years period 1996–2002 (P.A. Brivio et al., see page 154).

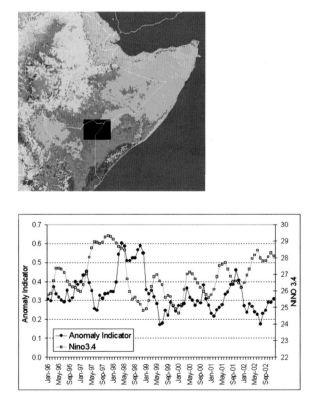

Figure 4.　Comparison between the sea surface temperature of Niño 3.4 and the Anomaly Indicator averaged over a region of interest (500 pixels) in Eastern Africa. High anomaly values are evident in correspondence with the 1997–98 El Niño event (P.A. Brivio et al., see page 155).

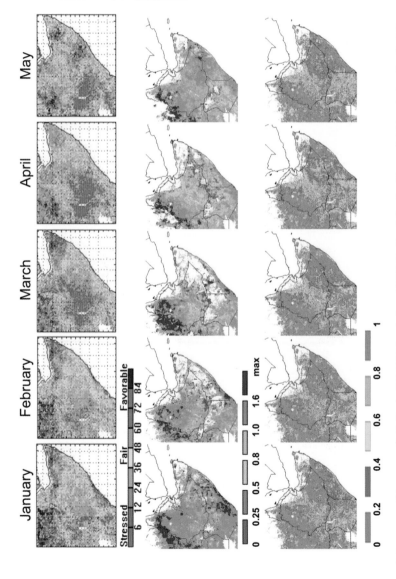

Figure 5. Comparison between monthly maps of *AI* (top), AN-NPP (centre) and VHI (bottom) for the Horn of Africa during the first five months of the year 2000 (P.A. Brivio et al., see page 156).

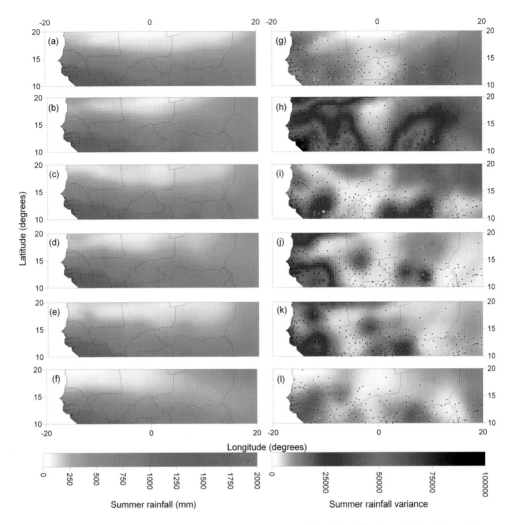

Figure 3. Maps of the per pixel (approximately 0.4°) average of the 300 realisations (a–f) and the variance for the conditional distributions (g–l) of summer rainfall in the West African Sahel for the years of 1935, 1945, 1955, 1965, 1975 and 1985, respectively. The rainfall stations (●) and the political boundaries for countries of the region are also shown using black lines (A. Chappell & C.T. Agnew, see page 168).

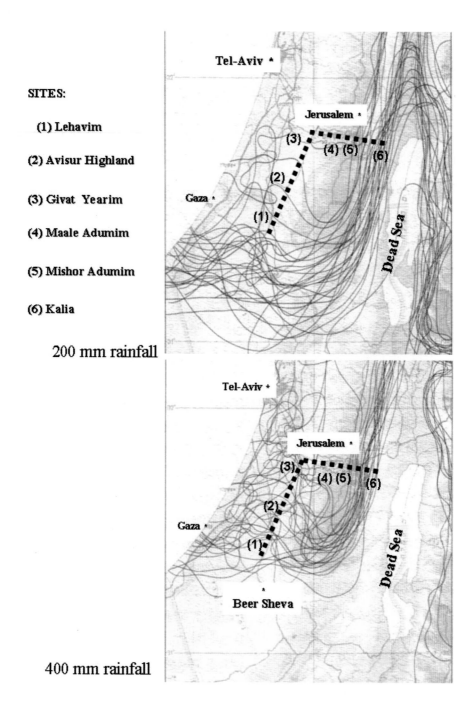

SITES:

(1) Lehavim

(2) Avisur Highland

(3) Givat Yearim

(4) Maale Adumim

(5) Mishor Adumim

(6) Kalia

200 mm rainfall

400 mm rainfall

Figure 1.   Location of the two transect areas with respect to the 200 mm and 400 mm Isohyets' fluctuations between 1930 and 1960 (Atlas of Israel 1964 Edition) (M. Shoshany, see page 190).

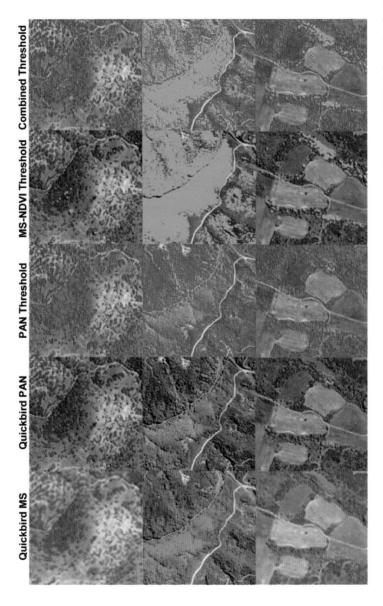

Figure 1. Threshold-based mapping of woody vegetation cover based on a combination of panchromatic and multispectral (NDVI) information for three locations: open shrubland (top), partially forested (middle), pastures with shrub edges (bottom) (J. Hill et al., see page 215).

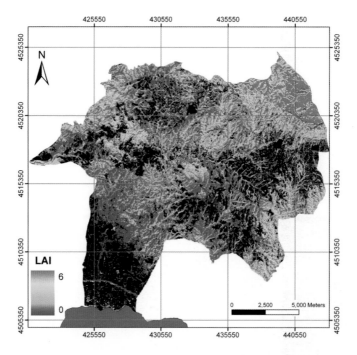

Figure 5. Map of site-LAI retrieved by combing constrained GeoSail-retrieved shrub-LAI with Quickbird-derived cover estimates (agricultural areas and settlements in black, parts of Lake Koronia visible in blue in the South) (J. Hill et al., see page 221).

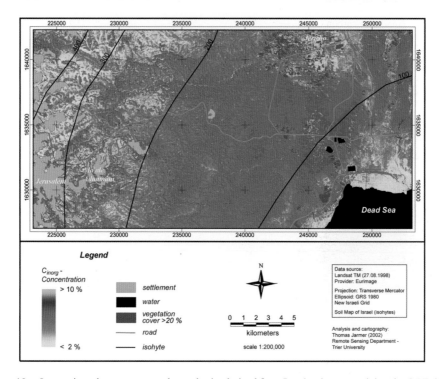

Figure 10. Inorganic carbon contents at the study site derived from Landsat image applying the C.I.E. based model (T. Jarmer et al., see page 237).

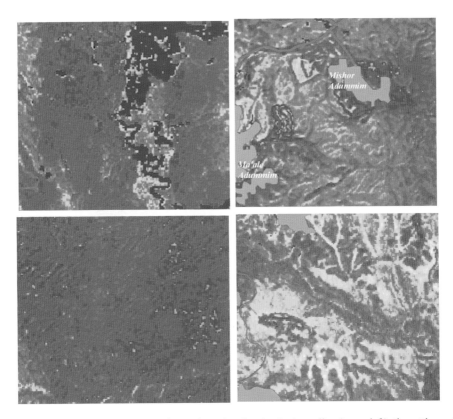

Figure 11. Inorganic carbon contents for a subset showing the Jordan valley (upper left), the settlement of Ma'ale Adummim and the industrial area of Mishor Adummim (upper right), the central part of the study area (lower left) and Wadis in the north-western part of the study site (lower right) (scale approx. 1:75.000; legend see figure 10) (T. Jarmer et al., see page 238).

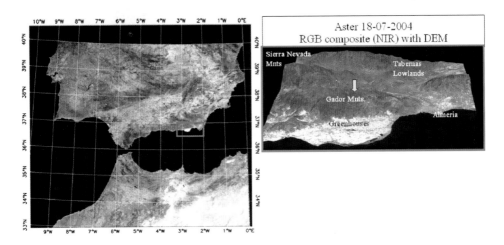

Figure 1. Location of the study site in South East Spain, Almeria province. The left panel shows MODIS True Color Composite corresponding to the study site and 18-07-2004. The right panel shows the study site with a false colour composite for ASTER 18-04-07 image with image relief. Location of the eddy covariance system is shown by the arrow (M. García et al., see page 263).

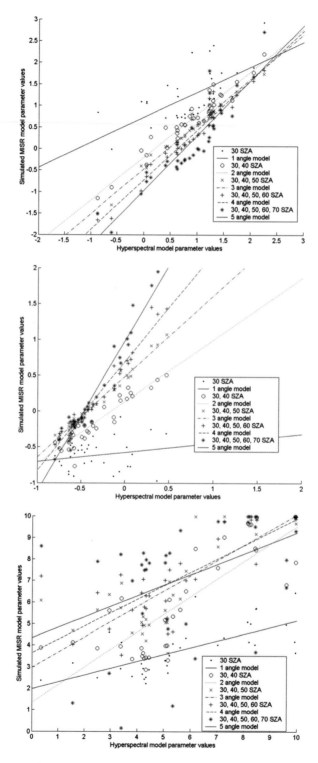

Figure 3. Linear regression models fitted to the values of bi-directional soil spectral reflectance model parameters b (a), c, (b), h (c), b' (d), c' (e) from hyperspectral data and simulated MISR view angles and solar zenith angles (SZA) (A. Chappell et al., see page 250). (*continued*)

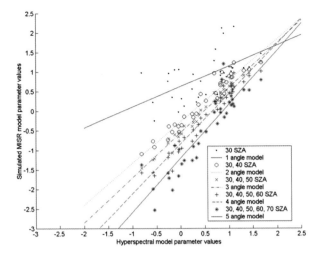

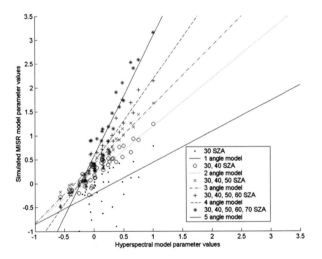

Figure 3. (*Continued*).

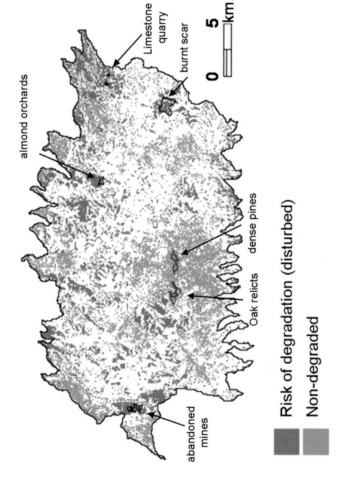

Figure 7. Classification map of areas with high risk of degradation (red) and low risk of degradation (green) in Sierra Gador (Almeria, Spain). Validation sites for non-degraded sites are composed of oak relicts with three density levels and pine forests (blue polygons), and disturbed sites considered at high risk of degradation are mapped as black polygons (burnt scar from 2002, limestone quarry, almond orchards, and abandonded mines) (M. Garcia et al., see page 274).

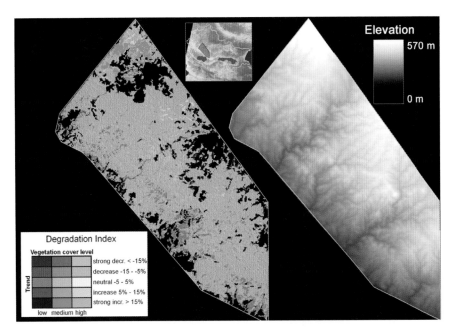

Figure 4. Degradation index for subset area and corresponding DEM-subset (interpretation in the text) (A. Röder, et al., see page 290).

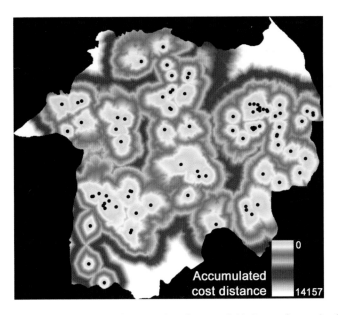

Figure 6. Accumulated cost distance surface based on integrated friction surface and calculated using a pushbroom algorithm starting from points of livestock concentration (PLC depicted as black dots; adapted from Röder et al., 2007) (A. Röder et al., see page 293).

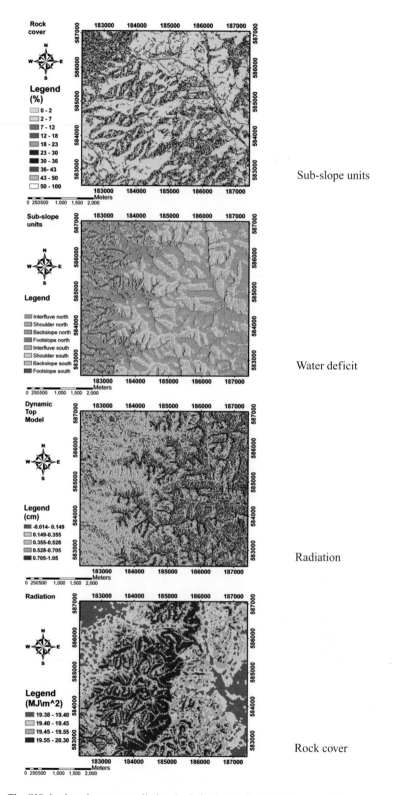

Figure 3. The GIS database layers as applied to the Lehavim study area (T. Svoray et al., see page 305).

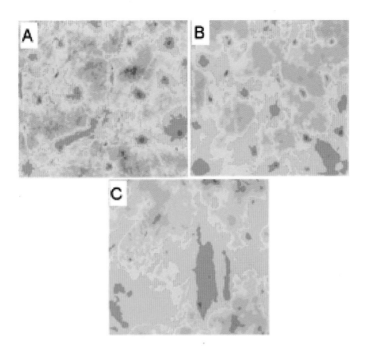

Figure 5. Kriging interpolation maps for images shown in Figure 2 based on the Brightness Index values. Dark tones indicate land degradation due to grazing and/or technological desertification. (A. Karnieli et al., see page 319).

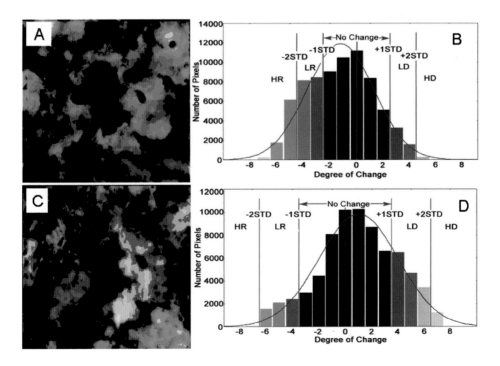

Figure 6. Change detection results shown as maps and histograms. HR = high rehabilitation; LR = low rehabilitation; HD = high degradation; LD = low degradation. (A, B) difference between the 1987 and 1975 images; (C, D) difference between the 2000 and 1987 images. (A. Karnieli et al., see page 320).

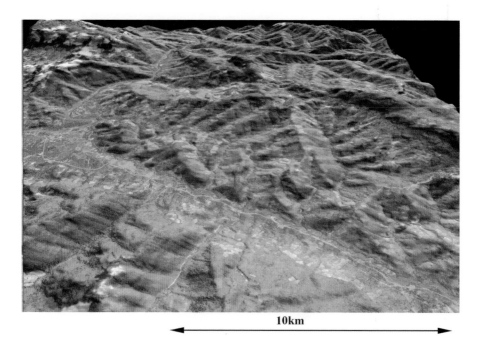

10km

Figure 4.   A 3D view of the intensive study area near Luang Prabang (HK-site) (Y. Inoue et al., see page 350). Note: The spectral image is from QuickBird on Oct. 18, 2003. The spatial resolution is 2.4 m for spectral images and 10 m for digital elevation map, respectively. The G, R, and NIR bands are assigned to B, G, and R colors, so that the bluish color represents the slash/burn land-use during the year, and reddish color indicates the densely vegetated areas, respectively. The area was approximately 320 km^2.

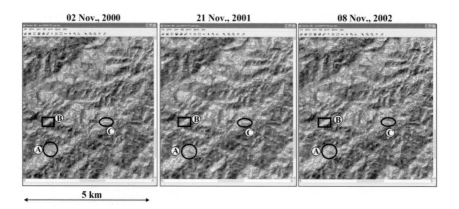

Figure 5.   Slash-and-burn patches identified on the composite image of 660 nm (B), 830 nm (G), and 1650 nm (R) in three consecutive years (Y. Inoue et al., see page 351). Note: Sample areas A, B and C indicate typical patches of slash-and-burning in 2000, 2001 and 2002, respectively.